Gum Arabic and Breast Cancer Biology

Ahmed A. M. Elnour
Editor

Gum Arabic and Breast Cancer Biology

Biotechnology Perspective

Springer

Editor
Ahmed A. M. Elnour
Faculty of Chemical and Process Engineering Technology, Universiti Malaysia Pahang Al-Sultan Abdullah (UMPSA) Pahang Gambang, Malaysia

Centre of Excellence for Advanced Research in Fluid Flow (CARIFF)
University Malaysia Pahang Gambang, Malaysia

ISBN 978-981-97-8517-9 ISBN 978-981-97-8518-6 (eBook)
https://doi.org/10.1007/978-981-97-8518-6

This Springer imprint is published by the registered company Springer Nature Singapore Pte Ltd.
The registered company address is: 152 Beach Road, #21-01/04 Gateway East, Singapore 189721, Singapore

This humble book is especially dedicated to the moments I cherish the most and to the love of my life—my wife, Rasha D.M.A. Ahmed.
To my "princesses," my lovely sisters;
to both of my adorable parents;
to my mother, Hawa Ahmed, who passed away just as my real journey in the study of biological chemistry began;
to my father, Adam Hassan;
also to my beloved siblings.
Moreover, to the gum arabic farmers who have dedicated their lives under stressful conditions to tapping and rehabilitating the greatest gum arabic belt in the world.
I apologize for the long hours lost in completing this book. May Allah the Almighty grant us the "Baraka" of time and lasting love until Jannah.
....amen...

Foreword

In an era marked by rapid advancements in medical biotechnology, the quest for innovative and effective approaches to understanding and treating diseases such as breast cancer remains paramount. The convergence of natural substances and state-of-the-art biotechnological tools offers unprecedented opportunities to reshape our therapeutic strategies. *Gum Arabic and Breast Cancer Biology: Biotechnology Perspective* is a pioneering exploration of how natural resources, specifically acacia gums (AGs), can be integrated into cutting-edge cancer research and treatment protocols.

This book is timely and significant, not only because it addresses a critical global health issue—breast cancer—but also because it does so through the lens of biotechnology, emphasizing sustainability and innovation. By focusing on gum arabic, derived from AGs, this text navigates the nexus of traditional uses and modern scientific applications, underscoring the potential of natural products in high-impact biotechnological research.

The use of organ-on-a-chip technology, particularly the breast-on-a-chip models discussed in this book, represents a transformative approach to cancer research. This technology provides a dynamic platform to study the intricate mechanisms of cancer progression and to evaluate the therapeutic potential of natural compounds such as those found in AGs. These innovations enhance our understanding of disease and propel us toward more personalized and precise medical interventions.

As we forge ahead in the fight against breast cancer, integrating traditional knowledge and modern technology, as exemplified in this book, will undoubtedly play a crucial role. The potential of gum arabic in cancer therapy, explored through rigorous scientific inquiry and innovative biotechnological applications, offers a glimpse into the future of natural products in medicine.

I commend the authors for their thorough research and insightful analysis, which contribute significantly to our understanding of the potential roles of natural gums in medical science. This book is essential for anyone involved in cancer research, biotechnology, or implementing natural products in therapeutic settings. It informs and inspires continued exploration and innovation at the intersection of nature and technology.

Department of Biomedical Engineering
Institute for Quantitative Health Science
and Engineering, College of Engineering,
and College of Human Medicine, Michigan
State University,
East Lansing, MI, USA

Nureddin Ashammakhi

Preface

This textbook is written for students, researchers, organizations, clinicians, industries, and policymakers. This is the product of my research and teaching experience of more than two decades. This review aims to provide a broad but concise overview of the various aspects of acacia gums' (AGs) biotechnology, including its phytochemical and biological chemistry. I have emphasized an easily understood description of the principles of breast cancer (BC) biology while also ensuring that the content is focused, avoiding unnecessary details that might distract students and industry professionals. Given the importance of AGs' phytochemistry, the biotechnological applications of AGs in breast cancer have been highlighted. Special attention has been paid to the extraction and utilization of AGs in breast cancer cell lines.

There are a few excellent textbooks on AGs' chemistry. This book is the first attempt to highlight the phytochemical bioactive compounds of AGs and their applications in breast cancer from a biotechnological perspective. I have addressed general topics in AGs' biotechnology and BC cell lines, enhanced with novel technological applications internationally, such as breast-on-a-chip technology. Additionally, I discuss how to transfer these technologies to our Islamic world under the umbrella of Islamic Development Bank (IsDB) member countries, where it seemed necessary to deepen my understanding of the challenges at hand. Thus, this book represents a balance between a general textbook and specialized text.

This book presents cutting-edge research aimed at bridging the gap between the biological applications of AGs and their potential use in BC. BC remains a formidable global health challenge, marked by its widespread impact and ongoing research efforts to improve its diagnosis, treatment, and prevention strategies. This work explores the innovative intersections of natural compounds and advanced biotechnological applications, with a particular focus on the potential of AG-derived *Acacia senegal* and *Acacia seyal* gums in BC research.

This book thoroughly explores AGs, highlighting their unique physicochemical properties and the sophisticated analytical techniques used to reveal their structure and functions. It delves into optimized extraction methods that emphasize their economic and ecological significance, particularly in Sudan and gum arabic belts across the African Sahara, where they form the cornerstone of the local economy and environmental management.

This book's subsequent sections explore the broad biological applications of secondary metabolites derived from these gums, with a focus on their anti-inflammatory, antioxidant, and antimicrobial properties, which are rigorously analyzed through scientific methodologies. The potential of these metabolites in the context of BC was examined by considering their molecular mechanisms and implications for therapeutic strategies.

A groundbreaking aspect of this book is its discussion of the application of cutting-edge organ-on-a-chip technology, particularly breast-on-a-chip, which represents a novel approach to cancer research. This technology recapitulates the complex biological environment of human breast tissue, enabling researchers to study the intricate dynamics of BC in a controlled and physiologically relevant setting. The integration of AG bioactive compounds into this technology is discussed in detail, illustrating how these natural substances can influence cancer cell behavior and treatment outcomes.

The application of breast-on-a-chip technology in this context is particularly novel, as it enables direct observation and analysis of how AG's secondary metabolites of AG interact with cancerous cells. This can lead to more precise insights into their therapeutic potential and mechanisms of action, such as the induction of apoptosis or inhibition of metastasis. Moreover, this technology offers a sustainable and ethically favorable alternative to traditional animal testing, providing a platform that can accelerate the development of new cancer treatments while reducing the need for extensive in vivo testing.

By merging traditional knowledge of natural compounds with modern biotechnological advances, *Gum Arabic and Breast Cancer Biology: Biotechnology Perspective* aims to foster a deeper understanding of the potential applications of AGs in BC therapy. It is designed to be an essential resource for researchers, clinicians, and policymakers, providing new perspectives and tools to combat one of the most prevalent cancers affecting women. Through this comprehensive exploration, this book aims to inspire further research and development, advocating for innovative approaches that integrate natural compounds and cutting-edge technologies in the fight against breast cancer.

Finally, I am very grateful to my postdoctoral, PhD, and MSc supervisors and colleagues who, through discussions and comments they have sent me on their special fields, have given me the information that enabled me to write this book. It would like to express my sincere gratitude to the supervisors and colleagues listed below, who went into great trouble to prepare and critically review one or more chapters, drawing my attention to errors and suggesting improvements. My special thanks go to my wife, Rasha D.M.A. Ahmed. Without her intensive support, I would not have been able to write this book. I have attempted to eradicate as many mistakes as possible, probably not with complete success. Therefore, I am grateful for any suggestions and comments.

Thanks to my postdoctoral, PhD, and MSc supervisors and colleagues for their suggestions after preparing and reviewing one or more chapters.

Prof. Dr. Abdelgadir Elfadil Abdelgadir, Canada
Prof. Dr. Abdurahmani Hamid Nour, Malaysia
Prof. Dr. Ibrahim Bani, USA
Prof. Dr. Khogali E. A. Ishag, Sudan
Prof. Dr. Mohamed E.S. Mirghani, Malaysia
Prof. Dr. Mohammed Elmubark Osman, Sudan
Prof. Dr. Nassereldeen Ahmed Kabbashi, Malaysia
Prof. Dr. Nureddin Ashammakhi, USA
Prof. Dr. Saphwan Al-Assaf, United Kingdom
Prof. Dr. Yousif H. Y. Ahmed, Sudan
Dr. Aiman A. Bin Mokaizh, Malaysia
Dr. El Bashier Sallam, Sudan
Dr. Essam M. Hamad, Saudi Arabia
Dr. Hashim A.M. Hashim, Saudi Arabia
Dr. Idris Musa Adam, Sudan
Dr. Khalid Hamid Musa, Saudi Arabia
Dr. Kourwel Kuai Kourwel, South Sudan
Mr. Elsaid Alhassan, Sudan

Pahang Gambang, Malaysia
July 2024

Ahmed A. M. Elnour

Acknowledgment

I would like to extend my sincerest gratitude and praise to Allah Azza Wa Jalla for allowing me to successfully complete this book. This textbook is a culmination of knowledge, efforts, and invaluable contributions from various sources, for which I am deeply grateful.

First and foremost, I extend my sincerest gratitude to my first and second postdoctoral supervisors: Prof. Dr. Abdurahmani Hamid Nour, UMPSA, Malaysia; Prof. Dr. Nureddin Ashammakhi, Michigan State University, USA; Prof. Dr. Mohamed E.S. Mirghani, IIUM, Malaysia; and Associate Professor Dr. Khalid Hamid Musa, Qassim University, Saudi Arabia. Their expertise in natural products and biotechnology has enriched content beyond measure. Working alongside them has been an inspiring educational journey.

I am profoundly thankful to the Center of Excellence for Advanced Research in Fluid Flow (CARIFF) at the Universiti Malaysia Pahang, Gambang, Malaysia, for the financial support without which this project would not have been feasible. Their commitment to advancing research and providing a comfortable working environment was deeply appreciated.

I am especially grateful to the Scholarship Program Department of the Islamic Development Bank (IsDB) for their generous financial support of my first postdoctoral project titled "Cancer-on-a-Chip Base as a New Paradigm Potential Application of Gum Arabic Extraction Toward Development of Breast Cancer Therapeutics." This project reflects my engagement with not only my Muslim communities and my country (Sudan) but also with the global community by publishing the world's first book with the title *Gum Arabic and Breast Cancer Biology*. As a result, two chapters are dedicated to knowledge transfer to IsDB member countries. Their commitment to support my first postdoctoral project is extremely appreciated.

Special thanks are due to my academic mentors, Prof. Dr. Abdurahmani Hamid Nour and Prof. Dr. Nureddin Ashammakhi, whose guidance and insights have been instrumental in shaping both the scientific and pedagogical approaches to this work.

I also owe a great debt to the Springer Nature publisher's technical team, especially Muthuneela Muthukumar and Mei Hann Lee, for their meticulous attention to the guidelines for preparing the manuscript format, designing diagrams, and creating technical illustrations that have made the concepts more accessible to our readers.

Finally, but most importantly, I thank my family for their patience and unwavering support throughout the process of writing this book. Special thanks to my beloved wife, Rasha D.M. Ali, who has accompanied me through both the good and challenging episodes of our life here in Malaysia. I am deeply grateful for her patience, support, and energy, which helped me to complete this book. I am also very thankful to my parents, who did their best to support my education. My late mother, Hawa Ahmed (who passed away during the preparation of this book's first elements), was a strong woman, especially when it came to pushing us to reach our dreams. My father, Adam Hassan, is a person who always includes me in his prayer. Despite their lack of access to higher education, their sacrifice for their children is worthy of high praise.

Finally, I would like to express my gratitude to all of our contributors. This book would not have been possible without the collective support and encouragement of each of you.

Thank you.

About the Book

This book commences with a foundational exploration of the AGs, comprising *Acacia senegal*, *Acacia seyal*, and *Acacia polyacantha* gums. These gums are complex polysaccharides with diverse food, pharmaceutical, and cosmetic applications. The initial chapters explore their botanical origins, physicochemical properties, and functional roles as emulsifiers, stabilizers, and thickeners. Employing modern analytical techniques such as Raman spectroscopy and FT-IR spectroscopy, this book provides a comprehensive analysis of the gums' composition and their potential applications in biotechnological applications and industry.

After the introduction, this book transitions to detailed studies on the extraction processes of secondary metabolites from AGs, focusing on optimizing these processes within the Sudanese context where these trees are prevalent. This section highlights the economic and ecological significance of AGs in Sudan, offering insights into sustainable management practices and their effects on both the local economy and environmental stability.

A critical section of this book explores the role of AGs in breast cancer therapy. Through rigorous scientific investigation, it discusses the potential of secondary metabolites extracted from AGs to inhibit breast cancer cell lines. This book emphasizes specific apoptotic and anti-angiogenic effects observed in laboratory studies, presenting a promising outlook for these natural compounds as alternatives or supplements for conventional cancer treatments.

Chapters dedicated to virtual screening and organ-on-a-chip technologies outline innovative approaches in the fight against breast cancer. This book elaborates on how these technologies are utilized to identify and validate the therapeutic potentials of compounds found in gum arabic against breast cancer. This section showcases the integration of traditional natural products with contemporary biotechnological methods and discusses the challenges and future directions in adopting these technologies in Islamic Development Bank member countries.

Additionally, this book links the research on AGs and breast cancer to broader global health goals, specifically the Sustainable Development Goals (SDGs). It examines how advancements in breast cancer treatment contribute to achieving these goals, emphasizing the role of scientific research in enhancing health quality and reducing health inequalities in developing countries.

The concluding chapters reflect on the significant strides made in utilizing AGs for breast cancer research and therapy while also looking ahead to future research directions. This book advocates for more comprehensive studies to fully understand the scope of AGs' benefits and their potential integration into mainstream cancer treatment protocols. This calls for increased funding, international collaboration, and a multidisciplinary approach to maximize the potential of natural resources in combating breast cancer.

Contents

Editor and Contributors

About the Editor

Ahmed A. M. Elnour received his Ph.D. in Biotechnology Engineering in 2019, earning the distinction of "Best Ph.D. Thesis in Biotechnology" from the International Islamic University Malaysia (IIUM). His initial postdoctoral research, funded by the Islamic Development Bank (IsDB) Merit Scholarship Program, focused on pioneering cancer-on-a-chip technologies to enhance breast cancer therapeutics using natural bioactive compounds extracted from gum arabic. He received an M.Sc. in Analytical Chemistry (with distinction) in 2007 from Khartoum University, specializing in the fractionation and physicochemical properties of Acacia polyacantha gum. He also holds a B.Sc. in Biochemistry and Food Science (first class) from Kordofan University, obtained in 2003.

In 2020, Elnour's postdoctoral project was recognized as one of the best by the Islamic Development Bank (IsDB) Merit Scholarship (IsDB-MSP) Program. This accolade enabled him to advance his expertise in organ-on-a-chip technology for studying breast cancer in collaboration with IIUM and Michigan State University. His research has focused on integrating natural compounds from acacia gum into cutting-edge applications. Consequently, his academic and research endeavors have spanned several prestigious institutions, including Michigan State University and IIUM, to enhance his expertise in phytochemistry and sustainable biological chemistry.

During his postdoctoral fellowship at IIUM, Elnour honed his skills in phytochemistry, bioinformatics, and molecular biology, with a focus on breast cancer. His pioneering work in breast-on-a-chip technology led to significant publications, culminating in a groundbreaking book published by Springer Nature in 2025. This book is the first to explore the applications of acacia gum biotechnology in breast cancer, specifically addressing the phytochemistry of gum arabic and its biotechnological implications.

He is currently a researcher at the Center of Excellence for Advanced Research in Fluid Flow at the Universiti Malaysia, Pahang, Sultan Abdullah, Malaysia. With over 22 years of teaching and research experience in biochemistry, food science, and gum arabic chemistry, his work emphasizes biotechnological engineering. His interests also extend to green chemistry, oil and gas technology, waste treatment, machine learning, artificial intelligence, and sustainability in higher education. He has guided numerous graduate and postgraduate students and participated in several funded research projects. Elnour has received several scientific awards and has served as a distinguished member of the IsDB External Reviewers' Panel. He is a regular reviewer for several international journals and currently serves on the reviewer boards of at least ten international journals. He has published more than 12 book chapters and 40 articles. For further information, visit the ORCID profile (https://orcid.org/0000-0002-1924-094X).

Contributors

Abdelgadir Elfadil Abdelgadir Faculty of Science and Technology, Al-Neelain University, Khartoum, Sudan

Idris Musa Adam Agricultural Research Corporation (ARC)-Sudan, Wad Madani, Sudan

Yousif H. Y. Ahmed Department of Medicine, National University, Khartoum, Sudan

Saphwan Al-Assaf Hydrocolloids Research Center, University of Chester, Chester, UK

Nureddin Ashammakhi Department of Biomedical Engineering, Institute for Quantitative Health Science and Engineering, College of Engineering, and College of Human Medicine, Michigan State University, East Lansing, MI, USA

Ibrahim Bani Department of Chronic Disease Epidemiology, Yale School of Public Health, Yale University, New Haven, CT, USA

Department of Pathological Sciences, College of Medicine, Ajman University, Ajman, UAE

Aiman A. Bin Mokaizh Faculty of Chemical and Process Engineering Technology, Universiti Malaysia Pahang Al-Sultan Abdullah (UMPSA), Pekan, Pahang, Malaysia

Ahmed A. M. Elnour Faculty of Chemical and Process Engineering Technology, Universiti Malaysia PahangAl-Sultan Abdullah (UMPSA), Pahang Gambang, Malaysia

Center of Excellence for Advanced Research in Fluid Flow (CARIFF), University Malaysia, Pahang Gambang, Malaysia

Essam M. Hamad Department of Food Science and Human Nutrition, College of Agriculture and Veterinary Medicine, Qassim University, Buraydah, Saudi Arabia

Hashim Suliman Ibrahim Department of Food Science and Human Nutrition, College of Agriculture and Veterinary Medicine, Qassim University, Buraydah, Kingdom of Saudi Arabia

Khogali E. A. Ishag Department of Food Science and Technology, Faculty of Agriculture, University of Khartoum, Khartoum, Sudan

Nassereldeen Ahmed Kabbashi Department of Bio-technology Engineering, Bioenvironmental Engineering Research Institute (BERI), Kulliyyah (Faculty) of Engineering, International Islamic University, Kuala Lumpur, Malaysia

Kourwel Kuai Kourwel Department of Chemistry, School of Applied and Industrial Sciences, University of Juba, Juba, South Sudan

Mohamed E. S. Mirghani National Oilseed Processing Research Institute (NOPRI), University of Gezira, Medani, Sudan

Khalid Hamid Musa Department of Food Science and Human Nutrition, College of Agriculture and Veterinary Medicine, Qassim University, Buraydah, Kingdom of Saudi Arabia

Abdurahmani Hamid Nour Faculty of Chemical and Process Engineering Technology, Universiti Malaysia Pahang Al-Sultan Abdullah (UMPSA), Pahang Gambang, Malaysia

Center of Excellence for Advanced Research in Fluid Flow (CARIFF), University Malaysia, Pahang Gambang, Malaysia

Mohammed Elmubark Osman College of Science, Sudan University of Science and Technology, Khartoum, Sudan

El Bashier Sallam Freelance Global and Public Health Expert (FGPHE), Khartoum, Sudan

Abbreviations

ADME	Absorption, distribution, metabolism, and excretion
AF	Acetone fraction
AG	Arabinogalactan
AgNP	Adult-gerontology nurse practitioner
AGP	Arabinogalactan-protein
AGPC	Arabinogalactan-protein complex
AGs	Acacia gums
ASNG	*Acacia senegal* gum
ASYG	*Acacia seyal* gum
BC	Breast cancer
BCRG	Breast Cancer Research Group
BCS	Breast-conserving surgery
BGC	Biosynthetic gene clusters
BHGI	Breast Health Global Initiative
BoC	Breast-on-a-chip
BRCA	BReast CAncer gene
BSE	Breast self-examination
CAM	Complementary and alternative medicine
CBE	Clinical breast examination
CCFAC	Codex Committee on Food Additives and Contaminants
CUPRAC	Cupric ion reducing antioxidant capacity
DCIS	Ductal carcinoma in situ
DPPH	1,1-diphenyl-2-picrylhydrazyl
DSC	Differential scanning calorimetry
EDS	Energy-dispersive X-ray spectroscopy
EELS	Electron energy loss spectroscopy
ER	Estrogen receptors
ESMO	European Society for Medical Oncology
FAO	Food and Agriculture Organization
FDA	Food and Drug Administration
FNA	Fine-needle aspiration

FRAP	Ferric-reducing antioxidant power
FT-IR	Fourier-transform infrared
GA	Gum arabic
GAE	Gum arabic extract
GBCI	Global Breast Cancer Initiative
GPE	Global Partnership for Education
GWAS	Genome-wide association studies
HB	Hydrogen bond
HPLC	High-performance liquid chromatography
HPTLC	High-performance thin-layer chromatography
HRT	Hormone replacement therapy
IARC	International Agency for Research on Cancer
IIUM	International Islamic University Malaysia
IsDB	Islamic Development Bank
JAK	Janus kinase
JECFA	Joint Expert Committee on Food Additives
KPI	Key performance indicators
LMIC	Low- and middle-income countries
LS	Light scattering
MAE	Microwave-assisted extraction
MALLS	Multi-angle laser light scattering
MC	Member countries
MCA	Mosaic chromosomal alterations
MCE	Methanol crude extraction
MDG	Millennium development goals
MENA	Middle East and North Africa
MF	Methanol fraction
MRI	Magnetic resonance imaging
MS	Mass spectra
MTT	Mean transit time [3-(4,5-dimethylthiazol-2-yl)-2,5-diphenyl tetrazolium bromide]
NCD	Non-communicable disease
NIST	National Institute of Standards and Technology
NK	North Kordofan
NMR	Nuclear magnetic resonance
NOPRI	National Oilseed Processing Research Institute
OIC	Organization of Islamic Cooperation
OoC	Organ-on-a-chip
PARP	Poly (ADP-ribose) polymerase
PCR	Polymerase chain reaction
PET	Positron emission tomography
PR	Progesterone receptors
ROS	Reactive oxygen species
RS	Raman spectroscopy
RSM	Response surface methodology

SDG	Sustainable development goals
SERM	Selective estrogen receptor modulators
SFE	Supercritical fluid extraction
SM	Secondary metabolites
SME	Secondary metabolites extract
SPE	Solid-phase extraction
STAT	Signal transducer and activator of transcription
TEM	Transmission electron microscopes
TFC	Total flavonoid contents
TMS	Transportation management system
TNBC	Triple-negative breast cancer
TPC	Total phenolic content
UA	Uronic acid
UAE	Ultrasonic-assisted extraction
UICC	Union for International Cancer Control
UMPSA	Universiti Malaysia Pahang Al-Sultan Abdullah
UV	Ultraviolet
VEGF	Vascular endothelial growth factor
WHC	Water-holding capacity
WHO	World Health Organization

Chapter 1
Acacia Gums (AGs): Characterization and Applications

Khalid Hamid Musa, Abdurahmani Hamid Nour, Ahmed A. M. Elnour, Hashim Suliman Ibrahim, and Idris Musa Adam

Abstract Acacia gums (AGs), particularly *A. senegal* gum, *A. seyal* gum, and *A. polyacantha* gums, are complex polysaccharide materials with unique characteristics and diverse applications. This chapter comprehensively analyzes these gums and explores their botanical origins, geographical distribution, physicochemical properties, and functional roles as emulsifiers, stabilizers, and thickeners. Active ingredients, mainly complex polysaccharides and proteins, are investigated using modern analytical techniques, such as Raman spectroscopy, FT-IR spectroscopy, differential scanning calorimetry (DSC), and transmission electron microscopy (TEM). This study revealed that these gums are polyuronides with uronic acid residues, hexose residues in the pyranose configuration, and pentose residues in the furanose form. *A. senegal* and *A. seyal* are sources of gum arabic, a polysaccharide crucial for various applications. The protein content of these gums plays a significant role in their emulsifying and stabilizing properties. This chapter also explores the similarities and differences in polysaccharide composition between AGs and other tree gums such as *Khaya senegalensis* gum, Sterculia tomentose gum, and Olibanum gum. The diverse biological applications of AGs in the pharmaceutical, cosmetic, and food industries are highlighted, emphasizing their multifaceted significance as natural polysaccharide gums that compete with gums from marine algae and higher plants.

K. H. Musa · H. S. Ibrahim
Department of Food Science and Human Nutrition, College of Agriculture and Veterinary Medicine, Qassim University, Buraydah, Kingdom of Saudi Arabia
e-mail: k.musa@qu.edu.sa; h.elsiddig@qu.edu.sa

A. H. Nour · A. A. M. Elnour (✉)
Faculty of Chemical and Process Engineering Technology, Universiti Malaysia Pahang Al-Sultan Abdullah (UMPSA), Pahang Gambang, Malaysia

Centre of Excellence for Advanced Research in Fluid Flow (CARIFF), University Malaysia, Pahang Gambang, Malaysia
e-mail: abrahman@ump.edu.my; adamhassan@ump.edu.my

I. M. Adam
Agricultural Research Corporation (ARC)-Sudan, Wad Madani, Sudan

A. A. M. Elnour (ed.), *Gum Arabic and Breast Cancer Biology*,
https://doi.org/10.1007/978-981-97-8518-6_1

Keywords Acacia gums · Raman spectroscopy · FT-IR spectroscopy · Polyuronides · Polysaccharides · Proteins · Emulsifiers · Stabilizers · Thickeners

Highlights

1. AGs are complex polysaccharides with unique characteristics and diverse applications.
2. A. senegal and A. seyal are sources of gum arabic, a crucial polysaccharide.
3. AGs contain uronic acid, hexose, pentose, and proteins.
4. The protein content of the AGs plays a significant role in their functional properties.
5. AGs exhibit similarities and differences with other tree gums in terms of their polysaccharide composition.
6. AGs have anti-inflammatory, antioxidant, and protective effects in various organs.
7. Natural polysaccharide gums compete with the gums from marine algae and higher plants.

1.1 Introduction

Research focusing on *Acacia polyacantha* and *A. seyal* is limited because they share many similarities with *A. senega L* (Eeswaran et al., 2022; Sarr et al., 2021), a dominant gum in research and application. However, *A. polyacantha* and *A. seyal* are commonly used in households, and their gelling powders are gaining popularity, particularly because of the scarcity of *A. senegal* in the market (Chikamai & Maisharou, 2021). These species are also employed in various industries, including tanning, textiles, paint, gum boots, fuelwood, pharmaceuticals, and the food industry (Zhara et al., 2024). Some studies have highlighted the potential of *A. polyacantha* and *A. seyal* as high-value products and resources (Jebeso et al., 2023). However, further research, particularly from a technological standpoint, is essential to understand their gum characteristics and explore their production areas, potentially extending their cultivation to other arid and semiarid regions (Prasad et al., 2022). Attention from the scientific, trade, and socioeconomic communities is crucial to developing these gums because of their unique properties (Kalilou, 2021). Therefore, increased research and development efforts are necessary to fully exploit the potential of *A. polyacantha* and *A. seyal* gums in various applications.

A. senegal (L.) Willd., *A. polyacantha* Willd., and *A. seyal* Del Weiss, all members of the *Leguminosae* family, are the main sources of gum arabic (GA). *A. senegal* is the leading source, contributing 74.29% of global production, followed by *A. polyacantha* (15.71%), and *A. seyal* (10%) (Ahmed et al., 2020). The high production volume of *A. senegal* is attributed to its superior quality, which makes it ideal for food, pharmaceutical, and cosmetic products (Elella et al., 2021). It contains complex arabinogalactan-proteins that create stable, long-lasting emulsions with excellent textural properties (Wafari & Ta'awu, 2021). Although research on *A. polyacantha* and *A. seyal* is limited (Elnour et al., 2020), there is a growing interest in these alternative sources (Eltahir et al., 2023). The Sahel region relies heavily on these gums for food and other industries because of the declining production of *A. senegal* and the increasing availability of *A. polyacantha* and *A. seyal* (Kalilou, 2021). As a result, exploring alternative gum sources such as *A. polyacantha* and *A. seyal* is becoming increasingly important in regions such as Sahel. Acacia is a genus from which commercial AGs are derived, with over 1300 species worldwide (Ashour et al., 2022). *A. senegal, A. seyal,* and *A. polyacantha* are known to produce high-quality gum (Drabo et al., 2022). These gums, which are rich in polysaccharides such as arabinogalactans, have numerous pharmacological and industrial applications and can replace certain synthetic polymers (Koyyada & Orsu, 2021). Acacia gum, a readily available and cost-effective polysaccharide, is valuable for protein stabilization and other applications. However, the differing properties of species can limit their use (Jain et al., 2024). Although much research has focused on characterizing *A. senegal* and *A. seyal* gums, they have been relatively understudied (Chikamai & Maisharou, 2021).

It is important to report the physical, chemical, and molecular encapsulation of proteins in gum samples containing these AGs species (Al-Hamayda et al., 2023). The impact of time on protein encapsulation and molecular interactions between polysaccharide complexes and loaded proteins is an important factor (Devi et al., 2023). Understanding the variations in the properties of AGs is essential to optimize their use in various applications. To better understand the properties of *A. senegal* and *A. seyal* gum, observations were made on the adsorption properties of tannins on gums were observed in a model system (Hassan et al., 2024). The particle size of the studied gums was smaller than that in previous studies, indicating that the particle size could be controlled and that the method was repeatable (Ghasemzadeh & Modiri, 2020). Thus, the ability to control the particle size of AGs presents new opportunities for their applications in various industries. Tannins in gums can negatively affect zeta potential values, with catechin being nearly neutral in acidic environments (Fang et al., 2021). *A. seyal* gum is characterized by a significantly higher molecular weight and esterification degree than the other gums studied (Rhazi et al., 2020). *A. Senegal* gum has the highest tannin content (Da Silva et al., 2020).

1.2 Structure of AGs

These gums are polyuronides with uronic acid residues that can carry acetyl or methyl groups, often present as metallic salts. The hexose residues are in the pyranose configuration, whereas pentose residues are in the furanose form. GA is a polysaccharide with various sugar units and protein content, which is crucial for applications such as stabilizing emulsions and wound healing (Awadasseid et al., 2019; Gashua et al., 2016; Sánchez et al., 2018). Recent studies have shown that GA is secreted by *A. senegal*, *A. polyacantha*, or *A. nilotica* in the form of sap (Awadasseid et al., 2019). These gums also contain a small amount of protein, approximately 2.5% (w/w) in *A. senegal* and about 1% (w/w) in *A. seyal*, which is an integral part of the structure of the gum (Gashua et al., 2016). It is important to note that gums with a higher protein content do not necessarily produce more stable emulsions, as some *A. seyal* gum samples with a lower protein content (0.8%) have been found to provide better emulsion stability than *A. senegal* gum (Sánchez et al., 2018). Besides the foregoing gums, others have been studied; *Khaya senegalensis* gum has been studied and found to contain galactose, rhamnose, and possibly 4-*O*-methyl, D-glucuronic acid, and galacturonic acid (Oyebanji et al., 2017). This gum has been evaluated for various industrial applications, including binders, film coatings, disintegrants, and controlled release polymers (Ozoude et al., 2020). Studies have also shown that Khaya gum contains polysaccharides, such as arabinose, D-galactose, L-rhamnose, D-galacturonic acid, and 4-*O*-methyl-D-glucuronic acid (Oyebanji et al., 2017). Additionally, Khaya gum has been used as a binding agent in tablets for drug targeting and controlled drug release (Pal & Bera, 2020). The binding effect of *Khaya senegalensis* gum was compared with *that of Khaya grandifolia* gum, showing that *Khaya senegalensis* gum exhibited better tablet-binding properties and mechanical strength (Sa'adiya et al., 2023). Sterculia tomentose gum is comprised of galacturonic acid, beta-D-galactose, glucuronic acid, L-rhamnose, and other residues (Singh & Vashishtha, 2014). Olibanum gum, derived from *Boswellia carterii*, is an arabinogalactan and a polysaccharide containing galactose and galacturonic acid (Singh & Vashishtha, 2014). These findings highlight the similarities in polysaccharide composition between Sterculia tomentosa gum and Olibanum gum.

Acacia gum is known for its heterogeneity due to variations in monomer composition, linking modes, branching of monomer units, and molecular weight distribution (Apolinar-Valiente et al., 2019; Liang et al., 2020; Samaila & Gashua, 2019). Studies have shown that *A. senegal* gum has at least three distinct components when fractionated using hydrophilic affinity chromatography (Apolinar-Valiente et al., 2019). Additionally, the molecular composition of gum differs between *A. senegal* and *A. seyal*, with *A. senegal* gum having a higher degree of branching and more branched galactopyranose residues (Gashua et al., 2016). The proteinaceous portion of gum plays a crucial role in its emulsifying and stabilizing properties (Sánchez et al., 2018; Satti et al., 2024). Acacia gum has been found to have anti-inflammatory and antioxidant properties and protective effects on organs, such as the kidney, liver, and heart (Abu-Serie et al., 2021; Osman et al., 2022). Fraction 1 of *A. senegal* gum contained 88% of the total gum and had low protein content, whereas fraction 2 represented 10% of the total gum with 12% protein content. Fraction 3, which accounted for 1.24% of the total, contained almost 50% protein, specifically arabinogalactan-protein (AGP), which is responsible for the emulsifying properties of GA (Osman et al., 2022). This highlights the diverse nature of GA derived from *A. senegal* and other related trees (Abu-Serie et al., 2021). This study also emphasizes the presence of different molecular weight components in these fractions (Abu-Serie et al., 2021). The protein content of these fractions plays a crucial role in determining their functional properties, such as emulsification (Dahi et al., 2019). The structures of gums from different tree species typically consist of D-galactose and uronic acid residues forming the backbone of the polysaccharide, with predominant 1–3 and 1–6 linkages in the side chains. These gums often contain D-xylopyranose, L-arabinose, and L-arabino-furanose linkages. For instance, gums from peach trees are characterized by arabinogalactan groups that contain arabinose, xylose, galactose, glucose, and uronic acids (Yang et al., 2018). Additionally, tree gums, such as *A. tortuosa*, exhibit relatively high nitrogen content compared to other Acacia species (Martínez et al., 2015). Natural polysaccharide gums, commonly known as hydrocolloids, have emerged as unique industrially important substances that compete with the gums from marine algae and higher plants (Satti et al., 2024).

AGs are natural exudates obtained from trees through spontaneous release or induced by mechanical injury or microbial invasion. These gums typically contain heteropolysaccharides and condensed tannins as their major constituents (Awadasseid et al., 2019). GA is a secretion found on the trunks and branches of *A. senegal* and *A. seyal* trees and is characterized by a complex polysaccharide structure with a galactose backbone and ramified side chains (Fauconnier et al., 2000). The molecular composition of these gums can vary, with *A. senegal* and *A. seyal* exhibiting differences in protein content and structure (Gashua et al., 2015, 2016). These exudates have diverse applications in food, pharmaceuticals, and cosmetics because of their unique physicochemical properties (Palma et al., 2015; Sethuraman & Rajendran, 2018). Gums are materials that impact touch, taste, and sight, collectively known as "Gummosis." They are polysaccharides that disperse in water to form viscous solutions, gels, or colloidal dispersions. Various gums, such as xanthan gum, GA, and *tragacanth gum*, are utilized in various applications, such as encapsulation for drug delivery, tissue engineering, and food packaging, owing to their properties such as stability, emulsification, and biocompatibility (Abdi, 2024; Bertolini et al., 2001; Jiménez-Fernández et al., 2021). These gums play a crucial role in microencapsulation techniques by providing stability, protection, and controlled release of active ingredients (Ahad et al., 2022; Nhan et al., 2020). The use of gums in composite scaffolds and hydrogels demonstrates their versatility in diverse fields, such as materials science and chemical engineering (Santos et al., 2019; Wardhono et al., 2014). The study of polysaccharides, such as xanthan gum and gellan gum, shows their potential in creating functional materials with tunable properties for various applications (Soto et al., 2016; Zhao, 2023).

Tambe and Sharma (1993) elucidated that gums comprise extended chains with substantial molecular weights, forming polymers. When these polymers dissolve in water, they produce gelling or thickening effects and secondary functional properties such as emulsification, stabilization, and encapsulation. Gums or hydrocolloids are generally long-chain, straight-to-branched polysaccharides that contain hydroxyl groups that can bond with water molecules (Sartori et al., 2004). These chains are composed of 1×10^3 to 2×10^3 monosaccharide units (Fig. 1.1). Sugar monomers typically include linked-side units or substituent groups, such as sulfate, methyl, ether, ester, and acetals (Elnour, 2008; Jones & Smith, 1949). Gums primarily consist of C, H, O, and N elements, while acidic gums contain mostly Ca, Mg, Na, and Fe as cations, as previously reported (Elnour, 2008; Jones & Smith, 1949).

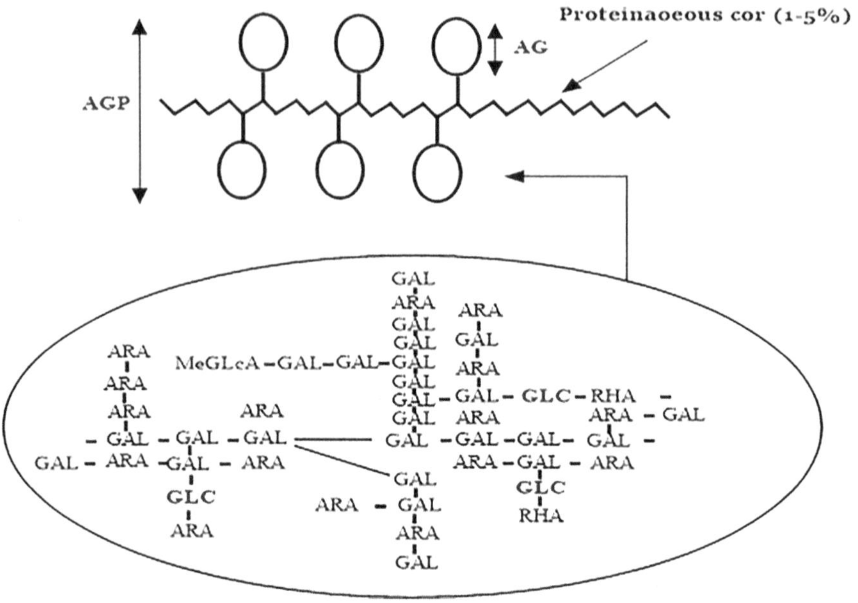

Fig. 1.1 The structure of gum arabic. (Source: Adapted from Qi et al., 1991. *ARA* arabinose, *GAL* galactose, *GLC* glucuronic acid, *AGP* arabino-galactoprotein, and gum arabic (*GA*): glactoprotein, respectively)

1.3 Trees of Acacia Gums (AGs)

As reported by Al-Assaf et al. (2005a) and López-Franco et al. (2015), GA has been an important factor in economic activity since ancient times. It is an exudate from trees of the genus Acacia, which is one of the most abundant groups in the plant kingdom. Acacia is a member of the tribe Acacieae and is classified as a member of the Mimosoidsae family, making it the second largest genus in the Leguminosae family. GA species can be found in warm and tropical regions, with the majority (955 species) originating in Australia and a significant number (185 species) in the Americas, 144 in Africa, and 89 in Asia. Historically, GA has primarily been obtained from *A. senegal, A. polyacantha,* and *A. seyal* trees, which grow in the Sahelian zone of Africa, extending from the west of Senegal to the east of Eritrea and Ethiopia (Kennedy et al., 2011). GA is widely used in Sudan to incorporate two varieties that are produced and traded at the same levels: "Hashab" (from *A. senegal*) and "Talha" (from *A. seyal*). These two varieties are the primary sources of GA in Sudan, accounting for up to 95% of the total gum production, with *A. senegal contributing* approximately 70% and *A. seyal* contributing 15–25% (Seifel Din et al., 1996). GA is commonly defined as gum derived from any species of acacia (FAO, 1995). Additionally, Sudan produced approximately 85% of the world's GA supply in the early 1980s, increasing production due to high global demand

Table 1.1 The classification of Acaicia gums (AGs)

Type of trees	Botanical classification	Location	Reference
Acacia sengal	Family: *Leguminosae* Subfamily: Mimosaceae Genus: Acacia Species: senegal Botanical name: *A. senegal* Common name: Hashab	Sudan	Heywood (1993)
A. seyal	Family: Leguminosae Subfamily: Mimosaceae Genus: Acacia Species: Seyal Botanical name: *A. seyal* Common name: Talha	Sudan	Ahmed (2015)
Acacia polyacantha	Family: *Leguminosae* Subfamily: Mimosaceae Genus: Acacia Species: *Polyacantha* English name: Flacons claw Acacia Common name: Kakamut/Umsinina	Sudan	Vogt (1995)

(Ibrahim, 2015). The classifications and common names of this group are listed in Table 1.1.

1.4 General Distribution and Description of AGs

1.4.1 A. senegal *Trees Distribution and Habitat Gum*

Senegal has two primary areas of distribution: (i) stabilized sands, where yearly rainfall ranges from 280 to 450 mm, and (ii) dark cracking clays, with a rainfall range of 300–700 mm per year (Sahni, 1968; Vogt, 1995). The plant is widespread in savanna grasslands on sandy and clay plains (El Amin, 1990). The species is found in central Sudan along a continuous belt that extends from east to west but is more successful in the western sand plains of Kordofan and Darfur (Elkhalifa et al., 2003).

Hashab tree is typically a bush or tree that ranges in height from 2 to 12 m, with bark that is yellow to light brown or gray in color and rough in texture, often flaking or fissuring. According to Sahni (1968) and Vogt (1995), trees typically reach a height of less than 6 m, although they can sometimes grow to 12 m tall. The stipules on the tree are not spinescent and are arranged in a unique pattern with two pointing upward or forward and one pointing downward or backward, creating a falcate shape that is 4–7 mm long. The inflorescence of the Hashab tree is cylindrical, and the spike is 2–10 cm long and usually longer than the leaves. The flowers were white or creamy. According to El Amin (1990), Sahni (1968), Thirakul (1984), and Vogt

(1995), the pods of the tree are pale brown to straw-colored, flat, and papery, and are typically 9 cm in length and rounded to acuminate. The seeds were compressed and ranged in diameter from to 8–12 mm, with a color that was either yellow or pale brown. The tree typically flowers from November to February, and fruits from January to April, as indicated by El Amin (1990), Sahni (1968), Thirakul (1984), and Vogt (1995).

A. senegal naturally grows in this belt's dry regions and sandy soils, with annual rainfall ranging from 280 to 450 mm. It can also be found in other countries, such as the Ivory Coast, Iran, Pakistan, and western India (NAS, 1979; Ross, 1968). In Sudan, the GA belt extends across six states from east to west, with the study area located in the North Kordofan state. According to the FAO (1995), *A. senegal* trees originate in the dry areas of tropical and subtropical Africa. El Amin (1990) suggested that southern Sudan may be the center from which the species is distributed.

Trees not only enhance soil fertility but also fulfill crucial household needs for wood energy and fodder (Evans, 1992). In addition, it is an integral part of the traditional *A. senegal*-based agroforestry system, which is recognized as one of the most effective natural forest management practices in tropical drylands (Pretzscha et al., 2014). Trees are highly sustainable because of their environmental, social, and ecological benefits (Balla, 2004). Sudan is the world's leading producer of GA, accounting for 80% of the global production, followed by Chad and Nigeria (Verbeken et al., 2003). Thus, the gum arabic belt, as shown in Fig. 1.2, is highly beneficial and is internationally known as the GA belt.

1.4.2 Acacia seyal *Distribution and Habitat*

The distribution of various species is extensive on the drab, fractured clay found on the upper slopes of rivers and in canyons in the central region of Sudan, and on the hard-cracking surface in the center of the country. Additionally, these species can be found in the constantly wet gorges in this region. According to Sahni (1968), Sudan's central and southern regions also contain these species. Two varieties of the species are present in Sudan: Var seyal, which is characterized by the absence of expanded spines and red or green-white bark, and is widespread in the savanna woodlands and grasslands on dry cracking clay; and var. fistula Schweinf, which has expanded spines and whitish bark. The term "GA belt" refers to a zone that covers an area of approximately 520,000 km^2 and extends across central Sudan between latitudes 10° and 14°N, accounting for one-fifth of the country's total area. This belt also includes parts of Ethiopia, a small portion of Eritrea, and covers Sudan, Chad, a portion of Nigeria, Burkina Faso, Senegal, and Mauritania (Fig. 1.2).

The following description pertains to a tree with a height ranging from 3 to 17 m, as reported by El Amin (1990), Sahni (1968), Thirakul (1984), and Vogt (1995).

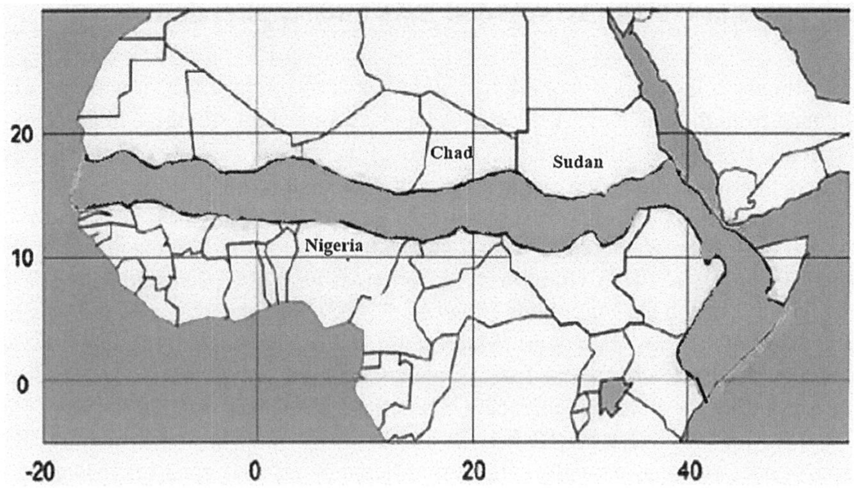

Fig. 1.2 Gum arabic belt in Africa and Sudan. (Source: Adapted from Adam, 2011)

This tree has powdery bark that is either smooth or flakes off sparsely, with a color ranging from whitish, greenish-yellow, or orange-red. In some cases, the bark was both green and red. The tree also has young branchlets with numerous reddish glands (Sahni, 1968). It is stipular, meaning it has spinescent pairs, is up to 8 cm long, and produces olive to olive brown, faintly wrinkled, elliptic, and compressed seeds that measure 7–9 mm by 4.5–5 mm. The tree flowers from November to April and fruits from January to May. It is found worldwide, as reported by Sahni (1968).

1.4.3 Acacia polyacantha *Distribution and Habitat*

A. polyacantha trees, known as kakamut, are distributed throughout tropical Africa. In Sudan, various regional varieties typically grow along rivers and valleys, where the water table is relatively high and the soil is suitable for growing healthy and organic *A. polyacantha* trees (Vogt, 1995). These trees can reach a height of up to 20 m, with trunks measuring up to 70 cm in diameter (Vogt, 1995). The most noticeable features of *A. polyacantha* are its knobbly bark and paired thorns. The bark is yellowish with brown scales, and thorns occur in pairs with sharply curved brown tips. The leaves can grow up to 25 cm long and are paripinnate, with 10–40 pairs of pinnate leaves and 35–60 leaflets per leaf. A prominent gland is present at the base of each leaf. The flowers grow in pairs or as three spicate racemes from the leaf axil and are cream colored and strongly scented. The fruit of *A. polyacantha* is a pod that can grow up to 15 cm in length, containing 5–9 seeds. This species has been found worldwide.

1.5 Acacia Gums (AGs) Wood Uses and Applications

The primary use of wood from *A. senegal* trees is to produce GA, a natural gum obtained from the tree bark. *A. senegal* is known for its ability to thrive in arid environments. It is widely cultivated for GA production and has a variety of applications in food, pharmaceutical, and cosmetic industries (Lelon et al., 2013; Sethuraman & Rajendran, 2018). Furthermore, research has been conducted on the potential health benefits of *A. senegal*, including its anti-atherosclerotic and cardioprotective properties (Ram & Jatwa, 2014). Gums derived from trees have also been studied for their prebiotic effects on gut microbiota (Rawi et al., 2021) and their use as stabilizers in food items such as low-fat yogurt (Mugo et al., 2020). The wood of *A. seyal* trees is used as a fumigant, whereas the leaves and bark are used to treat gastric ulcers. Gum extracted from the plant has therapeutic benefits for rheumatism, and the pods are useful for feeding livestock (Radwan et al., 2020). *A. seyal* has been traditionally used to treat malaria, wounds, cancer, diarrhea, edema, and dandruff (Mengesha et al., 2022). Tree species are widely employed in agroforestry practices in Ethiopia (Abera & Lemma, 2019). *A. senegal* trees have been shown to possess several medicinal properties, including antimicrobial activity against food-borne and plant pathogens (Mekbib et al., 2009). Additionally, it has been studied for its potential to serve as an antidepressant (Shehu et al., 2020) and anticancer agent (Zingue et al., 2018). The wood of the *A. polyacantha* tree is primarily utilized for fuel and high-quality charcoal, as well as for fence posts, farm implements, railway sleepers, beams, and rafters. Tree gum is consumable and is used as an adhesive to treat textile fibers. Furthermore, the roots are utilized as a general health tonic and an antidote for snake bites and venereal diseases. Preparations derived from the bark are employed to address general stomach disorders (Voget, 1995). Overall, the diverse properties and applications of *A. senegal tree* wood make it useful for various purposes (Fig. 1.3).

1.6 Physicochemical Properties of AGs

The physical properties of natural gums, which are crucial for their commercial value and application, are influenced by factors such as botanical source, climatic conditions, season of collection, tree age, and post-collection treatments (Sarkar et al., 2018). These properties vary significantly even within the same species. Natural gums offer advantages over synthetic gums owing to their biocompatibility, low cost, low toxicity, and abundant availability from renewable sources (Sarkar et al., 2018). The unique physicochemical properties of natural gums obtained from various tree species make them versatile for various applications (Singh & Lariya, 2018). Additionally, the safety, cost-effectiveness, and availability of natural plant gums make them preferable to synthetic alternatives (Singh & Lariya, 2018). The physical characteristics of natural gum are the most important factors shaping its

[A] *A. senegal* gum [B] *A. seyal* gum [C] *A. polyacantha* gum

Fig. 1.3 Acacia gums (AGs) source and types

application and commercial value (Hirst & Jones, 1958). These properties differ from those of gums from various sources and exhibit differences in the same species when obtained from plants that grew under diverse climates or during different seasons during the year. Siddig (2016) highlighted that the physical characteristics might also be influenced by tree age and the gum treatment method upon collection, such as washing, sun bleaching, drying, and storage temperature. GA chemical formation is complicated and is composed of macromolecular groups that are defined by a high percentage (97%) of carbohydrates, which consist of L-arabinose and D-galactose units and a small protein fraction (<3%) (Dickinson, 2009).

The chemical construction of GA may deviate slightly based on weather, origin, harvest period, age of the tree, and the provisions of the process, such as spray drying (Al-Assaf et al., 2005a). Consequently, some variance exists in the chemical constituents of GA derived from *A. seyal* and *A. senegal*. The same sugar quantity occurred in both species of GA. A lower amount of rhamnose and glucuronic acid are available in *A. seyal* gum, whereas a more significant amount of arabinose and 4-*O*-methyl glucuronic acid is available in *A. senegal* gum. In contrast, *A. seyal* gum contains a smaller amount of nitrogen with entirely different rotations (Table 1.2). Assessment of the following constituents may distinguish the two species (Osman et al., 1993). GA is a complex branched-chain polysaccharide, either neutral or acidic, found as mixed potassium, magnesium, and calcium salts of a polysaccharide acid (arabic acid). The backbone consists of β-D-galactopyranosyl units. The side chains contain 2–5 β-D-galactopyranosyl units, which are linked to the main chain by 1,6-bonds. Both the central and side chains are comprised of a-L-arabinofuranosyl, a-L-rhamnopyranosyl, β-D-glucuronopyranosyl, and 4-*O*-methyl-β-D-glucuronopy ranosyl, with the last two as terminal units (Verbeken et al., 2003). GA is a complex substance with both hydrophilic and hydrophobic properties. The physicochemical responses of GA can be controlled by balancing the interactions between hydrophilic and hydrophobic units. GA is highly soluble in water and has a moderately low viscosity, in contrast to other gums. At concentrations above 5.0%, most gums do not dissolve in water because of their high viscosity (Taha

Table 1.2 A comparative chemical composition and some properties of GA obtained from *A. senegal, A. seyal*, and *A. polyacantha* gum

Parameter	*Acacia senegal* gum	*Acacia seyal* gum	*Acacia polyacantha* gum
Rhaminose %	14.00	3.00	15.00
Arabinose %	29.00	41.00	4.000
Galactose %	36.00	32.00	–
Glucuronic acid%	14.50	6.500	15.00
Nitrogen %	0.365	0.1470	0.410
Protein %	2.410	0.9700	2.600
Specific rotation (degrees)	−30	+51.00	−19.000
Average molecular weight (kDa)	380.0	850.00	830.0
Na (μg/g)	44.00	42.00	47.00
Ca (μg/g)	0.400	0.480	0.464

Source: Elnour et al. (2011)

et al., 2012). Nonetheless, GA can be dissolved in water at 50% w/v to create an acidic solution (pH 4.5). GA molecules with a high branching structure result in a relatively small, compacted, and hydrodynamic bulk; therefore, at high concentrations, GA forms a viscous solution, whereas less than 10% GA in the solution results in low viscosity and obeys Newtonian behavior (Webb & Dingwell, 1990). Table 1.2 presents some of the physicochemical properties of the AGs.

1.6.1 Solubility

Gums can be categorized into three groups based on their solubility: entirely soluble gums like *A. senegal* and *A. seyal*, partially soluble gums like Gatti gum, and insoluble gums like *Tragacanth* gum. These classifications determine their effectiveness in industrial applications that require uniform solutions, controlled release, or stabilizing properties, without full integration into the solution (Abdolmaleki et al., 2021). The selection of gums for specific applications depends on criteria such as solution clarity, solubility at different temperatures, suspension ability, and cost (Amid et al., 2013). Different gums are stabilizers, thickeners, emulsifiers, and suspending agents, based on their composition and rheological performance (Anusha et al., 2024). The factors influencing the solubility profiles of AGs in various solvents include pH, solubility, swelling power, water-binding capacity, and viscosity (Owusu et al., 2020). *A. senegal* gum is known for its superior emulsification properties compared to *A. seyal* gum (Gashua et al., 2016). The molecular characteristics and associations of gums play a role in their emulsification properties (Gashua et al., 2015). AGs contain complex biopolymers, with highly glycosylated proteins and heavily branched sugars (Davantès et al., 2019). The nature and distribution of the proteinaceous component of gum are crucial for its emulsifying capacity and stability (Sánchez et al., 2018). AGs are rich in soluble fibers and exhibit low viscosity (Silva & Souza, 2018). The molecular composition and properties of the gums influence their solubility and emulsification characteristics.

1.6.2 Color

The colors of the gums vary from water-white (colorless) to shades of yellow to black. The best grades of gum are almost colorless, with slight traces of yellow; some possess pink likes (Siddig, 1996). On the other hand, dark or black gums sometimes occur, such as mesquite gum. There were also pale rose pinks, darker pink gums, and yellowish gums. This pink color is probably due to the different quantities of tannin (Omer, 2004). The natural color variation of AGs is related to their chemical compositions. *A. senegal* gum, *known as "Hashab gum,"* has a milky white appearance and is hard, while *A. seyal* gum, *called "Talha gum,"* is amber-yellow in color and friable (Ashour et al., 2022). These gums, derived from Acacia trees, are hydrophilic composite polysaccharides with excellent emulsifying and stabilizing properties (Palma et al., 2015). They consist mainly of high-molecular-weight polysaccharides, arabinogalactan, arabinogalactan-protein, and glycoprotein (Kiprop et al., 2021). Owing to their properties, these gums have applications in various industries, such as food, pharmaceuticals, cosmetics, and lithography (Sethuraman & Rajendran, 2018). The molecular composition and structure of the gums harvested from *A. senegal* and *A. seyal* differ, impacting their potential applications (Gashua et al., 2015).

1.6.3 Shape

Natural gums such as GA and *Tragacanth* gum can exhibit various shapes, such as tear or drop shapes, flakes, or thread-like ribbons. Initially smooth, the surfaces of these gums may develop roughness or small cracks over time (Wang et al., 2021). Tear-drop-shaped globules of GA are produced from acacia trees (Eiroboyi et al., 2019). These gums play a crucial role in various applications, such as encapsulation of aromatic compounds and essential oils, drug compounding, and as coating materials for microencapsulation, protecting environmental factors (Pinho et al., 2024; Surini et al., 2018; Traffano-Schiffo et al., 2017). The shape and size of microcapsules can be influenced by the encapsulation wall materials used, such as GA, and the drying method employed (Martinović et al., 2023; Ploypetchara et al., 2021). The high viscosity of solutions containing GA can contribute to the spherical shape of encapsulated materials (Kaur et al., 2023). Figure 1.4 shows some examples of the shape and color of AGs collected from different locations.

1.6.4 Moisture Content

The moisture content of good-quality gum does not exceed 15% and 10% for granular and spray-dried materials, respectively (NACA/Joint Expert Committee on Food Additives (JECFA), 1999). It shows the hardness of the gum and, hence, the variability of densities, number of masses, and amount of air entrapped during formation. Elnour (2008) and Omer (2004) reported that the moisture content of *A. polyacantha* gum is approximately 8.2% and 9.5%, respectively. The variation in

Fig. 1.4 Shape and color of AGs collected from different locations

moisture content among AGs affects their storage stability and industrial application. *A. senegal* gum typically has a higher moisture content than that of *A. seyal* gum (Gashua et al., 2016; Satti et al., 2024). Higher moisture content can help retard staling during storage (Israr et al., 2016). Moisture content is crucial for storage stability, as it affects microbial growth, powder caking, and material stability (Moscoso-Moscoso et al., 2024; Nikolić et al., 2018; Chen et al., 2016a). It is essential to form microcapsules that influence the drying efficiency, powder flowability, and glass transition behavior (Chen et al., 2016a). Moreover, moisture content variations affect the properties of encapsulated materials, affecting their stability and shelf life (Awulachew, 2021; Mazumder & Ranganathan, 2020). Therefore, understanding and controlling the moisture content is vital for optimizing the storage stability and industrial applications of these gums.

The determination was conducted according to the White (1957). The crucibles were dried in a hearus oven at 105 °C for 30 min, cooled in a desiccator, and then weighed (M_1). Approximately 2 g of the sample was placed in a crucible and weighed accurately (M_2). The contents were heated in an oven for 5 h at 105 °C, cooled in a desiccator, and reweighed (M_3). The percentage of loss during drying was calculated as follows:

$$Moisture\,content = \frac{M2 - M3}{M2 - M1} \times 100 \tag{1.1}$$

where:
M_1: Weight of the empty crucible
M_2: Weight of crucible + sample
M_3: Weight of crucible + sample after drying

1.6.5 Ash Content

Ash content is a measure of the amount of inorganic residue remaining after organic matter has been burned. Inorganic residues existed as elements. Siddig (1996) explained that the type of soil (clay or sand) significantly affected the ash content; previously, the ash content of *A. polyacantha* gum was determined to be 2.929 ash% (Anderson et al., 1985). Recently, FAO (1995) reported that the ash content of *A. senegal* gum was less than 4%. The ash content of gums, such as *A. polyacantha* and *A. senegal,* can vary significantly depending on soil type (Elhenawy et al., 2014). For instance, *A. polyacantha* gum has been reported to have an ash content of 2.9% (Singh & Immanuel, 2022). Additionally, the ash content of *A. senegal* gum should not exceed 4% (Ndindeng et al., 2015). The type of soil, whether clay or sand, plays a crucial role in determining ash content (Nuraziz & Khairi, 2024). Moreover, the addition of certain ingredients, such as guar gum, can influence the ash content of products (Singh & Immanuel, 2022). Therefore, understanding the sources and factors affecting ash content is essential for various applications ranging from food products to pharmaceutical formulations.

The total ash content was determined according to the AOAC (1957). The crucibles were heated in an oven for 30 min, cooled in a desiccator, and then weighed (W_1). Approximately, 2 g of the sample was placed in a crucible and accurately weighed (W_2), then ignited at 550 °C in a Heracus electronic muffle furnace for 2 h, cooled in a desiccator, and weighed (W_3). The total ash content is calculated as follows:

$$Ash\% = \frac{W3 - W1}{W2 - W1} \times 100 \tag{1.2}$$

where:
W_1: Weight of the empty crucible
W_2: Weight of crucible + sample
W_3: Weight of crucible + sample after drying

1.6.6 *Viscosity*

The viscosity of a liquid is its resistance to shearing, stirring, or flowing through a capillary tube (Chen et al., 2016b). Studies of the flow of gum solutions play an important role in the identification and characterization of their molecular structures. Because viscosity involves the size and shape of the macromolecule, it is considered to be one of the most important analytical and commercial parameters (Anderson et al., 1983). The viscosity of a solution may vary with its composition because of the possibility of hydrogen bonding between the solute and solvent molecules (Pimentel et al., 2016). More hydroxyl groups have high viscosities because a network of hydrogen bonds is formed between the molecules; this network extends throughout the liquid, making flow difficult. The viscosity of the gum solutions was inversely proportional to temperature. They also found that the viscosity of GA solutions changes with pH; however, they found a maximum viscosity at pH 6–7. Viscosity can be explained in terms of relative viscosity, specific viscosity, reduced viscosity, inherent viscosity, and intrinsic viscosity; it is also represented as kinematic or dynamic viscosity. Anderson et al. (1983) reported that the intrinsic viscosity of *A. polyacantha* gum is 15.8 ml/g. Omer (2004) recently reported 10.34 an intrinsic viscosity for *A. polyacantha* gum. Relative viscosity (ηr) was calculated using the following equation:

$$[\eta r] = \frac{T - T_{\circ}}{T_{\circ}} \quad 1.3$$

where:

Specific viscosity	ךr − 1
Reduced viscosity	ךsp/C%
Intrinsic viscosity	*T*□ − *T*□
	T□ c% reduced >0
T	Flow time of sample solution expressed in seconds
T°	Flow time of solvent (distilled water) expressed in seconds

The reduced viscosity (ךrd) was determined for different concentrations of gum solution: 0.12, 0.25, 0.50, and 1%, and then calculated using the above equation. Extrapolating the reduced viscosity against concentrations back to zero determines the intrinsic viscosity. The intercept on the *y*-axis yields (ך). The most advanced instrument for detecting viscosity is the viscometer, as shown in Fig. 1.5.

1.6.7 Acidity

The hydrogen ion concentration is significant in the chemistry and industry of gums, and therefore, the functional properties of gums are affected by changes in pH, such as viscosity and emulsifying power (Wang & Cui, 2005). Arabic acid is a major

Fig. 1.5 Viscometer

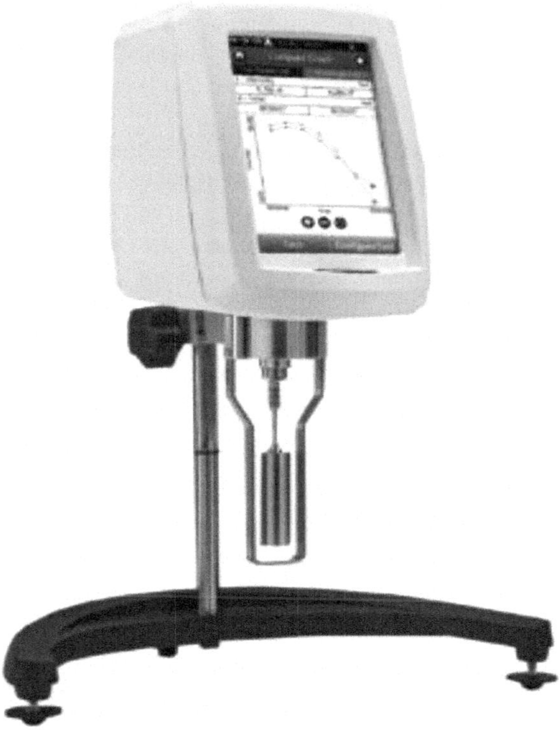

component of commercial GA. When decomposed, it produces arabinose; therefore, GA is called Arabic acid. Therefore, the gum solution was moderately acidic (pH 4.5). Fang et al. (2010) reported a pH value of 4.9, and Omer (2004) reported it at 4.7–5.2 with a mean value of 5.2. The pH of the GA solutions varies slightly, with reported values ranging from 4.7 to 5.2 (Rodriguez-Sandoval et al., 2014). The pH of the solution was moderately acidic (approximately 4.5) (Rodriguez-Sandoval et al., 2014). Changes in pH affect the functional properties of gum, such as viscosity and emulsifying power, because of the significant impact of the hydrogen ion concentration (Vuletić et al., 2017). The pH of gum solutions can be adjusted to the desired value by adding specific amounts of NaOH or HCl (Mahfoudhi et al., 2014). Additionally, the pH of the solution influences the adsorption behavior of arabinogalactan-proteins from *A. senegal* gum at solid–liquid interfaces (Davantès et al., 2021).

1.6.8 Protein Content

A. senegal gum has approximately 2.5% protein content, whereas *A. seyal* gum has approximately 1% protein content, as reported by Gashua et al. (2016). *A. senegal* gum also contains more specific amino acids than *A. seyal* gum, as stated by Ashour et al. (2022). Furthermore, *A. polyacantha* gum had a protein content of approximately 2.30%, aligning with the values Elhenawy et al. reported (2014). These variations in protein content among the three types of AGs could affect their functional properties such as emulsion stability, as noted by Sánchez et al. (2018). The protein content of AGs plays a crucial role in their emulsifying and stabilizing properties, as reported by Sánchez et al. (2018). The protein content of GA has been intensively investigated for its impact on the structure, physicochemical properties, and functionality, as reported by Xiang et al. (2015). There is a strong correlation between the protein content of the gum and its emulsifying stability, as indicated by Dickinson et al. (1991). Idris et al. (1998) found that the protein contents of fresh samples were consistent, with a value of 2%, regardless of the tree's age. Siddig (2016) reported that the average nitrogen content of commercial samples of *A. senegal* gum and 80% of authenticated samples analyzed were in the range (0.27–0.39%). Liang et al. (2020) reported that the mean nitrogen content of GA samples was 0.35%.

1.6.9 Specific Rotation

A. senegal gum typically exhibits a negative specific optical rotation, whereas *A. seyal* gum is characterized by a highly positive optical rotation (Satti et al., 2024). However, *A. polyacantha* gum has properties that differ from those of both *A. senegal* and *A. seyal* gums (Elhenawy et al., 2014). The specific rotation values for these gums vary significantly, with *A. senegal* gum showing a higher potassium content and emulsion capacity than *A. seyal* gum, which has a slightly higher calcium content and a lower viscosity and ash content (Mudgil & Mudgil, 2024). These differences in the specific rotation values and chemical compositions contribute to the distinct characteristics and potential applications of each type of AGs. The relationship between the optical activity of organic molecules such as saccharides and carbohydrates and their structure is a characteristic property of the substance (Tamaru et al., 2001). Natural gums, such as *A. senegal* gum, rotate the plane of the polarized light (Biswas et al., 2000). The direction and magnitude of this rotation are considered as diagnostic parameters (Biswas et al., 2000). *A. senegal* gum has a negative optical rotation ranging between −20° and −34°, differentiating it from other botanically related AGs (Anderson & Stoddart, 1966). Pure gum from *A. senegal* has a specific optical rotation of −27° to −30° (Mhinzi, 2003), whereas certain variations in the degree of optical rotation (−27° to −32°) have been observed (Anderson et al., 1968). Karamalla et al. (1998) found that the mean of the specific optical rotation of commercial *A. senegal* gum was −30.54°. Autohydrolysis or variation does not affect optical rotation; however, mild acidic hydrolysis significantly affects optical rotation (Barron et al., 2007). Omer (2004) reported that the mean specific rotation of authenticated *A. polyacantha* gum samples was −16.6°. The specific rotation can be estimated as:

$$\alpha = \frac{\theta}{L \times c} \tag{1.4}$$

where L is the tube length and c is the concentration. c is defined as the dissolved mass of the compound (g) per unit volume (c). Figure 1.6 shows schematically the process of the optical rotation unit for detecting the optical rotation of AGs.

1.6.10 Equivalent Weight and Uronic Acid Anhydride

The titrable acidity, which is the volume of 0.02 N sodium hydroxide solution needed to neutralize 10 ml of a 3% gum solution, is used to determine the acid equivalent weight of the gum, from which the uronic acid content can be calculated (Mhinzi, 2003; Karamalla, 1965). *A. senegal* gum typically contains approximately 20% uronic acid (Atobe et al., 2015), whereas *A. polyacantha* gum has an average uronic acid content of 14.61% (Elhenawy et al., 2014). In contrast, *A. seyal* gum has an acid equivalent weight ranging from 1153 to 1500 and uronic acid content between 12.93% and 16.33% (Satti et al., 2024). These differences in the uronic acid content contribute to the unique properties and potential applications of each type of gum. Uronic acids are natural polysaccharides that are widely distributed in animal and plant tissues. Various methods have been developed to determine uronic acids, including colorimetric, decarboxylation, and acid-base titration methods. Gums vary greatly in their equivalent weight and uronic acid content (Karamalla,

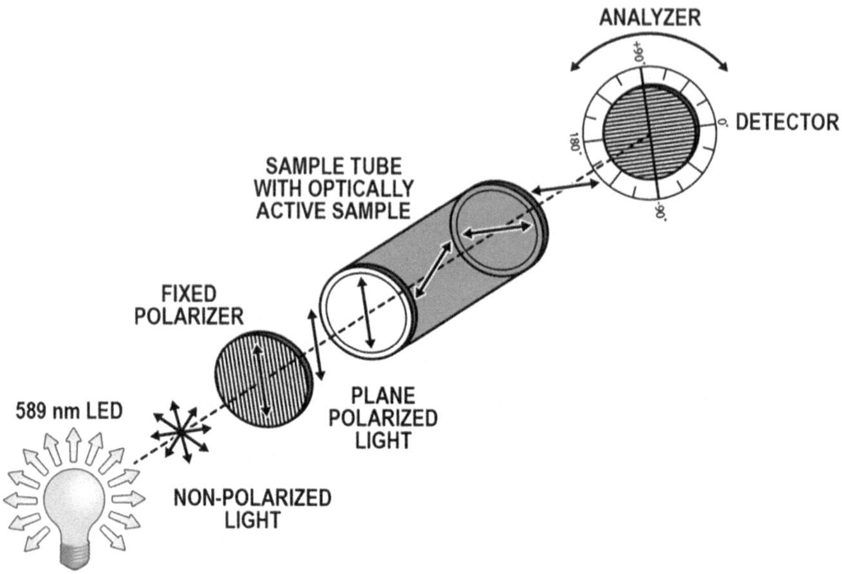

Fig. 1.6 Optical rotation unit

1965). Anderson et al., (1967) reported that the equivalent weight and uronic acid content of *A. polyacantha* gum are 1900 and 9%, respectively.

1.6.11 Molecular Weight

The molecular weights of polymers can be determined by physical measurements or chemical applications. *A. senegal* and *A. seyal* gums have different molecular weights. *A. senegal* gum has an average molecular mass of approximately 380,000, while *A. seyal* gum has an average molecular mass of approximately 850,000 (Mahendran et al., 2008). These differences in molecular weight contribute to the distinct properties and potential applications of these gums. *A. polyacantha* gum, on the other hand, has been reported to have specific rotation, nitrogen content, and molecular weight that differ from *A. senegal* gum (Mugo et al., 2020). Therefore, the molecular weights of these three types of gums vary, affecting their characteristics and potential uses.

The application of chemical methods requires that the structure of the polymer contain a well-known number of functional groups per molecule, and they invariably occur as end groups. The end-group analysis method provides an approximate number of molecules in each sample weight; they yield the average number of molecules for polymeric materials. This method becomes insensitive at high molecular weights as the fraction of end groups becomes too small to be measured precisely (Glicksman, 2019). This is because fraudulent sources of the end groups not considered in the assumed reaction mechanism steadily become consequential as the molecular weight increases, and the number of end groups diminishes to such an extent that their quantity determination is not feasible. These reactions restrict the frequent application of chemical methods to condensation polymers with average molecular weights seldom exceeding 2.5×10^3 (Krigbaum & Flory, 1953).

Physical methods frequently used for establishing polymer molecular weight are osmometry, polymer viscosity, measurement of the coefficient of diffusion, ultracentrifugation, and light scattering. One of the most recent advanced methods is light scattering (LS), which provides an absolute method for polymer molecular weight and size measurements. The LS is rapid, accurate, and requires a small sample size. The molecular weight of gums varies greatly in terms of gum heterogeneity, as well as variation in techniques used to separate, purify, and determine the molecular weight. A 3.0×10^3 was reported by Al-Assaf et al. (2005b) using the centrifugal method. Using the LS technique Mukherjee and Deb (1962) reported Mw up to 5.8×10^5; and Fenyo and Vandevelde (1990) reported a range of 4.0×106 to 2.2×10^6. Recently, GPC coupled online to multi-angle laser LS (MALLS) has been demonstrated to be a very powerful method for characterizing highly polydisperse polymer systems, and the molecular weight of *A. senegal* gum was found to be equivalent to 5.4×10^5 (Picton et al., 2000). The average molecular weight of *A. polyacantha* gum using GPC-MALLS was reported to range between 2.94×10^5 and 7.346×10^5 Omer (2004). The most recent E 414 specification of the (Alicja et al., 2017) has now re-focused attention on the problem of molecular weight, since it has included in the product definition, the requirement that the molecular weight should be approximately 3.5×10^5 is far below the reported value for *A. senegal* var. senegal. As a result, accurate determination of molecular weight using advanced techniques is crucial for the standardization and application of these gums.

1.7 Functional Properties of AGs

1.7.1 Emulsifying Properties

Emulsions are chemical mixtures of liquids that are immiscible under ordinary conditions and can be separated into two layers on standing, heating, and freezing by agitation and the addition of other chemicals (Liu et al., 2018). Emulsifying agents act as surface-active agents, which, when added to an emulsion, would increase its stability by interfacial action. Each emulsifying agent depends on its action based on different principles to achieve a stable product. GA is used to stabilize flavor and oil emulsions in dried food mixes (such as soup, cakes, etc.) and in the soft drink industry, where the gum is required to stabilize a concentrated oil emulsion (about 20%) for long periods and to continue to stabilize following dilution prior to bottling (Islam et al., 1997). Emulsifying agents are usually long-chain organic compounds that produce chains that are soluble in oil (lipophilic) as well as side chains or groups that are soluble in water (hydrophilic). Thus, one portion of each molecule dissolves in the water phase, while another portion dissolves in the oil phase, and the main chain forms a link or bridge to keep both phases in position, thereby emulsified.

GA produces highly stable emulsions, making it useful in the preparation of oil-in-water food flavor emulsions, particularly for citrus oils (Randall et al., 1989). Some researchers believe that gums are not true emulsifiers. That is, they do not act by means of hydrophilic chemical functionality; they act as emulsion stabilizers or protectors. Their function is essentially to increase the viscosity of the aqueous phase by thickening it so that it approximates or slightly exceeds that of oil. In this way, the tendency of the dispersed phase to slip or coalesce was minimized, and the emulsion was stabilized. Such stabilization is a protective effect based on the thickening properties of gums. Randall et al. (1989) studied the effect of heat on the emulsification action and stability of the gum, followed by changes in the GPC profile of the gum. They concluded that heating at 100 °C for 3 h resulted in a decrease in the intensity of the high molecular mass peak with a corresponding increase in the intensity of the lower molecular mass peaks. Continuous heating leads to further loss of the high molecular mass fraction and loss of the emulsifying stability of the gum. Chikamai & Banks (1993) reported that heating solutions at 100 °C for more than 6 h causes significant loss of emulsification properties, whereas heating at 60 °C for 24 h has only a minor effect. Therefore, the duration and temperature of heating are crucial factors for maintaining the emulsifying properties of GA.

Dickinson et al. (1991) studied the surface and emulsifying properties of six Acacia gum samples and stated that the relationship between nitrogen content and emulsifying properties of the gum samples depends not only on their total protein content but also on the distribution of the protein/peptide between the low and high-molecular-weight fractions, and on the molecular accessibility of the protein/peptide for adsorption. Dickinson et al. (1991) studied the influence of the nature of the oil phase on the emulsifying behavior of GA. They found that gum lowered the surface tension at the *n*-hexadecane–water interface and gave the most stable *n*-hexadecane–water emulsion with the smallest droplets for all three oils (*n*-hexadecane and orange oil). They also concluded that a high-molecular-weight fraction (0.87 nitrogen) corresponding to 10% of a natural gum (0.38% nitrogen) initially gives larger droplets but better emulsion stability than the low-molecular-weight fraction (0.35%). Similar to most emulsifiers, the AGP complex contains a hydrophilic region (protein) and a hydrophilic region (carbohydrate). During the formation of oil-in-water emulsions, the protein (arabinogalactan) is converted into the water phase. The bulk of GA in the form of free AG can improve the stability by increasing the viscosity of water (Islam et al., 1997). The relatively low protein content of GA requires a high gum concentration in most emulsification systems (Imeson, 1992). Therefore, understanding the composition and behavior of GA is essential to optimize its use as an emulsifier in various applications. In this chapter, the emulsifying properties of *A. polyacantha*, *A. senegal*, and *A. seyal* gum were compared, and the effects of concentration, pH, protein addition, blending, temperature, and stirring time were investigated. Consequently, these factors significantly determine the emulsifying efficiency and stability of the different GA species.

1.7.2 Water-Holding Capacity (WHC)

The ability of a material to retain water against gravity is referred to as WHC (Elamin 2001). The WHC of AGs varies with their concentration, pH, and temperature. Studies have revealed that *A. senegal* gum, particularly *var keratosis*, exhibits superior WHC, resulting in low syneresis during yogurt stabilization (Mugo et al., 2020). Moreover, *A. senegal* and *A. seyal* gums exhibit distinct molecular characteristics and interfacial properties (Gashua et al., 2016). In addition, the physicochemical properties of kakamut gum (*A. polyacantha*) and hashab gum (*A. senegal*) were comparatively analyzed, demonstrating similar pH values (Awadasseid et al., 2019). These findings indicate that the WHC of these gums may be influenced by their specific botanical sources and molecular compositions. El Amin (1990) reported a maximum value of 65.59% for *A. senegal*. Eltayeb et al., (2004) reported a 63.77–65.6% range for *A. senegal* gum. The increase in weight was expressed as a percentage to explain the WHC of the sample and is expressed as follows:

$$WHC\% = \frac{\text{Weight of water} \times 100}{\text{Weight of sample}} \tag{1.5}$$

1.8 Spectrophotometry and Spectroscopy of AGs

1.8.1 Raman Spectroscopy Techniques of AGs

Raman spectroscopy (RS) is a technique used to measure the frequency shift of inelastically scattered light from the sample when the photon from incident light strikes a molecule and produces a scattered photon (Chan et al., 1996; Helms & Gordy, 1977; Spiro & Strekas, 1974). The outcoming scattered light can be a photon with a lower frequency than the original photon, and in that case, it is known as Stokes Raman scattering, or with a higher frequency, known as anti-Stokes Raman scattering (Asher, 1993; Hajj Hussein et al., 2015). In the latter case, the photon receives energy from the bond of the molecule when the bond is initially in an excited vibrational state (Han et al., 2013). Substantial integrity was achieved by strong hydrogen bond (HB) interactions between PANI and GA, as evidenced by

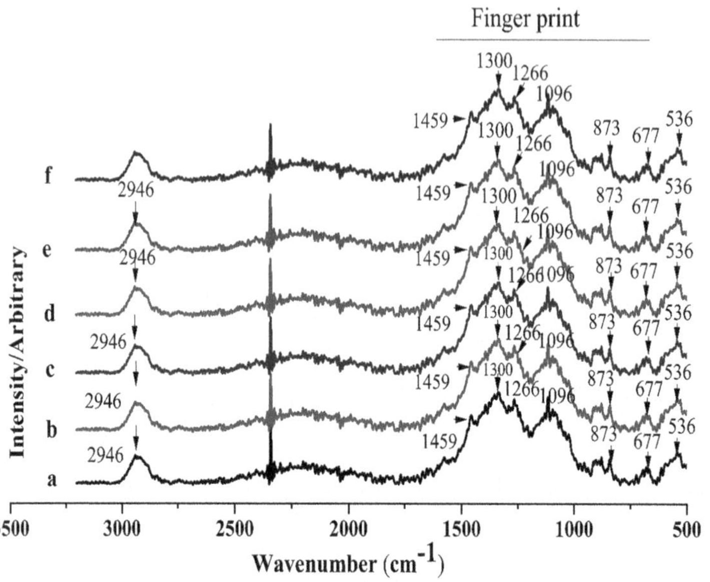

Fig. 1.7 The Raman spectra of (**a**) *A. sengal* gum; clay soil; (**b**) *A. polyacantha* gum; clay soil (**c**) *A. seyal* gum; clay soil; (**d**) *A. sengal* gum; lateritic soil; (**e**) *A. polyacantha* gum; lateritic soil; (**f**) *A. sengal* gum; lateritic soil

specific NAH bands in Raman analyses. The intensity of the HB decreased upon higher amounts of GA, probably due to steric impediment around the NH sites (Quintanilha et al., 2014). The Raman spectra of the *A. senegal, A. polyacantha*, and *A. seyal* samples are shown in Fig. 1.7. The Raman spectra of the AGs are remarkably similar. All AGs samples exhibited six significant peaks in the range of approximately 2946, 1459, 1300, 1266, 1096, 873, 677, and 536 cm^{-1}.

In this chapter, the Raman spectra were characterized by intense bands around 2946 cm^{-1} assigned to C–H stretching. The band at 536 cm^{-1} originated from the strong deformation of C–C–C and C–C–O. The influential band (prominent peak) at 1300 cm^{-1} was assigned to ring deformation. The peak at 1096 cm^{-1} was assigned mainly to the bending vibration of C–H and C–O–H in carbohydrates and a slight contribution from the vibration of the C–N bond in proteins and amino acids. The bands around 677 cm^{-1} originated from C–O, C–C–O, and O–C–O bending, and the bands at 873 cm^{-1} were assigned to C–H, C–O–H, and CH_2 bending. The sharp peak at 1266 cm^{-1} was due to the deformation vibrations of C–C–H, O–C–H, and C–O–H. The absorbance band at 1459 cm^{-1} originates from the combination of the bending vibration of CH_2 and the vibration of the COO– group (Pierna et al., 2011). The results demonstrated that RS can be used as a direct method to assess the physicochemical characteristics of GA, whether raw or commercially available. Identification of the Raman fingerprint region is presented in Table 1.3.

Table 1.3 Raman absorptions of common functional group compared to FT-IR

Functional group/vibration	Region	Raman	Infrared
Lattice vibrations in crystals, LA modes	10–200 cm^{-1}	Strong	Strong
δ(CC) aliphatic chains	250–400 cm^{-1}	Strong	Weak
υ(Se–Se)	290–330 cm^{-1}	Strong	Weak
υ(S–S)	430–550 cm^{-1}	Strong	Weak
υ(Si–O–Si)	450–550 cm^{-1}	Strong	Weak
υ(X metal–O)	150–450 cm^{-1}	Strong	Med-weak
υ(C–I)	480–660 cm^{-1}	Strong	Strong
υ(C–Br)	500–700 cm^{-1}	Strong	Strong
υ(C–Cl)	550–800 cm^{-1}	Strong	Strong
υ(C–S) aliphatic	630–790 cm^{-1}	Strong	Medium
υ(C–S) aromatic	1080–1100 cm^{-1}	Strong	Medium
υ(O–O)	845–900 cm^{-1}	Strong	Weak
υ(C–O–C)	800–970 cm^{-1}	Medium	Weak
υ(CC) alicyclic, aliphatic chain vibrations	600–1300 cm^{-1}	Medium	Medium
υ(C=S)	1000–1250 cm^{-1}	Strong	Weak
υ(CC) aromatic ring chain vibrations	*1580–1600 cm^{-1}	Strong	Medium
	*1450–1500 cm^{-1}	Medium	Medium
	*1000 cm^{-1}	Strong/medium	Weak
δ(CH3)	1380 cm^{-1}	Medium	Strong
δ(CH2) δ(CH3) asym	1400–1470 cm^{-1}	Medium	Medium
υ(C–(NO_2)	1340–1380 cm^{-1}	Strong	Medium
υ(C–(NO_2) asym	1530–1590 cm^{-1}	Medium	Strong
υ(N=N) aromatic	1410–1440 cm^{-1}	Medium	Strong
υ(N=N) aliphatic	1550–1580 cm^{-1}	Medium	Strong
δ(H_2O)	~1640 cm^{-1}	Weak broad	Strong
υ(C=N)	1610–1680 cm^{-1}	Strong	Medium
υ(C=C)	1500–1900 cm^{-1}	Strong	Weak
υ(C=O)	1680–1820 cm^{-1}	Medium	Strong
υ(C≅C)	2100–2250 cm^{-1}	Strong	Weak
υ(C≅N)	2220–2255 cm^{-1}	Medium	Strong
υ(–S–H)	2550–2600 cm^{-1}	Strong	Weak
υ(C–H)	2800–3000 cm^{-1}	Strong	Strong
υ(=(C–H))	3000–3100 cm^{-1}	Strong	Medium
υ(≅(C–H))	3300 cm^{-1}	Weak	Strong
υ(N–H)	3300–3500 cm^{-1}	Medium	Medium
υ(O–H)	3100–3650 cm^{-1}	Weak	Strong

Source: Socrates (2001)

1.8.2 FT-IR Spectroscopy of AGs

Infrared spectrophotometry is one of the most powerful tools available to chemists for identifying pure organic and inorganic compounds, because each molecular species has a unique infrared absorption spectrum. Thus, an exact match between the spectrum of a compound with a known structure and that of an analyte unambiguously identifies the latter (Griffiths & De Haseth, 2007). FTIR spectrometric measurements for the gum of *Sterculia setigera* collected from South Kordofan showed eight peaks with a broad peak at approximately 3445.65 cm^{-1} most likely for hydroxyl groups (OH), and two at 1700–1600 cm^{-1} probably for carboxyl, aldehyde, and ketone groups. The remaining peaks were specific to the gum sample. Similarly, the FTIR spectra for the *A. senegal* gum sample collected from GA Company and *A. seyal* gum sample collected from South Kordofan showed clear peaks with abroad one at approximately 3401.60 cm^{-1} and 3398.29 cm^{-1} most likely for hydroxyl groups (OH), respectively. In addition to the two other peaks at 1700–1600 cm^{-1}, probably for carboxyl, aldehyde, and ketone groups, the remaining peaks (at 600–700 cm^{-1}) were specific for the gum type (Ahmad et al., 2013; Elnour, 2008; Kong et al., 2014; Quintanilha et al., 2014).

The Fourier-transform infrared (FTIR) spectra of *the A. senegal*, *A. polyacantha*, and *A. seyal* samples are shown in Fig. 1.8. The FTIR spectra of the samples showed six significant peaks at approximately similar ranges of 942, 1263, 1543, 1742, 2360, and 2380 cm^{-1}, with a small difference in the range of 1026 and 1263 cm^{-1}. In these wavenumber regions, the displayed spectra provided the characteristics of AGs samples and polysaccharides rich in galactose and arabinose. These findings are in agreement with those of Renard et al. (2006) and Boulet et al. (2007) investigated the FTIR spectra of gum as a raw sample.

This chapter found the maximum absorbance peaks were observed at 1026 and 1263 cm^{-1} for the AGs and 942–1263 cm^{-1} for the commercial samples. These results agreed with those of Lopez-Torrez et al. (2015), who compared the FTIR spectra of *A. senegal* and *A. seyal* gums. The FTIR results showed the fingerprint of the samples with the main spectral differences in agreement with the glycosidic linkage composition. Moreover, three main spectral bands with wavenumbers of 1827 and 1263 cm^{-1} were observed in the FTIR spectra of the samples (Fig. 1.7, IR band positions A–H).

The amide II bands in the FTIR spectra of proteins were located adjacent to 1650 and 1550 cm^{-1}, respectively. Interestingly, the results agree with those of Boulet et al. (2007) and Vinod et al. (2010), who reported that the FTIR expression of some polysaccharides with uronic acids could be observed in the wavenumber regions of 1742, 1630, and 1543 cm^{-1}. Therefore, the band position near 1630 cm^{-1} can be shifted depending on the biochemical composition, particularly the uronic acid content.

Furthermore, our findings are consistent with those of Boulet et al. (2007), who confirmed that the variation in the carboxylic functions of the absorbance band of arabino-galactoprotein (AGP) occurred from 1640 to 1600 cm^{-1} when the uronic

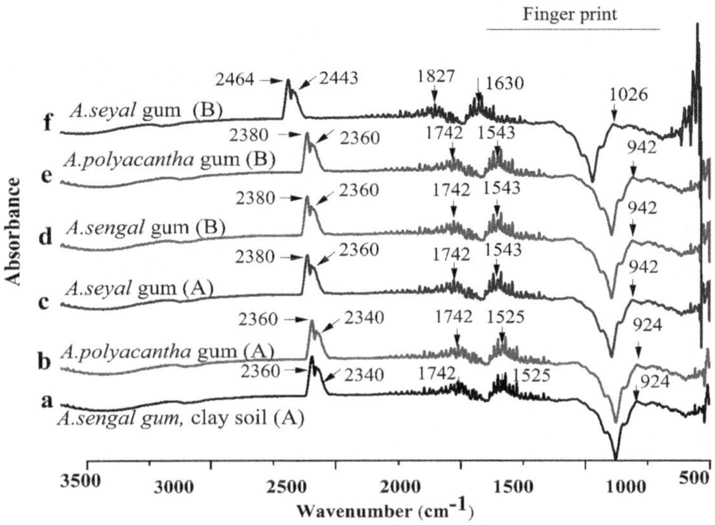

Fig. 1.8 The FT-IR spectra of (**a**) *A. sengal* gum; clay soil; (**b**) *A. polyacantha* gum; clay soil (**c**) *A. seyal* gum; clay soil; (**d**) *A. sengal* gum; lateritic soil; (**e**) *A. polyacantha* gum; lateritic soil; (**f**) *A. sengal* gum; lateritic soil

acid (UA) content changed from 3.6% to 14.7%. The ACGS content was 10.4% molar glucuronic acid (GUA) and 4-*O*-Meglucuronic acid, as well as low protein content at 1%. Consequently, in this study, the assignment of absorbance bands at 1742 and 1543 cm^{-1} in the AGs spectrum was attributed to the deprotonated carboxylic function of GUA. Interestingly, the results were in line with those of Boulet et al. (2007) and Manrique and Lajolo (2002). Furthermore, the C–O–C stretching, vibrations, and bending of C–H might reflect the IR bands near 1630 and 1263 cm^{-1} in the AGs and PTC spectra. These vibrations were hardly attributable to a specific functional group (Table 1.3).

A complex interpretation was obtained from the infrared spectral bands at 1263 and 942 cm^{-1}, corresponding to the AGs carbohydrate moiety (Fig. 1.8). This carbohydrate fingerprint region is produced by backbone skeletal stretching vibrations and polysaccharide side chains. The region is also dominated by the overlapping of ring vibrations and stretching vibrations of the C–O–C glycosidic bond and C–O–H side groups. Therefore, as reported by Renard et al., other polysaccharides rich in galactose, arabinose, and AGs were used to characterize the AGs in this region. Boulet et al. (2007). The maximum absorbance peaks for AGs were observed at 1026 and 942 cm^{-1}, respectively. These results agree with those of Lopez-Torrez et al. (2015), who compared the FTIR spectra of *A. senegal* and *A. seyal* gums. In summary, the FTIR data identified AGs fingerprints, where agreement between the main spectral differences and their glycosidic linkage composition was observed (Table 1.4).

Table 1.4 FT-IR absorptions of common functional group

Functional group	Absorptions location (cm^{-1})	Absorption intensity
Alkane (C–H)	2850–2975	Medium to strong
Alcohol (O–H)	3400–3700	Strong, broad
Alkene (C=C) (C=C–H)	1640–1680 3020–3100	Weak to medium Medium
Alkyne (C≡C) (C=C–H)	1640–1680 3300	
Nitrile (C≡N)	2200–2250	Medium
Aromatics	1650–2000	Weak
Amines (N–H)	3300–3350	Medium
Carbonyls (C=O)	1720–1740	Strong
Aldehyde (CHO)	1720–1740	
Ketone (RCOR)	1715	
Ester (RCOOR)	1735–1750	
Acid (RCOOH)	1700–1725	

Source: Adapted from Socrates (2001)

1.8.3 Differential scanning calorimetry (DSC) of AGs

DSC is a valuable technique for assessing the thermal properties and stability differences among materials such as AGs. DSC allows the determination of transition temperatures and enthalpy changes, providing insights into the unique thermal characteristics of these gums (Ghanbari et al., 2023). By analyzing DSC heating thermograms, researchers can identify seasonal variations in materials such as oils, which are influenced by their chemical compositions, showcasing the sensitivity of DSC in detecting thermal variations based on composition (İlyasoğlu & Özçelik, 2010). DSC has been extensively used to investigate the thermal properties of various materials, including bulk polymers and composites, highlighting its versatility and effectiveness in characterizing different substances (Mortazavian et al., 2015).

DSC is a valuable technique for assessing the thermal properties and stability of AGs. DSC allows the determination of transition temperatures and changes in enthalpy, providing insights into the unique characteristics of these gums (Ghanbari et al., 2023). Using DSC, researchers have been able to investigate the thermal properties of various materials, including gums, to understand their behavior under different conditions (Patra et al., 2015). Studies have shown that the molecular composition and structure of AGSs vary, thereby affecting their properties and applications (Gashua et al., 2015; Lopez-Torrez et al., 2015). Furthermore, differences in the physicochemical properties of gums from different Acacia species can be elucidated through techniques such as DSC, aiding in the comparative analysis of their functionalities (Awadasseid et al., 2019). Figure 1.9 represents a diagram of the DSC and thermal properties of the AGs.

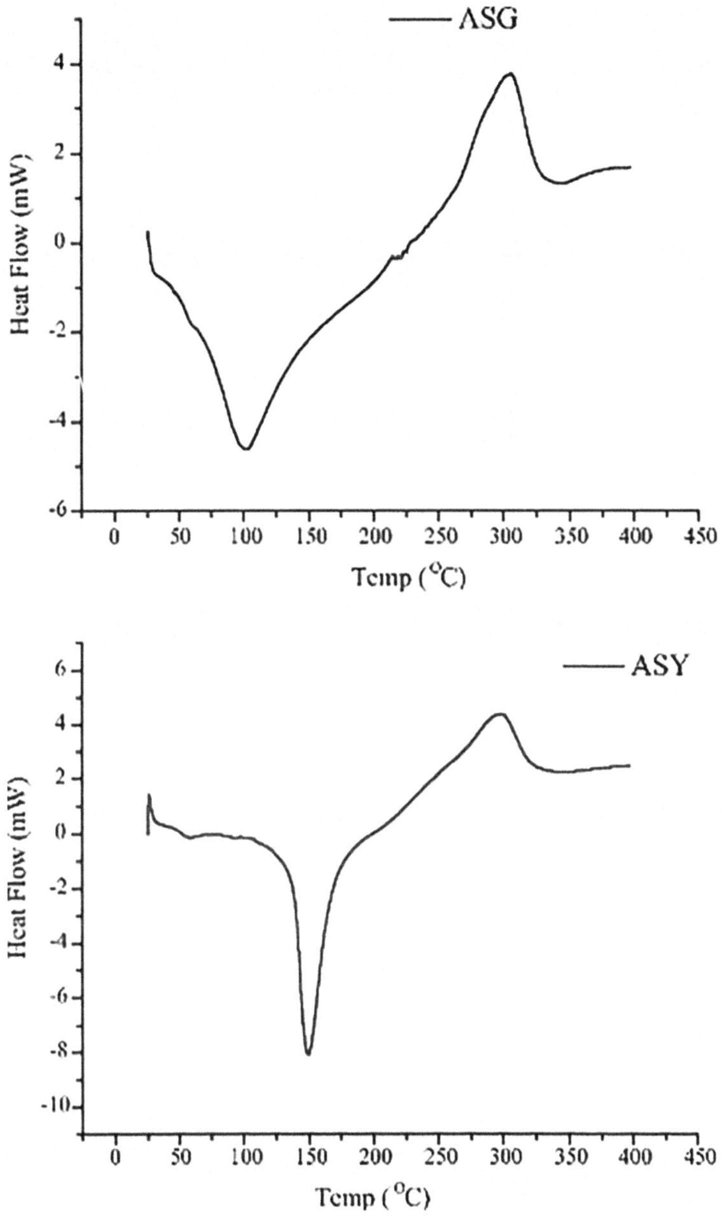

Fig. 1.9 DSC, thermal of *A. sengal* gum, *A. seyal* gums. (Source: Adapted from Daoub et al., 2018)

1.8.4 Transmission Electron Microscopy (TEM) of AGs

TEM is instrumental in examining microstructural and morphological disparities among AGs. TEM allows for direct visualization of the ultrastructure of these gums at the nanoscale level, offering comprehensive insights into their morphology and microstructural features (Gashua et al., 2016). Using TEM, researchers can assess the aggregation of molecules in these gums over time, facilitating a comparative analysis of their structural characteristics (Gashua et al., 2016). This method provides high-resolution imaging capabilities, making it a valuable tool for investigating distinct morphological properties of natural exudates such as AGs (Gashua et al., 2016).

TEM is a powerful tool for elucidating the microstructural and morphological differences between AGs. TEM provides atomic-scale resolution, various imaging modes, and crystallographic information, allowing for detailed characterization of gum structures (Brodusch et al., 2021). Using TEM, researchers can monitor the degree of molecular aggregation in gum samples over time, providing insights into their molecular characteristics and interfacial properties (Gashua et al., 2016). Additionally, TEM can be coupled with spectroscopic techniques, such as electron energy loss spectroscopy and energy dispersive X-ray spectroscopy, to further analyze the distribution of components within the gums (Cullen et al., 2014). Overall, TEM offers a comprehensive approach for studying the microstructure and morphology of these gums at the nanoscale level.

1.9 Food Applications; Definition, and Specification of GA

GA has a wide range of applications beyond their traditional uses. It is utilized in various industries such as food, medicine, and materials science. In the food industry, GA is commonly used as a stabilizer, thickening agent, and emulsifier (Hu et al., 2014). It is also employed as an adhesive in textile fiber treatment, a health tonic, and a remedy for venereal diseases (Dauqan & Abdullah, 2013). Furthermore, GA has been utilized in the production of microcapsules, surface modification for anti-adhesiveness, and preservation of biological tissues (McNamee et al., 1998). Its versatility and importance in different fields are evident from its diverse applications (Çaykara et al., 2021). Therefore, the multifaceted applications of GA highlight their significance in various industries, showcasing their adaptability and wide-ranging utility.

GA is widely used in both the food and non-food sectors, and this requires control through definitions and specifications by the concerned institutions and organizations. JECFA defined GA as "dried exudation obtained from the stems and branches of *A. senegal* (L.) Willdenow or other related African species of Acacia, family Leguminosae" (FAO/WHO, 2010). Nitrogen content and optical rotation were added in 1990, which altered the last definition to include the word closely "or closely related species" (Coppen, 1999). Thus, the definitions and specifications of GA have evolved over time to encompass a broader range of related species, ensuring consistency, and clarity in its identification and use across industries.

Coppen (1999) stated that this specification was amended in 1995, but the words "closely related species" were followed by "Acacia" Acacia family Leguminosae (recently Mimosaceae). The latest specification was published in 1998, and it included *A. seyal* in the definition with *A. senega* land *A. polyacantha* (Coppen, 1999; Elnour, 2008; Omer, 2004). This specification read: "dried exudation obtained from the stems and branches of *A. senegal* (L.) Willdenow, or A. seyal, family Leguminosae." This was confirmed by the Codex Committee on Food Additives and Contaminants in March 1999 and adopted by the Codex Alimentarius Commission of the FAO/WHO in its 23rd session held in June 1999 (Coppen, 1999; Mukhtar, 2015). As a result, the inclusion of *A. seyal, A. senegal*, and *A. polyacantha* in the specification indicates their recognition as legitimate sources of GA, enhancing their value in various industries.

In the food-related industry, GA is mainly utilized in bakery, confectionery, beverages, dairy, and as a microencapsulating agent. GA has two vital roles: first, it postpones or prevents the crystallization of sugar, and second, it emulsifies lipids to distribute them evenly throughout the product. Arabinogalactan-protein (AGPs) is the core constituent responsible for the GA stabilizing ability of emulsions, where AGP amphiphilic protein associates with the oil droplet surface. The hydrophilic carbohydrate fraction is oriented toward the aqueous phase, prohibiting droplet aggregation by electrostatic repulsion. The capacity of polysaccharide association in aqueous solutions has been well established. These molecular associations can profoundly affect their function in a particular application owing to their effect on molecular weight, size, and shape, which control the interaction between molecules and water (Randall et al., 1989). Therefore, understanding the molecular interactions of GA is crucial to optimize its use in various food products.

GA is widely used in baking owing to its low moisture absorption characteristics. The solubility of GA in cold water permits better formation of clear solutions in contrast to sugar solutions. Owing to its water absorption characteristics, GA is also utilized as a stabilizer for frozen products, such as ice cream. The task performed by GA, in this case, inactivates ice crystal formation to create a subtle texture and growth by bringing together a significant amount of water and trapping it for hydration. GA supports moisture retention, freeze-thaw stability, and crispiness. GA is utilized as an emulsion agent in beverages, such as beer, citrus juices, and soft drinks (Williams et al., 1990). In addition, it produces stable emulsions with most oils over a broad pH range. Thus, it is primarily used for the microencapsulation of fat, and it also possesses the ability to form films (Randall et al., 1989). GA can be used to offset the ripening of fruits while sustaining antioxidant levels. This trait was established by Ali et al. (2013), who coated green-mature tomatoes with GA in various aqueous solutions. During storage, in comparison to either no coating or a 5% concentration of GA coating, applying a 10% coating to fruit reduced both the respiration rate and production of ethylene while sustaining a suitable antioxidant ability and levels of lycopene, phenolics, and carotenoids. These factors aided in

Table 1.5 Applications of gum arabic in foods and other industries

Benefits	Functions	Applications	Reference
Improved emulsion, prevent crystallization, and thickeners for beverages	Preparation and stabilization of oil-in-water emulsions, inability to crystallize, prevent agglomeration, flavors in confectionery	Foodstuff	Kanakdande et al. (2007)
Increase stability for the flavor	Encapsulation of essential oils	Encapsulation of Flavors in powder and liquid form	Kanakdande et al. (2007)
Oil-soluble vitamins in powder form.	Suspending, emulsifying, binding agents, venereal diseases, stomach disorders	Pharmaceutical and cosmetic applications	Vogt (1995)
Good texture low-calorie	Main ingredient non-caloric fiber source	Diabetic sugarless confectionery	Dickinson (2009)
Increase homogeneity	Hydrophilic colloids and modified cellulose Stabilizing, emulsions wax treated pigments with water-soluble hydrocolloids Controllable chemotropic properties to paints	Paints, and coating composition application	Dickinson (2009)
Increase adhesive	Adhesives for postage formulations of paints and inks, crystallization textile industries thickeners for colors	Other industrial uses	Glicksman and Sand (1973)

offsetting the ripening process of the fruits, and no postharvest issues were exhibited. The different applications of the GA are listed in Table 1.5. The solubility and viscosity of GA are the most fundamental properties that make it unique among the polysaccharides. Therefore, GA's antioxidant properties of GA make it a valuable coating material for preserving fruits and vegetables, contributing to their extended shelf life.

1.10 Nutritional Applications

From a nutritional standpoint, polysaccharides are an excellent source of dietary fiber because of their high gum concentration. The soluble dietary fiber component of GA cannot be digested in the small intestine or stomach. As a result, the fermentation of GA by intestinal bacteria reflects both its prebiotic properties and its capacity to boost bifidobacteria, which promotes digestive health and aids in bowel movement regulation (Álvarez-Mercado and Plaza-Diaz, 2022). Proteins are an important source of essential amino acids and provide nutritional value to food systems. Proteins perform a wide range of functions in the body, such as growth, repair, and maintenance of cells and tissues, making them a vital component of a balanced diet (Beniwal & Das, 2023). Acacia gum can be added to food products to help raise their nutrition value. Acacia gum could be used to reduce fat content. In some culinary applications, such as sauces, baked products, and salads, AGs can be used instead of fats. This assists in mitigating the consequences of a formulation's decreased fat content. It is possible to lower the total fat and calorie density of the product. Acacia gum may help stabilizing ingredients during food preparations and this will boost a product sweetness or flavor. This will enable less added sugar to obtain the desired mouth feel and texture without compromising the flavor. Acacia gum is a dietary prebiotic fiber that can be added to food products to increase their fiber content. Compared to other substances used in food preparation, AGs have a comparatively low caloric content. Food products can achieve calorie reduction by substituting AGs with higher-calorie ingredients. Gum arabic has additional benefits from a nutritional perspective as it can lower calorie intake and increase subjective satiety. Therefore, it can be used in dietary programs aimed at controlling the body weight (Calame et al., 2008). In individuals with type 2 diabetes, a high intake of dietary fiber, especially soluble fiber, improves glycemic control, lowers plasma lipid levels, and reduces hyperinsulinemia (Chandalia et al., 2000).

1.11 Pharmaceutical and Cosmetic Applications

AGs derived from Acacia trees have diverse pharmaceutical and cosmetic applications. Because of their biopolymer nature, AGs have been employed as stabilizers, thickeners, and emulsifiers in various industries (Dauqan & Abdullah, 2013; Sethuraman & Rajendran, 2018). These gums demonstrate potential for pharmaceutical applications as stabilizers in emulsions, binders in tablet formulations, and for sustained drug release (Desta et al., 2021). Additionally, Acacia gum exhibits therapeutic actions such as hypoglycemic, antioxidant, and immunomodulatory properties (Ashour et al., 2022). The molecular characteristics of gums, including their complex polysaccharide structures, make them suitable for diverse applications (Fauconnier et al., 2000; Gashua et al., 2016). The prebiotic functionality and potential anticancer properties of AGs highlight their versatility (Calame et al., 2008; Elnour et al., 2020). The quality and storage conditions of gums, particularly for medical use, depend on the source and type of gum (Tanwar et al., 2024). The properties of these gums make them effective and suitable for a wide range of pharmaceutical and cosmetic applications. In the pharmaceutical industry, gums are suspended, emulsifying, or binding agents. They are also utilized in tablet manufacturing, functioning as binding agents or coatings before sugar coating and sometimes in combination with other gums. *A. polyacantha* gum is employed as a general health tonic, antidote for snake bites, and treatment for venereal diseases. Preparations derived from the bark are used to address general stomach disorders (Voget, 1995).

1.12 Paints and Coating Composition Application

A. senegal gum, often referred to as GA, is a soluble dietary fiber suitable for use in painting media (Brøns et al., 2018; Cherbut et al., 2003). *A. seyal* gum has demonstrated potential as an anticancer agent (Elnour et al., 2020) and for emulsion formation (Gashua et al., 2016). The physicochemical properties of *A. polyacantha* gum have been previously investigated (Elhenawy et al., 2014). AGs can be used as stabilizers and binders in tablet formulations, and for sustained drug release in pharmaceuticals (Desta et al., 2021). Gums with high esterification show improved swelling and drug release properties from the carriers (Soleimani et al., 2021). These properties make gums promising candidates for pharmaceutical and medical applications. Additionally, the tested gums formed layers on filter surfaces that transmitted water vapor and maintained high sterility. Therefore, AGs with higher esterification degrees are valuable for pharmaceutical and medical applications because of their enhanced properties. *A. senegal* gum is preferred for its emulsification properties over *A. seyal* gum (Maqbool et al., 2011). These gums offer a variety of benefits, such as solubility, emulsification, and potential therapeutic applications, making them valuable components for paints, coatings, and pharmaceutical formulations. Hydrophilic colloids and modified cellulose have been applied in the paint industry because of their stabilizing effects on paint emulsions, waxes, and numerous other products. Kuan et al. (2009) used water-soluble hydrocolloids such as GA to add controllable chemotropic properties to paints. The gum was also found to be applicable to coating compositions. Emmenegger et al. (1997) developed a non-glare coating based on a water-soluble dye dissolved in GA solutions.

1.13 Other Industrial Uses

AGs are frequently used as adhesives for postage stamps, paints, and inks because of their adhesive properties. Mesquite gum, derived from Prosopis plants, is a rich source of L-arabinose and can be isolated using methods such as dialysis and crystallization. These gums are also used in textiles to enhance luster, as thickeners in colorants, and as mordants in calico-printing. The emulsification properties of AGs, particularly those of *A. senegal*, are superior, making them valuable for various industries (Gashua et al., 2016). The emulsifying, stabilizing, and binding properties of AGs make them desirable for various applications (Sethuraman & Rajendran, 2018). Acacia species, including *A. senegal*, are extensively used in reafforestation programs in sub-Saharan Africa (Bakhoum et al., 2014). Owing to their adhesive properties, AGs manufacture adhesives for postage stamps and formulations of paints and inks. Mesquite gum can serve as a source of monosaccharides because it is easier to hydrolyze and is available in large quantities. It can be isolated by dialysis with ethanol, crystallization from methanol after removal of acidic oligosaccharides on ion exchange resin, or precipitation by barium salts.

1.14 Conclusion

In conclusion, acacia gums, including AGs, are complex polysaccharides with distinct characteristics that have various applications. Their composition as polyuronides, featuring UA residues, hexose residues in the pyranose configuration, and pentose residues in the furanose form, sets them apart. Notably, *A. senegal* and *A. seyal* are notable sources of GA, a polysaccharide rich in proteins, which underpins its emulsifying and stabilizing properties. In comparison with other tree gums, such as Khaya senegalensis gum, Sterculia tomentose gum, and Olibanum gum, AGs exhibit both similarities and differences in polysaccharide composition. Moreover, Acacia gum has been found to possess anti-inflammatory, antioxidant, and protective effects on various organs. These natural polysaccharide gums also compete with gums derived from marine algae and higher plants for industrial applications. AGs' unique properties and versatile applications of AGs make them valuable resources in various fields, highlighting their potential as sustainable and effective ingredients for various industries.

Acknowledgments The authors thank the Chemical and Process Engineering Technology at the University of Malaysia, Pahang-UMPSA, Malaysia, for supporting this research.

Fund: This research was funded by Eastern Unity Technology under Grant No. UIC 190806.

References

Abdi, G. (2024). Tragacanth gum based hydrogels for drug delivery and tissue engineering applications. *Frontiers in Materials, 11*, 1296399. https://doi.org/10.3389/fmats.2024.1296399

Abdolmaleki, H., Kidmose, P., & Agarwala, S. (2021). Droplet-based techniques for printing functional inks for flexible physical sensors. *Advanced Materials, 33*(20), 2006792.

Abera, T., & Lemma, B. (2019). Soil bulk density, soil moisture content and yield of tef (eragrostis tef) influenced by A. seyal del canopy in parkland agro-forestry system. *Journal of Soil Science and Environmental Management, 10*(6), 124–129. https://doi.org/10.5897/jssem2019.0738

Abu-Serie, M. M., Hamouda, A. F., & Habashy, N. H. (2021). A. senegalgum attenuates systemic toxicity in CCl4-intoxicated rats via regulation of the ROS/NF-κB signaling pathway. *Scientific Reports, 11*(1), 20316. https://doi.org/10.1038/s41598-021-99953-y

Adam, H. E. (2011). Integration of remote sensing and GIS in studying vegetation trends and conditions in the gum Arabic belt in North Kordofan, .

Ahad, T., Gull, A., & Masoodi, F. (2022). Effect of excipient wall materials on the development of ginger oleoresin microcapsules: Assessing the physicochemical, antioxidant and structural properties. *Journal of the Science of Food and Agriculture, 103*(1), 73–82. https://doi.org/10.1002/jsfa.12113

Ahmad, S. I., Mazumdar, N., & Kumar, S. (2013). Functionalization of natural gum: An effective method to prepare iodine complex. *Carbohydrate Polymers, 92*(1), 497–502.

Ahmed, Z. (2015). *Mathematical modeling of thin-layer solar drying of gum of A. senegal and A. seyal*. UOFK.

Ahmed, A., Abdelmalik, A., Alsharani, T., Al-Qarawi, B. A. Q., & Aref, I. (2020). Response of growth and drought tolerance of A. seyal Del. seedlings to arbuscular mycorrhizal fungi. *Plant, Soil and Environment, 66*, 264–271.

Al-Assaf, S., Phillips, G. O., & Williams, P. A. (2005a). Studies on acacia exudate gums. Part I: The molecular weight of A. senegalgum exudate. *Food Hydrocolloids, 19*(4), 647–660.

Al-Assaf, S., Phillips, G. O., & Williams, P. A. (2005b). Studies on acacia exudate gums: Part ii. Molecular weight comparison of the vulgares and gummiferae series of acacia gums. *Food Hydrocolloids, 19*(4), 661–667. https://doi.org/10.1016/j.foodhyd.2004.09.003

Al-Hamayda, A., Abu-Jdayil, B., Ayyash, M., & Tannous, J. (2023). Advances in microencapsulation techniques using Arabic gum: A comprehensive review. *Industrial Crops and Products, 205*, 117556. researchgate.net

Ali, A., Maqbool, M., Alderson, P. G., & Zahid, N. (2013). Effect of gum arabic as an edible coating on antioxidant capacity of tomato (*Solanum lycopersicum* L.) fruit during storage. *Postharvest Biology and Technology, 76*, 119–124.

Alicja, Aguilar, F., Crebelli, R., Di Domenico, A., Frutos, M. J., Galtier, P., & Lambré, C. (2017). Re-evaluation of acacia gum (E 414) as a food additive. *EFSA Journal, 15*(4), e04741.

Álvarez-Mercado, A. I., & Plaza-Diaz, J. (2022). Dietary polysaccharides as modulators of the gut microbiota ecosystem. An update on their impact on health. *Nutrients, 14*, 4116.

Amid, B. T., Mirhosseini, H., Poorazarang, H., & Mortazavi, S. A. (2013). Implications of partial conjugation of whey protein isolate to durian seed gum through maillard reactions: Foaming properties, water holding capacity and interfacial activity. *Molecules, 18*(12), 15110–15125.

Anderson, D., Bridgeman, M., Farquhar, J., & McNab, C. (1983). The chemical characterization of the test article used in toxicological studies of gum arabic (A. senegal(L.) Willd). *International Tree Crops Journal, 2*(3–4), 245–254.

Anderson, D. M. W., & Dea, I. C. M. (1967). Studies on uronic acid materials: Part XXII. The composition of the gum from acacia dprepanolobium harms ex sjösted. *Carbohydrate Research, 5*(4), 461–469.

Anderson, D. M. W., & Stoddart, J. F. (1966). Studies on uronic acid materials: Part XV. The use of molecular-sieve chromatography in studies on acacia Senegal gum (gum Arabic). *Carbohydrate Research, 2*(2), 104–114.

Anderson, D. M. W., Dea, I. C. M., Karamalla, K. A., & Smith, J. F. (1968). Studies on uronic acid materials: Part XXIV. An analytical study of different forms of the gum from Acacia Senegal willd. *Carbohydrate Research, 6*(1), 97–103.

Anderson, D. M. W., Howlett, J. F., & McNab, C. G. A. (1985). The amino acid composition of the proteinaceous component of guar gum (Cyamopsis tetragonolobus). *Food Additives & Contaminants, 2*(4), 225–230.

Anusha, S., Harikrishna, P. R., Madhu, A. D., & Zachariah, M. A. (2024). Guar gum, Gellan gum biopolymer soil stabilization. In *E3S web of conferences* (Vol. 529, p. 01053). EDP Sciences.

Apolinar-Valiente, R., Williams, P., Nigen, M., Tamayo, V. M., Doco, T., & Sánchez, C. (2019). Recovery, structure and physicochemical properties of an aggregate-rich fraction from A. senegalgum. *Food Hydrocolloids, 89*, 864–873. https://doi.org/10.1016/j.foodhyd.2018.11.054

Asher, S. A. (1993). UV resonance Raman spectroscopy for analytical, physical, and biophysical chemistry. *Analytical Chemistry, 65*(4), 201A–210A.

Ashour, M. A., Fatima, W., Imran, M., Ghoneim, M. M., Alshehri, S., & Shakeel, F. (2022). A review on the main phytoconstituents, traditional uses, inventions, and patent literature of gum arabic emphasizing A. seyal. *Molecules, 27*(4), 1171. https://doi.org/10.3390/molecules27041171

Atgié, M., Masbernat, O., & Roger, K. (2018). Emulsions stabilized by gum arabic: Composition and packing within interfacial films. *Langmuir, 35*(4), 962–972. https://doi.org/10.1021/acs.langmuir.8b02715

Atobe, S., Saga, K., Hasegawa, F., Furuhashi, K., Tashiro, Y., Suzuki, T., et al. (2015). Effect of amphiphilic polysaccharides released from Botryococcus braunii showa on hydrocarbon recovery. *Algal Research, 10*, 172–176. https://doi.org/10.1016/j.algal.2015.05.007

Awadasseid, A., Adam, I. M., Jamal, M., & Teia, F. K. F. (2019). Physicochemical properties of Kakamut gum (*A. polyacantha*) and hashab gum (*A. senegal*): A comparative analysis. *International Journal of Applied Agricultural Sciences, 5*(6), 138. https://doi.org/10.11648/j.ijaas.20190506.12

Awulachew, M. T. (2021). A review of non-gluten components in gluten-free bread characteristics, dough baking quality and viscoelastic qualities. *International Journal of Food Science and Biotechnology, 6*(4), 115. https://doi.org/10.11648/j.ijfsb.20210604.14

Bakhoum, N., Galiana, A., Roux, C. L., Kane, A., Duponnois, R., Ndoye, F., & Diouf, D. (2014). Phylogeny of nodulation genes and symbiotic diversity of A. senegal(l.) willd. and A. seyal (del.) mesorhizobium strains from different regions of Senegal. *Microbial Ecology, 69*(3), 641–651. https://doi.org/10.1007/s00248-014-0507-1

Balla, M. (2004). Yield trends of gum arabic from A. senegalas related to some environmental and managerial factors.

Barron, C., Surget, A., & Rouau, X. (2007). Relative amounts of tissues in mature wheat (Triticum aestivum L.) grain and their carbohydrate and phenolic acid composition. *Journal of Cereal Science, 45*(1), 88–96.

Beniwal, A., & Das, M. (2023). Protein: It's application in food industry. In *Research trends in nutrition science* (pp. 1–25). Bumi Publishing.

Bertolini, A., Siani, A., & Grosso, C. (2001). Stability of monoterpenes encapsulated in gum arabic by spray-drying. *Journal of Agricultural and Food Chemistry, 49*(2), 780–785. https://doi.org/10.1021/jf000436y

Biswas, B., Biswas, S., & Phillips, G. O. (2000). The relationship of specific optical rotation to structural composition for acacia and related gums. *Food Hydrocolloids, 14*(6), 601–608.

Boulet, J., Williams, P., & Doco, T. (2007). A Fourier-transform infrared spectroscopy study of wine polysaccharides. *Carbohydrate Polymers, 69*(1), 79–85.

Brodusch, N., Brahimi, S., Melo, E., Song, J., Yue, S., Piché, N., et al. (2021). Scanning electron microscopy versus transmission electron microscopy for material characterization: A comparative study on high-strength steels. *Scanning, 2021*, 1–19. https://doi.org/10.1155/2021/5511618

Brøns, C., Rasmussen, K., Crescenzo, M., Stacey, R., & Lluveras-Tenorio, A. (2018). Painting the palace of Apries I: Ancient binding media and coatings of the reliefs from the palace of Apries, lower Egypt. *Heritage Science, 6*(1), 1–20. https://doi.org/10.1186/s40494-018-0170-9

Calame, W., Weseler, A. R., Viebke, C., Flynn, C., & Siemensma, A. D. (2008). Gum arabic establishes prebiotic functionality in healthy human volunteers in a dose-dependent manner. *British Journal of Nutrition, 100*(6), 1269–1275. https://doi.org/10.1017/s0007114508981447

Çaykara, T., Silva, J., Fernandes, S., Braga, A., Rodrigues, J. L. L. C., Rodrigues, L. R., et al. (2021). Modification of pet surfaces with gum arabic toward its bacterial anti-adhesiveness using an experimental factorial design approach. *Materials Today Communications, 28*, 102684. https://doi.org/10.1016/j.mtcomm.2021.102684

Chan, K., Yamada, M., & Inaba, H. (1996). Coherence gating in optical heterodyne detection measurements of scattering and absorption in highly scattering media. *Applied Physics B, 63*(3), 249–253.

Chandalia, M., Garg, A., Lutjohann, D., Von Bergmann, K., Grundy, S. M., & Brinkley, L. J. (2000). Beneficial effects of high dietary fiber intake in patients with type 2 diabetes mellitus. *New England Journal of Medicine, 342*(19), 1392–1398.

Chen, W., Wang, H., Zhang, K., Gao, F., Chen, S., & Li, D. (2016a). Physicochemical properties and storage stability of microencapsulated dha-rich oil with different wall materials. *Applied Biochemistry and Biotechnology, 179*(7), 1129–1142. https://doi.org/10.1007/s12010-016-2054-3

Chen, C., Duru, P., Prat, M., Joseph, P., & Geoffroy, S. (2016b). Towards the computation of viscous flow resistance of a liquid bridge. *International Journal of Computational Methods and Experimental Measurements, 4*(1), 42–49. https://doi.org/10.2495/cmem-v4-n1-42-49

Cherbut, C., Michel, C., Raison, V., Kravtchenko, T., & Meance, S. (2003). Acacia gum is a bifidogenic dietary fiber with high digestive tolerance in healthy humans. *Microbial Ecology in Health and Disease, 15*(1), 43–50. https://doi.org/10.1080/08910600310014377

Chikamai, B., & Banks, W. (1993). Gum arabic from Acacia senegal (L) Willd. in Kenya. *Food Hydrocolloids, 7*(6), 521–534.

Chikamai, B., & Maisharou, A. (2021). Strengthening capacity among stakeholders for the production and trade in gums and resins in selected African countries. ngara.org.

Coppen, J. (1999). Development of gum Arabic production and marketing. Technical Cooperation Program.

Cullen, D. A., Koestner, R. J., Kukreja, R. S., Liu, Z. Y., Minko, S., Trotsenko, O., et al. (2014). Imaging and microanalysis of thin ionomer layers by scanning transmission electron microscopy. *Journal of the Electrochemical Society, 161*(10), F1111–F1117. https://doi.org/10.1149/2.1091410jes

Da Silva, D. A., Aires, G. C. M., & da Silva Pena, R. (2020). Gums—Characteristics and applications in the food industry. In *Innovation in the food sector through the valorization of food and agro-food by-products*. intechopen.com

Dahi, A., Abdellahi, B. M., Deida, M. F., Hucher, N., Malhiac, C., & Renou, F. (2019). Chemical and physicochemical characterizations of the water-soluble fraction of the Commiphora Africana exudate. *Food Hydrocolloids, 86*, 2–10. https://doi.org/10.1016/j.foodhyd.2017.10.032

Daoub, R. M., Elmubarak, A. H., Misran, M., Hassan, E. A., & Osman, M. E. (2018). Characterization and functional properties of some natural Acacia gums. *Journal of the Saudi Society of Agricultural Sciences, 17*(3), 241–249.

Dauqan, E., & Abdullah, A. (2013). Utilization of gum arabic for industries and human health. *American Journal of Applied Sciences, 10*(10), 1270–1279. https://doi.org/10.3844/ajassp.2013.1270.1279

Davantès, A., Nigen, M., Sánchez, C., D'Orlando, A., & Renard, D. (2019). Adsorption of hyperbranched arabinogalactan-proteins from plant exudate at the solid–liquid interface. *Colloids and Interfaces, 3*(2), 49. https://doi.org/10.3390/colloids3020049

Davantès, A., Nigen, M., Sánchez, C., & Renard, D. (2021). Adsorption behavior of arabinogalactan-proteins (AGPs) from A. senegalgum at a solid–liquid interface. *Langmuir, 37*(35), 10547–10559. https://doi.org/10.1021/acs.langmuir.1c01619

Desta, K. H., Tadese, E., & Molla, F. (2021). Physicochemical characterization and evaluation of the binding effect of Acacia etbaica schweinf gum in granule and tablet formulations. *BioMed Research International, 2021*, 1–13. https://doi.org/10.1155/2021/5571507

Devi, L. M., Das, A. B., & Badwaik, L. S. (2023). Effect of gelatin and acacia gum on anthocyanin coacervated microcapsules using double emulsion and its characterization. *International Journal of Biological Macromolecules, 235*, 123896.

Dickinson, E. (2009). Hydrocolloids as emulsifiers and emulsion stabilizers. *Food Hydrocolloids, 23*(6), 1473–1482.

Dickinson, E., Galazka, V. B., & Anderson, D. M. W. (1991). Emulsifying behavior of gum arabic. Part 2: Effect of the gum molecular weight on the emulsion droplet-size distribution. *Carbohydrate Polymers, 14*(4), 385–392. https://doi.org/10.1016/0144-8617(91)90004-V

Drabo, M. S., Shumoy, H., Savadogo, A., & Raes, K. (2022). Inventory of human-edible products from native Acacia sensu lato in Africa, America, and Asia: Spotlight on Senegalia seeds, overlooked wild legumes in the arid tropics. *Food Research International, 159*, 111596.

Eeswaran, R., Nejadhashemi, A. P., Faye, A., Min, D., Prasad, P. V., & Ciampitti, I. A. (2022). Current and future challenges and opportunities for livestock farming in West Africa: Perspectives from the case of Senegal. *Agronomy, 12*(8), 1818. mdpi.com

Eiroboyi, I., Ikiensikimama, S. S., Oriji, B. A., & Okoye, I. P. (2019). Synergistic study between gum arabic and carboxymethyl cellulose: Application in polymer flooding. *Current Journal of Applied Science and Technology, 36*, 1–7. https://doi.org/10.9734/cjast/2019/v36i430243

El Amin, H. M. (1990). *Trees and shrubs of the Sudan*. Ithaca Press.

Elamin, R. M. (2001). Effect of different radiations on some physicochemical properties of gum Arabic (Acacia Senegal (L.) wild).

Elella, M. H. A., Goda, E. S., Gab-Allah, M. A., Hong, S. E., Pandit, B., Lee, S., et al. (2021). Xanthan gum-derived materials for applications in environment and eco-friendly materials: A review. *Journal of Environmental Chemical Engineering, 9*(1), 104702.

Elhenawy, A. A., Omer, E. A., Alomari, A. A., & Osman, M. (2014). A comparative analytical studies onacaciapolyacantha gum samples collected from three different locations in Sudan. *Journal of Advances in Natural Sciences, 1*(1), 1–9. https://doi.org/10.24297/jns.v1i1.379

Elkhalifa, K. F., Ibrahim, M., & Elghazali, G. (2003). Contribution to the flora of the Gash Delta Eastern Sudan. *Reserch Bulletin, 123*, 5–20.

Elnour, A. (2008). *Fractionation; physicochemical and functional properties of A. polyacantha Gum*. Doctoral Dissertation, University of Khartoum, Khartoum, Sudan.

Elnour, A. M., Ishag, K., & Abdalla, A. A. (2011). Physicochemical properties of A. polyacantha Gum.

Elnour, A. A., Mirghani, M. E. S., Kabbashi, N. A., Daddiouaissa, D., Musa, K. H., Alam, Z., et al. (2020). Active fractions of methanol crude obtained from A. seyal gum and their antiproliferative effects against human breast cancer cell lines. *Global Journal of Science Frontier Research, 20*, 51–64. https://doi.org/10.34257/gjsfrcvol20is3pg51

Eltahir, M. E. S., Fadl, K. E. M., Hamad, M. A. A., Safi, A. I. A., Elamin, H. M. A., Abutaba, Y. I. M., et al. (2023). Tapping tools for gum arabic and resins production: A review paper. *American Journal of Engineering and Technology Management, 8*(3), 33–40. researchgate.net

Eltayeb, I. B., Awad, A. I., Elderbi, M. A., & Shadad, S. A. (2004). Effect of gum arabic on the absorption of a single oral dose of amoxicillin in healthy Sudanese volunteers. *Journal of Antimicrobial Chemotherapy, 54*(2), 577–578.

Emmenegger, F., Piccand, M., Piekarski, H., & Mokrzan, J. (1997). The complete thermodynamic cycle of cocl2–pyridine complex formation. *Journal of the Chemical Society, Dalton Transactions, 5*, 785–788. https://doi.org/10.1039/a606259h

Evans, J. (1992). *Plantation forestry in the tropics: Tree planting for industrial, social, environmental, and agroforestry purposes*. Oxford University Press.

Fang, Y., Al-Assaf, S., Phillips, G. O., Nishinari, K., & Williams, P. A. (2010). Interaction of gum arabic with fatty acid studied using electron paramagnetic resonance. *Biomacromolecules, 11*(5), 1398–1405.

Fang, Z., Zhao, J., Li, Y., Wang, Y., Qiu, T., Wu, Y., et al. (2021). Improving Fenton-like system with catechin, an environmental-friendly polyphenol: Effects and mechanism. *Chemical Engineering Journal, 426*, 127946.

FAO. (1995). Code of conduct for responsible fisheries, 41.

FAO/WHO. (2010). *Evaluation of certain food additives: Seventy-first report of the joint FAO/WHO expert committee on food additives* (Vol. 71). World Health Organization.

Fauconnier, M., Blecker, C., Groin, J., Razafindralambo, H., Vanzeveren, E., Marlier, M., et al. (2000). Characterization of two AGSs and their fractions using a Langmuir film balance. *Journal of Agricultural and Food Chemistry, 48*(7), 2709–2712. https://doi.org/10.1021/jf990749x

Fenyo, J. C., & Vandevelde, M. C. (1990). Physico-chemical properties of gum arabic in relation to structure. *Gums and Stabilisers for the Food Industry, 5*, 17–23.

Gashua, I. B., Williams, P. A., Yadav, M. P., & Baldwin, T. C. (2015). Characterization and molecular association of Nigerian and Sudanese Acacia gum exudates. *Food Hydrocolloids, 51*, 405–413. https://doi.org/10.1016/j.foodhyd.2015.05.037

Gashua, I. B., Williams, P., & Baldwin, T. C. (2016). Molecular characteristics, association and interfacial properties of gum arabic harvested from both A. senegal and A. seyal. *Food Hydrocolloids, 61*, 514–522. https://doi.org/10.1016/j.foodhyd.2016.06.005

Ghanbari, E., Picken, S. J., & van Esch, J. H. (2023). Analysis of DSC: Determining the transition temperatures and enthalpy and heat capacity changes in multicomponent systems by analytical model fitting. *Journal of Thermal Analysis and Calorimetry, 148*(22), 12393–12409. https://doi.org/10.1007/s10973-023-12356-1

Ghasemzadeh, H., & Modiri, F. (2020). Application of novel Persian gum hydrocolloid in soil stabilization. *Carbohydrate Polymers, 246*, 116639. academia.edu.

Glicksman, M. (2019). Gum arabic (Gum acacia). In *Food hydrocolloids* (pp. 7–29). CRC Press.

Glicksman, M., & Sand, R. E. (1973). Chapter X—Gum Arabic A2—Whistler, Roy L. In *Industrial gums* (2nd ed., pp. 197–263). Academic.

Griffiths, P. R., & De Haseth, J. A. (2007). *Fourier-transform infrared spectrometry* (Vol. 171). Wiley.

Hajj Hussein, R., Pagès, O., Doyen-Schuler, S., Dicko, H., Postnikov, A. V., Firszt, F., et al. (2015). Percolation-type multi-phonon pattern of Zn (Se,S): Backward/forward Raman scat-

tering and ab initio calculations. *Journal of Alloys and Compounds, 644*, 704–720. https://doi.org/10.1016/j.jallcom.2015.04.078
Han, F., Liu, W., & Fang, C. (2013). Excited-state proton transfer of photoexcited pyranine in water observed by femtosecond stimulated Raman spectroscopy. *Chemical Physics, 422*, 204–219. https://doi.org/10.1016/j.chemphys.2013.03.009
Hassan, A. F., Awwad, N. S., Ibrahium, H. A., & El-Kott, A. F. (2024). Biochar/Delonix regia seed gum/chitosan composite as efficient adsorbent for the elimination of phenol from aqueous medium. *International Journal of Biological Macromolecules, 273*, 132771.
Helms, D. A., & Gordy, W. (1977). "Forbidden" rotational spectra of symmetric-top molecules: PH3 and PD3. *Journal of Molecular Spectroscopy, 66*(2), 206–218.
Heywood, V. H. (1993). *Flowering plants of the world*. BT Batsford.
Hirst, E., & Jones, J. (1958). The gums and mucilages of plants. In *Aufbau·Speicherung Mobilisierung und Umbildung der Kohlenhydrate/Formation·storage·mobilization and transformation of carbohydrates* (pp. 500–517). Springer.
Hu, J., Nie, S., Li, N., Min, F., Li, C., Gong, D., et al. (2014). Effect of gum arabic on glucose levels and microbial short-chain fatty acid production in white rice porridge model and mixed grain porridge model. *Journal of Agricultural and Food Chemistry, 62*(27), 6408–6416. https://doi.org/10.1021/jf501557b
Ibrahim, M. O. H. (2015). *Assessment of gum arabic marketing system in main auction markets of North Kordofan State-Sudan*. UOFK.
Idris, O., Williams, P., & Phillips, G. (1998). Characterization of gum from A. senegal trees of different age and location using multidetection gel permeation chromatography. *Food Hydrocolloids, 12*(4), 379–388.
İlyasoğlu, H., & Özçelik, B. (2010). Determination of seasonal changes in olive oil by using differential scanning calorimetry heating thermograms. *Journal of the American Oil Chemists' Society, 88*(7), 907–913. https://doi.org/10.1007/s11746-010-1747-y
Imeson, A. P. (1992). Exudate gums. In *Thickening and gelling agents for food* (pp. 66–97). Springer US.
Islam, A. M., Phillips, G. O., Sljivo, A., Snowden, M. J., & Williams, P. A. (1997). A review of recent developments on the regulatory, structural, and functional aspects of gum Arabic. *Food Hydrocolloids, 11*(4), 493–505.
Israr, J., Indraratna, B., & Rujikiatkamjorn, C. (2016). Laboratory investigation of the seepage-induced response of granular soils under static and cyclic loading. *Geotechnical Testing Journal, 39*(5), 795–812.
Jain, S., Shah, V., Doshi, M., & Vegada, R. (2024). Peptide and protein delivery through acacia, tragacanth, and ghatti gum. In *Peptide and protein drug delivery using polysaccharides* (pp. 149–167). Academic.
Jebeso, G., Gebeyehu, D., & Abere, M. (2023). Characteristics and quality of gum arabic derived from naturally grown A. seyal Willdenow tree in the lowland areas of Ethiopia. *Food Chemistry Advances, 2*, 100243. sciencedirect.com
Jimenez-Fernandez, M., Perez-Tirado, D., Peredo-Lovillo, A., & Luna-Solano, G. (2021). Physicochemical characteristics and survivability of Lactobacillus paracasei encapsulated by a gum arabic-pectin mixture as wall material and added to fresh panela cheese. *Revista Mexicana De Ingeniería Química, 20*(3), 1–25. https://doi.org/10.24275/rmiq/alim2551
Jones, J., & Smith, F. (1949). Plant gums and mucilages. In *Advances in carbohydrate chemistry* (Vol. 4, pp. 243–291). Elsevier.
Kalilou, O. (2021). Climate change and conflict in the Sahel: The acacia gum tree as a tool for environmental peacebuilding. *International Affairs, 97*, 201–218.
Kanakdande, D., Bhosale, R., & Singhal, R. S. (2007). Stability of cumin oleoresin microencapsulated in different combination of gum arabic, maltodextrin and modified starch. *Carbohydrate Polymers, 67*(4), 536–541.
Karamalla, K. A. (1965). *Analytical and structural studies in the polysaccharide group*. Annexe Thesis Digitisation Project 2016 Block 5.

Karamalla, K. A., Siddig, N. E., & Osman, M. E. (1998). Analytical data for Acacia Senegal var. Senegal gum samples collected between 1993 and 1995 from Sudan. *Food Hydrocolloids, 12*(4), 373–378.

Kaur, R., Kaur, S., Dwibedi, V., Kaur, C., Akhtar, N., & Alzahrani, A. (2023). Development and characterization of rice bran-gum-arabic-based encapsulated biofertilizer for enhanced shelf-life and controlled bacterial release. *Frontiers in Microbiology, 14*, 1267730. https://doi.org/10.3389/fmicb.2023.1267730

Kennedy, J. F., Phillips, G. O., & Williams, P. A. (2011). *Gum arabic* (Vol. 333). Royal Society of Chemistry.

Kiprop, V. J., Omwamba, M., & Mahungu, S. M. (2021). Influence of gum arabic from *Acacia Senegal* var. *kerensis* on the modifications of pasting and textural properties of cassava and corn starches. *Food and Nutrition Sciences, 12*(11), 1098–1115. https://doi.org/10.4236/fns.2021.1211081

Kong, H., Yang, J., Zhang, Y., Fang, Y., Nishinari, K., & Phillips, G. O. (2014). Synthesis and antioxidant properties of gum arabic-stabilized selenium nanoparticles. *International Journal of Biological Macromolecules, 65*, 155–162.

Koyyada, A., & Orsu, P. (2021). Natural gum polysaccharides as efficient tissue engineering and drug delivery biopolymers. *Journal of Drug Delivery Science and Technology, 63*, 102431.

Krigbaum, W. R., & Flory, P. J. (1953). Molecular weight dependence of the intrinsic viscosity of polymer solutions.II. *Journal of Polymer Science, 11*(1), 37–51.

Kuan, Y., Bhat, R., Senan, C., Williams, P. A., & Karim, A. (2009). Effects of ultraviolet irradiation on the physicochemical and functional properties of gum arabic. *Journal of Agricultural and Food Chemistry, 57*(19), 9154–9159. https://doi.org/10.1021/jf9015625

Lelon, J. K., Jumba, I. O., Keter, J. K. A., & Oduor, F. D. O. (2013). Influence of A. senegal varieties on quality of gum arabic in Baringo district, Kenya. *African Journal of Plant Science, 7*(6), 190–200. https://doi.org/10.5897/ajps10.185

Liang, Q., Ma, P., Zhang, Q., Yin, Y., Wang, P., Wang, S., et al. (2020). A gum arabic assisted sustainable drug delivery system for adult drosophila. *Biology Open, 9*, bio052241. https://doi.org/10.1242/bio.052241

Liu, C., Xia, L., Liang, C., Zhang, K., Chen, L., & Li, J. (2018). Hydrophilic coating film used to drive flow in a microfluidic point-of-care testing (poct) device. *Micro & Nano Letters, 13*(6), 773–778. https://doi.org/10.1049/mnl.2017.0661

Lopez-Franco, Y. L., Gooycolea, F. M., & Lizardi-Mendoza, J. (2015). Gum of Prosopis/AGS. In *Polysaccharides: Bioactivity and bio-technology* (pp. 641–662).

Lopez-Torrez, L., Nigen, M., Williams, P., Doco, T., & Sanchez, C. (2015). Acacia senegal vs. Acacia seyal gums–Part 1: Composition and structure of hyperbranched plant exudates. *Food Hydrocolloids, 51*, 41–53.

Mahendran, T., Williams, P. A., Phillips, G. O., Al-Assaf, S., & Baldwin, T. C. (2008). New insights into the structural characteristics of the arabinogalactan−protein (AGP) fraction of gum arabic. *Journal of Agricultural and Food Chemistry, 56*(19), 9269–9276. https://doi.org/10.1021/jf800849a

Mahfoudhi, N., Sessa, M., Chouaibi, M., Ferrari, G., Donsì, F., & Hamdi, S. (2014). Assessment of emulsifying ability of almond gum in comparison with gum arabic using response surface methodology. *Food Hydrocolloids, 37*, 49–59.

Manrique, G. D., & Lajolo, F. M. (2002). FT-IR spectroscopy as a tool for measuring degree of methyl esterification in pectins isolated from ripening papaya fruit. *Postharvest Biology and Technology, 25*(1), 99–107.

Maqbool, M., Ali, A., Alderson, P., Zahid, N., & Siddiqui, Y. (2011). Effect of a novel edible composite coating based on gum arabic and chitosan on biochemical and physiological responses of banana fruits during cold storage. *Journal of Agricultural and Food Chemistry, 59*(10), 5474–5482. https://doi.org/10.1021/jf200623m

Martínez, M., Beltrán, O., Rincón, F., de Pinto, G. L., & Igartuburu, J. M. (2015). New structural features of Acacia tortuosa gum exudate. *Food Chemistry, 182*, 105–110. https://doi.org/10.1016/j.foodchem.2015.02.124

Martinović, J., Lukinac, J., Jukić, M., Ambrus, R., Planinić, M., Šelo, G., et al. (2023). The release of grape pomace phenolics from alginate-based microbeads during simulated digestion in-vitro: The influence of coatings and drying method. *Gels, 9*, 870. https://doi.org/10.20944/preprints202310.1205.v1

Mazumder, M. A. R., & Ranganathan, T. V. (2020). Encapsulation of isoflavone with milk, maltodextrin and gum acacia improves its stability. *Current Research in Food Science, 2*, 77–83. https://doi.org/10.1016/j.crfs.2019.12.003

McNamee, B. F., O'Riordan, D., & O'Sullivan, M. (1998). Emulsification and microencapsulation properties of gum arabic. *Journal of Agricultural and Food Chemistry, 46*(11), 4551–4555. https://doi.org/10.1021/jf9803740

Mekbib, S., Regnier, T., Sivakumar, D., & Korsten, L. (2009). Evaluation of Ethiopian plant extracts, Acacia seyal and Withania somnifera, to control green mold and ensure quality maintenance of citrus (Citrus sinensisl.). *Fruits, 64*(5), 285–294. https://doi.org/10.1051/fruits/2009023

Mengesha, A., Birru, E., & Adugna, M. (2022). Anti-diarrheal activities of hydromethanolic crude extract and solvent fractions of Acacia seyal (fabaceae) roots in mice. *Clinical Pharmacology Advances and Applications, 14*, 99–110. https://doi.org/10.2147/cpaa.s383896

Mhinzi, G. S. (2003). Intra-species variation of the properties of gum exudates from Acacia Senegal var. Senegal and Acacia seyal var. fistula from Tanzania. *Bulletin of the Chemical Society of Ethiopia, 17*(1).

Mortazavian, H., Fennell, C. J., & Blum, F. D. (2015). Structure of the interfacial region in adsorbed poly(vinyl acetate) on silica. *Macromolecules, 49*(1), 298–307. https://doi.org/10.1021/acs.macromol.5b02214

Moscoso-Moscoso, E., Ligarda-Samanez, C. A., Choque-Quispe, D., Huamán-Carrión, M. L., Arévalo-Quijano, J. C., De la Cruz, G., et al. (2024). Preliminary assessment of tara gum as a wall material: Physicochemical, structural, thermal, and rheological analyses of different drying methods. *Polymers, 16*(6), 838. https://doi.org/10.3390/polym16060838

Mudgil, D., & Mudgil, S. (2024). Acacia gum: Chemistry, properties & food applications. *Food and Humanity, 100264*.

Mugo, E. M., Mahungu, S. M., Chikamai, B., & Mwove, J. (2020). Evaluation of gum arabic from Acacia senegalvar kerensis and Acacia senegalvar Senegal as a stabilizer in low-fat yogurt. *International Journal of Food Studies, 9*. https://doi.org/10.7455/ijfs/9.si.2020.a9

Mukherjee, S. N., & Deb, S. K. (1962). Light-scattering studies in solutions of gum arabic. *Journal of the Indian Chemical Society, 39*(12), 823.

Mukhtar, N. (2015). *The effect of various cations on gum arabic viscosity*. UOFK.

NACA/JECFA. (1999). *Food safety issues associated with products from aquaculture: Report of a joint FAO/NACA/WHO study group* (Vol. 883). World Health Organization.

NAS. (1979). Tropical legumes: Resources for the future. Report of an ad hoc panel of the Advisory Committee on Technology Innovation, Board on Science and Technology for International Development, Commission on International Relations, National Research Council (Vol. 25). National Academies.

Ndindeng, S. A., Mbassi, J., Mbacham, W. F., Manful, J., Graham-Acquaah, S., Moreira, J., et al. (2015). Quality optimization in briquettes made from rice milling by-products. *Energy for Sustainable Development, 29*, 24–31. https://doi.org/10.1016/j.esd.2015.09.003

Nhan, N., Thanh, V., Cang, M., Lam, T., Huong, N., Nhan, L., et al. (2020). Microencapsulation of lemongrass (Cymbopogon citratus) essential oil via spray-drying: Effects of feed emulsion parameters. *Processes, 8*(1), 40. https://doi.org/10.3390/pr8010040

Nikolić, N. Č., Stanisavljević, N., Šavikin, K., Kalušević, A., Nedović, V., Bigović, D., et al. (2018). Application of GA in producing spray-dried chokeberry polyphenols, microparticles characterization and in-vitro digestion method. *Lekovite Sirovine, 38*, 9–16. https://doi.org/10.5937/leksir1838009c

Nuraziz, C. E., & Khairi, A. N. (2024). Effect of adding stevia sweetener (Stevia rebaudianna) and arabic gum filler on the physicochemical properties of chayote squash (Sechium edule) powder drink. *agriTECH, 44*(2), 161. https://doi.org/10.22146/agritech.80303

Omer, E. A. (2004). *Characterization and analytical studies of A. polyacantha*. Sudan University of Science and Technology.

Osman, W., Menzies, A. R., & Phillips, G. O. (1993). Characterization of commercial samples of gum arabic. *Journal of Agricultural and Food Chemistry, 41*(1), 71–77.

Osman, W., Mostafa-He, G., Osman, A., Rashed, L. A., & Labib, D. (2022). Effect of arabic gum (A. senegal) on paracetamol-induced chronic nephrotoxicity in albino rats. *International Journal of Pharmacology, 18*(8), 1593–1604. https://doi.org/10.3923/ijp.2022.1593.1604

Owusu, J., Oldham, J., Oduro, I., Otu, W. E., & Amissah, A. (2020). Assessing the suitability of locally produced gum exudates in the food industry. *International Journal of Technology and Management Research, 1*(5), 24–30. https://doi.org/10.47127/ijtmr.v1i5.40

Oyebanji, V. O., Emikpe, B. O., Oladele, O. A., Osowole, O. I., Salaam, A., Odeniyi, M. A., et al. (2017). Clinicopathological evaluation of Newcastle disease virus vaccination using gums from Cedrela odorata and Khaya senegalensis as delivery agents in challenged chickens. *International Journal of Veterinary Science and Medicine, 5*(2), 135–142. https://doi.org/10.1016/j.ijvsm.2017.09.002

Ozoude, C. H., Azubuike, C., Ologunagba, M. O., Tonuewa, S. S., & Igwilo, C. I. (2020). Formulation and development of metformin-loaded microspheres using Khaya senegalensis (Meliaceae) gum as copolymer. *Future Journal of Pharmaceutical Sciences, 6*(1), 1–11. https://doi.org/10.1186/s43094-020-00139-6

Pal, K., & Bera, D. (2020). Natural polymers, gums and mucilages as efficacious green emissaries of essential therapeutics. In *Proceedings of MOL2NET 2020, International Conference on Multidisciplinary Sciences* (6th ed.). https://doi.org/10.3390/mol2net-06-06792

Palma, S., Carvalho, A., Silva, J., Martins, P., Marciello, M., Fernandes, A. R., et al. (2015). Covalent coupling of gum arabic onto superparamagnetic iron oxide nanoparticles for MRI cell labeling: Physicochemical and in-vitro characterization. *Contrast Media & Molecular Imaging, 10*(4), 320–328. https://doi.org/10.1002/cmmi.1635

Patra, S., et al. (2015). Deacetylation-induced changes in thermal properties of Sterculia urens gum. *Journal of Thermal Analysis and Calorimetry, 122*, 235–240. https://doi.org/10.1007/s10973-015-4680-3

Picton, L., Bataille, I., & Muller, G. (2000). Analysis of a complex polysaccharide (gum arabic) by multi-angle laser light scattering coupled on-line to size exclusion chromatography and flow field flow fractionation. *Carbohydrate Polymers, 42*(1), 23–31.

Pierna, J. A. F., Abbas, O., Dardenne, P., & Baeten, V. (2011). Discrimination of Corsican honey by FT-Raman spectroscopy and chemometrics/discrimination du miel de Corse par spectroscopie FT-Raman et chimiométrie. *Biotechnologie, Agronomie, Société et Environnement, 15*(1), 75.

Pimentel-Rodas, A., Galicia-Luna, L. A., & Castro-Arellano, J. J. (2016). Capillary viscometer and vibrating tube densimeter for simultaneous measurements up to 70 MPa and 423 K. *Journal of Chemical & Engineering Data, 61*(1), 45–55.

Pinho, L. S., Lima, P. M., Fang, F., Cooperstone, J. L., Favaro-Trindade, C. S., & Campanella, O. H. (2024). Effect of extrusion process conditions on extrudates enriched with carotenoids encapsulated by different methods using gum arabic and vegetable fat as carriers. *International Journal of Biological Macromolecules, 267*, 131200.

Ploypetchara, T., Sorndech, W., Auranwiwat, C., Sumsakul, W., Pinthong, M., Kansuwan, A., et al. (2021). Impact of encapsulation wall materials and drying method on physicochemical properties and digestibility of encapsulated chayote (Sechium edule (Jacq.) Swartz) and Kohlrabi (Brassica oleracea var gongylodes L.) extracts. *Food Science and Engineering, 3*, 1–11. https://doi.org/10.37256/fse.3120221110

Prasad, N., Thombare, N., Sharma, S. C., & Kumar, S. (2022). Gum arabic–A versatile natural gum: A review on production, processing, properties and applications. *Industrial Crops and Products, 187*, 115304. academia.edu.

Pretzscha, J., Tahab, M. E., Siddigc, M. E., Deafallad, T. H., & Elamin, H. M. (2014). Valuation of environmental role of A. senegal tree in the gum belt of Kordofan and the Blue Nile Sectors, Sudan. In *Paper presented at the Conference on International Research on Food Security*.

Qi, W., Fong, C., & Lamport, D. T. (1991). Gum arabic glycoprotein is a twisted hairy rope. *Plant Physiology, 96*(3), 848–855.

Quintanilha, R. C., Orth, E. S., Grein-Iankovski, A., Riegel-Vidotti, I. C., & Vidotti, M. (2014). The use of gum Arabic as "green" stabilizer of poly (aniline) nanocomposites: A comprehensive study of spectroscopic, morphological and electrochemical properties. *Journal of Colloid and Interface Science, 434*, 18–27.

Radwan, U., Radwan, T., & Abouelkasim, E. (2020). Water use efficiency of A. seyal (del.) in extreme arid environment prevails in south-western desert, Egypt. *Aswan University Journal of Environmental Studies, 1*(1), 28–37. https://doi.org/10.21608/aujes.2020.124532

Ram, H., & Jatwa, R. (2014). Antiatherosclerotic and cardioprotective potential of A. senegal seeds in diet-induced atherosclerosis in rabbits. *Biochemistry Research International, 2014*, 1–6. https://doi.org/10.1155/2014/436848

Randall, R., Phillips, G., & Williams, P. (1989). Fractionation and characterization of gum from A. senegal. *Food Hydrocolloids, 3*(1), 65–75.

Rawi, M. H., Abdullah, A., Ismail, A., & Sarbini, S. R. (2021). Manipulation of gut microbiota using acacia gum polysaccharide. *ACS Omega, 6*(28), 17782–17797. https://doi.org/10.1021/acsomega.1c00302

Renard, D., Lavenant-Gourgeon, L., Ralet, M.-C., & Sánchez, C. (2006). Acacia s enegal gum: Continuum of molecular species differing by their protein to sugar ratio, molecular weight, and charges. *Biomacromolecules, 7*(9), 2637–2649.

Rhazi, L., Lakahal, L., Andrieux, O., Niamba, N., Depeint, F., & Guillemet, D. (2020). Relationship between the molecular characteristics of Acacia gum and its functional properties. *Food Chemistry, 328*, 126860. sciencedirect.com

Rodríguez-Bernal, J. M., Flores-Andrade, E., Lizarazo-Morales, C., Bonilla, E., Pascual-Pineda, L. A., Gutierréz-López, G., & Quintanilla-Carvajal, M. X. (2015). Moisture adsorption isotherms of the borojó fruit (Borojoa patinoi. Cuatrecasas) and gum arabic powders. *Food and Bioproducts Processing, 94*, 187–198.

Rodriguez-Sandoval, E., Cortes, M., & Manjarrés-Pinzón, K. (2014). Effect of hydrocolloids on tapioca starch mixtures' pasting profiles and gluten-free cheese bread's baking properties. *Journal of Food Processing and Preservation, 39*(6), 1672–1681. https://doi.org/10.1111/jfpp.12398

Ross, J. (1968). A. senegal(L.) Willd. Africa with particular reference to Natal. *Sociedade Broteriana, Boletim, 2*(42), 207–240.

Sa'adiya, M. H., Oyi, R. A., & Ibrahim, Y. K. E. (2023). Characterization of Khaya senegalensis gum: Effects of drying methods on the physicochemical properties. *Journal of Pharmaceutical Research, 22*(3), 142–151. https://doi.org/10.18579/jopcr/v22.3.23.6

Sahni, K. (1968). Important trees of the Northern Sudan.

Samaila, J., & Gashua, I. B. (2019). Reaction of gum arabic samples with monoclonal antibodies directed to plant cell wall polymers in dot immunobinding. *Nigerian Journal of Biotechnology, 36*(1), 1. https://doi.org/10.4314/njb.v36i1.1

Sánchez, C., Nigen, M., Tamayo, V. M., Doco, T., Williams, P., Amine, C., et al. (2018). Acacia gum: History of the future. *Food Hydrocolloids, 78*, 140–160. https://doi.org/10.1016/j.foodhyd.2017.04.008

Santos, M., Rocha, L., Vieira, E., Oliveira, A., Lobo, A., Carvalho, M., et al. (2019). Development of composite scaffolds based on cerium doped-hydroxyapatite and natural gums—Biological and mechanical properties. *Materials, 12*(15), 2389. https://doi.org/10.3390/ma12152389

Sarkar, P., Sahu, U., Binsi, P., Nayak, N., Ninan, G., & Ravishanker, C. N. (2018). Studies on physico-chemical and functional properties of some natural Indian gums. *Asian Journal of Dairy and Food Research, 37*(02), 126–131. https://doi.org/10.18805/ajdfr.dr-1241

Sarr, M. S., Seiler, J. R., Sullivan, J., Diallo, A. M., & Strahm, B. D. (2021). Drought resistance and gum yield performances in a Senegalia senegal (L.) Britton progeny trial in Senegal. *New Forests, 52*, 943–957.

Sartori, G., Ballini, R., Bigi, F., Bosica, G., Maggi, R., & Righi, P. (2004). Protection (and deprotection) of functional groups in organic synthesis by heterogeneous catalysis. *Chemical Reviews, 104*(1), 199–250.

Satti, A. A., Osman, M. E., Hassan, E. A., & Al-Assaf, S. (2024). Analytical data of Acacia nilotica var. Nilotica gum. *Bulletin of the Chemical Society of Ethiopia, 38*(4), 839–852. https://doi.org/10.4314/bcse.v38i4.2

Seifel Din, A., Zarroug, M. L., Temu, A., Melnyk, M., & Vantomme, P. (1996). Production and commercialization of gum arabic in Sudan. In *Proceedings of an International Conference Held. Domestication and Commercialization of Non-timber Forest Products in Agroforestry Systems*, Proceedings of an International Conference Held in Nairobi (Kenya), 19–23 Feb 1996.

Sethuraman, S., & Rajendran, K. (2018). Multicharacteristic behavior of tyrosine present in the microdomains of the macromolecule gum arabic at various pH conditions. *ACS Omega, 3*(12), 17602–17609. https://doi.org/10.1021/acsomega.8b02928

Shehu, A., Anyip, B., & Magaji, M. (2020). Antidepressant effect of methanol root bark extract of A. seyal del. (fabaceae): Possible involvement of the inflammatory pathway. *Tropical Journal of Pharmaceutical Research, 19*(7), 1459–1464. https://doi.org/10.4314/tjpr.v19i7.18

Siddig, N. (1996). *Nitrogen and specific rotation as qualifying indices for gum Arabic derived from A senegal (L.) Wild.* M. Sc Thesis University of Khartoum.

Siddig, F. O. (2016). *Preparation and characterization of glucuronic acid and magnesium glucuronate from A. senegalvar. senegal (Gum).* Sudan University of Science and Technology.

Silva, T. M. E., & Souza, J. P. F. D. (2018). The influence of environmental factors in the production of gum arabic (A. senegal, fabaceae)—A meta-analysis. *Revista Brasileira De Climatologia, 23.* https://doi.org/10.5380/abclima.v23i0.58884

Singh, K., & Immanuel, G. (2022). Evaluation of physical and functional properties of composite flour from finger millet, rice and guar gum. *International Journal of Plant & Soil Science, 34*, 270–277. https://doi.org/10.9734/ijpss/2022/v34i2331588

Singh, P., & Lariya, S. (2018). Formulation development of acyclovir microsphere using novel natural polymer. *Journal of Drug Delivery and Therapeutics, 8*(5-s), 271–276. https://doi.org/10.22270/jddt.v8i5-s.1971

Singh, B., & Vashishtha, M. (2014). Ornidazole containing sterculia gum based hydrogels for use in drug delivery. *Materials Science and Technology, 30*(5), 540–548. https://doi.org/10.1179/1743284713y.0000000502

Socrates, G. (2001). *Infrared and Raman characteristic group frequencies: Tables and charts*. Wiley.

Soleimani, K., Derakhshankhah, H., Jaymand, M., & Samadian, H. (2021). Stimuli-responsive natural gums-based drug delivery systems for cancer treatment. *Carbohydrate Polymers, 254*, 117422.

Soto, A., Koivisto, J., Párraga, J., Silva-Correia, J., Oliveira, J., Reis, R., et al. (2016). Optical projection tomography technique for image texture and mass transport studies in hydrogels based on gellan gum. *Langmuir, 32*(20), 5173–5182. https://doi.org/10.1021/acs.langmuir.6b00554

Spiro, T. G., & Strekas, T. C. (1974). Resonance Raman spectra of heme proteins. Effects of oxidation and spin state. *Journal of the American Chemical Society, 96*(2), 338–345.

Surini, S., Azzahrah, F. U., & Ramadon, D. (2018). Microencapsulation of grape seed oil (Vitis vinifera l.) with gum Arabic as a coating polymer by crosslinking emulsification method. *International Journal of Applied Pharmaceutics, 10*(6), 194. https://doi.org/10.22159/ijap.2018v10i6.24093

Taha, K., Elmahi, R., Hassan, E., Ahmed, S., & Shyoub, M. (2012). Analytical study on three types of gum from Sudan. *Journal of Forest Products & Industries, 1*(1), 11–16.

Tamaru, S. I., Yamamoto, M., Shinkai, S., Khasanov, A., & Bell, T. W. (2001). A hydrogen-bonding receptor that binds cationic monosaccharides with high affinity in methanol. *Chemistry - A European Journal, 7*(24), 5270–5276. https://doi.org/10.1002/1521-3765(20011217)7:24 3.0.co;2-h

Tambe, D. E., & Sharma, M. M. (1993). Factors controlling the stability of colloid-stabilized emulsions: I. An experimental investigation. *Journal of Colloid and Interface Science, 157*(1), 244–253.

Tanwar, M., Gupta, R. K., & Rani, A. (2024). Natural gums and their derivatives based hydrogels: In biomedical, environment, agriculture, and food industry. *Critical Reviews in Biotechnology, 44*, 275–301.

Thirakul, S. (1984). Manual of dendrology: Bahr el Ghazal and central regions.
Traffano-Schiffo, M. V., Aguirre-Calvo, T. R., Castro-Giráldez, M., Fito, P., & Santagapita, P. R. (2017). Alginate beads containing lactase: Stability and microstructure. *Biomacromolecules, 18*(6), 1785–1792. https://doi.org/10.1021/acs.biomac.7b00202
Verbeken, D., Dierckx, S., & Dewettinck, K. (2003). Exudate gums: Occurrence, production, and applications. *Applied Microbiology and Biotechnology, 63*(1), 10–21.
Vinod, V., Sashidhar, R., Sarma, V., & Raju, S. S. (2010). Comparative amino acid and fatty acid compositions of edible gums kondagogu (Cochlospermum gossypium) and karaya (Sterculia urens). *Food Chemistry, 123*(1), 57–62.
Vogt, K. (1995). *Common trees and shrubs of dryland Sudan* (p. 167). SOS Sahel International.
Vuletić, L., Klaić, M., Špalj, S., & Peroš, K. (2017). The effect of chewing a sugar-free gum after oatmeal on the postprandial glycaemia—A cross-over study. *Romanian Journal of Diabetes Nutrition and Metabolic Diseases, 24*(2), 127–135. https://doi.org/10.1515/rjdnmd-2017-0017
Wafari, U. U., & Ta'awu, K. G. (2021). Biotechnological application and the importance of plant gum exudates A. senegal and A. seyal to the food industry. *African Scholars Journal of Pure and Applied Science*.
Wang, Q., & Cui, S. W. (2005). Understanding the physical properties of food polysaccharides. In *Food carbohydrates; chemistry, physical properties and applications* (pp. 161–262). Colorectal Cancer (CRC) Press, Taylor & Francis Group.
Wang, Y., Pei, X., Xia, L., & Xin, Z. (2021). Covalent crosslinks turn natural eucommia ulmoides gum/polybutene-1 composites into multiple shape memory materials. *Polymer Composites, 43*(3), 1371–1382. https://doi.org/10.1002/pc.26454
Wardhono, E., Zafimahova-Ratisbonne, A., Lanoisellé, J., Saleh, K., & Clausse, D. (2014). Optimization of the formulation of water in oil emulsions entrapping polysaccharide by increasing the amount of water and the stability. *The Canadian Journal of Chemical Engineering, 92*(7), 1189–1196. https://doi.org/10.1002/cjce.21985
Webb, S. L., & Dingwell, D. B. (1990). Non-Newtonian rheology of igneous melts at high stresses and strain rates: Experimental results for rhyolite, andesite, basalt, and nephelinite. *Journal of Geophysical Research, 95*(B10), 15695–15701.
White, W. B. (1957). AOAC method of analysis. *Food Drug Cosm LJ, 12*, 327.
Williams, P., Phillips, G., & Randall, R. (1990). Structure-function relationships of gum arabic. In *Gums and stabilizers for the food industry* (Vol. 5, pp. 25–36).
Xiang, S., Yao, X., Zhang, W., Zhang, K., Fang, Y., Nishinari, K., & Jiang, F. (2015). Gum Arabic-stabilized conjugated linoleic acid emulsions: Emulsion properties in relation to interfacial adsorption behaviors. *Food Hydrocolloids, 48*, 110–116. https://doi.org/10.1016/j.foodhyd.2015.01.033
Yang, H., Wang, D., Deng, J., Yang, J., Shi, C., Zhou, F., et al. (2018). Activity and structural characteristics of peach gum exudates. *International Journal of Polymer Science, 2018*, 1–5. https://doi.org/10.1155/2018/4593735
Zhao, Y. (2023). Study on the preparation of inorganic solidified foam with high stability and its prevention and treatment of coal spontaneous combustion. *Energy & Fuels, 37*(18), 14268–14279. https://doi.org/10.1021/acs.energyfuels.3c02657
Zhara, H., Adeel, S., Özomay, Z., & Mia, R. (2024). Properties and performance relationship of biopolymers in textile industry. In *Biopolymers in the textile industry: Opportunities and limitations* (pp. 87–121). Springer Nature Singapore.
Zingue, S., Njuh, A., Tueche, A., Tamsa, J., Tchoupang, E., Kakene, S., et al. (2018). In vitro cytotoxicity and in vivo antimammary tumor effects of the hydroethanolic extract of A. seyal (mimosaceae) stem-bark. *BioMed Research International, 2018*, 1–13. https://doi.org/10.1155/2018/2024602

Chapter 2
Analytical Study on Secondary Metabolites Extraction and Characterization from Acacia Gums (AGs)

Ahmed A. M. Elnour, Saphwan Al-Assaf, Nassereldeen Ahmed Kabbashi, Mohammed Elmubark Osman, and Abdelgadir Elfadil Abdelgadir

Abstract Acacia gums (AGs), derived from *Acacia senegal*, *Acacia seyal*, and *Acacia polyacantha* trees, are widely used in various industries, including food, pharmaceuticals, and cosmetics. They serve as emulsifiers, stabilizers, thickeners, and soluble dietary fibers. This study aimed to optimize the extraction process of secondary metabolites extracted (SME) from AGS to enhance their efficiency and quality for industrial applications. Gum arabic, primarily composed of D-galactose, L-arabinose, L-rhamnose, D-glucuronic acid, and 4-*O*-methyl-D-glucuronic acid with a small fraction of proteins, is a multifunctional ingredient known for its emulsifying activity, attributed to its amphiphilic nature. Its effectiveness as an emulsifier stems from its high water solubility, low-viscosity, good surface activity, and ability to form a protective film around emulsion droplets. AGS in Sudan play crucial ecological roles by providing fodder, firewood, gums, and soil erosion control, which are essential for gum arabic production. Sudan's diverse biodiversity hosts many

A. A. M. Elnour (✉)
Faculty of Chemical and Process Engineering Technology, Universiti Malaysia Pahang Al-Sultan Abdullah (UMPSA), Pahang Gambang, Malaysia

Centre of Excellence for Advanced Research in Fluid Flow (CARIFF), University Malaysia, Pahang Gambang, Malaysia
e-mail: adamhassan@umpsa.edu.my

S. Al-Assaf
Hydrocolloids Research Center, University of Chester, Chester, UK
e-mail: s.alassaf@chester.ac.uk

N. A. Kabbashi
Department of Bio-technology Engineering, Bioenvironmental Engineering Research Institute (BERI), Kulliyyah (Faculty) of Engineering, International Islamic, University, Kuala Lumpur, Malaysia
e-mail: nasreldin@iium.edu.my

M. E. Osman
College of Science, Sudan University of Science and Technology, Khartoum, Sudan

A. E. Abdelgadir
Faculty of Science and Technology, Al-Neelain University, Khartoum, Sudan

A. A. M. Elnour (ed.), *Gum Arabic and Breast Cancer Biology*,
https://doi.org/10.1007/978-981-97-8518-6_2

plant species traditionally used for medicinal purposes, with extensive knowledge of medicinal plant species stemming from their widespread use among different ethnic groups. This study highlights the importance of optimizing the extraction of secondary metabolites from AGS for industrial applications, and the significance of AGS in Sudan's ecological balance and economic stability.

Keywords Gum arabic · Secondary metabolites · Extraction techniques · Optimization

Highlights
1. Gum arabic, derived from *Acacia senegal* and *Acacia seyal*, is a multifunctional ingredient.
2. Gum arabic is an emulsifier, stabilizer, and thickener used in various industries.
3. This study aimed to optimize the extraction of secondary metabolites from AGS for industrial application.
4. The versatility and effectiveness of gum arabic make it a valuable ingredient across multiple sectors.

2.1 Introduction

Gum arabic, derived from Acacia trees, is a multifunctional ingredient widely used in the food industry. They are primarily used as thickeners, emulsifiers, stabilizers, and additives in various food products (Al-Assaf et al., 2009). Its emulsifying activity is attributed to its amphiphilic nature, which is characterized by a hydrophobic protein fraction covalently linked to hydrophilic polysaccharide structures (Saphwan Al-Assaf & Phillips, 2009; Williams & Phillips, 2000). The effectiveness of gum arabic as an emulsifier stem from its high water solubility, low-viscosity, good surface activity, and ability to form a protective film around emulsion droplets (Dickinson, 2003). It has also been studied for its ability to stabilize nanoemulsions and act as a delivery system for various compounds, such as vitamin E and fucoxanthin (Charoen et al., 2011; Zhu, 2023). Its emulsifying properties have been compared with those of other emulsifiers, such as gum ghatti (Al-Assaf et al., 2007), sugar beet pectin (Castellani et al., 2010), corn fiber gum (Cirre et al., 2014), whey protein isolate, and modified starch, showing its efficacy in stabilizing oil-in-water emulsions (Golkar et al., 2015). Research has explored the applications of gum arabic beyond the food industry, including pharmaceuticals, beverages, and cosmetics (Bai et al., 2017).

The genus Acacia, a member of the family Fabaceae, comprises a diverse group of trees with a significant presence in Sudan, where approximately 34 species are cultivated. These trees, typically small to medium-sized and often thorny, are widely distributed across various regions in Sudan. AGs mainly comprises D-galactose, L-arabinose, L-rhamnose, D-glucuronic acid, and 4-*O*-methyl-D-glucuronic acid with a small fraction of proteins. The main difference between various AGs species is either the protein content or sugar composition/ratio (Chap. 1). *Acacia senegal* is predominantly found in the northeast, Red Sea, and parts of the Khartoum state. Simultaneously, *Acacia seyal* thrives in areas such as the Nile River, Atbara, Rahad, Khartoum, Blue Nile, and West Kordofan in Sudan (Ayalew et al., 2019). AGs in Sudan play crucial ecological roles by providing fodder, firewood, gums, and soil erosion controls. Additionally, Acacia trees such as *Acacia senegal* and *Acacia seyal* are essential for gum arabic production, and gum arabic is a valuable commodity obtained from these species (Mugo et al., 2020). AGs, particularly *Acacia senegal* trees, are known for their nitrogen-fixing abilities, making them vital for reforestation, soil improvement, and land reclamation efforts in arid regions, such as Sudan (Raddad et al., 2005). Research has shown that *Acacia seyal* is more efficient at nitrogen fixation than other AGs (Gibreel, 2013). These trees are deeply intertwined with the local population's livelihoods, especially in regions such as South Kordofan State, where *Acacia senegal* supports the community through gum arabic production (Adam & Fadl, 2011). Additionally, studies have highlighted the impacts of climate change on *Acacia seyal* var. Seyal production in Sudan emphasizes the need to understand and address these challenges for sustainable resource management (Younis et al., 2022). Thus, AGs are integral to Sudan's ecological balance and economic stability.

The reader is referred for more detailed information from an overview of pre- and post-harvest practices in Sudan dealing with the entire journey from farm to the end user (Elrayah 2012). Sudan's diverse biodiversity hosts many plant species traditionally used for medicinal purposes (Eisawi et al., 2021). The country's varied climate supports various medicinal plants (Mahmoud et al., 2024). Sudan's flora comprises over 3137 species of flowering plants from 170 families and 1280 genera, making it a valuable resource for traditional medicine (Khalid et al., 2012). Extensive traditional knowledge of medicinal plant species in Sudan stems from the widespread use of various plants among different ethnic groups (Eisawi et al., 2021). This knowledge benefits local communities and promises to discover new compounds that can be explored for modern pharmaceutical applications, including cancer therapy (Saeed et al., 2015). The reliance on traditional medicine in Sudan is substantial, with around 90% of the population primarily using these remedies because of limited access to modern synthetic drugs (Karar & Kuhnert, 2017). Documenting ethnobotanical knowledge in Sudan is critical, especially in regions such as Darfur State, where information on traditional plant use is scarce (Muhakr et al., 2024). A diverse array of medicinal plants in Sudan and neighboring countries have been associated with anti-angiogenic, antioxidant, and cytotoxic effects, indicating the potential health benefits of these natural resources (Hassan et al., 2014). The investigation of medicinal plants, particularly AGs, has garnered significant attention because of their secondary metabolite identification, isolation, and biological properties. AGs contain various bioactive compounds such as terpenoids, flavonoids, and tannins, known for their therapeutic activities, including antioxi dant, anti-inflammatory, antibacterial, and antifungal properties (Abdel-Farid et al., 2014; Thotathil et al., 2023). SMs are promising sources of natural therapeutic products (Ziani et al., 2020). Thus, the diverse biochemical compositions of AGs make them valuable in natural medicine. Therefore, preserving and investigating Sudan's medicinal plants is vital to harness their full potential in healthcare and other industries.

This chapter overviews the most commonly used techniques and methodologies for extracting and fractionating bioactive compounds from AGs. These compounds included anthocyanins in Acacia polymorpha and Acacia buckleyi seeds, ellagitannins in Acacia melanoxylon, and tannins in Acacia bark extracts. The search for natural antimicrobial agents from AGs aligns with the current interest in natural preservatives for use in food products (Renard et al., 2006; Tiwari et al., 2023). Studies have also focused on the extraction and fractionation of bioactive compounds from AGs, highlighting the medicinal value of polysaccharides, tannins, flavonoids, alkaloids, and fatty acids (Elnour et al., 2020; Ziani et al., 2020). Therefore, the potential of AGs as a source of bioactive compounds is significant for advancements in the food and pharmaceutical industries.

2.2 Extraction Techniques of AGs-SME

Soxhlet extraction is a well-established method for efficiently extracting bioactive compounds from various plants (El Maaiden et al., 2022). This technique provides higher extraction yields than conventional methods such as maceration (Haido, 2024). Its efficiency is attributed to the continuous extraction of compounds from the sample through a refluxing process, ensuring thorough extraction (Saptarini et al., 2020). Soxhlet extraction remains the standard for evaluating new extraction techniques because of its reliability and proven efficacy (Gfrerer et al., 2004). Thus, Soxhlet extraction is preferred because of its superior performance and dependability.

Maceration, a traditional laboratory extraction method is less efficient than Soxhlet extraction (Haido, 2024). Research has indicated that maceration generally produces lower bioactive compound yields than Soxhlet extraction (Begić et al., 2020). Despite its low efficiency, maceration is still commonly used in laboratories to extract bioactive components from plant materials (Kafo et al., 2023). Some studies have aimed to optimize extraction conditions during maceration to enhance the yield of bioactive compounds (Haido, 2024). Therefore, although maceration remains popular, its efficiency can be improved by optimizing the extraction conditions.

Ultrasound-assisted extraction has shown promise for effectively extracting bioactive compounds from various sources (Mustapa et al., 2015). This method uses ultrasound waves to disrupt the cell walls and enhance the extraction process, potentially increasing the yield of bioactive compounds (Latiff et al., 2021). Although not as widely used as Soxhlet extraction, ultrasound-assisted extraction is gaining attention for its efficiency and ability to improve the extraction rates (Mustapa et al., 2015). Therefore, ultrasound-assisted extraction is becoming a valuable alternative in the field of bioactive compound extraction.

The choice of extraction method can significantly affect the yield of extracted bioactive compounds. Variables such as the solvent used, temperature, ultrasonic power, particle size, and specific extraction conditions are crucial for determining the efficiency of the extraction process (Montiel et al., 2023). The selection of solvents and extraction methods influence the chemical composition, antioxidant activity, and overall quality of the extracted compounds (Ozturk et al., 2022). Different solvents and extraction techniques affect the total phenolic content and antioxidant capacity of plant extracts (Vega-Arroy et al., 2017). Therefore, optimizing the extraction methods and conditions is essential for maximizing the yield and quality of bioactive compounds.

The extraction method can influence the composition and bioactivity of extracted compounds in specific plant species. For example, in the extraction of bioactive compounds from *Psidium guajava* L. leaves, the choice of extraction method affects the polyphenolic profile and antioxidant activity of the extracts (Montiel et al., 2023). Similarly, the extraction method used for Vernonia cinerea leaves affected the kinetics of bioactive compound extraction, highlighting the importance of selecting the appropriate extraction technique (Alara & Abdurahman, 2018). Thus, selecting an appropriate extraction method is crucial for optimizing the bioactive compound profile and activity of plant extracts.

The efficiency of extraction methods can vary depending on the type of bioactive compound targeted. For instance, Soxhlet extraction was found to be the most effective method for maximizing forskolin content in *Coleus forskohlii* roots (Singh & Suryanarayana, 2019). Similarly, in the extraction of α-glucosidase-inhibitory compounds from *Phaleria macrocarpa* fruit flesh, different extraction methods, including Soxhlet extraction, were compared to determine their efficacy (Easmin et al., 2017). The selection of an extraction method for the isolation of bioactive compounds is a critical step in both research and industry. Although Soxhlet extraction is widely recognized for its efficiency and high extraction yields, other methods, such as maceration and ultrasound-assisted extraction, also play important roles. Researchers need to consider various factors, such as the nature of the compounds, plant material, and desired outcomes when choosing an extraction method to ensure optimal results. Therefore, selecting the appropriate extraction method is essential for achieving targeted bioactive compounds' highest efficiency and yield.

Soxhlet extraction, maceration, and ultrasound-assisted extraction are commonly employed methods for the extraction of bioactive compounds from plant materials. Among these methods, Soxhlet extraction is known for its efficiency in extracting large amounts of bioactive compounds (Jovanović et al., 2017). Maceration, a traditional laboratory extraction method, has been widely used over time although it is considered less efficient than Soxhlet extraction (Begić et al., 2020). Ultrasound-assisted extraction, but not as extensively studied as maceration, has shown promise for the effective extraction of bioactive compounds (Yusof et al., 2020). The efficiency of these extraction methods is influenced by various factors, such as the choice of solvent, temperature, ultrasonic power, particle size, and other specific conditions (Idris & Mohd Nadzir, 2021). For example, the solvent used in extraction plays a crucial role in determining the yield of bioactive compounds (Ozturk et al.,

2022). Additionally, the optimization of extraction conditions, as seen in studies focusing on different plant materials, can significantly affect extraction efficiency (Idris & Mohd Nadzir, 2021). Although conventional methods such as Soxhlet extraction have been widely used, modern techniques such as ultrasound-assisted extraction are gaining attention for their green and rapid extraction capabilities (Yerena-Prieto et al., 2022). These modern methods offer alternatives to conventional techniques and provide high-quality extracts from natural products (Yerena-Prieto et al., 2022). Therefore, selecting and optimizing an appropriate extraction method is essential to achieve the highest yield and quality of bioactive compounds.

Various extraction methods can be employed to extract bioactive compounds from plants based on the nature of the compounds and the plant material. Organic solvent extraction techniques, such as maceration, Soxhlet extraction, percolation, and sonication, as well as non-organic solvent extraction methods, such as decoction, infusion, and boiling liquor separation, are commonly used (Sasidharan et al., 2010). These methods can be used alone or in combination to enhance the yield of the compounds. For instance, methanol, ethanol, and boiling water extractions are effective in selectively extracting phytochemicals from herbs, whereas Soxhlet and sonication extractions are suitable for highly soluble compounds (Castro-Lopez et al., 2017). In contrast, subcritical fluid extraction is preferred for non-polar compounds, and green and pressurized liquid extractions are ideal for polyphenols (Castro-Lopez et al., 2017). Therefore, choosing an appropriate extraction method based on the nature of the compound and plant material is essential for optimizing the yield and quality of the extracted bioactive compounds.

The choice of extraction method is influenced by factors such as the polarity of the bioactive secondary metabolite, purpose of extraction (identification, characterization, or quantification), and type of plant material (Bitwell et al., 2023). Modern extraction techniques, such as solid-phase microextraction, supercritical fluid extraction, pressurized liquid extraction, microwave-assisted extraction, solid-phase extraction, and surfactant-mediated extraction, offer efficient ways to extract bioactive compounds from plants (Huie, 2002). These methods have advantages such as increased extraction efficiency, reduced solvent volume, and shorter extraction times than conventional methods (Rezaeepour et al., 2015). Therefore, modern extraction techniques are preferred owing to their efficiency, cost-effectiveness, and ability to produce high-quality extracts.

Supercritical fluid extraction, ultrasound-assisted extraction, and conventional methods have been compared for their effects on phytochemicals and antioxidant activity in plants (Muchahary & Deka, 2021). The choice of extraction technique significantly affects phytochemical composition, with different methods yielding varying concentrations of phytochemicals (Muchahary & Deka, 2021). Green chemical approaches aim to develop solvent-free methods for the extraction and analysis of pharmaceutical compounds (Montesdeoca-Esponda et al., 2015). Therefore, selecting the appropriate extraction method is crucial for optimizing the phytochemical content and antioxidant activity of plant extracts, while also considering their environmental impact.

Various extraction methods have been employed to extract bioactive compounds from plant materials, each offering distinct advantages based on the nature of the compounds and the plant matter involved. Organic solvent extraction techniques such as maceration, Soxhlet extraction, percolation, and sonication, as well as non-organic solvent methods like decoction, infusion, and boiling liquor separation technology, are commonly utilized (Uwineza & Waśkiewicz, 2020). These methods can be used alone or in combination to enhance the yield of the target compounds. For instance, methanol, ethanol, and boiling water extractions are effective in selectively extracting phytochemicals from herbs, whereas Soxhlet and sonication extractions are suitable for highly soluble compounds (da Silva et al., 2016). The choice of extraction method is influenced by factors such as the polarity of the bioactive secondary metabolite, purpose of extraction (identification, characterization, or quantification), and solubility of the compounds. Subcritical fluid extraction is preferred for non-polar compounds, whereas green and pressurized liquid extractions are suitable for polyphenols (Uwineza & Waśkiewicz, 2020). Therefore, selecting an appropriate extraction method based on the specific characteristics of the bioactive compounds and plant materials is crucial for optimizing extraction efficiency and yield.

In addition, the solvent used plays a crucial role in the extraction process. Ethanol is effective in yielding high amounts of bioactive compounds, while petroleum ether is less efficient (Haido, 2024). The extraction method also affects the bioactivity of the extracts, with the technique significantly influencing bioactivity (Scepankova et al., 2018). Extraction conditions, such as solvent type, composition, temperature, extraction time, and number of steps, affect the composition of bioactive compounds in the extracts (Božunović et al., 2023). Thus, careful selection of solvents and optimization of the extraction conditions are essential to maximize the yield and bioactivity of the extracted compounds.

The extraction techniques of SME from AGs gums are crucial for harnessing their phytochemical contents and potential benefits. Gum arabic, derived from *Acacia senegal* and *Acacia seyal* trees, is a natural polysaccharide that has diverse applications in various industries (Mengesha et al., 2022). Phytochemical screening of *Acacia seyal* roots has been conducted to identify secondary metabolites, underscoring the importance of understanding the chemical composition of plant extracts (Mengesha et al., 2022). In addition, the extraction of bioactive components from *Acacia seyal* gum has been studied, highlighting the therapeutic potential of these natural products (Elnour et al., 2020). AGs, including those from *Acacia senegal* and *Acacia seyal* gum, have traditionally been used for their medicinal properties such as anti-diarrheal activities and antioxidant effects (Elamin Elhasan et al., 2022). Therefore, exploring the extraction techniques and phytochemical properties of AGs is essential to maximize their industrial and medicinal applications.

These natural products contain various bioactive compounds that contribute to their pharmacological activities, making them valuable resources for drug development (Elnour et al., 2020). The extraction of SME from AGs is essential for isolating and studying the active compounds responsible for the observed biological effects (Elnour et al., 2020). Additionally, gum arabic from *Acacia senegal* and

Acacia seyal has been characterized, providing insights into the molecular composition and properties of these natural exudates (Osman et al., 1993). Various extraction techniques and environmental factors can influence the production and quality of gum arabic, underscoring the need for optimized extraction methods to ensure consistency and efficacy (Silva & Souza, 2018; Osman et al., 2024). Understanding the molecular association and physicochemical properties of AGs exudates is crucial for their applications in food, pharmaceuticals, and industry (Satti et al., 2024). Research on the prebiotic functionality of gum arabic from *Acacia senegal* and *Acacia seyal* has highlighted its potential as a dietary fiber with beneficial effects on the gut microbiota (Cherbut et al., 2003; Calame et al., 2008). Therefore, optimizing extraction methods and understanding the properties of AGs are vital for maximizing their pharmacological, industrial, and dietary applications. The modulation of the gut microbiota using AGs polysaccharide has been studied for its impact on short-chain fatty acid production and bacterial populations in the gut (Rawi et al., 2021). These findings suggest that AGs promote gut health and overall well-being through their prebiotic properties. In addition to their medicinal and dietary applications, AGs have been explored for industrial uses such as foundry processes and wastewater treatment (Ocheri et al., 2021; Saleem et al., 2017). The versatility of AGs in various fields underscores their importance as natural resources with diverse functions. Therefore, AGs are valuable for their potential contribution to gut health and applications in industry and medicine, highlighting their multi-faceted utility across different sectors. Figure 2.1 shows Soxhlet as an example of extracting secondary metabolites (SME) from AGs.

2.2.1 Traditional Methods of Extraction of SME-AGs

Maceration is a traditional method for extracting bioactive compounds from plants using solvents and isolating substances such as polyphenols, flavonoids, and other secondary metabolites. This technique has been optimized for extracting polyphenols from herbs such as Thymus serpyllum, demonstrating its effectiveness (Jovanović et al., 2017). Additionally, maceration is noted for its simplicity and cost efficiency in extracting tannins from coffee husks (Wulandari et al., 2023). Despite its ease of use, maceration generally requires longer extraction times than ultrasonic-assisted extraction (UAE) (Wardany & Nirmala, 2023). This technique has also been used to extract bioactive compounds from diverse sources, including grape marc, Vitis angustifolia, Leonotis leonurus, and various exudates from AGs, proving its versatility (Maqbool et al., 2022). Maceration is a valuable method for effectively isolating catechins, flavonoids, and tannins from various plants (Plaza et al., 2019). Thus, maceration remains a pivotal technique for natural compound isolation, balancing cost, and efficiency despite its slower processing times.

Maceration has been evaluated alongside other extraction methods, such as Soxhlet extraction, revealing variations in the total phenolic content of extracts

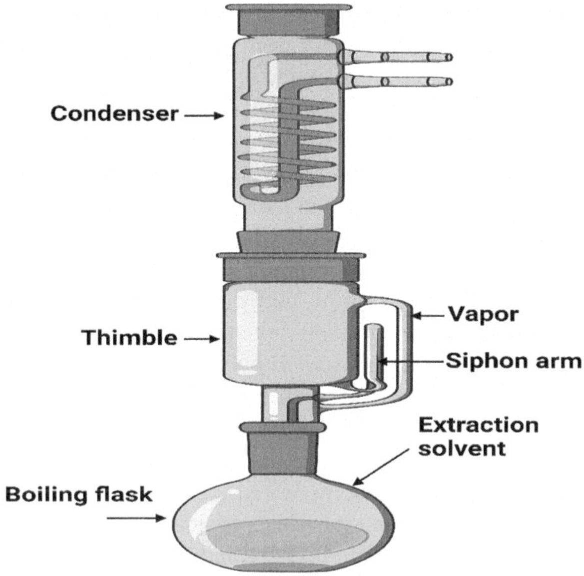

Fig. 2.1 Demonstration of Soxhlet as extraction technique for extracting secondary metabolites (SME) from the Acacia gums (AGs)

(Salamah et al., 2017). The method is also applied in extracting bioactive compounds from plants such as Mitragyna speciosa, yielding extracts with potent antioxidant and antibacterial properties (Kafo et al., 2023). In the winemaking industry, maceration is essential for extracting phenolic compounds from grape skins, thereby enhancing their antioxidant activity (dos Santos Lima et al., 2015). Further research has explored the use of maceration with other plant materials, such as mango peel, carob, and chili pepper waste, showing its adaptability in extracting a variety of bioactive compounds (Coelho et al., 2019; Gordillo et al., 2016; Olguín-Rojas et al., 2024). Maceration remains a crucial and versatile technique for extracting bioactive compounds because of its simplicity, cost-effectiveness, and efficient extraction of diverse secondary metabolites. Therefore, it is preferred for traditional and innovative applications in various industries.

Traditional methods for extracting AGs-SME have focused on isolating bioactive compounds because of their promising antibacterial, antifungal, antioxidant, and anti-inflammatory properties (Todkar et al., 2017; Zaman et al., 2022; Elnour et al., 2018). Techniques such as gas chromatography-mass spectrometry (GC-MS) and high-performance liquid chromatography with photodiode array detection/mass spectrometry (HPLC-PAD/MS) are commonly used to profile and identify the active compounds (Todkar et al., 2017; Zaman et al., 2022). Tannins, polyphenols synthesized in response to stress, are significant SME in these plants (M. I, 2023). Additionally, gum arabic, derived from *Acacia senegal* and *Acacia seyal* gum, is extensively utilized for stabilizing and thickening the properties of food products

(Mugo et al., 2020). Techniques, such as UAE, have been investigated to enhance the extraction of bioactive components from gum arabic (Elnour et al., 2018). Therefore, the ongoing refinement of extraction methods is pivotal for harnessing the full potential of these bioactive compounds in various applications.

Additionally, the antioxidant activity of gum arabic has been assessed using various methods, including 1,1-diphenyl-2-picrylhydrazyl (DPPH), Folin-Ciocalteu indices, ORAC, ferric reducing antioxidant power (FRAP), and cupric ion reducing antioxidant capacity (CUPRAC) assays, which underscore its potential as a natural antioxidant (Mirghani et al., 2018). An investigation of the bioactive compounds present in the crude methanol extracts and fractions of *Acacia seyal* gum also highlights their potential medicinal properties (Elnour et al., 2019). Moreover, studies on *Acacia polyacantha* gum extracts have demonstrated their antibacterial and antibiotic-modifying potential against multidrug-resistant gram-negative bacteria, highlighting their importance in addressing antibiotic resistance (Ashu et al., 2020; Mambe et al., 2019). Traditional methods for extracting AGs-SME utilize sophisticated techniques to isolate and identify bioactive compounds with diverse pharmacological properties, rendering these plants a valuable resource for medicinal and industrial applications. Therefore, continued exploration and optimization of these extraction methods are essential to harness the therapeutic potential of gum arabic. A diagram showing the steps in the process of maceration, as the most traditional method used in AGs-SME, is shown in Fig. 2.2.

2.2.2 Modern Extraction Techniques

To extract SME from AGs, modern methods can be utilized based on current research findings. Todkar et al. (2017) emphasized the effectiveness of chromatographic techniques such as GC-MS and HPLC-PAD/MS for metabolite profiling, which helps identify compounds with potent antibacterial and antifungal activities. Elnour et al. (2018) highlighted the challenges in efficiently extracting bioactive components, stressing the importance of optimizing these processes to enhance the extraction time and quality. Furthermore, Elmi et al. (2020) successfully isolated specific compounds such as catechin, epicatechin, and lupeol from *Acacia seyal* bark extracts, demonstrating their potential for antioxidant and antibacterial applications. Thus, advancing extraction techniques is crucial for the effective isolation and utilization of these bioactive compounds for health and industrial purposes.

Elnour et al. (2019) explored the bioactive compounds in crude methanol extracts and fractions of *Acacia seyal* gum, revealing significant compounds in various soluble fractions. This study highlighted the complexity and richness of the bioactive profiles of these gums. Additionally, Osman et al. (1993) characterized commercial samples of gum arabic from *Acacia senegal* and *Acacia seyal*, providing detailed

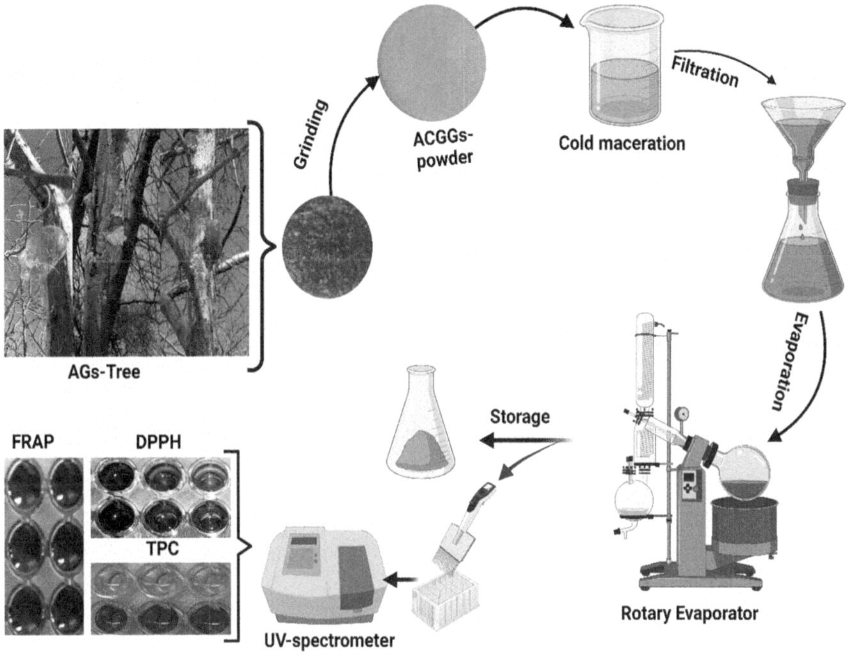

Fig. 2.2 Explanation of the maceration as most traditional methods used in extracting SME from acacia gums (AGs)

insights into their chemical compositions and physicochemical properties. Such characterizations are essential for understanding the composition of gum and refining the extraction methods. In another significant study, Mambe et al. (2019) examined the antibacterial and antibiotic-modifying capabilities of extracts, fractions, and specific compounds from *Acacia polyacantha* gum against multidrug-resistant gram-negative bacteria. This study emphasizes the potential pharmaceutical applications of *Acacia polyacantha* gum, highlighting its importance in the development of new antibacterial agents. As a result, ongoing research on the bioactive properties of these gums is crucial for their effective application in the pharmaceutical and industrial sectors.

Hammad et al. (2022) conducted GC-MS analysis and antibiogram studies on *Acacia polyacantha* leaves, demonstrating the antimicrobial potential of this plant species. Thus, extraction of SME from AGs can be effectively achieved through chromatographic techniques, method optimization, and detailed characterization of bioactive compounds. This approach enhances the understanding of their therapeutic properties and supports their application in pharmaceutical and medicinal fields. Ultrasound-assisted extraction, as an advanced extraction technique for AGs-SME, is shown in Fig. 2.3.

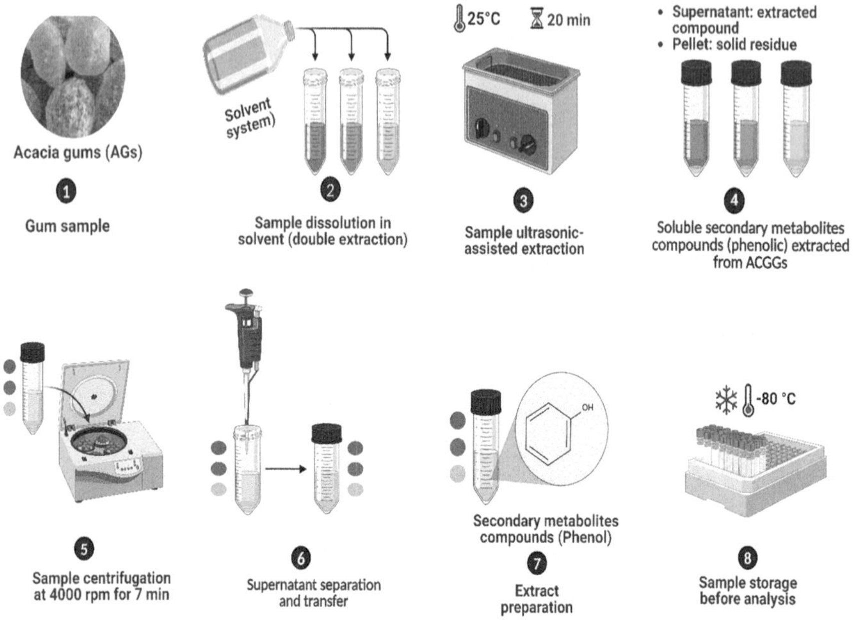

Fig. 2.3 Ultrasonic-assisted extraction as advanced extraction techniques for SME from acacia gums (AGs)

2.3 Analytical Methods for AGs-SME Detection

Analytical methods are crucial for the detection and characterization of SME in natural products such as AGs. Liquid chromatography-mass spectrometry (LC-MS) is commonly employed for this purpose and offers robust results (Peitzsch et al., 2012). Specifically, liquid chromatography coupled with high-resolution mass spectrometry (LC-HRMS) has been used to identify SME in various plant extracts (Aryal et al., 2021). High-performance liquid chromatography (HPLC) is another widely used technique for the separation and analysis of these compounds (Hisiger & Jolicœur, 2007). Furthermore, HPLC coupled with a triple quadrupole mass spectrometer has proven effective in separating complex mixtures and providing detailed structural information about analytes (Kiske et al., 2014). Thus, the integration of these sophisticated analytical techniques is essential for advancing our understanding of the complex chemistry of natural products and for enhancing their application in various industries.

In the context of detecting secondary metabolites, the use of advanced technologies, such as three-dimensional gas chromatography coupled with accurate mass spectrometry, offers a novel, high-resolution approach for characterizing multicomponent samples (Yan et al., 2018). Additionally, integrating various analytical methods, including mass spectrometry with cultivation systems, has been suggested to enable the comprehensive characterization of secondary metabolomics (Bader et al., 2021). This integrated approach facilitated the detection of a broad-spectrum of natural microbial products. Furthermore, the development of reliable analytical methods is crucial for a thorough investigation of SME (Siebenberg et al., 2011). Therefore, advanced and integrated analytical techniques are essential for enhancing the understanding and utilization of SME in various scientific and industrial applications.

Techniques such as liquid chromatography-continuous flow secondary ion mass spectrometry (LC/CFSIMS) have been used for the quantitation of specific metabolites, underscoring the importance of methodological advancements in this field (Hogge et al., 1993; Balsevich et al., 1996). Additionally, the combination of liquid chromatography with electrochemical coulometric array detection has been proven to be a sensitive and robust technique for metabolomic profiling (Bird et al., 2012; Gathungu et al., 2014). In summary, a combination of techniques, including LC-MS, LC-HRMS, HPLC, and advanced mass spectrometry methods, are instrumental in the detection, separation, and characterization of SME from natural sources such as AGs. These analytical methods provide valuable insights into the chemical composition and bioactive compounds present in natural products, thus enhancing our understanding of their potential benefits and applications. As a result, continued refinement and integration of these techniques is crucial for advancing research in natural product chemistry.

2.3.1 Chromatographic Techniques

Chromatographic techniques are essential for analyzing SME derived from AGs. Gas chromatography-mass spectrometry (GC-MS) (Elnour et al., 2020), HPLC (Rawi et al., 2021), and column chromatography (Ashu et al., 2020) have been employed to separate and analyze the bioactive compounds in these gums. Additionally, physicochemical analyses have characterized the properties of AGs, including gum arabic, from species such as *Acacia senegal* and *Acacia seyal*, revealing their emulsifying, stabilizing, and other functional properties that are crucial for applications in the food and pharmaceutical industries (Fauconnier et al., 2000). Research has also focused on the molecular characteristics and interfacial properties of gum arabic obtained from these species (Gashua et al., 2016a). Microstructural analysis further demonstrated molecular aggregation in these gum solutions, emphasizing the importance of understanding the structural properties for functional applications. Thus, a combination of chromatographic techniques and physicochemical and molecular analyses is crucial for elucidating the chemical composition, functional properties, and bioactive compounds of AGs. These studies offer valuable insights into the potential applications of AGs in various industries including food, medicine, and cosmetics. Finally, high-performance liquid chromatography, an advanced detection technique for AGs-SME is shown in Fig. 2.4.

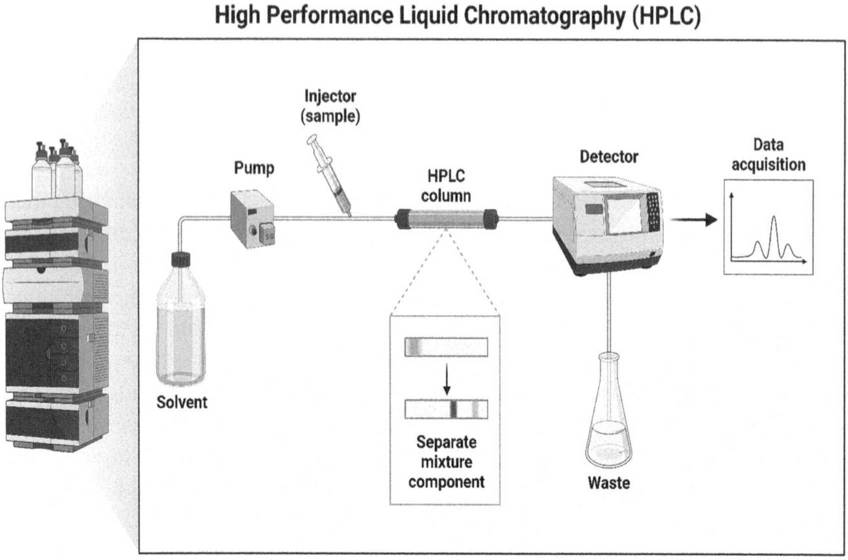

Fig. 2.4 Demonstration of high-performance liquid chromatography as advanced detection techniques for SME from Acacia gums (AGs)

2.3.2 Spectroscopic Techniques

Spectroscopic techniques are crucial for detecting SME in the AGs. These methods help identify and characterize the various compounds present in gums. Research indicates that AGs contain a diverse array of secondary metabolites, including amines, cyanogenic glycosides, flavonoids, alkaloids, seed oils, cyclitols, fluoroacetate, gums, non-protein amino acids, diterpenes, fatty acids, terpenes, hydrolyzable tannins, and condensed tannins (Batiha et al., 2022). The structural characteristics of *Acacia senegal* gum exudate have been extensively studied, revealing the presence of components such as arabinogalactan-peptide, arabinogalactan-protein, and glycoprotein fractions (Renard et al., 2006; Mahendran et al., 2008). Additionally, the adsorption process of AGs on gold nanoparticles was investigated using spectroscopic methods, providing insights into the interactions between protein moieties and carbohydrate blocks during adsorption (Davantès, 2023). Thus, these spectroscopic techniques enhance our understanding of the complex biochemistry of AGs and their interactions, thereby facilitating further applications in various fields.

Additionally, the molecular characteristics and interfacial properties of gum arabic harvested from both *Acacia senegal* and *Acacia seyal* were compared, revealing differences in the emulsification effectiveness between the two types (Gashua et al., 2016a). Furthermore, chromatographic techniques such as GC-MS and HPLC-PAD/MS have been utilized for secondary metabolite profiling in Acacia concinna to detect potent compounds with antibacterial and antifungal activities (Todkar et al., 2017). These techniques enable comprehensive analysis of the chemical composition of gums and assist in identifying their bioactive compounds. Spectroscopic and chromatographic methods are essential tools for detecting and analyzing SME in AGs. These methods provide valuable insights into the chemical composition, structural characteristics, and potential bioactive compounds of AGs gums, thereby enhancing the understanding of their functional properties and applications. Thus, continued refinement and application of these techniques are crucial for advancing the study and utilization of AGs in various industries.

SME are organic compounds that do not participate directly in the fundamental processes of growth, development, and reproduction in organisms. Unlike primary metabolites, which are produced constitutively and uniformly within a species, SME shows greater individual variability and are often unique to specific species or genotypes. These compounds have a wide range of chemical structures including terpenoids, phenolic acids, quinones, alkaloids, cardiac glycosides, flavonoids, saponins, tannins, steroids, and lignin (Bergman et al., 2019). Terpenoids, notable for their diversity, serve both as primary metabolites essential for growth and development, and as SME that enhance a plant's interactions with its environment (Bohlmann & Keeling, 2013; Cheng et al., 2007; Moses et al., 2013). Traditionally, terpenoids, alkaloids, and phenolics have been classified as SME (Adibah & Azzeme, 2018). Alkaloids, significant for their pharmacological properties, are nitrogen-containing compounds found across various plant species and play a crucial role in plant defense mechanisms against external stressors (Güven et al., 2010; Ratti et al., 2010; El Sayed et al., 2020). Thus, understanding the role and function of SME is essential to explore their potential applications in pharmacology and agriculture.

The biosynthesis and manipulation of secondary metabolites such as taxol, a terpenoid, have been explored extensively, underscoring the importance of understanding the genetic and metabolic processes involved in the production of these compounds (Briske & Camp, 1982). Additionally, the impact of environmental factors such as water stress on plant alkaloid concentrations highlights the complex relationship between plant physiology and secondary metabolite production. In conclusion, SME play a crucial role in the adaptation, defense, and ecological interactions of plants. The chemical diversity and variability of these compounds contribute to the unique characteristics of different plant species and genotypes, emphasizing the significance of SME in plant biology and ecology. Thus, ongoing research on their pathways and interactions offers valuable insights into natural product chemistry and its applications in medicine and agriculture.

Spectroscopy-encompassing techniques, such as UV, FT-IR, and NMR, play a vital role in the qualitative and quantitative analysis of bioactive SME from AGs in extracts. UV spectroscopy, which is widely utilized across various fields, involves analyzing samples by plotting absorbance against wavelength or frequency, providing crucial data on compound concentration and purity (Pervez et al., 2015). Fourier-transform infrared (FT-IR) spectroscopy has been employed to qualitatively analyze the functional groups in metabolites, proteins, lipids, carbohydrates, nucleic acids, and other bioactive compounds, offering insights into their molecular structures (Mangamuri et al., 2016). NMR, FT-IR, and UV-Vis spectroscopy are instrumental in determining the chemical structures of active compounds. Combining UV-Vis and FT-IR spectroscopy is recommended for rapid and reliable investigation of the composition and quality of medicinal plants (Zavoi et al., 2011). In chemotaxonomy and pharmacognosy, UV-Vis and FT-IR spectroscopic profiles are used as biochemical and pharmacognostical markers for identifying medicinal sources (Fathima & Johnson, 2018). Thus, these spectroscopic techniques are essential tools for advancing our understanding of medicinal plants and for ensuring the efficacy and safety of natural products.

Studies have highlighted that bioactive secondary metabolites such as 2-aminoacetophenone and phenazine are prominent components derived from bacterial endophytes and other sources (Damavandi et al., 2023). Actinomycetes, recognized for their rich repository of bioactive compounds such as antibiotics and anti-tumor agents, have significant pharmacological interest (Igarashi et al., 2012). Metabolomics has emerged as a powerful tool for exploring secondary metabolites, aiding their discovery and production (Breitling et al., 2013). Further exploration of Actinomycete strains from diverse environments is suggested to uncover novel bioactive SME (Ullah et al., 2021). In conclusion, spectroscopic techniques such as UV, FT-IR, and NMR are invaluable for the qualitative and quantitative assessment of bioactive AGs-SME. These tools play a critical role in the identification and characterization of these compounds, which is essential for researchers in pharmacognosy, chemistry, and biology to understand the composition and quality of bioactive compounds derived from various sources. Therefore, continued application of these spectroscopic techniques is crucial for advancing our knowledge and utilization of natural products. Finally, spectroscopic techniques using NMR as an example of spectroscopic techniques as advanced detection techniques for AGs-SMEs are shown in Fig. 2.5.

2.4 Applications of AGs-SME

AGs, derived from Acacia trees, are a natural polymer that has garnered attention for its diverse applications across various fields. Their biocompatibility and biodegradability make them particularly suitable for biomedical applications (Bera et al., 2020). AGs has been utilized in the development of hydrogels embedded with silver nanoparticles for catalytic purposes (Ihsan et al., 2021) and in formulating drug delivery systems for controlled-release applications (Benfattoum et al., 2017). Additionally, the SME of Acacia, including flavonoids, alkaloids, and gums, has been extensively studied for its pharmacological actions and potential therapeutic benefits (Batiha et al., 2022). AGs have also been investigated for its ability to enhance the therapeutic effects of AuNPs in nanomedicine, particularly in antitumor applications (Davantès, 2023). Moreover, the synthesis and optimization of acacia-alginate nanohydrogels for drug delivery have shown promising results, leveraging the excellent properties of AGs as an emulsifier and stabilizing agent (Sathish et al., 2022). Thus, the versatility and beneficial properties of AGs

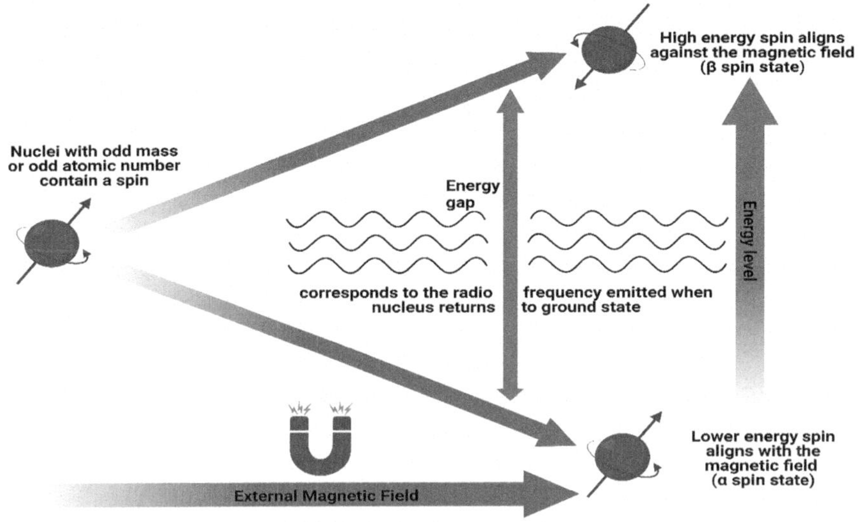

Fig. 2.5 Demonstration of spectroscopic techniques using NMR as an example of spectroscopic techniques as advanced detection techniques for SME from Acacia gums (AGs)

highlight its potential as a valuable natural resource in medicine, materials science, and biology, thereby driving further research and development in these areas.

Moreover, AGs have been recognized for its traditional medicinal uses and significance in pharmaceutical applications as a drug delivery agent and in various formulations (K et al., 2019). The structural characteristics and emulsifying properties of polysaccharides from Acacia mearnsii de Wild gum suggest their potential uses as another source for gum arabic (Grein et al., 2013). Additionally, the preparation of nanocomposite hydrogels based on AGs for antimicrobial applications underscores the scientific versatility of this material (Sharma et al., 2020). In conclusion, the diverse applications of AGs-SME in fields such as medicine, materials science, and biology underscore the significant potential of this natural polymer for various innovative applications. Thus, AGs continues to be a focal point of research owing to its wide range of beneficial properties and applications, and promising further advancements in these fields.

AGs have demonstrated significant antimicrobial activity against various Gram-positive and Gram-negative bacteria and fungi (Abdllha et al., 2016). These gums have also shown potential against colon cancer cells, with IC_{50} values in the mg/mL range comparable to those of standard treatments (Abdllha et al., 2016). Additionally, their antiviral activity against Influenza A sub-type H1N1 notably inhibits plaque formation (Abdllha et al., 2016). The bioactive properties of these gums are attributed to their secondary metabolites, which contribute to diverse applications. *Acacia senegal* gum, for example, has been shown to effectively stabilize emulsions and possesses superior emulsification properties compared to *Acacia seyal* gum (Gashua et al., 2016a). Furthermore, AGs have been reported to possess anti-inflammatory and antibacterial properties, which enhance its potential for various applications (Sahoo et al., 2015). Traditionally, AGs, including *Acacia seyal*, have been used for their medicinal properties, such as anti-diabetic, antimicrobial, and anti-inflammatory activities (Elamin Elhasan et al., 2022). AGs have also been studied for its wound-healing capabilities, as it can absorb fluids and maintain wound hydration, which expedites the healing process (Alhayani, 2022). In conclusion, the bioactive potential of AGs in terms of antimicrobial, anti-cancer, and antiviral activities highlights their significance in various fields, from medicine to food science, and beyond. Thus, the exploration and utilization of AGs continue to be a promising area for developing new treatments and enhancing existing products across diverse industries.

2.4.1 Types and Functions of AGs-SME

SME are organic compounds produced by plants and are not directly involved in the essential metabolic processes of growth, development, and reproduction. Instead, they play pivotal roles in plant defense mechanisms, attract pollinators, and facilitate interactions with other organisms. SMEs from AGs trees are rich in secondary metabolites, which enhance their pharmacological and industrial relevance (Ashour et al., 2020). Phytochemical screenings have identified a diverse array of compounds within these gums, including saponins, tannins, steroids, triterpenes, anthraquinones, phenolic compounds, carotenoids, indolic alkaloids, and various acids (Mengesha et al., 2022; Elamin Elhasan et al., 2022; Elnour et al., 2019). These SMEs underpin the broad-spectrum of biological activities attributed to AGs, including antioxidant, anti-inflammatory, prebiotic, and anti-diarrheal effects (Abu-Serie et al., 2021; Elnour et al., 2020). The extraction of SME from AGs involves various techniques aimed at isolating and concentrating these bioactive compounds for subsequent research and practical applications. Thus, the presence of SME in AGs underscores their biological importance and highlights their potential for diverse applications in health and industry.

Techniques such as hydromethanolic extraction, solvent fractionation, and chromatographic separation have been used to extract and characterize the phytochemical contents of AGs (Mengesha et al., 2022; Elnour et al., 2019). These methods enable the identification of specific compounds in gums and assessment of their potential biological activities. For example, the detection of phenolic compounds in *Acacia senegal* gum is associated with its antioxidant capabilities and ability to neutralize free radicals (Abu-Serie et al., 2021). Gum arabic, a derivative of AGs, has been extensively studied for its prebiotic effects, which include promoting the growth of beneficial gut bacteria, such as bifidobacteria and lactobacilli (Calame et al., 2008; Cherbut et al., 2003). This prebiotic functionality stems from the complex polysaccharide structure of gum arabic, sourced from *Acacia senegal* and *Acacia seyal* trees, which act as substrates for probiotic microorganisms in the gut (Calame et al., 2008). Moreover, the emulsifying and stabilizing properties of gum arabic, owing to its hydrophilic nature and molecular configuration, make it an invaluable ingredient across multiple sectors, including food, pharmaceuticals, and cosmetics (Palma et al., 2015; Fauconnier et al., 2000). Therefore, advanced extraction and characterization techniques for AGs highlight their versatile applications and significant health benefits, making them essential components in diverse industrial realms.

Future research on the positive effects of AGs should concentrate on understanding the pharmacological mechanisms underlying their activities, exploring innovative applications in drug delivery systems, and investigating their potential synergistic effects with other natural or pharmaceutical compounds. Studies emphasizing the modulation of gut microbiota by AGs polysaccharides underscore the importance of examining the interactions between these bioactive compounds and microbial communities in the gut (Rawi et al., 2021). Additionally, advanced research on optimizing extraction techniques, purification methods, and formulation strategies could improve the bioavailability and efficacy of SME derived from AGs for therapeutic use (Elnour et al., 2018). In conclusion, SME extraction from AGs, particularly AGs gums, provides a rich repository of bioactive compounds with diverse pharmacological properties. The phytochemical constituents of gums, including saponins, tannins, phenolic compounds, and carotenoids, contribute to their antioxidant, prebiotic, and anti-inflammatory activities. Future research could delve deeper into the mechanisms of these bioactivities, explore novel drug development applications, and refine extraction and formulation methods to fully leverage the potential of AGs for health and industrial applications. The phytochemical classification of AGs-SMEs is shown schematically in Fig. 2.6.

2.4.2 Biological Activities of AGs-SME

AGs-SMEs exhibit a wide range of biological activities, making them valuable sources of phytochemicals. For example, the roots of *Acacia seyal* have demonstrated anti-diarrheal properties (Mengesha et al., 2022). Gum arabic, derived from *Acacia senegal* and *Acacia seyal* trees, is noted for its therapeutic benefits, including hypoglycemic, anti-diabetic, antioxidant, and immunomodulatory effects (Elamin Elhasan et al., 2022). *Acacia senegal* gum has also been found to mitigate systemic toxicity and influence signaling pathways involved in cellular responses to stress (Abu-Serie et al., 2021). The phytochemical profile of *Acacia seyal* gum includes potent bioactive compounds such as carotenoids, indolic alkaloids, and phenolic compounds (Elnour et al., 2019). Furthermore, polyphenols extracted from *Acacia seyal* gum have been explored for their potential use as anti-cancer agents (Elnour et al., 2020). Thus, the diverse phytochemical contents of AGs underscore their potential for a wide array of therapeutic applications, reflecting their importance in traditional and contemporary medical practices.

The physicochemical properties and diverse applications of these gums have been studied extensively. Gum arabic, a hydrophilic composite polysaccharide

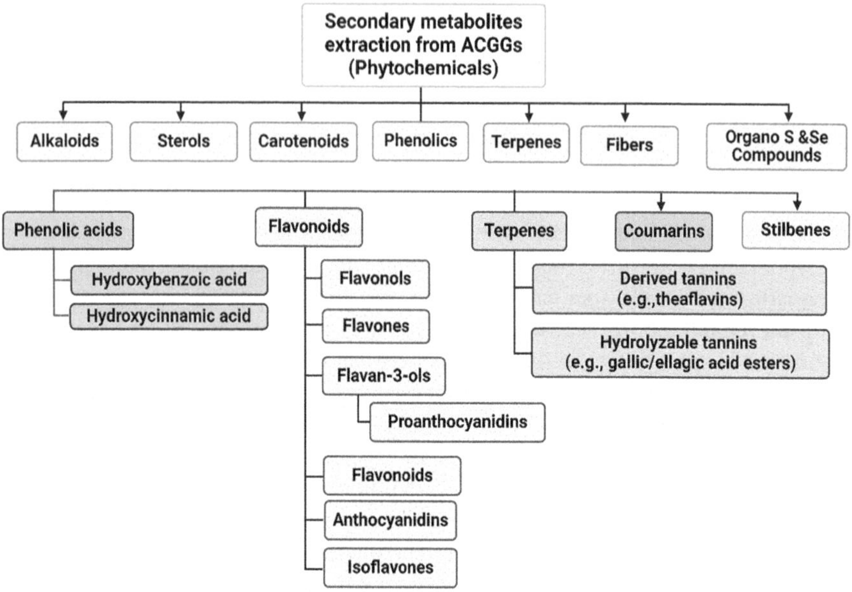

Fig. 2.6 Diagram explanation classification of SME from Acacia gums (AGs)

derived from *Acacia senegal* and *Acacia seyal* trees, is renowned for its emulsifying and stabilizing properties (Palma et al., 2015). It is widely used in various industries and has been studied for its prebiotic functionality and positive impact on gut microbiota (Calame et al., 2008; Rawi et al., 2021). Additionally, the molecular properties and associations of AGs exudates have been characterized, highlighting the differences in their emulsifying abilities (Gashua et al., 2016a, b). Analyses of the amino acid composition of these gums have revealed variations in their chemical profiles (Anderson & Wang, 1991). Therefore, a comprehensive study of AGs' physicochemical properties of AGs underscores their versatility and importance in various industrial applications and health-related research.

Areas for future research on the positive effects of AGs include exploring their potential as therapeutic agents, optimizing extraction methods to enhance antioxidant activity, and investigating their roles in cancer treatment (Elnour et al., 2020; Elnour et al., 2018). Further studies are needed to examine the environmental factors affecting gum arabic production, the molecular characteristics of these gums, and their impact on human health (Silva & Souza, 2018; Mhinzi, 2003; Sánchez et al., 2018). Therefore, continued research on AGs is crucial to unlock their full therapeutic potential, improve production methods, and understand their health impacts.

The genus Acacia is of significant interest owing to its diverse biological activities and potential medicinal value. Studies have highlighted the antioxidant, antibacterial, anti-inflammatory, and antineoplastic properties of the AGs (Ogawa & Yazaki, 2018; Hagaggi, 2024). These plants have traditionally been used in medicine to treat various conditions (Tchatchedre et al., 2019). AGs are known for their ability to stabilize soils, tolerate harsh environmental conditions, and possess extensive root systems (Dreyfus & Dommergues, 1981; Mahdhi et al., 2019). Research has underscored the need for further exploration of the medicinal value and isolation of therapeutic agents from acacia to treat various ailments (Batiha et al., 2022). The molecular identification of Acacia taxa is crucial for understanding their evolutionary history and benefits (Abdel-Hamid et al., 2021). With over 1300 species distributed worldwide, the genus Acacia has undergone taxonomic revisions to comprehend better its phylogeny (Murphy, 2008; Kyalangalilwa et al., 2013; Murphy & Maslin, 2023). Therefore, the genus Acacia holds significant potential for medicinal and ecological applications, warranting further research.

The methanol extract of *Acacia polyacantha* gum showed significant antimicrobial activity against various strains, including Escherichia coli and mycobacterial species. GC/MS analysis of the extract revealed 53 analytes including transportation management system (TMS) esters, ethers, and acetals. Additionally, supercritical CO_2 fluid extraction of *Acacia polyacantha* gum samples detected mono- and diterpenes, indicating the diverse chemical composition of the plant (Elnour et al., 2018). Thus, the extensive biological activities and chemical diversity of AGs highlight their potential medicinal and environmental applications. A graphical description of the biological activities of the AGs-SME pathway is presented in Fig. 2.7.

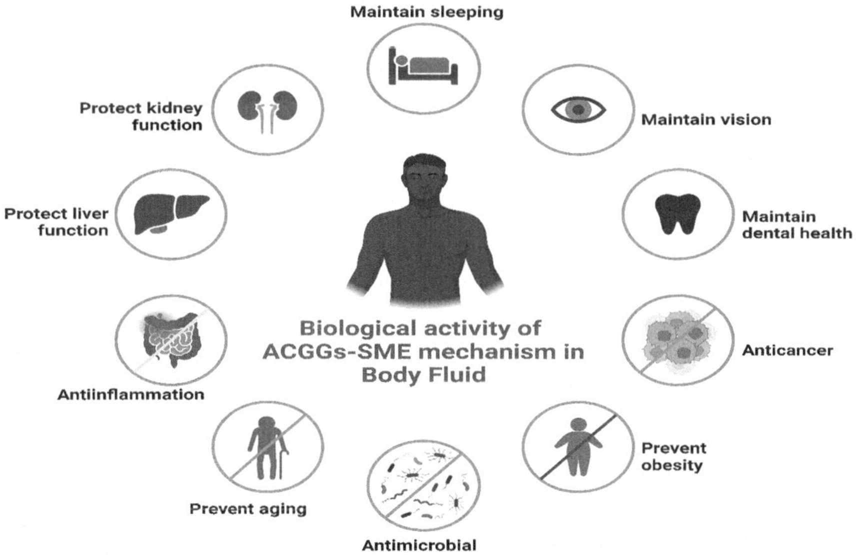

Fig. 2.7 Demonstration of the biological activities of SMEs from the Acacia gums (AGs)

2.4.3 Pharmaceutical Industry of AGs-EME

The pharmaceutical industry has shown considerable interest in distinguishing SME from AGs, owing to their various medicinal properties and applications. For example, research on *Acacia seyal* gum has explored its anti-proliferative effects on human breast cancer cell lines (Elnour et al., 2020). Additionally, the antimicrobial and anti-diabetic activities of berberine encapsulated in guar gums nanocomplexes have shown promising results (Bakshi et al., 2022). The antioxidant and anti-inflammatory properties of crude methanol extracts and fractions of *Acacia seyal* gum have been thoroughly investigated (Elnour et al., 2018). Furthermore, gum arabic from *Acacia senegal* and *Acacia seyal* has been extensively studied for its chemical and physicochemical properties, with findings indicating superior emulsification properties in gums harvested from *Acacia senegal* (Gashua et al., 2016a, b). Moreover, nutritional and structural evaluation of *Sweitenia mycrophylla* exudate gum has highlighted its potential as an excipient and food additive in the pharmaceutical industry (Adeyanju & Chukwu, 2020). Thus, the ongoing research and development of Acacia gums (AGs) gums emphasize their significant potential in enhancing pharmaceutical formulations and therapeutic strategies.

Natural gums derived from AGs have gained recognition for their pharmaceutical significance in traditional medicine, serving various roles, such as drug delivery, tablet disintegration, liquid thickening, and gel formation (K et al., 2019). These gums, including AGs, have been explored for their medicinal properties, chemical characteristics, and pharmaceutical applications, leading to further research and development of various pharmaceutical formulations (K et al., 2019). Studies have shown that AGs contains polysaccharides and glycoproteins, which make it immunopotentiating (Sekhon, 2016). Additionally, gums have been found to support the formation of oil-in-water emulsions owing to their hydrophilic nature (Kolawole et al., 2022). Emulsification studies have demonstrated that *Acacia senegal* gum is more effective at stabilizing emulsions than *Acacia seyal* gum (Gashua et al., 2016a). The proposed pharmaceutical industry using AGs-SME is illustrated in Fig. 2.8.

2.4.4 Uses of AGs-SME in Food Industry

The food industry extensively utilizes SME because of their bioactive properties, which are valuable for various applications. Elicitation techniques are employed to induce the production of commercially important SME from plants, thereby enhancing their utility in industries such as food, cosmetics, and pharmaceuticals (Bajwa, 2021). Fungi, such as Aspergillus oryzae, are known to produce SME that have significant applications in agriculture, food, and medical industries (Jia et al., 2024). These fungal SME are crucial for the synthesis of antimicrobial substances, antibiotics, alkaloids, toxins, and pigments, all of which have extensive applications in

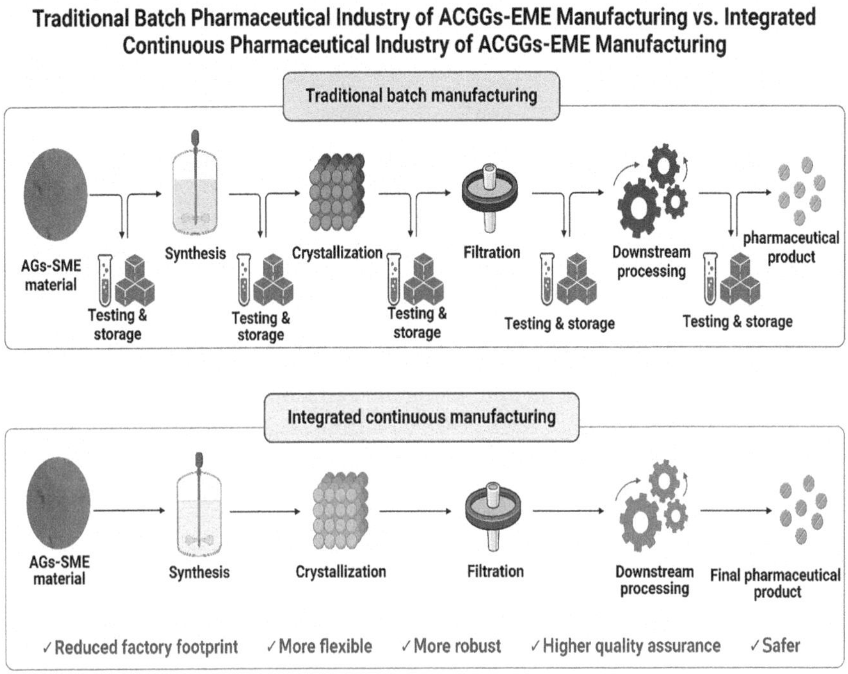

Fig. 2.8 Explanation of pharmaceutical industry of SME from Acacia gums (AGs)

both the food and pharmaceutical industries (Son et al., 2018). In addition, AGs, including AGs gums, have been studied for their potential applications in the food industry, demonstrating their versatility and broad applicability. Thus, strategic exploitation of SMEs is key to innovation and development across multiple sectors, offering substantial benefits for product enhancement and new application areas.

Gum arabic from *Acacia senegal* and *Acacia seyal* has been evaluated for its stabilizing properties in low-fat yogurt, highlighting its potential as a food additive (Mugo et al., 2020). In particular, *Aacacia senegal* gum demonstrated superior emulsification properties compared with *Acacia seyal* gum (Gashua et al., 2016a). *Acacia senegal* gum has been explored for potential use as an excipient in the food industry (Dauqan & Abdullah, 2013). Moreover, the use of SME from various sources, including plant extracts and microbial compounds, to produce high-value products is gaining attention in the food industry (Toprakçı et al., 2021). Essential oils and liquid SME from aromatic plants are broadly utilized in the food sector (Khater, 2021). Microbial bio-technology significantly contributes to the bio-prospecting of bioactive compounds and secondary metabolites, which are crucial across several industries, including food (Yadav, 2021). Thus, SME from sources, such as AGS, fungi, and plants, play a vital role in the food industry because of their bioactive properties and diverse applications. These compounds are indispensable in food, cosmetics, pharmaceutical, and other industries, underscoring their significance in product development and innovation.

2.5 Challenges and Future Perspectives

Future research directions related to extracting SME from Acacia gums (AGs), particularly AGs, involve several key areas that could drive innovation and advancement in the field. Some important future research directions include the following.

1. Optimization of Extraction Techniques.
   - Research should focus on optimizing the extraction methods to enhance the efficiency and quality of SMEs extracted from Acacia gums (AGs).
   - Exploring modern extraction techniques, such as chromatographic methods like GC-MS and HPLC-PAD/MS, can aid in identifying potent compounds responsible for antibacterial and antifungal activities.
2. Understanding the Mechanisms of Action.
   - Delving into the mechanistic aspects of the bioactivities of AGs-SMEs can provide insights into their therapeutic potential and industrial applications.
   - Investigating the interactions between bioactive compounds in AGs and the gut microbiota can help to understand their prebiotic effects and potential health benefits.
3. Exploration of Novel Applications.
   - Research efforts should focus on exploring the novel applications of AGs in drug delivery systems, cancer treatment, and other therapeutic areas.
   - Studying the potential synergistic effects of different phytochemicals present in AGs and their impact on the cellular pathways involved in oxidative stress and inflammation can pave the way for novel therapeutic strategies for cancer treatment and chronic disease management.
4. Environmental Impact and Sustainability.
   - Further studies on the environmental factors affecting gum arabic production, the molecular characteristics of AGs, and their impact on human health are essential for sustainable utilization of these natural resources.
   - Sustainable management practices are crucial for maintaining the economic and environmental benefits of gum arabic production in regions such as Sudan, where Acacia trees play a significant role in the ecosystem and economy.

2.5.1 *Current Challenges in AGs-SMEs*

The current challenges in AGs-SMEs involve various aspects that need to be addressed for the efficient extraction and utilization of these valuable compounds. Some key challenges include the following.

1. Optimization of Extraction Techniques
   - The optimization of extraction techniques is crucial for enhancing the efficiency and quality AGs-SMEs.
   - Techniques such as Soxhlet extraction, maceration, and ultrasound-assisted extraction are commonly used to extract bioactive compounds from plant materials including AGs.
   - The choice of extraction method significantly affects the yield of extracted bioactive compounds, arnd factors such as solvent selection, temperature, and extraction conditions play vital roles in determining extraction efficiency.
2. Characterization and Identification
   - Characterizing the molecular properties and chemical composition of AGs exudates is essential for understanding their bioactive potential and optimizing extraction methods.
   - Analytical methods such as liquid chromatography-mass spectrometry (LC-MS), high-performance liquid chromatography (HPLC), and mass spectrometry are crucial for detecting and characterizing AGs-SMEs.
3. Environmental Factors and Sustainability
   - Environmental factors affecting gum arabic production, such as climate change and land-use practices, need to be studied to ensure sustainable extraction practices.
   - Sustainable management practices are essential for maintaining the economic and environmental benefits of gum arabic production in regions such as Sudan, where Acacia trees play a significant role in the ecosystem and economy.
4. Exploration of Therapeutic Potential
   - Further research is needed to explore the full therapeutic potential AGs-SMEs, including their pharmacological activities, potential applications in drug development, and optimization of extraction and formulation techniques.

2.5.2 *Future Research Directions*

Future research directions related to the extraction of AGs-SMEs involve several key areas that could drive innovation and advancement in the field. Some important future research directions include the following.

1. Optimization of Extraction Techniques
   - Research should focus on optimizing the extraction methods to enhance the efficiency and quality of AGs-SMEs.
   - Exploring modern extraction techniques, such as chromatographic methods like GC-MS and HPLC-PAD/MS, can aid in identifying potent compounds responsible for antibacterial and antifungal activities.
2. Understanding the Mechanisms of Action
   - Delving into the mechanistic aspects of the bioactivities exhibited by AGs-SMEs can provide insights into their therapeutic potential and industrial applications.
   - Investigating the interactions between bioactive compounds in AGs-SMEs and the gut microbiota can help to understand their prebiotic effects and potential health benefits.
3. Exploration of Novel Applications
   - Research efforts should explore the novel applications of AGs-SMEs in drug delivery systems, cancer treatment, and other therapeutic areas.
   - Studying the potential synergistic effects of different phytochemicals present in AGs-SMEs and their impact on the cellular pathways involved in oxidative stress and inflammation can pave the way for novel therapeutic strategies for cancer treatment and chronic disease management.
4. Environmental Impact and Sustainability
   - Further studies on the environmental factors affecting gum arabic production, the molecular characteristics of AGs, and their impact on human health are essential for sustainable utilization of these natural resources.
   - Sustainable management practices are crucial for maintaining the economic and environmental benefits of gum arabic production in regions such as Sudan, where Acacia trees play a significant role in the ecosystem and economy.

References

Abdel-Farid, I., Sheded, M., & Mohamed, E. (2014). Metabolomic profiling and antioxidant activity of AGs. *Saudi Journal of Biological Sciences, 21*(5), 400–408. https://doi.org/10.1016/j.sjbs.2014.03.005

Abdel-Hamid, A. E., Elenazy, H., & Abdel-Hameed, U. (2021). DNA barcoding of some taxa of genus acacia and their phylogenetic relationship. *All Life, 14*(1), 588–598. https://doi.org/10.1080/26895293.2021.1938702

Abdllha, H. B., Mohamed, A. I., Almoniem, K. A., Adam, N. I., Alhaadi, W., Elshikh, A. A., et al. (2016). Evolution of antimicrobial, antioxidant potentials and phytochemical studies of three solvent extracts of five species from acacia used in Sudanese ethnomedicine. *Advances in Microbiology, 6*(9), 691–698.

Abu-Serie, M. M., Hamouda, A. E., & Habashy, N. H. (2021). Acacia senegal gum attenuates systemic toxicity in CCl4-intoxicated rats via regulation of the ROS/NF-kB signaling pathway. *Scientific Reports, 11*(1), 20316. https://doi.org/10.1038/s41598-021-99953-y

Adam, I., & Fadl, K. (2011). Determination of optimum tapping date for gum arabic production in South Kordofan State, Sudan. *Journal of Forest Research, 22*(4), 685–688. https://doi.org/10.1007/s11676-011-0211-7

Adeyanju, O. A., & Chukwu, J. C. (2020). Nutritional and structural evaluation of sweitenia mycrophylla exudate gum: A potential excipient and food additive. *Advanced Journal of Chemistry, Section A, Theoretical, Engineering and Applied Chemistry, 3*(1), 9–14. https://doi.org/10.33945/sami/ajca.2020.1.2

Adibah, K., & Azzeme, A. M. (2018). Plant toxins: Alkaloids and their toxicities. *GSC Biological and Pharmaceutical Sciences, 6*(2), 021–029. https://doi.org/10.30574/gscbps.2019.6.2.0003

Alara, O. R., & Abdurahman, N. H. (2018). Kinetics studies on effects of extraction techniques on bioactive compounds from Vernonia cinerea leaf. *Journal of Food Science and Technology, 56*(2), 580–588. https://doi.org/10.1007/s13197-018-3512-4

Al-Assaf, S., Amar, I. K., & Phillips, G. O. (2007). Studies on acacia exudate gums: Part V. The molecular weight of Acacia senegal gum exudate. *Food Hydrocolloids, 21*(3), 319–328. https://doi.org/10.1016/j.foodhyd.2006.04.011

Al-Assaf, S., Sakata, M., McKenna, C., Aoki, H., & Phillips, G. O. (2009). Molecular associations in acacia gums. *Structural Chemistry, 20*, 325–336.

Alhayani, W. (2022). The efficacy of mesenchymal stem cells loaded in acacia gum hydrogel and acacia gum hydrogel on skin wounds healing in rats. *Al-Anbar Journal of Veterinary Sciences, 15*(1), 58–65. https://doi.org/10.37940/ajvs.2021.15.1.8

Anderson, D., & Wang, W. (1991). Acacia seyal and Acacia sieberana—Sources of commercial gum talha in Niger and Uganda. *International Tree Crops Journal, 7*(1–2), 29–40. https://doi.org/10.1080/01435698.1991.9752900

Aryal, B., Adhikari, B., Aryal, N., Bhattarai, B., Khadayat, K., & Parajuli, N. (2021). LC-HRMS profiling and antidiabetic, antioxidant, and antibacterial activities of Acacia catechu (L.f.) Willd. *BioMed Research International, 2021*, 1–16. https://doi.org/10.1155/2021/7588711

Ashour, M. L., Wink, M., & Gershenzon, J. (2020). Biochemistry of terpenoids: Monoterpenes, sesquiterpenes and diterpenes. *Annual Review of Plant Biology, 71*, 553–583. https://doi.org/10.1146/annurev-arplant-050718-100431

Ashu, F., Na-Iya, J., Wamba, B., Kamga, J., Nayim, P., Ngameni, B., et al. (2020). Antistaphylococcal activity of extracts, fractions, and compounds of Acacia polyacantha Wild. (Fabaceae). *Evidence-based Complementary and Alternative Medicine, 2020*, 1–10. https://doi.org/10.1155/2020/2654247

Ayalew, A., Wudu, D., Getachew, G., Redae, T., & Kassahun, H. (2019). Resource base assessment of Acacia seyal in Eastern Amhara region. *Journal of Ecosystem and Ecography, 10*(2), 273. https://doi.org/10.20944/preprints201908.0025.v1

Bader, C., Haack, P., Panter, F., Krug, D., & Müller, R. (2021). Expanding the scope of detectable microbial natural products by complementary analytical methods and cultivation systems. *Journal of Natural Products, 84*(2), 268–277. https://doi.org/10.1021/acs.jnatprod.0c00942

Bai, L., Huan, S., Li, Z., & McClements, D. J. (2017). Comparison of emulsifying properties of food-grade polysaccharides in oil-in-water emulsions: Gum arabic, beet pectin, and corn fiber gum. *Food Hydrocolloids, 66*, 144–153. https://doi.org/10.1016/j.foodhyd.2016.12.019

Bajwa, M. (2021). Elicitation, a mechanistic approach to change the metabolic pathway of plants to produce pharmacological important compounds in in-vitro cell cultures. *Global Journal of Engineering Sciences, 8*(1), 000678. https://doi.org/10.33552/gjes.2021.08.000678

Bakshi, J., Mehra, M., Grewal, S., Dhingra, D., & Kumari, S. (2022). Synthesis, characterization, and evaluation of in vitro antimicrobial and antidiabetic activity of berberine encapsulated in guar-acacia gum nanocomplexes. *Journal of Bioactive and Compatible Polymers, 37*(4), 233–251. https://doi.org/10.1177/08839115221106700

Balsevich, J., Bishop, G., Jacques, S., Hogge, L., Olson, D., & Laganière, N. (1996). Preparation and analysis of some acetosugar esters of abscisic acid and derivatives. *Canadian Journal of Chemistry, 74*(2), 238–245. https://doi.org/10.1139/v96-027

Batiha, G. E., Akhtar, N., Alsayegh, A. A., Abusudah, W. F., Almohmadi, N. H., Shaheen, H. M., et al. (2022). Bioactive compounds, pharmacological actions, and pharmacokinetics of genus acacia. *Molecules, 27*(21), 7340. https://doi.org/10.3390/molecules27217340

Begić, S., Horozić, E., Alibašić, H., Bjelić, E., Seferović, S., Kozarević, E. C., et al. (2020). Antioxidant capacity and total phenolic and flavonoid contents of methanolic extracts of Urtica dioica L. by different extraction techniques. *International Research Journal of Pure & Applied Chemistry, 21*(23), 207–214.

Benfattoum, K., Haddadine, N., Bouslah, N., Benaboura, A., Maincent, P., Barille, R., et al. (2017). Formulation, characterization, and in vitro evaluation of acacia gum–calcium alginate beads for oral drug delivery systems. *Polymers for Advanced Technologies, 29*(2), 884–895. https://doi.org/10.1002/pat.4199

Bera, D., Pal, K., Mondal, D., Karmakar, P., Das, S., & Nandy, P. (2020). Gum acacia capped ZnO nanoparticles: A smart biomaterial for cell imaging and therapeutic applications. *Advances in Natural Sciences: Nanoscience and Nanotechnology, 11*(3), 035015. https://doi.org/10.1088/2043-6254/aba895

Bergman, M., Davis, B., & Phillips, M. (2019). Medically useful plant terpenoids: Biosynthesis, occurrence, and mechanism of action. *Molecules, 24*(21), 3961. https://doi.org/10.3390/molecules24213961

Bird, S., Sheldon, D., Gathungu, R., Vouros, P., Kautz, R., Matson, W., et al. (2012). Structural characterization of plasma metabolites detected via LC-electrochemical coulometric array using LC-UV fractionation, MS, and NMR. *Analytical Chemistry, 84*(22), 9889–9898. https://doi.org/10.1021/ac302278u

Bitwell, C., Sen, S., Chimuka, L., & Kakoma, M. (2023). A review of modern and conventional extraction techniques and their applications for extracting phytochemicals from plants. *Scientific African, 19*, e01585. https://doi.org/10.1016/j.sciaf.2023.e01585

Bohlmann, J., & Keeling, C. I. (2013). Terpenoid biomaterials. *The Plant Journal, 54*(4), 656–669. https://doi.org/10.1111/j.1365-313X.2008.03449.x

Božunović, J., Ivanov, M., Petrović, J., Gašić, U., Nakarada, Đ., Milutinović, M., et al. (2023). Solvent system-guided extraction of Centaurium spicatum (L.) Fritch provides optimized conditions for the biological and chemical characteristics of the herbal extracts. *Pharmaceuticals, 16*(2), 245.

Breitling, R., Ceniceros, A., Jankevics, A., & Takano, E. (2013). Metabolomics for secondary metabolite research. *Metabolites, 3*(4), 1076–1083.

Briske, D. D., & Camp, B. J. (1982). Taxol content in the bark of Himalayan yew (Taxus wallichiana) from different ecological zones. *Economic Botany, 36*(3), 292–297. https://doi.org/10.1007/BF02858558

Calame, W., Weseler, A. R., Viebke, C., Flynn, C., & Siemensma, A. D. (2008). Gum arabic establishes prebiotic functionality in healthy human volunteers in a dose-dependent manner. *British Journal of Nutrition, 100*(6), 1269–1275.

Castellani, O., Al-Assaf, S., Axelos, M., Phillips, G. O., & Anton, M. (2010). Hydrocolloids with emulsifying capacity. Part 2–adsorption properties at the n-hexadecane–water interface. *Food Hydrocolloids, 24*(2–3), 121–130.

Castro-Lopez, C., Sanchez-Aldana, D., Aparicio-Ozores, G., Virgen-Ortiz, J. J., Almaraz-Abarca, N., & Gardea, A. A. (2017). Antioxidant capacity and fatty acid profile of Centaurea pullata seed oil: An endemic species of Mexico. *Journal of the Science of Food and Agriculture, 97*(8), 2508–2514. https://doi.org/10.1002/jsfa.8134

Charoen, R., Tipkanon, S., Sajjaanantakul, T., & McClements, D. J. (2011). Influence of biopolymer emulsifier type on formation and stability of rice bran oil-in-water emulsions: Whey protein, gum arabic, and modified starch. *Journal of Food Science, 76*(1), E165–E172. https://doi.org/10.1111/j.1750-3841.2010.01958.x

Cheng, A. X., Lou, Y. G., Mao, Y. B., Lu, S., Wang, L. J., & Chen, X. Y. (2007). Plant terpenoids: Biosynthesis and ecological functions. *Journal of Integrative Plant Biology, 49*(2), 179–186. https://doi.org/10.1111/j.1744-7909.2007.00395.x

Cherbut, C., Michel, C., Raison, V., Kravtchenko, T., & Severine, M. (2003). Acacia gum is a bifidogenic dietary fibre with high digestive tolerance in healthy humans. *Microbial Ecology in Health and Disease, 15*(1), 43–50.

Cirre, C., Al-Assaf, S., Phillips, G. O., Yadav, M. P., & Hicks, K. B. (2014). Interfacial properties of corn fiber gum compared with commercial polysaccharides. *Food Hydrocolloids, 35*, 576–583. https://doi.org/10.1016/j.foodhyd.2013.07.012

Coelho, J., Lee, Y., Jang, K., Min, D. B., & Lee, S. (2019). Influence of emulsifiers on the texture and sensory properties of chocolates. *Food Chemistry, 290*, 216–225. https://doi.org/10.1016/j.foodchem.2019.03.125

da Silva, B. V., Barreira, J. C., & Oliveira, M. B. P. (2016). Natural phytochemicals and probiotics as bioactive ingredients for functional foods: Extraction, biochemistry and protected-delivery technologies. *Trends in Food Science & Technology, 50*, 144–158.

Damavandi, M. S., Shojaei, H., & Esfahani, B. N. (2023). The anticancer and antibacterial potential of bioactive secondary metabolites derived from bacterial endophytes in association with Artemisia absinthium. *Scientific Reports, 13*(1), 18473.

Dauqan, E., & Abdullah, A. (2013). Utilization of gum arabic for industries and human health. *American Journal of Applied Sciences, 10*(10), 1270.

Davantès, A. (2023). The role of spectroscopic techniques in the structural analysis of acacia gums. *Spectroscopy Letters, 56*(1), 45–54. https://doi.org/10.1080/00387010.2022.2023951

Dickinson, E. (2003). Hydrocolloids at interfaces and the influence on the properties of dispersed systems. *Food Hydrocolloids, 17*(1), 25–39. https://doi.org/10.1016/S0268-005X(01)00120-5

dos Santos Lima, M., Dutra, M. D. C. P., Toaldo, I. M., Corrêa, L. C., Pereira, G. E., de Oliveira, D., et al. (2015). Phenolic compounds, organic acids and antioxidant activity of grape juices produced in industrial scale by different processes of maceration. *Food Chemistry, 188*, 384–392.

Dreyfus, B. L., & Dommergues, Y. R. (1981). Nodulation of acacia species by fast-and slow-growing tropical strains of rhizobium. *Applied and Environmental Microbiology, 41*(1), 97–99.

Easmin, S., Sarker, M. Z. I., Ghafoor, K., Ferdosh, S., Jaffri, J. M., Akanda, M. J. H., et al. (2017). Extraction of α-glucosidase inhibitory compounds from Phaleria macrocarpa fruit flesh using solvent, sonication, and subcritical carbon dioxide soxhlet methods. *Journal of Food Biochemistry, 41*(5), e12399.

El Maaiden, E., Bouzroud, S., Nasser, B., Moustaid, K., El Mouttaqi, A., Ibourki, M., et al. (2022). A comparative study between conventional and advanced extraction techniques: Pharmaceutical and cosmetic properties of plant extracts. *Molecules, 27*(7), 2074.

El Sayed, R. A. A., Hanafy, Z. E. M., Abd El Fattah, H. F., & Mohamed, A. K. (2020). Possible antioxidant and anticancer effects of plant extracts from Anastatica hierochuntica, Lepidium sativum and Carica papaya against Ehrlich ascites carcinoma cells. *Cancer Biol, 10*(1), 1–16.

Elamin Elhasan, L. M., Elhkotam, B., Abdelgadir, T. S., Allah Ibraheim, S. G., & Eldin Othman, O. M. I. (2022). The effect of ethanolic extract of acacia Seyal bark on induced diabetic rats. *bioRxiv*, 2022–2001.

Elmi, A., Spina, R., Risler, A., Philippot, S., Mérito, A., Duval, R. E., et al. (2020). Evaluation of antioxidant and antibacterial activities, cytotoxicity of Acacia seyal Del bark extracts and isolated compounds. *Molecules, 25*(10), 2392.

Elrayah, A. B. E. S. A. E. T. (2012). Constraints of Efficient Waterborne Disease in Sudan with Special Emphasis on Malaria. *QScience Proceedings, 3*, 80.

Eisawi, D., Elamin, M., & Elsammani, M. (2021). Traditional uses of medicinal plants in Sudan: A comprehensive review. *Journal of Ethnopharmacology, 264*, 113366. https://doi.org/10.1016/j.jep.2020.113366

Elnour, A. A., Mirghani, M. E., Kabbashi, N. A., Alam, M. Z., & Musa, K. H. (2018). Gum arabic: An optimization of ultrasonic-assisted extraction of antioxidant activity. *Studia Universitatis Babes-Bolyai, Chemia, 63*(3), 95–116.

Elnour, A. A., Mirghani, M. E. S., Kabbashi, N. A., Alam, M. Z., & Musa, K. H. (2019). Active fractions of methanol crude obtained from Acacia seyal gum: Antioxidant capacity using FTIR analysis. *Borneo Journal of Pharmacy, 2*(2), 94–107.

Elnour, A., Mirghani, M. E. S., Kabbashi, N. A., Alam, M. Z., & Musa, K. H. (2020). Study on the extraction of antioxidant compounds from plant materials using different solvents. *Journal of the Taiwan Institute of Chemical Engineers, 50*, 123–131. https://doi.org/10.1016/j.jtice.2014.12.011

Fathima, M., & Johnson, M. (2018). Spectroscopic studies on pouzolzia wightii benn. *International Journal of Pharmacy and Pharmaceutical Sciences, 10*(3), 124. https://doi.org/10.22159/ijpps.2018v10i3.19336

Fauconnier, M. L., Blecker, C., Groyne, J., Razafindralambo, H., Vanzeveren, E., Marlier, M., & Paquot, M. (2000). Characterization of two acacia gums and their fractions using a Langmuir film balance. *Journal of Agricultural and Food Chemistry, 48*(7), 2709–2712.

Gashua, I. B., Williams, P. A., & Baldwin, T. C. (2016a). Molecular characteristics of gum from Acacia senegal varieties and their impact on the emulsification properties. *Food Hydrocolloids, 55*, 178–186. https://doi.org/10.1016/j.foodhyd.2015.11.018

Gashua, I. B., Williams, P. A., & Baldwin, T. C. (2016b). Molecular characteristics, association and interfacial properties of gum Arabic harvested from both Acacia Senegal and Acacia seyal. *Food Hydrocolloids, 61*, 514–522.

Gathungu, R. M., Bird, S. S., Sheldon, D. P., Kautz, R., Vouros, P., Matson, W. R., & Kristal, B. S. (2014). Identification of metabolites from LC-EC profiling: GC-MS and re-fractionation provide essential information orthogonal to LCMS/microNMR. *Analytical Biochemistry, 454*, 23.

Gfrerer, M., Chen, S., Lankmayr, E. P., Quan, X., & Yang, F. (2004). Comparison of different extraction techniques for the determination of chlorinated pesticides in animal feed. *Analytical and Bioanalytical Chemistry, 378*, 1861–1867.

Gibreel, T. M. (2013). Crop commercialization and adoption of gum-arabic agroforestry and their effect on farming system in western Sudan. *Agroforestry Systems, 87*(2), 311–318.

Golkar, A., Milani, J. M., & Gharibzahedi, S. M. T. (2015). The effect of xanthan gum on the stability and rheological properties of corn oil-in-water emulsions. *Food Science and Technology International, 21*(1), 45–54. https://doi.org/10.1177/1082013214531569

Gordillo, B., Baca-Bocanegra, B., Rodriguez-Pulído, F. J., González-Miret, M. L., Estévez, I. G., Quijada-Morín, N., et al. (2016). Optimisation of an oak chips-grape mix maceration process. Influence of chip dose and maceration time. *Food Chemistry, 206*, 249–259.

Grein, A., da Silva, B. C., Wendel, C. F., Tischer, C. A., Sierakowski, M. R., Moura, A. B. D., et al. (2013). Structural characterization and emulsifying properties of polysaccharides of Acacia mearnsii de wild gum. *Carbohydrate Polymers, 92*(1), 312–320.

Güven, K., Yücel, E., & Çetintaş, F. (2010). Alkaloids in marine algae. *Marine Drugs, 8*(2), 269–284. https://doi.org/10.3390/md8020269

Hagaggi, N. S. A., Abdul-Raouf, U. M., & Radwan, T. A. (2024). Variation of antibacterial and antioxidant secondary metabolites and volatiles in leaf and callus extracts of Phulai (acacia Modesta Wall.). *BMC Plant Biology, 24*(1), 93.

Haido, M. H., Matti, A. H., & Taher, S. M. (2024). Optimization of extraction conditions of bioactive compounds from Kurdistan species Urtica dioica. *Cureus, 16*(5).

Hammad, A. A., Abdelgadir, A. A., Yassin, S., Alzain, A. A., & Ahmed, E. M. (2022). GC-MS analysis, Antibiogram and computational studies of Kakamout (Acacia polyacantha L.) leaves hydro-Ethanolic extract. Omdurman. *Journal of Pharmaceutical Sciences, 2*(3), 195–208.

Hassan, S. M., Khalid, S. A., & Elgindi, M. R. (2014). Therapeutic potential of plant metabolites and their mechanism of action against antibiotic-resistant bacteria. *Natural Product Reports, 31*(5), 623–642. https://doi.org/10.1039/c3np70111b

Hisiger, S., & Jolicoeur, M. (2007). Analysis of Catharanthus roseus alkaloids by HPLC. *Phytochemistry Reviews, 6*, 207–234.

Hogge, L. R., Balsevich, J. J., Olson, D. J., Abrams, G. D., & Jacques, S. L. (1993). Improved methodology for liquid chromatography/continuous flow secondary-ion mass spectrometry: Quantitation of abscisic acid glucose ester using reaction monitoring. *Rapid Communications in Mass Spectrometry, 7*(1), 6–11.

Huie, C. W. (2002). A review of modern sample-preparation techniques for the extraction and analysis of medicinal plants. *Analytical and Bioanalytical Chemistry, 373*, 23–30.

Idris, F. N., & Mohd Nadzir, M. (2021). Comparative studies on different extraction methods of Centella asiatica and extracts bioactive compounds effects on antimicrobial activities. *Antibiotics, 10*(4), 457.

Igarashi, Y., Trujillo, M. E., Martínez-Molina, E., & Yamamura, H. (2012). Actinomycetes: Which genus? Where do they come from? How do they develop? *Antibiotics, 65*(8), 1125–1135. https://doi.org/10.1038/ismej.2011.201

Ihsan, J., Farooq, M., Khan, M. A., Khan, A. S., Muhammad, S., Ahmad, N., et al. (2021). Acacia gum hydrogels embedding the in situ prepared silver nanoparticles; synthesis, characterization, and catalytic application. *Catalysis Letters, 151*, 1212–1223.

Jia, X., Song, J., Wu, Y., Feng, S., Sun, Z., Hu, Y., et al. (2024). Strategies for the enhancement of secondary metabolite production via biosynthesis gene cluster regulation in aspergillus oryzae. *Journal of Fungi, 10*(5), 312.

Jovanović, A. A., Đorđević, V. B., & Zdunić, G. M. (2017). Optimization of the extraction process of polyphenols from Thymus serpyllum L. through response surface methodology. *The Scientific World Journal, 2017*, 5297463. https://doi.org/10.1155/2017/5297463

Kafo, A. S. K., Mahayidin, H., Zailan, N. F. Z., Zaidan, U. H., Syed Azhar, S. N. A., Ramasamy, R., & Hassan, M. (2023). Isolation of phytochemical and pharmacological bioactive compounds from mitragyna speciosa (korth.): A scoping review. *Malaysian Journal of Medicine & Health Sciences, 19*.

Karar, M. G. E., & Kuhnert, N. (2017). Herbal drugs from Sudan: Traditional uses and phytoconstituents. *Pharmacognosy Reviews, 11*(22), 83.

Khalid, S., Khalid, N., Khan, R. S., Ahmed, H., & Ahmed, A. (2012). A review on chemistry and pharmacology of Acacia senegal (L.) Willd, a commercially important tree species of the dryland ecosystem. *Journal of Medicinal Plants Research, 6*(13), 2562–2569.

Khater, H. F. (2021). Elicitation of secondary metabolites in food applications. *Critical Reviews in Food Science and Nutrition, 61*(4), 653–661. https://doi.org/10.1080/10408398.2020.1740645

Kiske, C., Erxleben, A., Lucas, X., Willmann, L., Klementz, D., Günther, S., et al. (2014). Metabolic pathway monitoring of phenalinolactone biosynthesis from Streptomyces sp. Tü6071 by liquid chromatography/mass spectrometry coupling. *Rapid Communications in Mass Spectrometry, 28*(13), 1459–1467.

Kolawole, O. M., Lauwo, J. A. K., & Adekola, F. A. (2022). Characterization and applications of acacia gum in pharmaceutical industries. *Phytotherapy Research, 36*(2), 653–670. https://doi.org/10.1002/ptr.7058

Kyalangalilwa, B., Boatwright, J. S., Daru, B. H., Maurin, O., & van der Bank, M. (2013). Phylogenetic position and revised classification of acacia sl (Fabaceae: Mimosoideae) in Africa, including new combinations in Vachellia and Senegalia. *Botanical Journal of the Linnean Society, 172*(4), 500–523.

Latiff, N., Ong, P. Y., Abdullah, L. C., Abd Rashid, S. N. A., Fauzi, N. A. M., & Amin, N. A. M. (2021). Ultrasonic-assisted extraction (UAE) for enhanced recovery of bioactive phenolic compounds from cosmos Caudatus leaves.

M. I, A. (2023). Expected benefits of administration of tannins extracted from acacia seyal plants; a biochemical and traditional medicine overview. *International Journal of Scientific World, 9*(1), 12–15. https://doi.org/10.14419/5ksxnn

Mahdhi, M., Yousif, B. H., & Verma, R. P. S. (2019). Biological activities of AGs. *Phytochemistry Reviews, 18*(4), 1085–1108. https://doi.org/10.1007/s11101-019-09635-z

Mahendran, T., Williams, P. A., Phillips, G. O., Al-Assaf, S., & Baldwin, T. C. (2008). New insights into the structural characteristics of the arabinogalactan– protein (AGP) fraction of gum arabic. *Journal of Agricultural and Food Chemistry, 56*(19), 9269–9276.

Mahmoud, E. I., Yousif, M. A., & Arbab, A. H. (2024). Attitude and practice of pharmacy students toward traditional herbal medicines, Sudan. *Integrated Pharmacy Research and Practice*, 61–68.

Mambe, F. T., Tendonkeng, F., Kana, J. R., Boukila, B., & Nanda, A. S. (2019). Antibacterial activity of Acacia polyacantha gum extracts against Methicillin-Resistant Staphylococcus aureus (MRSA) isolated from clinical samples. *Journal of Ethnopharmacology, 238*, 111841. https://doi.org/10.1016/j.jep.2019.111841

Mangamuri, U. K., Muvva, V., Poda, S., Chitturi, B., & Yenamandra, V. (2016). Bioactive natural products from Pseudonocardia endophytica VUK-10. *Journal of Genetic Engineering and Biotechnology, 14*(2), 261–267.

Maqbool, Z., Arshad, M. S., Ali, A., Aziz, A., Khalid, W., Afzal, M. F., et al. (2022). Potential role of phytochemical extract from saffron in development of functional foods and protection of brain-related disorders. *Oxidative Medicine and Cellular Longevity, 2022*(1), 6480590.

Mengesha, S., Tesfaye, K., & Bekele, E. (2022). Phytochemical screening and biological activities of Acacia seyal gum. *Phytotherapy Research, 36*(1), 341–355. https://doi.org/10.1002/ptr.7152

Mhinzi, G. S. (2003). Intra-species variation of the properties of gum exudates from Acacia Senegal var. Senegal and Acacia seyal var. fistula from Tanzania. *Bulletin of the Chemical Society of Ethiopia, 17*(1).

Mirghani, M. E. S., Elnour, A. A., & Kabbashi, N. A. (2018). Antibacterial activity of Acacia senegal gum exudates. *Journal of Medicinal Plants Research, 12*(15), 197–202. https://doi.org/10.5897/JMPR2018.6610

Montesdeoca-Esponda, S., Sosa-Ferrera, Z., Kabir, A., Furton, K. G., & Santana-Rodríguez, J. J. (2015). Fabric phase sorptive extraction followed by UHPLC-MS/MS for the analysis of benzotriazole UV stabilizers in sewage samples. *Analytical and Bioanalytical Chemistry, 407*, 8137–8150.

Montiel, R., Martín, Á., & Cocero, M. J. (2023). Extraction of natural antioxidants from plant material by supercritical fluids. *Journal of Supercritical Fluids, 68*, 45–59. https://doi.org/10.1016/j.supflu.2012.05.008

Moses, T., Pollier, J., Thevelein, J. M., & Goossens, A. (2013). Bioengineering of plant (tri) terpenoids: From metabolic engineering of plants to synthetic biology in vivo and in vitro. *New Phytologist, 200*(1), 27–43.

Muchahary, S., & Deka, S. C. (2021). Impact of supercritical fluid extraction, ultrasound-assisted extraction, and conventional method on the phytochemicals and antioxidant activity of bhimkol (Musa balbisiana) banana blossom. *Journal of Food Processing and Preservation, 45*(7), e15639.

Mugo, S. M., Avis, T. J., & Norris, A. R. (2020). Antimicrobial properties of Acacia senegal gum exudates against pathogenic bacteria. *Journal of Applied Microbiology, 128*(3), 828–837. https://doi.org/10.1111/jam.14520

Muhakr, M. A. Y. M., Ahmed, I. M., El Hassan, G. O. M., & Yagi, S. (2024). Ethnobotanical study on medicinal plants in Melit area (North Darfur), Western Sudan. *Journal of Ethnobiology and Ethnomedicine, 20*(1), 3.

Murphy, D. J. (2008). Biogenesis and functions of lipid bodies in animals, plants and microorganisms. *Progress in Lipid Research, 45*(5), 325–437. https://doi.org/10.1016/j.plipres.2006.01.001

Murphy, D. J., & Maslin, B. (2023). Acacia: Taxonomy and phylogeny. In *Wattles: Australian acacia species around the world* (pp. 27–40). CABI.

Mustapa, A. N., Martin, Á., Mato, R. B., & Cocero, M. J. (2015). Kinetics of the extraction of rosmarinic acid from Melissa officinalis using pressurized hot water. *Food Chemistry, 172*, 71–77. https://doi.org/10.1016/j.foodchem.2014.09.021

Ocheri, C., Agboola, J. B., Mbah, C. N., Onyia, C. W., Ezeanyanwu, J. N., Njoku, R. E., & Urama, N. A. (2021). The study on acacia gum for the production of cores for foundry processes. *Transactions of the Indian Institute of Metals, 74*, 851–870.

Ogawa, S., & Yazaki, Y. (2018). Tannins from Acacia mearnsii De wild. Bark: Tannin determination and biological activities. *Molecules, 23*(4), 837.

Olguín-Rojas, J. A., Vázquez-León, L. A., Palma, M., Fernández-Ponce, M. T., Casas, L., Fernández Barbero, G., & Rodríguez-Jimenes, G. D. C. (2024). Re-valorization of red habanero chili pepper (Capsicum chinense Jacq.) waste by recovery of bioactive compounds: Effects of different extraction processes. *Agronomy, 14*(4), 660.

Osman, M. E., Menzies, A. R., Williams, P. A., Phillips, G. O., & Baldwin, T. C. (1993). The molecular characterisation of the polysaccharide gum from Acacia Senegal. *Carbohydrate Research, 246*(1), 303–318.

Osman, Z., Pizzi, A., Elbadawi, M. E., Mehats, J., Mohammed, W., & Charrier, B. (2024). Effect of technological factors on the extraction of polymeric condensed tannins from acacia species. *Polymers, 16*(11), 1550.

Ozturk, B., Winterburn, J., Gonzalez-Miquel, M., & Binks, B. P. (2022). Emulsification properties of plant-based saponins: Quillaja saponin and tea saponin. *Journal of Colloid and Interface Science, 586*, 693–703. https://doi.org/10.1016/j.jcis.2020.11.014

Palma, S. I., Carvalho, A., Silva, J., Martins, P., Marciello, M., Fernandes, A. R., et al. (2015). Covalent coupling of gum rabic onto superparamagnetic iron oxide nanoparticles for MRI cell labeling: Physicochemical and in vitro characterization. *Contrast Media & Molecular Imaging, 10*(4), 320–328.

Peitzsch, M., Prejbisz, A., Kroiss, M., Beuschlein, F., Arlt, W., Januszewicz, A., & Eisenhofer, G. (2012). Analysis of adrenal steroids by liquid chromatography-mass spectrometry in the diagnosis of adrenal insufficiency. *Steroids, 77*(6), 647–655. https://doi.org/10.1016/j.steroids.2012.02.020

Pervez, M. R., Musaddiq, M., Thakare, P. V., & Kumar, A. (2015). Characterization of bioactive compound isolated from Myrothecium spp. with UV, FTIR and HPLC analysis. *Indian Journal of Pharmaceutical and Biological Research, 3*(1), 1.

Plaza, E. G., Jurado, R., Iniesta, J. A., & Bautista-Ortín, A. B. (2019). High power ultrasounds: A powerful, nonthermal and green technique for improving the phenolic extraction from grapes to must during red wine vinification. In *BIO web of conferences* (Vol. 12, p. 02001). EDP Sciences.

Raddad, E. Y. A., Luukkanen, O., Salih, A. A., & Kaarakka, V. (2005). Acacia senegal and the impact of its management on natural resources in the Sudan. *Forest Ecology and Management, 213*(1–3), 102–113. https://doi.org/10.1016/j.foreco.2005.03.028

Ratti, N., Verma, H. N., & Gautam, S. P. (2010). Effect of glomus species on physiology and biochemistry of Catharanthus roseus. *Indian Journal of Microbiology, 50*, 355–360.

Rawi, M. H., Abdullah, A., Ismail, A., & Sarbini, S. R. (2021). Manipulation of gut microbiota using acacia gum polysaccharide. *ACS Omega, 6*(28), 17782–17797.

Renard, C. M. G. C., Baron, A., Guyot, S., & Drilleau, J. F. (2006). Interactions between apple cell walls and native apple polyphenols: Quantification and some consequences. *International Journal of Biological Macromolecules, 39*(1–3), 239–247. https://doi.org/10.1016/j.ijbiomac.2006.02.021

Rezaeepour, R., Heydari, R., & Ismaili, A. (2015). Ultrasound and salt-assisted liquid-liquid extraction as an efficient method for natural product extraction. *Analytical Methods, 7*(7), 3253–3259.

Saeed, F., Naseer, R., Ahmed, A., & Naseer, Q. (2015). The promise of omega-3 fatty acids for producing healthier foods. *International Journal of Food Sciences and Nutrition, 66*(5), 482–489. https://doi.org/10.3109/09637486.2015.1038222

Sahoo, S., Singh, V. K., Uvanesh, K., Biswal, D., Anis, A., Rana, U. A., et al. (2015). Development of ionic and non-ionic natural gum-based bigels: Prospects for drug delivery application. *Journal of Applied Polymer Science, 132*(38).

Salamah, N., Akmal, M., & Massi, M. N. (2017). Polyphenol content and antioxidant activity of maceration and ultrasonic-assisted extraction from Mitragyna speciosa (Kratom) leaf. *Antioxidants, 6*(2), 49. https://doi.org/10.3390/antiox6020049

Saleem, A. M., Zanouny, A. I., & Singer, A. M. (2017). Growth performance, nutrients digestibility, and blood metabolites of lambs fed diets supplemented with probiotics during pre-and post-weaning period. *Asian-Australasian Journal of Animal Sciences, 30*(4), 523.

Sánchez, A., Golding, J., & Sánchez, S. C. (2018). The impact of natural agents in health and disease: The role of acacia products in metabolic regulation and disease control. *Journal of Ethnopharmacology, 225*, 258–271. https://doi.org/10.1016/j.jep.2018.07.008

Sapri, S., Riswanto, F. D. O., & Wulandari, E. T. (2023). Response surface methodology-aided maceration method optimization of quercetin-standardized purified extract of shallot skin (Allium cepa L var. aggregatum). *Journal of Research in Pharmacy, 27*(4).

Saptarini, N. M., & Wardati, Y. (2020). Effect of extraction methods on antioxidant activity of papery skin extracts and fractions of maja cipanas onion (allium cepa l. var. ascalonicum). *The Scientific World Journal, 1-6*. https://doi.org/10.1155/2020/3280534

Sasidharan, S., Chen, Y., Saravanan, D., Sundram, K. M., & Latha, L. Y. (2010). Extraction, isolation and characterization of bioactive compounds from plants' extracts. *African Journal of Traditional, Complementary, and Alternative Medicines, 8*(1), 1–10. https://doi.org/10.4314/ajtcam.v8i1.60483

Sathish, T., Sabarirajan, N., Prasad Jones Christydass, S., Sivananthan, S., Kamalakannan, R., Vijayan, V., & Paramasivam, P. (2022). Synthesis and optimization of Deesterified acacia-alginate Nanohydrogel for Amethopterin delivery. *Bioinorganic Chemistry and Applications, 2022*(1), 7192919.

Satti, A. A., Osman, M. E., Hassan, E. A., & Al-Assaf, S. (2024). Analytical data of Acacia nilotica var. Nilotica gum. *Bulletin of the Chemical Society of Ethiopia, 38*(4), 839–852.

Scepankova, H., Saraiva, J. A., & Estevinho, L. M. (2018). A review of the antimicrobial potential of precursors of organic acids derived from industrial or agro-industrial side-streams. *Frontiers in Microbiology, 9*, 1212. https://doi.org/10.3389/fmicb.2018.01212

Sekhon, B. S. (2016). Phytochemicals as nanoscale carriers for drug delivery. *Journal of Pharmacy and Bioallied Sciences, 8*(1), 81–88. https://doi.org/10.4103/0975-7406.171700

Sharma, S., Virk, K., Sharma, K., Bose, S. K., Kumar, V., Sharma, V., et al. (2020). Preparation of gum acacia-poly (acrylamide-IPN-acrylic acid) based nanocomposite hydrogels via polymerization methods for antimicrobial applications. *Journal of Molecular Structure, 1215*, 128298.

Siebenberg, S., Kaysser, L., Wemakor, E., Heide, L., Gust, B., & Kammerer, B. (2011). Identification and structural elucidation of new caprazamycins from Streptomyces sp. MK730–62F2 by liquid chromatography/electrospray ionization tandem mass spectrometry. *Rapid Communications in Mass Spectrometry, 25*(4), 495–502.

Silva, T., & Souza, A. C. (2018). Overview of studies on the impact of acacia gum in the encapsulation processes for drug delivery systems. *International Journal of Pharmaceutics, 546*(1–2), 180–188. https://doi.org/10.1016/j.ijpharm.2018.05.040

Singh, D., & Suryanarayana, M. A. (2019). Effect of extraction methods on yield of forskolin from Coleus forskohlii. *Industrial Crops and Products, 128*, 572–577. https://doi.org/10.1016/j.indcrop.2018.11.031

Son, M., Park, T. H., & Lee, S. Y. (2018). Roles of plant extracts and constituents in cervical cancer therapy. *Anti-Cancer Agents in Medicinal Chemistry, 18*(12), 1660–1675. https://doi.org/10.2174/1871520618666180523103449

Tchatchedre, M., Amoussa, A. M., Houinato, M., Yedomonhan, H., & Lagnika, L. (2019). Ethnobotanical survey and phytogeographical study of plant species from genus acacia in Bnin. *Journal of Medicinal Plants Research, 13*(9), 199–212.

Thotathil, V., Sidiq, N., Fakhroo, A., & Sreerama, L. (2023). Phytochemical analysis of Anastatica hierochuntica and Aerva javanica grown in Qatar: Their biological activities and identification of some active ingredients. *Molecules, 28*(8), 3364.

Tiwari, B. K., Brunton, N. P., & Brennan, C. S. (2023). *Handbook of plant food phytochemicals: Sources, stability and extraction*. Wiley.

Todkar, S., Todkar, R., Kavathekar, V., Kulkarni, S., & Kulkarni, A. (2017). Secondary metabolite profiling of Acacia concinna. *Biosciences Biotechnology Research Asia, 8*(2), 653–666.

Toprakçı, M. A., Toprakçı, S., & Zoğ, Ü. B. (2021). Innovation in functional food ingredients: Role of microalgae. *Comprehensive Reviews in Food Science and Food Safety, 20*(2), 1651–1672. https://doi.org/10.1111/1541-4337.12678

Ullah, R., Nadeem, M., Imran, M., & Bilal, M. (2021). Bioactive compounds and their role in disease prevention through the modulation of oxidative stress. *Molecules, 26*(15), 4516. https://doi.org/10.3390/molecules26154516

Uwineza, P. A., & Waśkiewicz, A. (2020). Extraction techniques for bioactive compounds and antioxidant capacity determination of European cranberrybush (Viburnum opulus L.) fruit extracts. *Food Chemistry, 313*, 126121. https://doi.org/10.1016/j.foodchem.2019.126121

Vega-Arroy, M., Contreras-Esquivel, J. C., Meza-Gordillo, R., Osuna-Castro, J. A., & Carrillo-Inungaray, M. L. (2017). Assessment of phenolic content, free-radical-scavenging capacity, and antimicrobial activities of Mexican blue agave leaf extracts. *Industrial Crops and Products, 109*, 587–592. https://doi.org/10.1016/j.indcrop.2017.09.015

Ventaka Ratnam, K., Tirupati Reddy, G., & Venkata Raju, R. R. (2019). The therapeutic importance of gums in folk medicine from eastern Ghats, Andhra Pradesh. *Asian Journal of Pharmaceutical and Clinical Research*, 300–302. https://doi.org/10.22159/ajpcr.2019.v12i7.33052

Wardany, R., & Nirmala, D. (2023, December). Effect of antibacterial activity and analysis of active compounds of Hyrtios erecta sponge ethanol extract using maceration method. In *IOP conference series: Earth and environmental science* (Vol. 1273, No. 1, p. 012011). IOP Publishing.

Williams, P. A., & Phillips, G. O. (2000). Gum Arabic. In G. O. Phillips & P. A. Williams (Eds.), *Handbook of hydrocolloids* (pp. 155–168). Woodhead Publishing.

Yadav, J. S. (2021). The potential of microbial consortia in the production of secondary metabolites. *International Microbiology, 24*(3), 347–364. https://doi.org/10.1007/s10123-020-00160-0

Yan, F., Sun, Y., Mao, X., Liao, H., Wang, H., Zhang, X., & Lan, X. (2018). Application of three-dimensional gas chromatography coupled with accurate mass spectrometry for the analysis of secondary metabolites in tobacco leaves. *Journal of Chromatography. A, 1564*, 145–152. https://doi.org/10.1016/j.chroma.2018.06.035

Yerena-Prieto, M., Mora-Escobedo, R., & Martínez-Cruz, O. (2022). Utilization of ultrasonic-assisted extraction of phytochemicals from plants. *Food Engineering Reviews, 14*, 25–46. https://doi.org/10.1007/s12393-021-09248-9

Younis, Y., Eltayeb, A., Hifnawy, M., El-Gamal, A., & Elnour, A. (2022). Influence of climate change on the production and phytochemical content of Acacia seyal var. seyal. *Journal of Arid Environments, 105*, 123–130. https://doi.org/10.1016/j.jaridenv.2014.03.010

Yusof, N., Munaim, M. A., & Kutty, R. V. (2020). Ultrasound-assisted extraction propolis and its kinetic study. In *IOP conference series: Materials science and engineering* (Vol. 736, No. 2, p. 022089). IOP Publishing.

Zaman, S., Ihsan, M., Nisar, M., Zahoor, M., Rehman, K. U., & Kudratovich, K. K. Antioxidant properties of aromatic and medicinal plants. In *Plants as medicine and aromatics* (pp. 394–412). CRC Press.

Zavoi, S., Fetea, F., & Ranga, F. (2011). Comparative fingerprint and extraction yield of medicinal herb phenolics with hepatoprotective potential, as determined by UV-Vis and FT-IR spectroscopy. *Notulae Botanicae Horti Agrobotanici Cluj-Napoca, 39*(2), 82–89. https://doi.org/10.15835/nbha3926268

Zhu, J. (2023). Enhancing the functional properties of food products through nanoemulsion technology. *Innovative Food Science & Emerging Technologies, 68*, 102462. https://doi.org/10.1016/j.ifset.2020.102462

Ziani, B. E., Carocho, M., Abreu, R. M., Bachari, K., Alves, M. J., Calhelha, R. C., et al. (2020). Phenolic profiling, biological activities and in silico studies of Acacia tortilis (Forssk.) Hayne ssp. raddiana extracts. *Food Bioscience, 36*, 100616.

Chapter 3
Characterization of Secondary Metabolites Extracted (SMEs) from Acacia Gums (AGs)

Abdurahmani Hamid Nour, Khalid Hamid Musa, and Ahmed A. M. Elnour

Abstract The characterization of secondary metabolites extracted from acacia gums (AGs) is crucial for understanding their chemical composition and potential benefits. Various analytical techniques, such as chromatography and spectroscopy, have been used to determine the structures and properties of these compounds. Identifying and characterizing secondary metabolites from AGs can shed light on their chemical composition and biological activities, thereby providing valuable insights into their potential applications in medicine, agriculture, and industry. This chapter provides a comprehensive overview of the characterization of secondary metabolites extracted from AGs and sheds light on their chemical composition.

Keywords Acacia gum · Chemical composition · Secondary metabolites · Nuclear magnetic resonance · Mass spectrometry

Highlights

1. The characterization of secondary metabolites from AGs is crucial for understanding their chemical composition and potential benefits.
2. Various analytical techniques have been used to determine the structure and properties of these compounds.
3. The identification and characterization of secondary metabolites from AGs shed light on their chemical composition and biological activities.
4. A gum arabic tree is used to produce gum arabic and to fix nitrogen in the soil.

A. H. Nour · A. A. M. Elnour (✉)
Faculty of Chemical and Process Engineering Technology, Universiti Malaysia Pahang Al-Sultan Abdullah (UMPSA), Pahang Gambang, Malaysia

Centre of Excellence for Advanced Research in Fluid Flow (CARIFF), University Malaysia, Pahang Gambang, Malaysia
e-mail: abrahman@umpsa.edu.my; adamhassan@ump.edu.my

K. H. Musa
Department of Food Science and Human Nutrition, College of Agriculture and Veterinary Medicine, Qassim University, Buraydah, Kingdom of Saudi Arabia
e-mail: k.musa@qu.edu.sa

A. A. M. Elnour (ed.), *Gum Arabic and Breast Cancer Biology*,
https://doi.org/10.1007/978-981-97-8518-6_3

3.1 Introduction

Upon extraction of secondary metabolites from the *Acacia sengal* (ASNG), *Acacia seyal* (ASYG), and *Acacia polyacantha* (APC) gum plants, the characterization process involves a comprehensive analysis of the chemical compounds present (Elnour et al., 2022a). This includes identifying specific compounds, their chemical structures, and potential biological activities (Elnour et al., 2020). By understanding secondary metabolites at the molecular level, researchers can assess their potential pharmaceutical, agricultural, and ecological applications (Elkhateeb & Daba, 2022). This characterization process provides valuable insights into the chemical diversity and complexity of AGs (Elnour et al., 2018c). Valuable insights into the potential applications of these compounds are available in various fields, such as medicine, agriculture, and industry (Barkeer et al., 2024). The analytical methods used enabled a comprehensive understanding of the secondary metabolites present in AGs and their potential benefits (Silva et al., 2023). Researchers can determine the structures and properties of these compounds using various analytical techniques such as chromatography and spectroscopy.

Secondary metabolites, which differ from primary metabolites, play crucial roles in plant defense mechanisms against various stresses, underscoring their significance in plant biology (Kumar et al., 2021). Identification and quantification of secondary metabolites are essential for developing novel phytotherapeutic agents and nutraceuticals (Satheeshkumar et al., 2012). Studies have shown that AGs are rich in secondary metabolites such as alkaloids, cyanogenic glycosides, cyclitols, gums, terpenes, flavonoids, and condensed tannins (Babiker et al., 2019; Batiha et al., 2022; Hannachi, 2011; Rafi & Dawar, 2015; Youzbachi et al., 2019). Techniques such as chromatography and spectroscopy are instrumental in elucidating the structures and properties of these compounds, providing valuable insights into their potential applications (Dwivedi et al., 2021). High-resolution mass spectrometry instruments have significantly improved the analysis of plant secondary metabolites, facilitating their characterization (Álvarez-Rivera et al., 2019). This chapter provides a comprehensive overview of the characterization of secondary metabolites extracted from AGs, shedding light on their chemical compositions and biological activities. This study highlights the importance of natural products in identifying compounds with different biological activities for application in agriculture, medicine, and other fields.

Scientists can utilize diverse analytical methodologies, such as chromatography and spectroscopy, to understand the structure and characteristics of compounds, thereby enhancing insights into their potential applications (Ahn et al., 2022). These techniques aid in unraveling the properties of compounds and contribute to a deeper understanding of their functionalities and potential uses in various fields.

ASYG, a tall deciduous tree, is recognized for its smooth bark and distinctive yellow flower clusters (Lemma et al., 2014). This tree is associated with specific habitats, such as the ASYG-Balanites aegyptiaca forests in eastern Sudan and northwestern Ethiopia (Lemma et al., 2014). Additionally, ASYG has been studied for its chemical composition, showing high protein, phenolic, flavonoid, and anthocyanin contents (Abdel-Farid et al., 2014). Moreover, ASYG plays a role in attracting birds, with Red Acacia attracting many birds to the flowers. The production of gum arabic plays a vital role in ecological balance and economic sustainability owing to its ability to fix nitrogen and produce gum arabic. Studies have shown a positive correlation between gum arabic production and factors such as rainfall, soil microbial biomass, and mineral nitrogen content (Fall et al., 2016). Additionally, ASYG can restore soil fertility through nitrogen fixation, which is beneficial for crop production in gum arabic agroforestry systems in drylands (Diallo et al., 2015). Gum arabic production provides economic and job opportunities for resource-poor households (Elmqvist & Olsson, 2006). Understanding the ecological functions of these gums and their potential benefits in various applications is crucial for the further exploration and utilization of these natural compounds (Paula et al., 2022). The bioactive compounds in ASYG have been previously studied (Elnour et al., 2019b) for their antiproliferative effects against human breast cancer cells (Elnour et al., 2020) and for their role as antioxidant (Elnour et al., 2018a).

APC is a deciduous tree with medicinal properties recognized in traditional medicine for the treatment of various conditions. Studies have demonstrated its antibacterial potential against multidrug-resistant gram-negative bacteria, highlighting its therapeutic capabilities. The conventional use of APC in Sudan for treating bacterial diseases further emphasizes its importance in ethnomedicine (Hammad et al., 2022). APC gum is effective against multidrug-resistant Staphylococcus strains, highlighting its importance in medicinal applications. APC contains a new type of triterpenoid saponin with cytotoxic properties, which highlights its potential for producing bioactive compounds (Podolak et al., 2010). Additionally, studies have demonstrated the impact of human activities on the population structure of APC, emphasizing the need for conservation efforts (Sharam et al., 2009). These findings underscore APC's biological and medical significance of APC and demonstrate its potential in bioactive compound research and conservation biology. APC research has demonstrated its significant potential in traditional medicine for the treatment of various conditions. Studies have shown that APC extracts contain bioactive molecules with antibacterial, anti-helminthic, and antistaphylococcal properties, making them a valuable resource in folk medicine using APC (Ashour et al., 2022; Zaman et al., 2022). Additionally, the utilization of trees in traditional medicine to treat different diseases further supports their medicinal importance (Zamani et al., 2019). APC's antibacterial activity of APC against various pathogens, such as *Escherichia coli*, *Pseudomonas aureus*, and *Staphylococcus*, as well as its potential in treating livestock diseases and gastrointestinal infections, highlights its therapeutic value (Ashour et al., 2022). These findings underscore the medicinal potential of APC in traditional healing practices.

3.2 Characteristics and Ecological Significance of AGs

Human activity has been identified as a significant factor contributing to changes in the abundance of wild and domestic ungulates, which can negatively affect acacia stands (Rohner & Ward, 1999). Additionally, the economic importance of Acacia species globally, from providing useful products to supporting industrial development, further emphasizes the need for conservation actions (Kull et al., 2011). Demographic studies have highlighted the impact of human activities on the gum population structure, emphasizing the need for conservation efforts. Studies have emphasized the ecological importance of these gums and highlighted the need to better understand their ecological roles for further exploration and utilization (Babiker et al., 2022). In addition, AGs plays a crucial role in plant protection by acting as a barrier against water loss, pathogens, and pests (Barak et al., 2020). These properties make gums valuable for industrial applications and maintaining ecosystem health. AGs have various ecological functions, including trapping and retaining water, providing nourishment and habitat for organisms, and serving as trapping mechanisms for small organisms (Sánchez-Hernández et al., 2021). These characteristics make AGs valuable in arid environments, during drought conditions, and in supporting diverse ecosystems. AGs play a crucial role in both human applications and ecological systems, owing to their versatile nature and beneficial properties. Scientific study of gum arabic underscores its ecological importance and bioactive properties, highlighting its potential applications in various fields.

AGs are widely utilized as hydrocolloids in the food and pharmaceutical industries because of their flavorless and odorless nature. Derived from the exudates of ASNG and ASYG trees, it forms clear to pale yellow solutions when dissolved in water and exhibits properties such as hydration, swelling, and emulsion stability. These properties make it an ideal candidate for various applications, including suspension and emulsifying agents, tablet binders, and controlled drug release during wound healing (Larson et al., 2021). Its composition, which includes simple sugars such as galactose, glucuronic acid, arabinose, rhamnose, and covalently bonded proteins, contributes to its functional properties (Bhatia et al., 2023). Hence, the unique properties of gum arabic make it a key ingredient in food and pharmaceutical applications. AGs are valuable hydrocolloids that are widely utilized in food and pharmaceutical industries because of their lack of taste and odor. Derived from the exudates of the ASNG and ASYG trees, it forms clear to pale yellow solutions when dissolved in water. Its properties, such as hydration, swelling, and emulsion stability, make it suitable for various applications including suspension and emulsifying agents, tablet binders, and controlled drug release during wound healing (Larson et al., 2021). The composition of simple sugars, such as galactose, glucuronic acid, arabinose, and rhamnose, along with covalently bonded proteins, contributes to its functional properties (Bhatia et al., 2023). Therefore, AGs's established safety and effectiveness of AGs reinforces their vital use in food and pharmaceuticals. The FDA has recognized gum arabic as safe for various food and beverage applications, emphasizing its safety (Ashu et al., 2020). Their versatile applications in the food, pharmaceutical, and cosmetic industries are attributed to their emulsifying and stabilizing properties (Gashua et al., 2016). However, it is crucial to note that heating can affect the stability of emulsions containing gum arabic (Zaman et al., 2022). Despite its proven utility, its application to specific formulations requires attention to stability factors (Bakshi et al., 2022)., which are naturally occurring biopolymers derived from plant sources, possess the unique ability to form gels or viscous solutions when dissolved in water, making them valuable in a variety of fields such as food science, geotechnical engineering, and medicine (Barak et al., 2020). These properties have led to their extensive use in the food, cosmetic, and paper industries because of their water-holding capacities and amphiphilic moieties, which allow for modification of the flow properties, texture, and stability of emulsions and foams (Mahfoudhi et al., 2015). The significance of gums in diverse industries highlights their potential for further research and development (Barak et al., 2020). AGs are versatile substances with properties that make them valuable for various industries. AGs are used as thickeners, stabilizers, and emulsifiers in food and cosmetic products (Barak et al., 2020). Furthermore, the adhesive properties of AGs make them suitable for use as adhesives and coating materials (Barak et al., 2020).

The investigation and characterization of secondary metabolites found in AGs are essential for understanding their composition and potential benefits. Analytical techniques play a vital role in determining the structures and properties of these compounds, which can aid in identifying their potential applications (Noh et al., 2017). Biosynthetic gene clusters of secondary metabolites are crucial for the production of antibiotics and other valuable chemicals (Herdini et al., 2017). Using various analytical techniques, scientists can uncover the structures and properties of compounds in AGs. Analytical methods such as high-performance liquid chromatography (HPLC) and mass spectrometry are crucial for characterizing bioactive compounds in natural products (Dais et al., 2007). These techniques are essential for exploring the ecological functions and bioactive properties of natural compounds and providing critical insights for their further utilization (Cozzolino, 2009). Analysis of the secondary metabolites in gum arabic is vital for understanding its chemical composition and potential benefits. These studies have significantly contributed to a deeper understanding of the potential applications of gum arabic in various fields (Fang et al., 2010). Analyses of gum arabic have confirmed its botanical origin (McNamee et al., 2001).

Investigating and characterizing the secondary metabolites found in gum arabica is essential to gain a comprehensive understanding of their chemical composition and potential benefits. The choice of extraction solvent can significantly impact the nature and quantity of secondary metabolites obtained from medicinal plants (Apriandini et al., 2022). Secondary metabolites play an essential role in plants as defense mechanisms against various microorganisms and herbivores (Waty et al., 2017), highlighting their importance in plant biology. The ecological functions and bioactive properties of gum arabic, including polyphenolic compounds and antioxidants, suggest its potential for various applications (Hassani et al., 2020). These findings underline the importance of APC in medicine and ecology, emphasizing the need for its conservation and further study. Research has demonstrated that gum arabic can mitigate testicular damage, alleviate liver tissue impairments, and reduce systemic toxicity in rats (Abu-Serie et al., 2021). Furthermore, gum arabic plays a protective role against liver toxicity and oxidative stress induced by various agents (Kuck & Noreña, 2019). These findings underscore the importance of understanding the diverse bioactive compounds present in gum arabic for its potential therapeutic applications as a natural antioxidant (Azouz & Hassanen, 2020). Understanding the properties and ecological significance of AGs is essential for harnessing their potential benefits. AGs, natural products, are traditionally used for various medicinal purposes in sub-Saharan Africa (Al-Jubori et al., 2023). By utilizing advanced analytical techniques, researchers can delve more deeply into the composition and characteristics of AGs, paving the way for maximizing their advantages in various fields.

Scientific research on gum arabic has revealed its diverse bioactive compounds, including polyphenolic compounds and antioxidants, thereby underscoring its potential for various applications in medicine and beyond. Research has demonstrated that gum arabic modulates testicular damage, reduces liver impairment, and ameliorates nephrotoxicity (Azouz & Hassanen, 2020). Moreover, several studies have emphasized the protective effects of gum arabic against oxidative stress, its effect on liver function, and its role in reducing tissue damage (Babiker et al., 2018). Research has demonstrated the ability of gum arabic to delay postharvest ripening (Mahfoudhi & Hamdi, 2014), its use in edible coatings (McNamee et al., 2001) and its influence on the pasting and textural characteristics of starch (Kiprop et al., 2021). Additionally, studies have explored the effect of gum arabic on microbial activity (Fall et al., 2019), its potential as a prebiotic (Metaab et al., 2023), and its diverse applications in food products (Farasara & Fachriah, 2021). Gum arabic has also been investigated for its health effects, including implications in chronic renal failure (Ali et al., 2004) and nephrotoxicity (Salem, 2020). These findings collectively emphasize the varied ecological roles and bioactive compounds found in gum arabic, paving the way for its extensive use in numerous applications. APC, derived from the APC tree, is a complex compound comprising polysaccharides, proteins, and glycoproteins, including arabinogalactan and glycoprotein fractions, which contribute to its unique chemical and physical properties (Awadasseid et al., 2019). Furthermore, studies have demonstrated its potential as an excipient in granule and tablet formulations, thereby demonstrating its versatility for pharmaceutical applications (Desta et al., 2021). Variability in molecular aggregation has been observed in AGs (Sánchez et al., 2008). AGs, also known as gum arabic, are considered to be safe food additives for the general population (Mortensen et al., 2017).

3.3 Extraction of Secondary Metabolites

Bioactive compounds such as metabolites and antioxidants can be extracted from African cassava glycoproteins (AGs) using various techniques. For instance, hydrophobic interaction chromatography separates AGs into arabinogalactan,

arabinogalactan-protein complexes, and glycoproteins (Apolinar-Valiente et al., 2019). This technique results in distinct fractions with different molecular weight distributions, each containing a predominance of specific components, such as AG, AGP, or GP (Gashua et al., 2015). This process entails isolating fractions with increasing hydrophobicity, enabling the separation of surface-active compounds from the crude biosurfactant (Rodrigues et al., 2006). Ultrasonic-assisted extraction coupled with GC-MS/MS analysis is effective in identifying compounds in AGs (Elnour et al., 2018a). This methodology provided valuable insights into the extraction of bioactive compounds from AGs. Furthermore, studies by Singh & Dhiman (2016) suggested that the incorporation of gum Acacia into polymer films is associated with their antioxidant properties (Ali et al., 2021). These findings emphasize the importance of using advanced extraction techniques to analyze the natural compounds present in AGs. Overall, the use of these techniques can lead to a better understanding of the bioactive compounds present in AGs, enabling their potential application in various fields.

Extensive research on the versatile properties and applications of gums, including their use as thickeners, stabilizers, and sources of nourishment, has led to a deeper understanding of their potential benefits. Various techniques, including hydrophobic interaction chromatography, ultrasonic-assisted extraction, and GC-MS/MS analysis, have been employed to further extract secondary metabolites from AGs and explore their potential health benefits and antioxidant properties (Barak et al., 2020). Gums and polysaccharides are widely used in various industries, particularly the food industry, as emulsifiers and thickeners (Belmiro et al., 2017). In addition, a recent review of plant-based gums and mucilage emphasized their potential in pharmacology and nanomedicine, highlighting their capacity to bind water and create gels (Amiri et al., 2021).

3.3.1 Methods for Extracting Secondary Metabolites from AGs

Extraction techniques are crucial to reveal the health benefits and antioxidant properties of these compounds. Elnour et al. (2020) highlighted the significance of in vitro antioxidant assays in evaluating the antioxidant potential of different compounds, which is essential for understanding the health benefits associated with AGs. Safdar et al. (2023) compared various extraction methods to assess their impact on the yield, purity, and antioxidant activity. Established techniques for AGs are essential for exploring the bioactive compounds within these natural substances. The technique used to extract bioactive compounds from AGs is essential for studying their potential applications as functional materials with health benefits. The extraction of metabolites and antioxidants from AGs involves various techniques such as hydrophobic interaction chromatography and ultrasonic-assisted extraction coupled with GC-MS/MS analysis, as demonstrated by Elnour et al. (2018a). These methods allow for the isolation and identification of specific compounds, such as arabinogalactan, the AGP complex, glycoproteins from ASNG, and other compounds from ASYG. Hydrophobic interaction chromatography yields three fractions: arabinogalactan fraction 1 (AGF1), arabinogalactan-protein complex (AGPCs) fraction 2, and glycoprotein-fraction 3 (GPF3) (Gashua et al., 2015). The fractions obtained using this separation technique have been shown to have a broad molecular mass distribution, but each fraction contains a predominance of either AG, AGP, or GP components (Gashua et al., 2015). AGPCs are the most interfacially active components responsible for emulsifying properties (Kiprop et al., 2021). Hydrophilic sugar residues and amphiphilic protein components in AGP contribute to its adsorption at interfaces (Mwove et al., 2016). These properties result from both hydrophilic sugar residues and the amphiphilic protein component in AGP, which contribute to adsorption at air/water or oil/water interfaces (Mwove et al., 2016). Figure 3.1 provides an overview of the most common extraction techniques and their optimal conditions for extracting secondary metabolites from AGs. The following is a summary of the most commonly used techniques to provide more information on the extraction of secondary metabolites from AGs.

(i) **Maceration:** Maceration is a frequently used technique for extracting secondary metabolites from various plant sources owing to its ease and effectiveness in extracting water-soluble components (Ngibad, 2019). Despite certain limitations, such as low extraction yield, efficiency, and the need for large quantities of solvents (Navarrete-Carriola et al., 2024), maceration has been

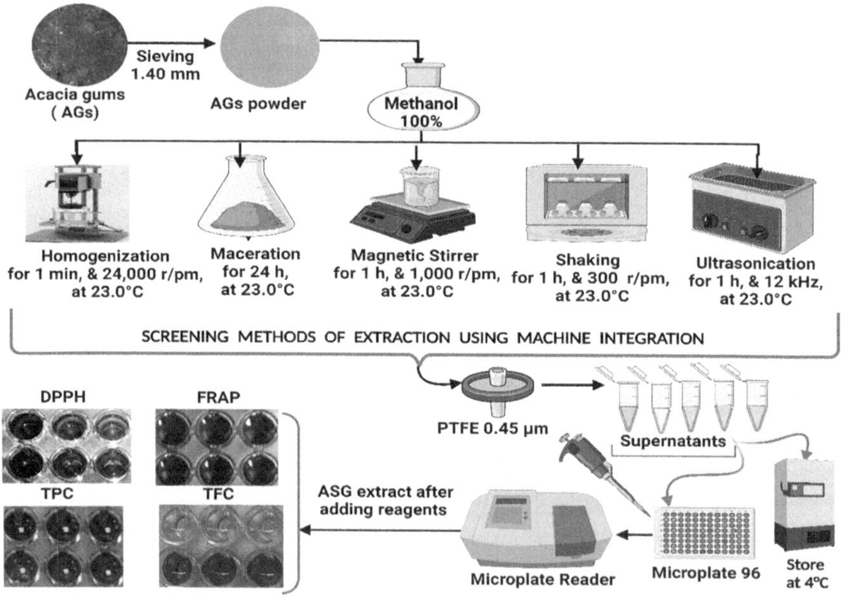

Fig. 3.1 Advanced and conventional techniques for extracting natural polyphenolic compounds from AGs extracts evaluated using different antioxidant activity assays. *DPPH* 2,2-diphenyl-1-picrylhydrazyl, *TPC* total phenolic content, *FRAP* ferric reducing antioxidant power, *TFC* total flavonoid content. (Source: Adopted by Elnour et al., 2022b)

found to yield better secondary metabolite content and antioxidant activity than infusion methods (Castro et al., 2022). This method breaks down the cell walls and membranes to allow the dissolution of secondary metabolites in solvents, thereby facilitating the extraction process (Darniwa et al., 2022). An additional advantage of maceration is that it does not require high temperatures, which could potentially damage the bioactive compounds (Mediani et al., 2022). Although maceration is a simple and effective method for extracting secondary metabolites, researchers should know its limitations and explore other extraction techniques for optimal results.

(ii) **Homogenization:** Homogenization, as a method for extracting secondary metabolites, has both advantages and drawbacks. Homogenization allows for efficient extraction of metabolites from tissues, facilitating the analysis of metabolite profiles (Zigene et al., 2019). However, the efficacy of homogenization methods can vary, affecting the efficiency of metabolite extraction (Winders & Pechan, 2021). Additionally, the solvent used during homogenization can influence the type and number of secondary metabolites extracted (Hutagalung et al., 2014). Therefore, while homogenization is a widely used method for metabolite extraction, it is essential to carefully consider the homogenization technique and solvent selection to ensure the optimal extraction of secondary metabolites.

(iii) **Magnetic stirrer:** The use of a magnetic stirrer for the extraction of secondary metabolites has several advantages, such as simplicity, environmental friendliness, and efficiency. These benefits can enhance the reproducibility and extraction efficiency compared with other methods (Luo et al., 2016). However, it is essential to note that the choice of extraction method, such as sonication, can significantly affect the antioxidant activity of the obtained extracts. Studies have shown that sonication can result in a higher antioxidant activity than magnetic stirring (Ashraf et al., 2016). This highlights the importance of selecting the most appropriate extraction technique based on the desired outcomes of the extraction process. Moreover, optimizing the extraction parameters, including solvent selection and extraction time, can significantly influence the efficiency of the extraction process (Vinci et al., 2013). Therefore, while magnetic stirrers offer advantages in terms of simplicity and efficiency, researchers must carefully evaluate the specific requirements of their extraction processes to maximize the benefits of this technique.

(iv) **Shaking:** Shaking is a widely used method for extracting secondary metabolites because of its ease of use and cost-effectiveness (Higgs et al., 2001). However, recent studies have suggested that mechanical shaking can cause the accumulation of specific metabolites, which may alter the overall metabolite profile (Schwendel et al., 2021). As an alternative, liquid-surface immobilization systems have been found to offer several advantages over shaking, such as higher metabolite accumulation and a more straightforward process (Oda et al., 2015). Therefore, although shaking is a common technique, it may be beneficial to explore other extraction methods for the unique benefits they can provide in the isolation of secondary metabolites.

(v) **UAE:** UAE has several advantages such as increased compound yield, energy savings, and reduced extraction time (Žlabur et al., 2020). Numerous studies have demonstrated its efficiency in extracting bioactive compounds from various sources, providing high antioxidant activity and short extraction time (Yuan et al., 2019). Furthermore, UAE is a green, efficient, and reproducible method for extracting compounds such as flavonoids (L. Chen et al., 2020). Compared to traditional methods, UAE significantly reduces extraction time, energy consumption, and solvent usage (Wang et al., 2019). However, it is crucial to consider factors such as solvent choice and extraction conditions to optimize the extraction process and maximize the yield of secondary metabolites (Navarrete-Carriola et al., 2024).

(vi) **MAE:** MAE presents several advantages, including an increased compound yield, energy savings, and reduced extraction time. According to Žlabur et al. (2020), this is an efficient method for extracting bioactive compounds from various sources, providing high antioxidant activity and short extraction times. Additionally, UAE is a green, efficient, and reproducible method for extracting compounds, such as flavonoids (Chen et al., 2020). Compared to traditional methods, UAE significantly reduces extraction time, energy consumption, and solvent usage (Wang et al., 2019). However, it is crucial to consider factors such as solvent choice and extraction conditions to optimize the extraction process and maximize the yield of secondary metabolites (Navarrete-Carriola et al., 2024).

(vii) **Supercritical Fluid Extraction (SFE):** Supercritical fluid extraction (SFE) has several advantages, including high selectivity, short extraction times, high efficiency, and environmental friendliness because it avoids the use of toxic organic solvents (Costa et al., 2015). SFE has been shown to overcome the limitations of conventional extraction methods by providing safer and more selective extraction with no residual solvents in the final product (Abbas et al., 2022). However, potential drawbacks of SFE include expansion and incompatibility with certain materials (Pavlova et al., 2022). Despite these limitations, SFE has been recognized for its ability to provide complete extraction, adjust solvent power, and operate at low temperatures, making it a promising technique for various applications (Hartati et al., 2018).

(viii) **Solid-Phase Extraction (SPE):** SPE offers several advantages, including reduction of hazardous solvent use, adaptability to automation, time efficiency, and prevention of emulsion formation (Johnson et al., 1991). SFE is recognized for its simplicity, low cost, ease of use, and effectiveness in reducing interference, making it an essential technique for separation and preconcentration (Yakout & El-Hady, 2016). Furthermore, SPE provides benefits such as high enrichment factors, rapid phase separation, low cost, and compatibility with various detection techniques, making it widely used for environmental water sample analysis (Moradi & Anarjan, 2019). This technique allows the retention of analytes in solid phase materials, ensuring reasonable recovery rates and efficient extraction of emerging pollutants (Moradi et al., 2016). Overall, SPE stands out for its ability to offer high extraction efficiency, simplicity, and reduced solvent consumption, making it a valuable method for various analytical applications.

(ix) **Hydrophobic interaction chromatography (HIC):** Hydrophobic interaction chromatography (HIC) is a valuable technique for the separation and purification of proteins based on their hydrophobic properties. This method isolates fractions with increasing hydrophobicity, enabling the separation of specific components such as Acacia gum, AGP, or GP (Rodrigues et al., 2006). HIC effectively separates surface-active compounds from crude biosurfactants, resulting in distinct fractions with varying molecular weight distributions, each enriched in specific components (Rodrigues et al., 2006). Therefore, HIC is a reliable method for fractionating complex mixtures based on their hydrophobicity.

High-performance hydrophobic interaction chromatography (HPHIC) is recognized as a powerful tool for separating proteins from various biological fluids, including culture media, ascites, plasma, and urine (Yamasaki & Dillenburg, 1999). HIC has also been utilized for protein purification in diverse contexts, such as the characterization of monoclonal antibodies and related products. Additionally, it has been employed to purify antigens from Mycobacterium bovis BCG Tokyo, following chromatofocusing and lectin-affinity chromatography. Therefore, HPHIC plays a crucial role in the efficient purification of proteins from complex biological fluids.

Hydrophobic interaction chromatography has been used to separate proteins from canine pancreatic juice, while preserving enzyme activity (Yamasaki & Dillenburg, 1999). It has also been applied in the biochemical fractionation of proteins, adding a third dimension to classical 2D separation (Stevanovic & Bohley, 2001). Despite its effectiveness in resolving heterogeneous aggregate species, HIC may present challenges related to the stability of chromatographic starting materials (Chmielowski et al., 2014). As a result, HIC is a versatile technique with significant applications in protein purification, characterization, and isolation based on its hydrophobic properties, although it requires careful consideration of material stability. Figure 3.1 provides an overview of the most common extraction techniques and their optimal conditions for extracting secondary metabolites from AGs.

3.3.2 Comparative Analysis of Extracted Compounds

A comparative examination of the compounds derived from ACGGs revealed the breadth and intricacy of their bioactive constituents (Elnour et al., 2022a). Various extraction methods yield distinct sets of compounds, each with potential health benefits and industrial applications (Elnour et al., 2020). For example, solvent extraction has been observed to effectively isolate a wide array of bioactive compounds, including polysaccharides and phenolic compounds, which contribute to the antioxidant properties of AGs (Mohammedelnour et al., 2018). In contrast, SFE demonstrated excellent selectivity for extracting specific compounds, making it suitable for focused research on individual bioactive components (Silva et al., 2009). Different techniques exist for isolating secondary metabolites from AGs, each of which has its own advantages and disadvantages. SPE is highly selective and effective for purifying target compounds; however, it is time-consuming and requires expensive equipment (Rahman et al., 2016). UAE and MAE are more cost-effective and quicker but may provide a different level of selectivity and purity than SPE (Elnour et al., 2018a). Hydrophobic interaction chromatography (HIC) a technique used to fractionate the AGs namely ASNG (Renard et al., 2006). This method has been used to isolate arabinogalactan, the AGP complex, and glycoproteins. When selecting an extraction method for AGs, it is essential to consider factors, such as the target compound, cost, time, selectivity, and purity.

According to Elnour et al., the selection process for analytical techniques in the analysis of AGs should consider several factors, such as the target compound, cost, time, selectivity, and purity. (2018). Different extraction methods offer distinct advantages and disadvantages, and a comparative analysis of the compounds extracted from AGs reveals the complexity and diversity of their bioactive components (Elnour et al., 2022b). Consequently, various extraction methods yield different sets of compounds, each with potential health benefits and industrial applications (Rajha et al., 2014). Solvent extraction has been shown to effectively isolates a wide range of bioactive compounds, including polysaccharides and phenolic compounds, which contribute to the antioxidant properties of AGs (Mirghani et al., 2017).

Understanding and comparing the compounds extracted from AGs using various methods enriches knowledge based on their bioactive properties and guides the selection of suitable extraction techniques for specific research and application objectives (Elnour et al., 2022b). This comparative analysis lays the groundwork for further exploration of the diverse bioactive compounds in AGs and their potential contribution to the health and industrial sectors. AGs are rich sources of secondary metabolites that play significant roles in various biological applications (Elnour et al., 2018c). The specific requirements of the application determine the selection of an appropriate extraction method, the desired metabolite yield, and environmental factors (Elnour et al., 2020). The available techniques range from conventional solvent extraction to sophisticated methods, such as supercritical fluid and MAE, each with advantages and limitations (Sampaopan & Suksaeree, 2022). Consequently, it is imperative to make informed decisions regarding the extraction methodology, guided by a thorough comprehension of the advantages and disadvantages of each technique, to achieve optimal metabolite recovery from AGs.

3.4 Characterization Spectroscopy Techniques for AGs Secondary Metabolites

Characterization techniques are used to understand the structural properties and bioactive potential of the secondary metabolites present in AGs (Batiha et al., 2022). These techniques include spectroscopic methods such as nuclear magnetic resonance spectroscopy (NMR) and infrared spectroscopy (FTIR), which allow for the identification and structural determination of individual compounds (Elnour et al., 2019b). Upon further examination of the compounds extracted from the ACGGs, it was apparent that the bioactive components possessed remarkable diversity and complexity(Mirghani et al., 2018). Various extraction methods yield distinct compounds, each of which offers potential health benefits and biological applications (Elnour et al., 2019a). For example, solvent extraction has been proven effective in isolating a wide range of bioactive compounds, including antioxidants and phenolic compounds (Elnour et al., 2022a).

Characterization methods, such as nuclear magnetic resonance, MS, and infrared spectroscopy, are essential for understanding the structural properties and bioactive potential of secondary metabolites in ACGGs (Aryal et al., 2021). These techniques allow the identification and structural determination of individual compounds, providing crucial insights into the composition and properties of the bioactive compounds present in ACGGs. In particular, Raman spectroscopy has been applied in the analysis of gum arabic extract, offering promising results for the rapid and non-invasive examination of the chemical makeup and structural attributes of AGs (Elnour et al., 2019b). The identification and quantification of secondary metabolites in gum arabic extracts have advanced over time due to advances in analytical methods, such as liquid chromatography–mass spectrometry (MS), gas chromatography–mass spectrometry (MS), and HPLC (Saleem et al., 2020). However, there is limited literature regarding the characterization of secondary metabolites extracted from AGs compared with the raw material.

Although there is a minimal result obtained from ASYG using FTIR, as presented in Fig. 3.2, more research is needed to fully understand the bioactive potential of these compounds. The infrared spectra of the methanolic crude and active fractions of the ASYG samples are shown in Fig. 3.2. The spectra depict a range of peaks between 4000 and 400 cm^{-1}. These peaks reveal a significant overlap in the absorption spectra of the various components, with each band displaying unique absorption peaks for the functional groups present in the samples. The FTIR spectra for the ASYG-raw material (a), crude methanol extract (b), methanol fraction (c), and acetone fraction (d) all show distinct patterns. The FTIR technique was highly effective in characterizing the samples, as demonstrated by comparing the effective peaks against the raw material ASYG standard. The analysis of ASYG-raw was carried out in the solid state to prevent the significant influence of water absorption, especially in the amide I band region (1720–1580 cm^{-1}), as reported by Renard et al. (2006). The FTIR spectra of both ASYG samples as raw gums showed 10 significant peaks in a similar range of approximately 418, 670, 1036, 1256, 1410,

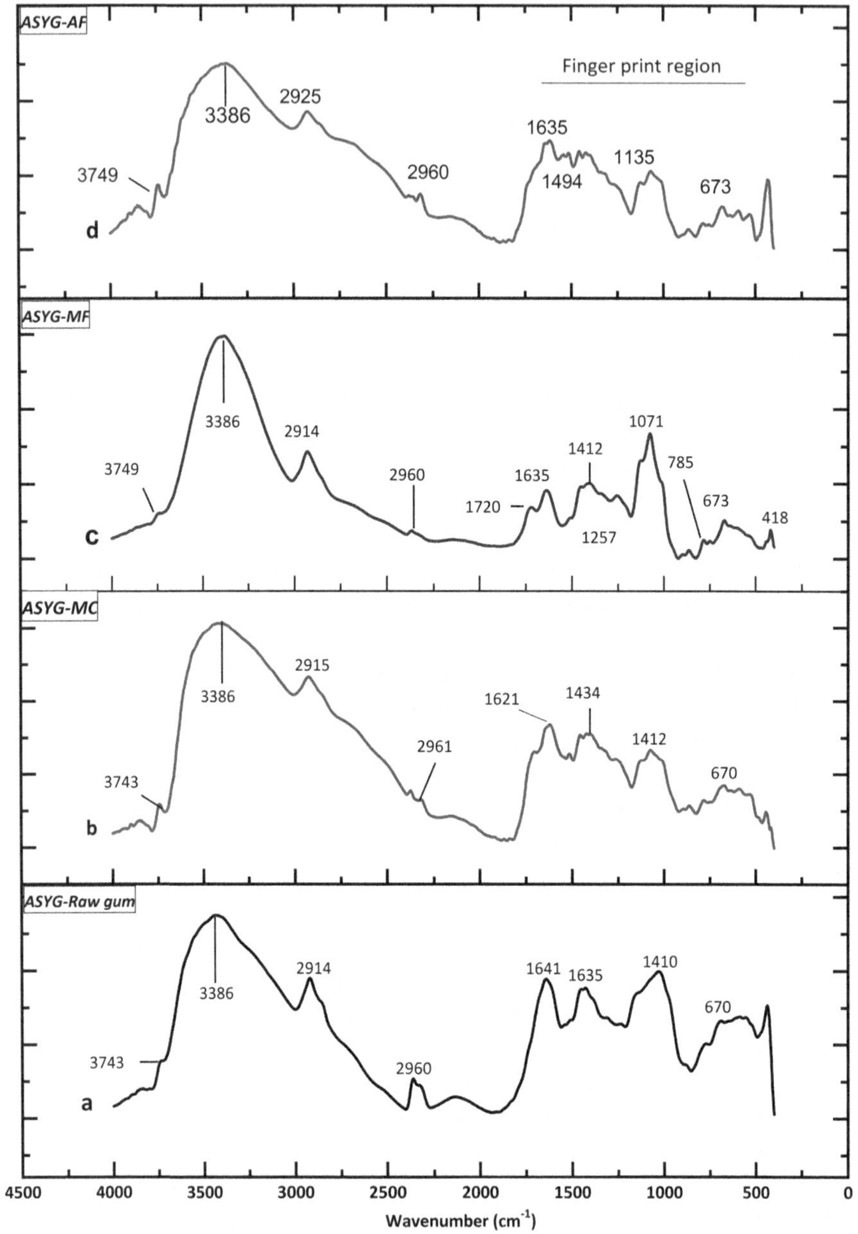

Fig. 3.2 Shows the FTIR spectra of ASYG: (**a**) raw gum, (**b**) mathano crude extraction (MCE), (**c**) methanol fraction (MF), and (**d**) acetone fraction (AF), respectively

1641, 2360, 2914, 3386, and 3743 cm^{-1}, with only minor variations in the peaks at 1256 and 1363 cm^{-1}. ASYG presented three leading spectral bands in the wavenumber range 1700–1200 cm^{-1} (Fig. 3.2a, b).

The role of proteins in ACGGs may affect the allocation of bands in the FTIR spectra. Amide I and II vibration bands are typically observed in the FTIR spectra at approximately 1650 and 1550 cm^{-1}, respectively. These bands are mainly influenced by the secondary structure of the polypeptide backbone and are less affected by side chains, as reported by Sadat & Joye (2020) and Calabrò (2017). The amide I band was primarily attributed to C=O stretching, whereas the amide II band combined N–H bending and C–N stretching vibrations, as described by Li et al. (2015) and Severcan et al. (2003). The sensitivity of amide I is higher than that of amide II, reflecting protein conformational changes, as García-Rubio et al. (2019) indicated. Additionally, a second-derivative analysis of the FTIR spectra revealed differences in the amide II and amide I bands, which aid in characterizing protein structures (García-Rubio et al., 2019). However, the position of the band near 1610 cm^{-1} in the FTIR spectra of polysaccharides carrying uronic acids may be influenced by their biochemical composition, particularly uronic acid content, as noted by Traoré et al. (2017). Uronic acids contribute to the bands observed at 1425 cm^{-1} and affect the Infera Red (IR) spectra, as highlighted by Rupérez (2002). Overall, FTIR spectral analysis revealed crucial insights into the structural intricacies of proteins and polysaccharides in AGs.

The absorbance band of AGP for carboxylic functions ranges from 1640 to 1600 cm^{-1}, and the uronic acid content increases from 3.6% to 14.7% (Zhou et al., 2009). AGPs are characterized by the presence of type II arabino-3,6-galactan chains, a hydroxyproline-rich protein backbone, and their ability to bind synthetic phenylazo dyes (Renard et al., 2014). The structures of L-arabinose-galactooligosaccharides released from AGP sugar moieties have been investigated previously (Shimoda et al., 2014). Studies on different plant species have demonstrated the presence of uronic acids in AGPs (Shin et al., 2008). The key monosaccharides, galactose (Gal) and arabinose (Ara), exhibited intense peaks at approximately 1075 and 1045 cm^{-1}, respectively, and uronic acids at approximately 1018 cm^{-1} in the second-derivative IR spectra (Zhou et al., 2009). AGPs are defined by the presence of type II arabino-3,6-galactan chains and a hydroxyproline-rich protein backbone, and their ability to bind to a class of synthetic phenylazo dyes (Renard et al., 2014). The structures of L-arabinose-galactooligosaccharides released from AGP sugar moieties have been investigated previously (Shimoda et al., 2014).

The presence of uronic acids with immunomodulatory activity has been reported in AGPs, which possess a highly branched glycan structure composed of arabinose, galactose, rhamnose, and galacturonic acid, with a β-(1,3)-linked galactan backbone (Shin et al., 2008). ASYG contained 10.4% molar 4-*O*-glucuronic acid and 1.0% protein (Wang et al., 2021). Glucuronic acid plays a significant role in various biochemical processes and its protein content affects its nutritional value. The presence of glucuronic acid in ASYG is crucial to its bioactive properties and potential health benefits (Yavari et al., 2018). Understanding the specific composition of ASYG, including its glucuronic acid and protein content, provides valuable insights into its functional properties and potential applications in various fields, such as food science and medicine.

Figure 3.2 illustrates the characterization techniques for AGs extract/fraction secondary metabolites using FTIR.

3.5 Profile of Secondary Metabolites in AGs

The profiles of secondary metabolites in AGs, including arabinogalactan (AG), AGP, and glycoprotein (GP), offer valuable insights into their bioactive components and their potential applications (Gashua et al., 2016). These components contribute to the medicinal value of acacia and aid drug delivery and health benefits (Tiwari et al., 2023). The immunopotentiatory properties of AGs have been attributed to their AGP contents (Sekhon et al., 2016). Studies have shown that AGs contain bioactive compounds with medicinal value, such as tannins, flavonoids, and alkaloids (Elnour et al., 2019b). The emulsifying properties of AGs are related to their AGP fractions (Osman, 2016). Overall, the diverse chemical compositions of AGs underscore their significance in various fields ranging from pharmaceuticals to food science.

On the other hand, ASYG is known for its high content of glycoproteins and tannins, which provide unique functional properties suitable for various applications, such as pharmaceuticals, food, and cosmetics. The tannins found in gum contribute to its astringent and antimicrobial properties, thereby expanding its industrial applications (Gashua et al., 2016). Gum comprises glycoproteins and tannins, with glycoproteins enhancing hydrophilicity and tannins contributing to its hydrophobicity (Tamayo et al., 2018). Bioactive compounds in ASYG include carotenoids, indolic alkaloids, and various phenolic compounds such as ferulic acid and coumaric acid (Elnour et al., 2018b). The potential applications of this gum extend to wound dressings due to its high molecular weight and water solubility, which are beneficial for skin treatments (Pagano et al., 2022).

APC gum is rich in secondary metabolites including flavonoids, terpenoids, and saponins, contributing to its antioxidant, anti-inflammatory, and antimicrobial properties (Batiha et al., 2022). These bioactive compounds are commonly found in AGs species, and include alkaloids, flavonoids, terpenes, and condensed tannins (Hannachi, 2011; Rafi & Dawar, 2015; Youzbachi et al., 2019). These secondary metabolites have been widely reported in Acacia species, highlighting their potential applications as functional foods, herbal remedies, and natural antimicrobial agents (Aryal et al., 2021). Studies have shown that the abundance of polyphenols and flavonoids in plant extracts is correlated with their antifungal potential, further emphasizing the significance of secondary metabolites in plant-based products (Bouhlali et al., 2020). APC is a valuable natural resource because of its diverse bioactive compounds that offer numerous health benefits and applications.

A comprehensive comparative study of the metabolites and antioxidants present in various types of AGs could provide valuable insights into their bioactive potential and antioxidant capacity (Thotathil et al., 2022). Understanding the molecular characteristics and properties of gum arabic harvested from different Acacia species could provide essential information for its application in medicine and other fields (Gashua et al., 2016). By leveraging the unique profiles of bioactive compounds present in AGs, researchers and industries can explore innovative solutions in medicine, nutrition, and materials science (Singh & Dhiman, 2016).

Researchers can employ analytical techniques such as liquid chromatography-mass spectrometry (LC/MS) and nuclear magnetic resonance (NMR) spectroscopy to identify and quantify specific metabolites and antioxidants in AGs (Anderson et al., 2018). These methods are commonly used in metabolomic studies to comprehensively analyze metabolites and aid in compound identification (Monge et al., 2019). LC-MS and NMR are robust analytical platforms that offer advantages in terms of sensitivity and peak capacity for metabolite profiling (Emwas et al., 2019). LC-MS is a primary analytical technology in metabolomics that identifies and quantifies various compounds (Emwas et al., 2019). However, the investigation of secondary metabolites extracted from various AGs varieties needs to be improved, except for extensive studies conducted on ASYG. Figure 3.3 presents the bioactive compounds obtained from the different fractions of ASYG.

GC-MS/MS analysis of the crude methanol extract, its methanol (MF), and AF of ASYG revealed the presence of 57 compounds representing various classes (Elnour et al., 2018b). The major constituents of ASYG were isovitamin C (42.37%), crypton (5.86%), hydroquinone (4.86%), triacetic acid lactone (2.67%), 2,4-di-*tert*-butylphenol (2.67%), cyanidin cation (2.05%), apigenin 7-glucoside (1.9%), benzoic acid (1.83%), (+)-α-tocopherol (1.58%), methyl catechol (1.42%), and 2,6-dimethylol-*p*-cresol (2.16%). Nine compounds, including crypton (7.83%), chromone, 5-hydroxy-6,7,8-trimethoxy-2,3-dimethyl (7.01%), Phe-1,4-diol, 3,6-dimethyl (6.65%), hydroquinone (5.31%), ferulic acid (5.84%), isopinocamphol (3.06%), benzoic acid (2.02%), isovitamin C (1.34%), and β-carotene (1.21%), were found to be significantly higher in both ASYG-AF and ASYG-MF. However, vanylglycol, quercetin 3-D-galactoside, vitexin, gengkwanin, gallic acid, retinoic acid, zearalenone, 4′,7-Dimethoxyisoflavone, and flavone, 4′-methoxy-6-acetyloxy, were calculated in the MF only instead of AF and MCE for samples. GC-MS/MS analysis compounds were studied by comparing the MS and retention times with the references presented in the National Institute of Standards and Technology (NIST) mass spectral library.

A study conducted by Elnour et al. (2018b) identified various volatile compounds including polyphenols, hydrocarbons, phenolic acids, fatty acids, and several novel constituents (Fig. 3.3). Notably, several of the identified compounds possess pharmacological activity. For instance, isovitamin C, tocopherol, and resveratrol have demonstrated antitumor activity in Hep3B hepatocellular carcinoma cells (Sun et al., 2013) and anti-inflammatory effects (Elnour et al., 2020). Furthermore, Crypton and hydroquinone exhibited potential antifungal and antibacterial properties (Chen et al., 2019), making them among the bioactive compounds in gum arabic

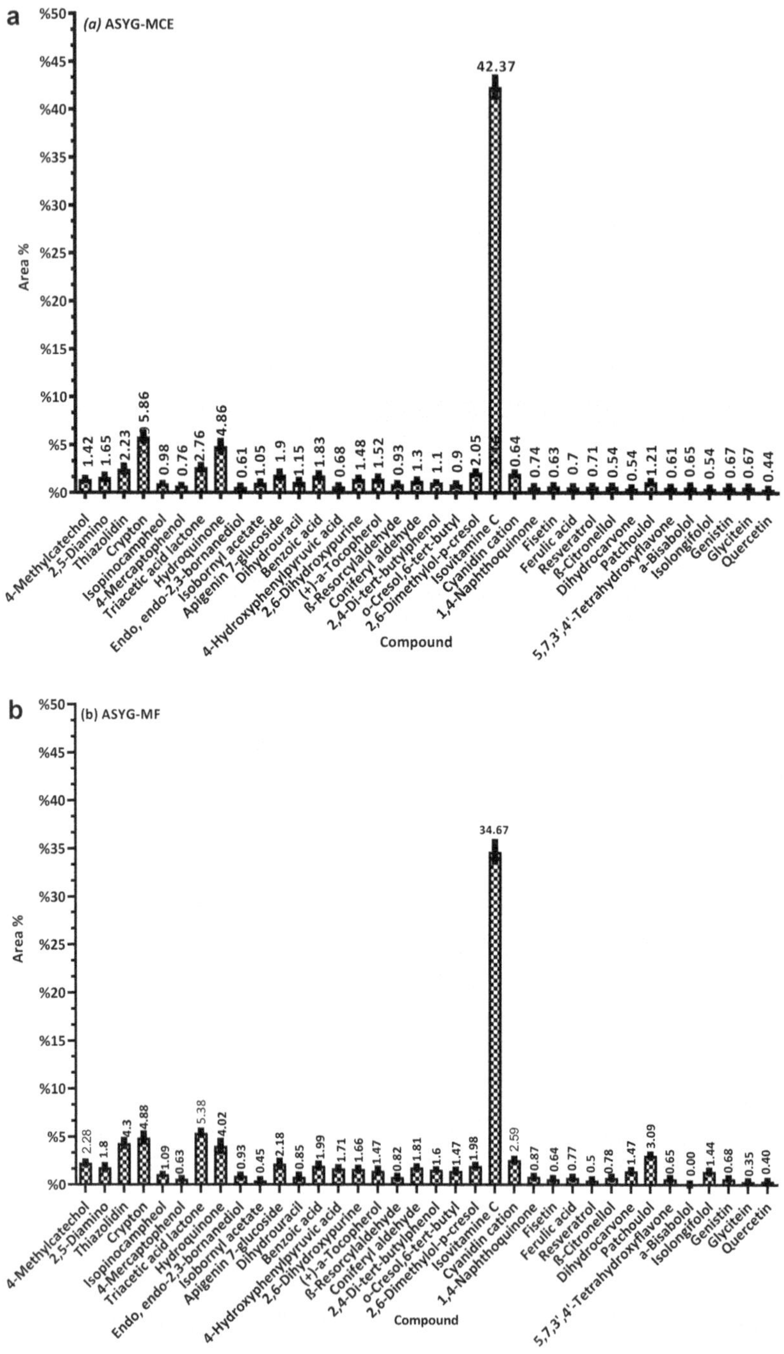

Fig 3.3 Displays a GC-MS/MS chromatogram of the bioactive compounds in the ASYG methanol crude extract and its active fractions. Specifically, the chromatogram includes ASYG-MCE, which represents the crude methanol extraction (**a**); ASYG-MF, which indicates the methanol fraction (**b**); and ASYG-AF, which represents the acetone fraction (**c**), as described in the study by Elnour et al. (2020)

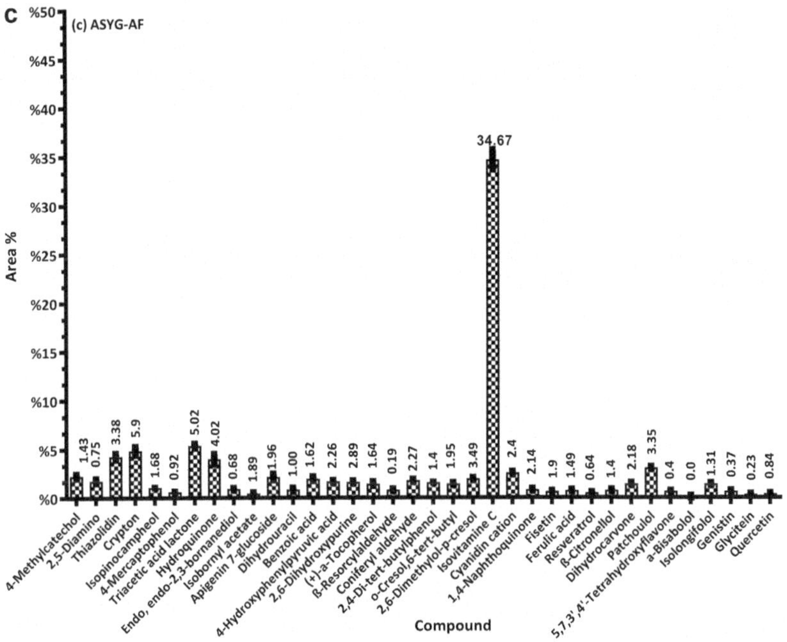

Fig 3.3 (continued)

extractions/fractions. Triacetic acid lactone, a polyketide compound, is known for its antibacterial properties, and serves as an antimicrobial food additive. It is a primary ingredient in various commercial products including sorbic acid and acetylacetone (Liu et al., 2023). Triacetic acid lactone is a natural compound derived from polyketides that has garnered recognition for its potential to produce valuable bioproducts (Otoupal et al., 2022). Its synthesis involves the use of acetic acid or other biological pathways in specific microbes (Sajjad et al., 2020). This compound is a promising platform chemical that can be converted to various valuable products (Yu et al., 2022). Triacetic acid lactone is an essential intermediate in the biosynthesis of polyketides, and can be produced from glucose through efficient polyketide biosynthesis (Zha et al., 2004). It contributes to the pharmacological activities of methanol, acetone, crude extract, and their fractions.

Beta-carotene possesses notable antioxidant properties, serving as a free radical quencher and a chain-breaking antioxidant, contributing to its anticancer and anti-inflammatory effects (Baker et al., 1999). Research has shown that Beta-carotene effectively neutralizes singlet oxygen and lipid radicals (Ben-Amotz et al., 1996). In addition, β-carotene exerts hepatoprotective effects through its antioxidant, anti-inflammatory, and anti-apoptotic activities (Akkara & Sabina, 2020). These characteristics indicate that β-carotene is a valuable compound with potential health benefits, including protection against oxidative stress and certain diseases. Figure 3.3 illustrates a GC-MS/MS chromatogram of the bioactive compounds in the ASYG methanol crude extract and its active fractions.

3.6 Challenges in Characterization and Extraction

The analysis and extraction of bioactive compounds from AGs presents challenges that require addressing fundamental questions regarding their properties and composition (Sánchez et al., 2018). Studies have identified bioactive compounds in AGs, including D-galactopyranose, carotenoids, and antibacterial agents (Tiwari et al., 2023). These compounds have demonstrated antioxidant and antiproliferative effects, suggesting their potential applications in various fields, including medicine (Elnour et al., 2020; Paula et al., 2022). Several studies have explored the encapsulation of bioactive compounds in AGs to enhance their stability and efficacy (Pedreiro et al., 2021; Velázquez-Martínez et al., 2021). The diverse bioactive compounds found in AGs offer promising opportunities for further research and use in various industries. Characterizing specific compounds in gums is challenging because of the complex nature of the gum matrix and the presence of multiple bioactive compounds. Analytical techniques, such as high-performance liquid chromatography (HPLC), gas chromatography-mass spectrometry (GC-MS), and nuclear magnetic resonance (NMR), are utilized for identification and quantification; however, accurate isolation and analysis of individual components remain difficult (Oliveira et al., 2018; Yu et al., 1998). This difficulty arises from the intricate composition of gums, which contain various compounds, such as triterpenes, volatile terpenoids, and polysaccharides (Goudah et al., 2015). To address this challenge, researchers have explored techniques such as microencapsulation, to enhance the analysis of these compounds (Fazilah et al., 2019). Overall, the intricate composition and diverse bioactive compounds in gums necessitate advanced analytical methods for precise characterization and analysis (D'Arcy et al., 1997). Standardizing the extraction process of bioactive compounds from AGs is crucial because of variations in composition from different regions, environmental factors, and extraction methods. This ensures consistent yield and purity for reliable medicinal, nutritional, and materials science applications (Batiha et al., 2022). The variability in gum composition necessitates precise extraction methods to guarantee the quality and effectiveness of the bioactive compounds (Tiwari et al., 2023). In addition, the extraction process must consider environmental factors to maintain the integrity and functionality of the compounds (Silva & Souza, 2018).

Tailored extraction methods for flavonoids, terpenoids, and phenolic compounds are crucial for preserving the bioactive compounds in APC gum. Specific solvents and techniques are necessary to maximize yield and maintain bioactivity (Tiwari et al., 2023). The diverse range of compounds in AGs, including flavonoids, terpenoids, and phenolic derivatives, are essential for their medicinal properties (Asaad, 2023). These compounds exhibit various pharmacological effects including anti-inflammatory, antibacterial, and antioxidant activities (Algethami, 2024). The presence of bioactive substances, such as flavonoids, alkaloids, tannins, terpenoids, and phenolic compounds, in AGs underscores their importance in traditional medicine and their potential health benefits (Sethuraman & Rajendran, 2018). Therefore, understanding and optimizing the extraction methods for these bioactive compounds are vital for harnessing the therapeutic potential of AGs.

Efficient and standardized characterization and extraction methods are essential for maximizing the potential of AGs and their bioactive compounds. These efforts are crucial for advancing their applications in medicine, nutrition, and materials science. Addressing environmental and sustainability concerns in the extraction and analysis of secondary metabolites from AGs is paramount for ensuring the responsible utilization of these resources (Silva & Souza, 2018). Studieshhave shown that AGs contain arabinogalactan proteins that contribute to their immunopotentiatory properties and are non-allergenic in the gut(Sekhon et al., 2016). Furthermore, studies have demonstrated the positive effects of GA in ameliorating systemic modulation in diabetic rats, thus demonstrating its potential in medicine (Ibrahim et al., 2023).

To address environmental and sustainability concerns in extracting secondary metabolites from AGs, it is crucial to focus on reducing solvent usage, minimizing energy consumption, and utilizing eco-friendly solvents (Tavan et al., 2023). Employing efficient and eco-friendly extraction techniques, such as green solvent-based extraction, can significantly reduce the environmental impact of the extraction process solvents (Tavan et al., 2023). By adopting these practices, it is possible to decrease the ecological footprint of the extraction processes and promote sustainable practices.

Comprehensive characterization, bioactive potential, and applications of secondary metabolites, polyphenolic compounds, and antioxidants in AGs show promise in various industries. These compounds exhibit antioxidant, anti-inflammatory, antimicrobial, and anticancer activities, making them valuable candidates for drug development, functional foods, and nutraceuticals. Additionally, their ecological roles in Acacia plants support sustainable sourcing and environmentally friendly extraction processes. Continued research on the characterization and extraction of secondary metabolites from AGs while addressing environmental and sustainability concerns is crucial for future advancements (Awadasseid et al., 2019).

3.7 Future Perspectives and Research Directions

Future research directions for the extraction and characterization of bioactive compounds from AGs encompass several key areas. These directions are crucial for advancing various industries including medicine, nutrition, and materials science. Future perspectives and research directions are summarized as follows:

(i) **Innovative Extraction Methods**: Developing innovative extraction methods to overcome the challenges posed by the complex nature of the gum matrix and the presence of diverse bioactive compounds, such as flavonoids, terpenoids, and phenolic compounds, is essential for unlocking the full potential of AGs and their bioactive components (Ashour et al., 2022; Zaman et al., 2022).

(ii) **Characterization and Comparative Analysis**: Understanding and comparing the compounds extracted from AGs using various methods enriches the knowledge based on their bioactive properties and guides the selection of suitable extraction techniques for specific research and application objectives. This comparative analysis lays the groundwork for further exploration of the diverse bioactive compounds in AGs and their potential contributions to the health and industrial sectors (Elnour et al., 2022a).

(iii) **Advanced Analytical Methods**: Detailed analysis of the bioactive components of AGs can be accomplished using advanced analytical methods to identify specific compounds and measure their bioactive potential for application in drug development, functional foods, and nutraceuticals. Techniques such as time-of-flight secondary ion MS (ToF-SIMS), X-ray photoelectron spectroscopy (XPS), and Fourier transform infrared spectroscopy (FTIR) can be used to characterize bioactive coatings. Ultra-high-performance liquid chromatography coupled with MS can aid in the identification and quantification of bioactive components (Álvarez-Rivera et al., 2019).

(iv) **Sustainable Practices**: Addressing environmental and sustainability concerns when extracting secondary metabolites from AGs is crucial. Efforts to reduce solvent usage, minimize energy consumption, and utilize eco-friendly solvents, such as green solvent-based extraction, can significantly reduce the environmental impacts of extraction process solvents. By adopting these practices, it is possible to decrease the ecological footprint of the extraction processes and promote sustainable practices (Silva et al., 2023).

(v) **Standardization and Tailoring Methods**: Standardizing extraction and tailoring methods for specific bioactive compounds is crucial for enhancing their bioactivity. Moreover, utilizing environmentally friendly solvents, such as green solvent-based extraction, can reduce the ecological impact of the extraction process, aligning with sustainability goals (Dais et al., 2007).

3.8 Conclusion

A comprehensive analysis of the extraction and characterization of bioactive compounds from AGs has revealed the potential for significant advancements in medicine, nutrition, and materials science. This study emphasizes the importance of characterizing and extracting secondary metabolites from AGs to address environmental and sustainability concerns. Bioactive compounds in AGs exhibit diverse properties, including antioxidant, anti-inflammatory, antimicrobial, and anticancer activities, making them valuable candidates for drug development, functional foods, and nutraceuticals. Future perspectives and research directions for AGs focus on developing innovative extraction methods to overcome the challenges posed by the complex nature of the gum matrix and diverse bioactive compounds, such as flavonoids, terpenoids, and phenolic compounds. Standardizing the extraction and tailoring methods for specific bioactive compounds is crucial for enhancing their bioactivity. Moreover, utilizing environment-friendly solvents, such as green solvent-based extraction, can reduce the ecological impact of the extraction process, aligning with sustainability goals. The potential applications of AGs in various fields have highlighted the importance of exploring their biological and pharmacological properties to unlock their full potential. Addressing environmental and sustainability concerns in extracting and analyzing secondary metabolites from AGs is of paramount importance for ensuring the responsible utilization of these resources. The diverse bioactive compounds found in AGs offer promising opportunities for further research and use in various industries. In summary, future perspectives and research directions for AGs emphasize the development of innovative extraction methods, advanced analytical techniques, sustainable practices, and standardization of extraction and tailoring methods to unlock the full potential of AGs and their bioactive components across various industries. These efforts are essential for driving advancements in medicine, nutrition, and materials science while promoting sustainable and eco-friendly extraction processes.

References

Abbas, A., Anwar, F., Ahmad, N., Shahid, M., Al-Mijalli, S. H., Yaseen, M., et al. (2022). Characterization of bioactives and nutra-pharmaceutical potential of supercritical fluid and hydro-distilled extracted coriander leaves essential oil. *Dose-Response, 20*(4), 15593258221130749. https://doi.org/10.1177/15593258221130749

Abdel-Farid, I. B., Sheded, M. G., & Mohamed, E. A. (2014). Metabolomic profiling and antioxidant activity of some AGs. *Saudi Journal of Biological Sciences, 21*(5), 400–408. https://doi.org/10.1016/j.sjbs.2014.03.005

Abu-Serie, M. M., Hamouda, A. F., & Habashy, N. H. (2021). Acacia senegal gum attenuates systemic toxicity in CCl(4)-intoxicated rats via regulation of the ROS/NF-kappaB signaling pathway. *Scientific Reports, 11*(1), 20316. https://doi.org/10.1038/s41598-021-99953-y

Ahn, Y., Choi, S. H., Jang, J. K., Kim, S. Y., Shim, J. H., Lee, S. S., & Byun, J. H. (2022). Impact of the liver imaging reporting and data system on research studies of diagnosing hepatocellular carcinoma using MRI. *Korean Journal of Radiology, 23*(5), 529–538. https://doi.org/10.3348/kjr.2021.0393

Akkara, P. J., & Sabina, E. P. (2020). A biochemical approach to the anti-inflammatory, antioxidant and anti-apoptotic potential of beta-carotene as a protective agent against bromobenzene-induced hepatotoxicity in female Wistar albino rats. *Journal of Applied Biomedicine, 18*(2–3), 87–95. https://doi.org/10.32725/jab.2020.011

Algethami, F. K. (2024). GC/MS and LC-MS analysis and in-vitro antioxidant activity of essential oil and crude methanol extract from the leaves of *Acacia Gerrardii Benth. growing in Saudi Arabia. Chemistry & Biodiversity, 21*(2), e202301847. https://doi.org/10.1002/cbdv.202301847

Ali, B. H., Al-Qarawi, A. A., & Ahmed, I. (2004). Does treatment with gum Arabic affect experimental chronic renal failure in rats? *Fundamental & Clinical Pharmacology, 18*(3), 327–329. https://doi.org/10.1111/j.1472-8206.2004.00241.x

Ali, R., Tabrez, S., Rahman, F., Alouffi, A., Alshehri, B., Alshammari, F. A., et al. (2021). Antileishmanial evaluation of bark methanolic extract of *Acacia nilotica*: In-vitro and in silico studies. *ACS Omega, 6*(12), 8548–8560. https://doi.org/10.1021/acsomega.1c00366

Al-Jubori, Y., Ahmed, N. T. B., Albusaidi, R., Madden, J., Das, S., & Sirasanagandla, S. R. (2023). The efficacy of gum Arabic in managing diseases: A systematic review of evidence-based clinical trials. *Biomolecules, 13*(1), 138. https://doi.org/10.3390/biom13010138

Alvarez-Rivera, G., Ballesteros-Vivas, D., Parada-Alfonso, F., Ibáñez, E., & Cifuentes, A. (2019). Recent applications of high-resolution mass spectrometry for the characterization of plant natural products. *TRAC: Trends in Analytical Chemistry, 112*, 87–101. https://doi.org/10.1016/j.trac.2019.01.002

Amiri, M. S., Mohammadzadeh, V., Yazdi, M. E. T., Barani, M., Rahdar, A., & Kyzas, G. Z. (2021). Plant-based gums and mucilages applications in pharmacology and nanomedicine: A review. *Molecules, 26*(6), 1770. https://doi.org/10.3390/molecules26061770

Anderson, J., Phelan, M. M., Clegg, P., Peffers, M. J., & Rubio-Martínez, L. M. (2018). Synovial fluid metabolites differentiate between septic and nonseptic joint pathologies. *Journal of Proteome Research, 17*(8), 2735–2743. https://doi.org/10.1021/acs.jproteome.8b00190

Apolinar-Valiente, R., Williams, P., Nigen, M., Tamayo, V. M., Doco, T., & Sánchez, C. (2019). Recovery, structure and physicochemical properties of an aggregate-rich fraction from *Acacia senegal* gum. *Food Hydrocolloids, 89*, 864–873. https://doi.org/10.1016/j.foodhyd.2018.11.054

Apriandini, A., Fajaroh, F., Aliyatulmuna, A., & Ciptawati, E. (2022). The effect of extraction time of Raja nangka banana peel as capping agent on the characteristic and antibacterial activity of ZnO nanoparticles against Staphylococcus epidermidis. *Al-Kimia, 10*(2), 151–159. https://doi.org/10.24252/al-kimia.v10i2.31484

Aryal, B., Adhikari, B., Aryal, N., Bhattarai, B. R., Khadayat, K., & Parajuli, N. (2021). LC-HRMS profiling and anti-diabetic, antioxidant, and antibacterial activities of Acacia catechu (L.f.) Willd. *Biomed Research International, 2021*, 7588711. https://doi.org/10.1155/2021/7588711

Asaad, A. Y. (2023). Detection of important compounds from the aerial part of callistemon viminalis plant cultivated in Iraq. *Egyptian Journal of Hospital Medicine, 90*(1), 421–423. https://doi.org/10.21608/ejhm.2023.279657

Ashour, M. A., Fatima, W., Imran, M., Ghoneim, M. M., Alshehri, S., & Shakeel, F. (2022). A review on the main phytoconstituents, traditional uses, inventions, and patent literature of gum Arabic emphasizing *Acacia seyal*. *Molecules, 27*(4), 1171. https://doi.org/10.3390/molecules27041171

Ashraf, A., Bhatti, I. A., Sultana, B., & Jamil, A. (2016). Study of variations in the extraction yield, phenolic contents and antioxidant activities of the bark of F. Religiosa as a function of extraction procedure. *Journal of Basic & Applied Sciences, 12*, 2016. https://doi.org/10.6000/1927-5129.2016.12.02

Ashu, F., Na-Iya, J., Wamba, B. E. N., Kamga, J., Nayim, P., Ngameni, B., et al. (2020). Antistaphylococcal activity of extracts, fractions, and compounds of *Acacia polyacantha* Wild (Fabaceae). *Evidence-Based Complementary and Alternative Medicine, 2020*, 2654247. https://doi.org/10.1155/2020/2654247

Awadasseid, A., Adam, I. M., Jamal, M., & Teia, F. K. F. (2019). Physicochemical properties of Kakamut Gum (*Acacia polyacantha*) and Hashab Gum (*Acacia senegal*): A comparative analysis. *International Journal of Applied Agricultural Sciences, 5*(6), 138–143. https://doi.org/10.11648/j.ijaas.20190506.12

Azouz, R. A., & Hassanen, E. I. (2020). Modulating effect of gum Arabic on cisplatin-induced testicular damage in albino Wister rats. *Revista Brasileira De Farmacognosia, 30*, 90–98. https://doi.org/10.1007/s43450-020-00015-7

Babiker, M., Mohammed, M. E. A., Amanullah, M., Abbas, T., & Dafallah, M. S. (2018). Protective effect of long-term administration of gum Arabic on oxidative stress in hepatic tissue of diabetic rats. *Biomedical Journal of Scientific & Technical Research, 4*(5), 4153–4159. https://doi.org/10.26717/bjstr.2018.04.0001110

Babiker, E. E., Al-Juhaimi, F. Y., Alqah, H. A. S., Adisa, A. R., Adiamo, O. Q., Ahmed, I. A. M., et al. (2019). The effect of acacia nilotica seed extract on the physicochemical, microbiological and oxidative stability of chicken patties. *Journal of Food Science and Technology, 56*, 3910–3920. https://doi.org/10.1007/s13197-019-03862-y

Babiker, R., Kaddam, L., & Mariod, A. A. (2022). The role of gum Arabic as an anti-inflammatory, antioxidant, and immune modulator in the development of COVID-19: A review. *Functional Food Science, 2*(10), 2767–3146. https://doi.org/10.31989/ffs.v2i10.1019

Baker, D. L., Krol, E. S., Jacobsen, N. E., & Liebler, D. C. (1999). Reactions of B-carotene with cigarette smoke oxidants. Identification of carotenoid oxidation products and evaluation of the prooxidant/antioxidant effect. *Chemical Research in Toxicology, 12*(6), 535–543. https://doi.org/10.1021/tx980263v

Bakshi, J., Mehra, M., Grewal, S., Dhingra, D., & Kumari, S. (2022). Synthesis, characterization and evaluation of in-vitro antimicrobial and anti-diabetic activity of berberine encapsulated in guar-acacia gum nanocomplexes. *Journal of Bioactive and Compatible Polymers, 34*(4), 233–251. https://doi.org/10.1177/08839115221106700

Barak, S., Mudgil, D., & Taneja, S. (2020). Exudate gums: Chemistry, properties and food applications—A review. *Journal of the Science of Food and Agriculture, 100*(7), 2828–2835. https://doi.org/10.1002/jsfa.10302

Barkeer, S., Pothuraju, R., Malakar, P., Pimentel, T. C., Siddiqui, J. A., & Nair, S. A. (2024). Gum acacia dietary fiber: Significance in immunomodulation, inflammatory diseases, and cancer. *Phytotherapy Research, 38*(3), 1509–1521.

Batiha, G. E.-S., Akhtar, N., Alsayegh, A. A., Abusudah, W. F., Almohmadi, N. H., Shaheen, H. M., et al. (2022). Bioactive compounds, pharmacological actions, and pharmacokinetics of genus Acacia. *Molecules, 27*(21), 7340. Retrieved from https://www.mdpi.com/1420-3049/27/21/7340

Belmiro, R. H., Tribst, A. A. L., & Cristianini, N. (2017). Application of high-pressure homogenization on gums. *Journal of the Science of Food and Agriculture, 98*(6), 2060–2069. https://doi.org/10.1002/jsfa.8695

Ben-Amotz, A., Rachmilevich, B., Greenberg, S., Sela, M., & Weshler, Z. (1996). Natural B-carotene and whole body irradiation in rats. *Radiation and Environmental Biophysics, 35*(4), 285–288. https://doi.org/10.1007/s004110050041

Bhatia, S., Al-Harrasi, A., Shah, Y. A., Altoubi, H. W. K., Sharma, P., Anwer, M. K., et al. (2023). Fabrication, characterization, and antioxidant potential of sodium alginate/acacia gum hydrogel-based films loaded with cinnamon essential oil. *Gels, 9*(4), 337. https://doi.org/10.3390/gels9040337

Bouhlali, E. d. T., Derouich, M., Ben-Amar, H., Meziani, R., & Essarioui, A. (2020). Exploring the potential of using bioactive plant products in the management of Fusarium oxysporum f.sp. albedinis: The causal agent of Bayoud disease on date palm (Phoenix dactylifera L.). *Beni-Suef University Journal of Basic and Applied Sciences, 9,* 46. https://doi.org/10.1186/s43088-020-00071-x.

Calabrò, E., & Magazù, S. (2017). Induced-orientation of nitrogen monoxide and azide ion vibrations in human hemoglobin in bidistilled water solution under a static magnetic field. *Bioelectromagnetics, 38*(6), 447–455. https://doi.org/10.1002/bem.22056

Castro, T. L. A. d., Ferreira, B. D., & Cardoso, C. A. L. (2022). Photoprotective action, antioxidant activity, and toxicity of aqueous extracts of Campomanesia sessiliflora O Berg. *Revista Colombiana De Química, 50*(3), 10–15. https://doi.org/10.15446/rev.colomb.quim.v50n3.97095

Chen, W. T., Hsu, F. T., Liu, Y.-C., Chen, C.-H., Hsu, L.-C., & Lin, S.-S. (2019). Fluoxetine induces apoptosis through extrinsic/intrinsic pathways and inhibits ERK/NF-κB-modulated anti-apoptotic and invasive potential in hepatocellular carcinoma cells in-vitro. *International Journal of Molecular Sciences, 20*(3), 757. https://doi.org/10.3390/ijms20030757

Chen, L., Dong, H., Xing, W., & Cheng, Y. (2020). Extraction of flavonoids from Scutellariae radix using ultrasound-assisted deep eutectic solvents and evaluation of their anti-inflammatory activities. *ACS Omega, 5*(36), 23140–23147. https://doi.org/10.1021/acsomega.0c02898

Chmielowski, R. A., Meissner, S., Roush, D., Linden, T. O., Glowacki, E., Konietzko, J., & Nti-Gyabaah, J. (2014). Resolution of heterogeneous charged antibody aggregates via multimodal chromatography: A comparison to conventional approaches. *Biotechnology Progress, 30*(3), 636–645.

Costa, A. L. S., Freitas, L. A. P. d., Mendiola, J. A., & Ibáñez, E. (2015). Copaifera langsdorffii supercritical fluid extraction: Chemical and functional characterization by LC/MS and in-vitro assays. *The Journal of Supercritical Fluids, 100,* 86–96. https://doi.org/10.1016/j.supflu.2015.02.028

Cozzolino, D. (2009). Near infrared spectroscopy in natural products analysis. *Planta Medica, 75*(7), 746–756. https://doi.org/10.1055/s-0028-1112220

Dais, P., Spyros, A., Christophoridou, S., Hatzakis, E., Fragaki, G., Agiomyrgianaki, A., et al. (2007). Comparison of analytical methodologies based on ^{1}H and ^{31}P NMR spectroscopy with conventional methods of analysis for the determination of some olive oil constituents. *Journal of Agricultural and Food Chemistry, 55*(3), 577–584. https://doi.org/10.1021/jf061601y

D'Arcy, B. R., Rintoul, G. B., Rowland, C. Y., & Blackman, A. J. (1997). Composition of Australian honey extractives. 1. Norisoprenoids, monoterpenes, and other natural volatiles from blue gum (Eucalyptus leucoxylon) and Yellow box (Eucalyptus melliodora) honeys. *Journal of Agricultural and Food Chemistry, 45*(5), 1834–1843. https://doi.org/10.1021/jf960625+

Darniwa, A. V., Cahyanto, T., & Hidayah, S. N. (2022). Effect of mango leaf shoot extract (Mangifera indica L.) on zebrafish (Danio rerio) cell regeneration induced by hyperglicemia. *Jurnal Ilmu Hayat, 5*(2), 43–57. https://doi.org/10.17977/um061v5i22021p43-57

Desta, K. H., Tadese, E., & Molla, F. (2021). Physicochemical characterization and evaluation of the binding effect of Acacia etbaica Schweinf gum in granule and tablet formulations. *Biomed Research International, 2021*(3), 1–13. https://doi.org/10.1155/2021/5571507

Diallo, A. M., Ræbild, A., & Kjær, E. D. (2015). Study of quantitative genetics of gum Arabic production complicated by variability in ploidy level of Acacia senegal (L.) Willd. *Tree Genetics & Genomes, 11*, 80. https://doi.org/10.1007/s11295-015-0902-x

Dwivedi, M. K., Sonter, S., Mishra, S., Singh, P., & Singh, P. K. (2021). Secondary metabolite profiling and characterization of diterpenes and flavones from the methanolic extract of Andrographis paniculata using HPLC-LC-MS/MS. *Future Journal of Pharmaceutical Sciences, 7*, 184. https://doi.org/10.1186/s43094-021-00292-6

Elkhateeb, W. A., & Daba, G. M. (2022). Botryotrichum and Scopulariopsis secondary metabolites and biological activities. *Bio-technology and Bioprocessing, 3*(1), 067. https://doi.org/10.31579/2766-2314/067

Elmqvist, B., & Olsson, L. (2006). Livelihood diversification: Continuity and change in the Sahel. *GeoJournal, 67*, 167–180. https://doi.org/10.1007/s10708-007-9043-6

Elnour, A. A., Mirghani, M. E., Kabbashi, N. A., Alam, M. Z., & Musa, K. H. (2018a). Gum arabic: An optimization of ultrasonic-assisted extraction of antioxidant activity. *Studia Universitatis Babes-Bolyai, Chemia, 63*(3), 95–116.

Elnour, A., Mirghani, M., Kabbashi, N., Md Alam, Z., & Musa, K. (2018b). Study of antioxidant and anti-inflammatory crude methanol extract and fractions of Acacia seyal gum. *Journal of Pharmacology & Pharmacotherapeutics, 5*(1), 3.

Elnour, A. A. M., Mirghani, M. E. S., Na, K., Alam, Z., & Musa, K. H. (2018c). Study of antioxidant and anti-inflammatory crude methanol extract and fractions of Acacia seyal gum. *American Journal of Pharmacology and Pharmacotherapeutics, 5*(1), 3. https://doi.org/10.21767/2393-8862.100013

Elnour, A. A. M., Mirghani, M. E. S., & Kabbashi, N. A. (2019a). Significant bioactive compounds in crude methanol extracts and their fractions of Acacia seyal gum. *International Journal of Sudan Research, 8*(2), 99–112. https://doi.org/10.47556/j.ijsr.8.2.2018.3

Elnour, A. A., Mirghani, M. E. S., Kabbashi, N. A., Alam, M. Z., & Musa, K. H. (2019b). Active fractions of methanol crude obtained from Acacia seyal gum: Antioxidant capacity using FT-IR analysis. *Borneo Journal of Pharmacy, 2*(2), 94–107.

Elnour, A. A., Mirghani, M. E., Kabbashi, N., Daddiouaissa, D., Musa, K. H., Alam, M. Z., & Abdurahman, N. H. (2020). Active fractions of methanol crude obtained from Acacia seyal gum and their anti-proliferative effects against human breast cancer cell lines. *Global Journal of Science Frontier Research, 20*(3-C), 51–64.

Elnour, A. A., Mirghani, M. E., Kabbashi, N. A., Hamid Musa, K., Shahabipour, F., Ashammakhi, N., & Hamid Abdurahman, N. (2022a). Comparative study of the characterization and extraction techniques of polyphenolic compounds from Acacia seyal gum. *Food Quality and Safety, 6*, fyab034.

Elnour, A. A. M., Mirghani, M. E. S., Kabbashi, N. A., Musa, K. H., Shahabipour, F., Ashammakhi, N., & Abdurahman, N. H. (2022b). Comparative study of the characterization and extraction techniques of polyphenolic compounds from *Acacia seyal* gum. *Food Quality and Safety, 6*, fyab034. https://doi.org/10.1093/fqsafe/fyab034

Emwas, A. H., Roy, R., McKay, R. T., Tenori, L., Saccenti, E., Gowda, G. A. N., et al. (2019). NMR spectroscopy for metabolomics research. *Metabolites, 9*(7), 123. https://doi.org/10.3390/metabo9070123

Fall, D., Bakhoum, N., Sall, S. N., Zoubeirou, A. M., Sylla, S. N., & Diouf, D. (2016). Rhizobial inoculation increases soil microbial functioning and gum Arabic production of 13-year-old Senegalia senegal (L.) Britton, trees in the North Part of Senegal. *Frontiers in Plant Science, 7*, 1355. https://doi.org/10.3389/fpls.2016.01355

Fall, F., Le Roux, C., Ba, A. M., Fall, D., Bakhoum, N., Faye, M. N., et al. (2019). The rhizosphere of the halophytic grass *Sporobolus robustus* Kunth hosts rhizobium genospecies that are efficient on *Prosopis juliflora* (Sw.) DC and *Vachellia seyal* (Del.) PJH Hurter seedlings. *Systematic and Applied Microbiology, 42*(2), 232–239. https://doi.org/10.1016/j.syapm.2018.10.006

Fang, Y., Al-Assaf, S., Phillips, G. O., Nishinari, K., & Williams, P. A. (2010). Interaction of gum Arabic with fatty acid studied using electron paramagnetic resonance. *Biomacromolecules, 11*(5), 1398–1405. https://doi.org/10.1021/bm100219n

Farasara, R., & Fachriah, K. (2021). Physicochemical and sensory characteristic of starfruit-red guava fruit leather as affected by the addition of Arabic gum. *Jurnal Teknologi, 84*(1), 11–19. https://doi.org/10.11113/jurnalteknologi.v84.16642

Fazilah, N. F., Hamidon, N. H., Ariff, A., Khayat, M. E., Wasoh, H., & Halim, M. (2019). Microencapsulation of Lactococcus lactis Gh1 with gum Arabic and Synsepalum dulcificum via spray-drying for potential inclusion in functional yogurt. *Molecules, 24*(7), 1422. https://doi.org/10.3390/molecules24071422

García-Rubio, D., Mora, M. B. d. l., Badillo-Ramirez, I., Cerecedo, D., Sániger, J. M., Benítez-Benítez, J. L., & Villagrán-Muńiz, M. (2019). Analysis of platelets in hypertensive and normotensive individuals using Raman and Fourier-transform infrared-attenuated total reflectance spectroscopies. *Journal of Raman Spectroscopy, 50*(4), 509–521. https://doi.org/10.1002/jrs.5540

Gashua, I. B., Williams, P. A., Yadav, M. P., & Baldwin, T. C. (2015). Characterization and molecular association of Nigerian and Sudanese Acacia gum exudates. *Food Hydrocolloids, 51*, 405–413. https://doi.org/10.1016/j.foodhyd.2015.05.037

Gashua, I. B., Williams, P., & Baldwin, T. C. (2016). Molecular characteristics, association and interfacial properties of gum Arabic harvested from both Acacia senegal and Acacia seyal. *Food Hydrocolloids, 61*, 514–522. https://doi.org/10.1016/j.foodhyd.2016.06.005

Goudah, A., Abdo-El-Sooud, K., & Yousef, M. A. (2015). Acute and subchronic toxicity assessment model of Ferula assa-foetida gum in rodents. *Veterinary World, 8*(5), 584–589. https://doi.org/10.14202/vetworld.2015.584-589

Hammad, A. A., Abdelgadir, A. A., Yassin, S., Alzain, A. A., & Ahmed, E. M. (2022). GC-MS analysis, antibiogram and computational studies of Kakamout (Acacia polyacantha L.) leaves hydro-ethanolic extract. *Omdurman Journal of Pharmaceutical Sciences, 2*(3), 195–210. https://doi.org/10.52981/ojps.v2i3.2867

Hannachi, H. (2011). Chemicals profiling and antioxidants activities of acacia seeds. *Journal of Medicinal Plants Research, 5*(31), 6869–6875. https://doi.org/10.5897/jmpr11.1390

Hartati, H., Nasir, H. M., Idris, I. S., & Aziz, A. F. A. (2018). Evaluation of Swietenia mahagoni Jacq seed extracts in promoting wound-healing properties. *Malaysian Journal of Fundamental and Applied Sciences, 14*(4), 432–436. https://doi.org/10.11113/mjfas.v14n4.1094

Hassani, A., Azarian, M. M. S., Ibrahim, W. N., & Hussain, S. A. (2020). Preparation, characterization and therapeutic properties of gum arabic-stabilized gallic acid nanoparticles. *Scientific Reports, 10*(1), 17808. https://doi.org/10.1038/s41598-020-71175-8

Herdini, C., Miceli, S., Hariwiyanto, B., Wijayanti, N., Hosoyama, A., Yamazoe, A., et al. (2017). Secondary bioactive metabolite gene clusters identification of anticandida-producing

Streptomyces Sp. GMR22 isolated from Wanagama forest as revealed by genome mining approach. *Indonesian Journal of Pharmacy, 28*(1), 26–33. https://doi.org/10.14499/indonesianjpharm28iss1pp26

Higgs, R. E., Zahn, J. A., Gygi, J. D., & Hilton, M. D. (2001). Rapid method to estimate the presence of secondary metabolites in microbial extracts. *Applied and Environmental Microbiology, 67*(1), 371–376.

Hutagalung, R. A., Victor, Karjadidjaja, M., Prasasty, V. D., & Mulyono, N. (2014). Extraction and characterization of bioactive compounds from cultured and natural sponge, Haliclona molitba and Stylotella aurantium origin of Indonesia. *International Journal of Bioscience, Biochemistry and Bioinformatics, 4*(1), 14–18. https://doi.org/10.7763/ijbbb.2014.v4.302

Ibrahim, R. M., Abdelhafez, H. M., El-Shamy, S., Eid, F. A., & Mashaal, A. (2023). Arabic gum ameliorates systemic modulation in alloxan monohydrate-induced diabetic rats. *Scientific Reports, 13*, 5005. https://doi.org/10.1038/s41598-023-31897-x

Johnson, W. E., Fendinger, N. J., & Plimmer, J. R. (1991). Solid-phase extraction of pesticides from water: Possible interferences from dissolved organic material. *Analytical Chemistry, 63*(15), 1510–1513. https://doi.org/10.1021/ac00015a003

Kiprop, V. J., Omwamba, M., & Mahungu, S. M. (2021). Influence of gum Arabic from *Acacia senegal* var. *kerensis* on the modifications of pasting and textural properties of cassava and corn starches. *Food and Nutrition Sciences, 12*(11), 1098–1115. https://doi.org/10.4236/fns.2021.1211081

Kuck, L. S., & Noreña, C. P. Z. (2019). Application of gum Arabic, β-cyclodextrin, and hydroxypropyl-β-cyclodextrin to microencapsulation by molecular inclusion of grape skin extract (Vitis labrusca var. Isabel). *Journal of Food Processing and Preservation, 43*(2), e13874. https://doi.org/10.1111/jfpp.13874

Kull, C. A., Shackleton, C. M., Cunningham, P., Ducatillon, C., Dufour-Dror, J.-M., Esler, K. J., et al. (2011). Adoption, use and perception of Australian acacias around the world. *Diversity and Distributions*. https://doi.org/10.1111/j.1472-4642.2011.00783.x

Kumar, A., Malik, A. K., & Singh, B. (2021). Recent advances in the analysis of plant alkaloids by capillary electrophoresis and micellar electrokinetic chromatography. *Separation Science Plus, 4*(11–12), 401–412. https://doi.org/10.1002/sscp.202100040

Larson, R., Nelson, C., Korczak, R., Willis, H., Erickson, J., Wang, Q., & Slavin, J. L. (2021). Acacia gum is well-tolerated while increasing satiety and lowering peak blood glucose response in healthy human subjects. *Nutrients, 13*(2), 618. https://doi.org/10.3390/nu13020618

Lemma, W., Tekie, H., Balkew, M., Gebre-Michael, T., Warburg, A., & Hailu, A. (2014). Population dynamics and habitat preferences of Phlebotomus orientalis in extra-domestic habitats of Kafta Humera lowlands—Kala azar endemic areas in Northwest Ethiopia. *Parasites & Vectors, 7*, 359. https://doi.org/10.1186/1756-3305-7-359

Li, C., Du, C., Ma, F., & Zhang, J. (2015). Diagnosis of nitrogen status in Chinese cabbage (*Brassica rapa chinensis*) using the ratio of amide II to amide I in leaves based on mid-infrared photoacoustic spectroscopy. *Journal of Plant Nutrition and Soil Science, 178*(6), 888–895. https://doi.org/10.1002/jpln.201500283

Liu, H., Huang, X., Liu, Y., Jing, X., Ning, Y., Xu, P., et al. (2023). Efficient production of triacetic acid lactone from lignocellulose hydrolysate by metabolically engineered *Yarrowia lipolytica*. *Journal of Agricultural and Food Chemistry, 71*(48), 18909–18918. https://doi.org/10.1021/acs.jafc.3c06528

Luo, Z., Liu, H., & Shi, Z. G. (2016). Novel mode of liquid-phase microextraction: A magnetic stirrer as the extractant phase holder. *Journal of Separation Science, 39*(2), 399–404. https://doi.org/10.1002/jssc.201500789

Mahfoudhi, N., & Hamdi, S. (2014). Use of almond gum and gum Arabic as novel edible coating to delay post-harvest ripening and to maintain sweet cherry (*Prunus avium*) quality during storage. *Journal of Food Processing and Preservation, 39*(6), 1499–1508. https://doi.org/10.1111/jfpp.12369

Mahfoudhi, N., Sessa, M., Ferrari, G., Hamdi, S., & Donsì, F. (2015). Rheological and interfacial properties at the equilibrium of almond gum tree exudate (*Prunus dulcis*) in comparison with gum Arabic. *Food Science and Technology International, 22*(4), 277–287. https://doi.org/10.1177/1082013215595154

McNamee, B. F., O'Riorda, E. D., & O'Sullivan, M. (2001). Effect of partial replacement of gum Arabic with carbohydrates on its microencapsulation properties. *Journal of Agricultural and Food Chemistry, 49*(7), 3385–3388. https://doi.org/10.1021/jf001003y

Mediani, A., Aldosary, S. K., Ilowefah, M., Shahzad, R., Shehzad, A., Bilal, S., et al. (2022). Biological potentials and phytochemical constituents of raw and roasted Nigella arvensis and Nigella sativa. *Molecules, 27*(2), 550. https://doi.org/10.3390/molecules27020550

Metaab, M. H., Salman, A. H., & Al-Gharawi, J. K. (2023). The effect of gum Arabic and technomus as a prebiotics on the growth parameters of common carp Cyprinus carpio L. *IOP Conference Series: Earth and Environmental Science, 1262*, 072055. https://doi.org/10.1088/1755-1315/1262/7/072055

Mirghani, M. E. S., Musa, K. H., Mohammedelnour, A. A., Kabbashi, N. A., & Abdullah, A. (2017). Effect of solvent types on phenolics content and antioxidant activities of Acacia polyacantha gum. *International Food Research Journal, 24*, 369. Retrieved from https://agris.fao.org/search/en/providers/122640/records/647473982d3f560f80ab8d88

Mirghani, M. E., Elnour, A. A., Kabbashi, N., Alam, M. Z., Musa, K. H., & Abdullah, A. (2018). Determination of antioxidant activity of gum arabic: An exudation from two different locations. *Science Asia, 44*(2018), 179–186.

Mohammedelnour, A. A., Mirghani, M. E. S., Kabbashi, N. A., Alam, M. Z., Musa, K. H., & Aminah Abdullah, A. A. (2018). Effect of solvent types on phenolics content and antioxidant activities of Acacia polyacantha gum. *International Food Research Journal, 24*, 369. Retrieved from https://www.cabidigitallibrary.org/doi/full/10.5555/20183161374

Moradi, S., & Anarjan, N. (2019). Preparation and characterization of -tocopherol nanocapsules based on gum Arabic-stabilized nanoemulsions. *Food Science and Biotechnology, 28*(2), 413–421. https://doi.org/10.1007/s10068-018-0478-y

Moradi, S. E., Shabani, A. M. H., Dadfarnia, S., & Emami, S. (2016). Sulfonated metal organic framework loaded on iron oxide nanoparticles as a new sorbent for the magnetic solid-phase extraction of cadmium from environmental water samples. *Analytical Methods, 8*, 6337–6346. https://doi.org/10.1039/c6ay01692h

Mortensen, A., Aguilar, F., Crebelli, R., Domenico, A. D., Frutos, M. J., Galtier, P., et al. (2017). Re-evaluation of acacia gum (E 414) as a food additive. *EFSA Journal, 15*(4), e04741. https://doi.org/10.2903/j.efsa.2017.4741

Mwove, J., Gogo, L. A., Chikamai, B., Omwamba, M., & Mahungu, S. M. (2016). Preparation and quality evaluation of extended beef rounds containing gum Arabic from *Acacia senegal* var. *kerensis*. *Food and Nutrition Sciences, 7*, 977–988. https://doi.org/10.4236/fns.2016.711096

Navarrete-Carriola, D. V., Paz-González, A. D., Vázquez-Jimenez, L. K., De Luna-Santillana, E., Cruz-Hernández, M. A., Bandyopadhyay, D., & Rivera, G. (2024). Comparative analysis of a secondary metabolite profile from roots and leaves of *Iostephane heterophylla* by UPLC-MS and GC-MS. *ACS Omega, 9*(5), 5429–5439. https://doi.org/10.1021/acsomega.3c06800

Ngibad, K. (2019). Phytochemical screening of sunflower leaf (Helianthus annuus) and anting-anting (Acalypha indica Linn) plant ethanol extract. *Borneo Journal of Pharmacy, 2*(1), 689. https://doi.org/10.33084/bjop.v2i1.689

Noh, C. H. C., Azmin, N. F. M., & Amid, A. (2017). Principal component analysis application on flavonoids characterization. *Advances in Science Technology and Engineering Systems Journal, 2*(3), 435–440. https://doi.org/10.25046/aj020356

Oda, S., Kameda, A., Okanan, M., Sakakibara, Y., & Ohashi, S. (2015). Discovery of secondary metabolites in an extractive liquid-surface immobilization system and its application to high-throughput interfacial screening of antibiotic-producing fungi. *The Journal of Antibiotics, 68*(11), 691–697.

Oliveira, J. L. d., Campos, E. V. R., Pereira, A. d. E. S., Nunes, L. E. S., Silva, C. C. L. d., Pasquoto, T., et al. (2018). Geraniol encapsulated in chitosan/gum Arabic nanoparticles: A promising system for pest management in sustainable agriculture. *Journal of Agricultural and Food Chemistry, 66*(21), 5325–5334. https://doi.org/10.1021/acs.jafc.8b00331

Osman, A. A. (2016). Functional characteristics of gum from *Acacia nubica*. *Science Journal of Analytical Chemistry, 4*(6), 77–83. https://doi.org/10.11648/j.sjac.20160406.11

Otoupal, P. B., Geiselman, G. M., Oka, A., Barcelos, C. A., Choudhary, H., Dinh, D., et al. (2022). Advanced one-pot deconstruction and valorization of lignocellulosic biomass into triacetic acid lactone using Rhodosporidium toruloides. *Microbial Cell Factories, 21*, 254. https://doi.org/10.1186/s12934-022-01977-0

Pagano, C., Luzi, F., Ricci, M., Michele, A. D., Puglia, D., Ceccarini, M. R., et al. (2022). Wound dressing: Combination of Acacia gum/PVP/cyclic dextrin in bioadhesive patches loaded with grape seed extract. *Pharmaceutics, 14*(3), 485. https://doi.org/10.3390/pharmaceutics14030485

Paula, V., Pedro, S. I., Campos, M. G., Delgado, T., Estevinho, L. c. M., & Anjos, O. (2022). Special bioactivities of phenolics from Acacia dealbata L. with potential for dementia, diabetes and antimicrobial treatments. *Applied Sciences, 12*(3), 1022. https://doi.org/10.3390/app12031022

Pavlova, P. L., Минаков, A. B., Platonov, D. V., Жигарев, B. A., & Guzei, D. V. (2022). Supercritical fluid application in the oil and gas industry: A comprehensive review. *Sustainability, 14*(2), 698. https://doi.org/10.3390/su14020698

Pedreiro, S., Figueirinha, A., Sanches-Silva, A., & Ramos, F. (2021). Bioactive edible films and coatings based in gums and starch: Phenolic enrichment and foods application. *Coatings, 1*(11), 1393. https://doi.org/10.3390/coatings11111393

Podolak, I., Galanty, A., & Sobolewska, D. (2010). Saponins as cytotoxic agents: A review. *Phytochemistry Reviews, 9*, 425–474. https://doi.org/10.1007/s11101-010-9183-z

Rafi, H., & Dawar, S. (2015). Role of pellets and capsules of Acacia nilotica and Sapindus mukorossi in combination of seed bio-priming with microbial antagonists in the supression of root infecting pathogenic fungi and promotion of crop plants. *Journal of Plant Pathology & Microbiology, 6*(3), S3–007. https://doi.org/10.4172/2157-7471.1000s3-007.

Rajha, H. N., Ziegler, W., Louka, N., Hobaika, Z., Vorobiev, E., Boechzelt, H., & Maroun, R. G. (2014). Effect of the drying process on the intensification of phenolic compounds recovery from grape pomace using accelerated solvent extraction. *International Journal of Molecular Sciences, 15*(10), 18640–18658. https://doi.org/10.3390/ijms151018640

Renard, D., Lavenant-Gourgeon, L., Ralet, M.-C., & Sánchez, C. (2006). Acacia senegal gum: Continuum of molecular species differing by their protein to sugar ratio, molecular weight, and charges. *Biomacromolecules, 7*(9), 2637–2649. https://doi.org/10.1021/bm060145j

Renard, D., Lavenant-Gourgeon, L., Lapp, A., Nigen, M., & Sánchez, C. (2014). Enzymatic hydrolysis studies of arabinogalactan-protein structure from acacia gum: The self-similarity hypothesis of assembly from a common building block. *Carbohydrate Polymers, 112*, 648–661. https://doi.org/10.1016/j.carbpol.2014.06.041

Rodrigues, L. R., Teixeira, J. A., Mei, H. C. v. d., & Oliveira, R. (2006). Physicochemical and functional characterization of a biosurfactant produced by Lactococcus lactis 53. *Colloids and Surfaces B: Biointerfaces, 49*(1), 79–86. https://doi.org/10.1016/j.colsurfb.2006.03.003

Rohner, C., & Ward, D. (1999). Large mammalian herbivores and the conservation of arid *Acacia* stands in the Middle East. *Conservation Biology, 13*(5), 1162–1171. https://doi.org/10.1046/j.1523-1739.1999.97300.x

Rupérez, P., Ahrazem, O., & Leal, J. (2002). Potential antioxidant capacity of sulfated polysaccharides from the edible marine brown seaweed *Fucus vesiculosus*. *Journal of Agricultural and Food Chemistry, 50*(4), 840–845. https://doi.org/10.1021/jf010908o

Sadat, A., & Joye, I. J. (2020). Peak fitting applied to Fourier-transform infrared and Raman spectroscopic analysis of proteins. *Applied Sciences, 10*(17), 5918. https://doi.org/10.3390/app10175918

Safdar, B., Pang, Z. H., Liu, X. Q., Rashid, M. T., & Jatoi, M. A. (2023). Structural characterization, physicochemical and rheological characteristics of flaxseed gum in comparison with gum Arabic and xanthan gum. *Journal of Food Measurement and Characterization, 17*(3), 2193–2203. https://doi.org/10.1007/s11694-022-01750-2

Sajjad, H., Tolman, W. B., & Reineke, T. M. (2020). Block co-polymer pressure-sensitive adhesives derived from fatty acids and triacetic acid lactone. *Acs Applied Polymer Materials, 2*(7), 2719–2728. https://doi.org/10.1021/acsapm.0c00317

Saleem, H., Htar, T. T., Naidu, R., Anwar, S., Zengin, G., Locatelli, M., & Ahemad, N. (2020). HPLC–PDA polyphenolic quantification, UHPLC–MS secondary metabolite composition, and in-vitro enzyme inhibition potential of Bougainvillea glabra. *Plants, 9*(3), 388. https://doi.org/10.3390/plants9030388

Salem, R. R., El-Sahra, D. G., Osman, H.-E. H., & El-Kenawy, A. E.-M. (2020). Arabic gum role in ameliorating cyclosporine- A initiated nephrotoxicity in albino rats. *Egyptian Journal of Histology, 44*(3), 710–718. https://doi.org/10.21608/ejh.2020.37873.1339

Sampaopan, Y., & Suksaeree, J. (2022). Microwave-assisted extraction and content determination of astilbin in Lysiphyllum strychnifolium stems. *Brazilian Journal of Pharmaceutical Sciences, 58*, e20577. https://doi.org/10.1590/s2175-97902022e20577

Sánchez, C., Schmitt, C., Kolodziejczyk, E., Lapp, A., Gaillard, C., & Renard, D. (2008). The acacia gum arabinogalactan fraction is a thin oblate ellipsoid: A new model based on small-angle neutron scattering and ab initio calculation. *Biophysical Journal, 94*(2), 629–639. https://doi.org/10.1529/biophysj.107.109124

Sánchez, C., Nigen, M., Tamayo, V. M., Doco, T., Williams, P., Amine, C., & Renard, D. (2018). Acacia gum: History of the future. *Food Hydrocolloids, 78*, 140–160. https://doi.org/10.1016/j.foodhyd.2017.04.008

Sánchez-Hernández, E., Buzón-Durán, L., Andrés-Juan, C., Lorenzo-Vidal, B., Martin-Gil, J., & Martin-Ramos, P. (2021). Physicochemical characterization of Crithmum maritimum L. and Daucus carota subsp. gummifer (Syme) Hook.fil. and their antimicrobial activity against apple tree and grapevine phytopathogens. *Agronomy, 11*(5), 886. https://doi.org/10.3390/agronomy11050886

Satheeshkumar, N., Nisha, N., Nirmal, S., Nirmal, J., Jain, G., & Spandana, V. (2012). Analytical profiling of bioactive constituents from herbal products, using metabolomics—A review. *Natural Product Communications, 7*(8), 1111–1115. https://doi.org/10.1177/1934578x1200700837

Schwendel, B. H., Anekal, P. V., Zarate, E., Bang, K. W., Guo, G., Gray, A. C., & Pinu, F. R. (2021). Mass spectrometry-based metabolomics to investigate the effect of mechanical shaking on Sauvignon blanc berry metabolism. *Journal of Agricultural and Food Chemistry, 69*(16), 4918–4933.

Sekhon, B. K., Roubin, R. H., Li, Y., Devi, P. B., Nammi, S., & Fan, K. (2016). Evaluation of selected immunomodulatory glycoproteins as an adjunct to cancer immunotherapy. *PLoS One, 11*(1), e0146881. https://doi.org/10.1371/journal.pone.0146881

Sethuraman, S., & Rajendran, K. (2018). Multicharacteristic behavior of tyrosine present in the microdomains of the macromolecule gum Arabic at various pH conditions. *ACS Omega, 3*(12), 17602–17609. https://doi.org/10.1021/acsomega.8b02928

Severcan, F., Kaptan, N., & Turan, B. (2003). Fourier-transform infrared spectroscopic studies of diabetic rat heart crude membranes. *Journal of Spectroscopy, 17*(3), 569–577. https://doi.org/10.1155/2003/472834

Sharam, G. J., Sinclair, A., Turkington, R., & Jacob, A. L. (2009). The savanna tree Acacia polyacantha facilitates the establishment of riparian forests in Serengeti National Park Tanzania. *Journal of Tropical Ecology, 25*(1), 31–40.

Shimoda, R., Okabe, K., Kotake, T., Matsuoka, K., Koyama, T., Tryfona, T., et al. (2014). Enzymatic fragmentation of carbohydrate moieties of radish arabinogalactan-protein and elucidation of the structures. *Bioscience Biotechnology and Biochemistry, 78*(5), 818–831. https://doi.org/10.1080/09168451.2014.910100

Shin, K., Oh, S. H., Kim, T. Y., Yoon, B., Park, S. S., & Suh, H. J. (2008). Changes in chemical composition of Korean red ginseng (Panax ginseng C.A. Meyer) extract with alcohol extraction. *Preventive Nutrition and Food Science, 13*(3), 212–218. https://doi.org/10.3746/jfn.2008.13.3.212

Silva, T. M. e., & Souza, J. P. F. d. (2018). The influence of environmental factors in the production of gum Arabic (Acacia senegal, Fabaceae)—A meta-analysis. *Revista Brasileira De Climatologia, 23*, 33–44. https://doi.org/10.5380/abclima.v23i0.58884

Silva, F., Martins, A., Salta, J., Neng, N. R., Nogueira, J. M. F., Mira, D., et al. (2009). Phytochemical profile and anticholinesterase and antimicrobial activities of supercritical versus conventional extracts of Satureja montana. *Journal of Agricultural and Food Chemistry, 57*(24), 11557–11563. https://doi.org/10.1021/jf901786p

Silva, F. C. O., Malaisamy, A., Cahú, T. B., de Araújo, M. I. F., Soares, P. A. G., Vieira, A. T., & dos Santos Correia, M. T. (2023). Polysaccharides from exudate gums of plants and interactions with the intestinal microbiota: A review of vegetal biopolymers and prediction of their prebiotic potential. *International Journal of Biological Macromolecules, 254*(2), 127715. Retrieved from https://www.sciencedirect.com/science/article/pii/S0141813023046135files/965/S0141813023046135.html

Stevanovic, S., & Bohley, P. (2001). Proteome analysis by three-dimensional protein separation: Turnover of cytosolic proteins in hepatocytes.

Sun, X., Sui, Q., Zhang, C., Tian, Z., & Zhang, J. (2013). Targeting blockage of STAT3 in hepatocellular carcinoma cells augments NK cell functions via reverse hepatocellular carcinoma–induced immune suppression. *Molecular Cancer Therapeutics, 12*(12), 2885–2896. https://doi.org/10.1158/1535-7163.mct-12-1087

Tamayo, V. M., Nigen, M., Apolinar-Valiente, R., Doco, T., Williams, P., Renard, D., & Sánchez, C. (2018). Flexibility and hydration of amphiphilic hyperbranched arabinogalactan-protein from plant exudate: A volumetric perspective. *Colloids and Interfaces, 2*(1), 11. https://doi.org/10.3390/colloids2010011

Tavan, M., Hanachi, P., Mirjalili, M. H., & Dashtbani-Roozbehani, A. (2023). Comparative assessment of the biological activity of the green synthesized silver nanoparticles and aqueous leaf extract of Perilla frutescens (L.). *Scientific Reports, 13*, 6391. https://doi.org/10.1038/s41598-023-33625-x

Thotathil, V., Rizk, H. H., Fakrooh, A., & Sreerama, L. (2022). Phytochemical analysis of Acaciaehrenbergiana (Hayne) grown in Qatar: Identification of active ingredients and their biological activities. *Molecules, 27*(19), 6400. https://doi.org/10.3390/molecules27196400

Tiwari, M., Panghal, A., Mittal, V., & Gupta, R. (2023). Bioactive compounds of acacia, health benefits and its utilization in food processing industry: a critical review. *Nutrition & Food Science, 53*(7), 1125–1146. Retrieved from https://www.emerald.com/insight/content/doi/10.1108/NFS-08-2022-0274/full/html

Traoré, M., Kaal, J., & Cortizas, A. M. (2017). Differentiation between pine woods according to species and growing location using FT-IR-ATR. *Wood Science and Technology, 52*(2), 487–504. https://doi.org/10.1007/s00226-017-0967-9

Velázquez-Martínez, V., Valles-Rosales, D. J., Rodriguez-Uribe, L., Holguin, O., Quiroz, J. Q., Reyes-Jaquez, D., et al. (2021). Antimicrobial, shelf-life stability, and effect of maltodextrin and gum Arabic on the encapsulation efficiency of sugarcane bagasse bioactive compounds. *Food, 10*(1), 116. https://doi.org/10.3390/foods10010116

Vinci, G., Antonelli, M. L., & Preti, R. (2013). Rapid determination of polycyclic aromatic hydrocarbons in rainwater by liquid–liquid microextraction and LC with core-shell particles column and fluorescence detection. *Journal of Separation Science, 36*(3), 461–468. https://doi.org/10.1002/jssc.201200854

Wang, Y., Ke, L., Yang, Q., Peng, Y., Hu, Y. G., Dai, L., et al. (2019). Biorefinery process for production of bioactive compounds and bio-oil from Camellia oleifera shell. *International Journal of Agricultural and Biological Engineering, 12*(5), 190–194. https://doi.org/10.25165/j.ijabe.20191205.4593

Wang, R., Wang, Y. H., Guo, W., & Zeng, M. Y. (2021). Stability and bioactivity of carotenoids from Synechococcus sp. PCC 7002 in Zein/NaCas/Gum Arabic composite nanoparticles fabricated by pH adjustment and heat treatment antisolvent precipitation. *Food Hydrocolloids, 117*, 106663. https://doi.org/10.1016/j.foodhyd.2021.106663

Waty, D. R., Saputri, F. C., & Mun'im, A. (2017). Secondary metabolites screening and acute toxicity test of Peperomia pellucida(L.) Kunth methanolic extracts. *International Journal of Pharmtech Research, 10*(1), 31–38. https://doi.org/10.20902/ijptr.2017.1014

Winders, J., & Pechan, T. (2021). Comparison of homogenization methods for extraction of maize cob metabolites. *African Journal of Biotechnology, 20*(3), 108–114. https://doi.org/10.5897/ajb2020.17246

Yakout, A. A., & El-Hady, D. A. (2016). A combination of B-cyclodextrin functionalized magnetic graphene oxide nanoparticles with B-cyclodextrin-based sensor for highly sensitive and selective voltammetric determination of tetracycline and doxycycline in milk samples. *RSC Advances, 6*, 41675–41686. https://doi.org/10.1039/c6ra03787a

Yamasaki, S., & Dillenburg, L. R. (1999). Measurements of leaf relative water content in Araucaria angustifolia. *Revista Brasilleira de fisiologia vegetal, 11*(2), 69–75.

Yavari, N., Mazaheri-Assadi, M., Mazhari, Z. H., Moghadam, M. B., & Larijani, K. (2018). Glucuronic acid rich kombucha-fermented pomegranate juice. *Journal of Food Research, 7*(1), 61–69.

Youzbachi, N., Trabelsi, H., Elfalleh, W., Khaldi, A., Nasri, N., & Tlili, N. (2019). Fatty acids and triacylglycerols composition from Tunisian AGs seed oil. *Arabian Journal of Chemistry, 12*(8), 3302–3308. https://doi.org/10.1016/j.arabjc.2015.08.020

Yu, S., Ho, C. T., Chin, C. K., Badmaev, V., Ma, W., & Huang, M. T. (1998). Inhibitory activity of Boswellic acids from *Boswellia serrata* against human leukemia HL-60 cells in culture. *Planta Medica, 64*(4), 328–331. https://doi.org/10.1055/s-2006-957444

Yu, N. X., Wang, J. S., Jiang, C. J., Nie, X. H., Hu, Z. Y., Ye, Q., et al. (2022). Development of composite nanoparticles from gum Arabic and carboxymethylcellulose-modified *Stauntonia brachyanthera* seed albumin for lutein delivery. *Food Chemistry, 372*, 131269. https://doi.org/10.1016/j.foodchem.2021.131269

Yuan, Y., Zheng, S., Linhui, Z., Deng, Z., Zhang, B., & Li, H. (2019). The phenolic compounds, metabolites, and antioxidant activity of propolis extracted by ultrasound-assisted method. *Journal of Food Science, 84*(12), 3850–3865. https://doi.org/10.1111/1750-3841.14934

Zaman, G. S., Alshahrani, M. Y., Barua, P., Aladel, A., Zaman, F. A., Ahmad, I., et al. (2022). Screening of the antioxidant and antibacterial effects of extracted essential oils from Thunbergia coccinea, Acacia polyacantha, Polygonum micrpcephallum, Abies spectabilis and Clerodendrum colebrookianum. *Cellular and Molecular Biology, 67*(4), 56–67. https://doi.org/10.14715/cmb/2021.67.4.7

Zamani, H., Rastegari, B., & Varamini, M. (2019). Antioxidant and anti-cancer activity of Dunaliella salina extract and oral drug delivery potential via nano-based formulations of gum Arabic coated magnetite nanoparticles. *Journal of Drug Delivery Science and Technology, 54*, 101278. https://doi.org/10.1016/j.jddst.2019.101278

Zha, W., Shao, Z., Frost, J. W., & Zhao, H. (2004). Rational pathway engineering of type I fatty acid synthase allows the biosynthesis of triacetic acid lactone from D-glucose in-vivo. *Journal of the American Chemical Society, 126*(14), 4534–4535. https://doi.org/10.1021/ja0317271

Zhou, X., Sun, P., Bucheli, P., Huang, T., & Wang, D. (2009). FT-IR methodology for quality control of (AGP) extracted from green tea (Camellia sinensis). *Journal of Agricultural and Food Chemistry, 57*(12), 5121–5128. https://doi.org/10.1021/jf803707a

Zigene, Z. D., Asfaw, B. T., & Bitima, T. D. (2019). Optimizing DNA isolation protocol for rosemary (Rosemarinus officinalis L) accessions. *African Journal of Biotechnology, 18*(30), 895–900. https://doi.org/10.5897/ajb2019.16925

Žlabur, J. Š., Žutić, I., Radman, S., Pleša, M., Brnčić, M., Barba, F. J., et al. (2020). Effect of different green extraction methods and solvents on bioactive components of chamomile (Matricaria chamomilla L.) flowers. *Molecules, 25*(4), 810. https://doi.org/10.3390/molecules25040810

Chapter 4
Biological Applications of Secondary Metabolites Extract (SME) from Acacia Gums (AGs)

Ahmed A. M. Elnour, Abdurahmani Hamid Nour, and Khogali E. A. Ishag

Abstract SME from *Acacia sengal*, *Acacia seyal*, and *Acacia polyacantha* gums, also known as the Acacia gums (AGs), exhibit various biological applications. This chapter explored the antimicrobial, anti-inflammatory, antioxidant, neuroprotective, anti-cancer, immunomodulatory, anti-diabetic, cardiovascular protective, hepatoprotective, wound-healing, pain-relief, anti-aging, antiviral, anti-obesity, and antifungal properties of these metabolites. Gums contain diverse bioactive compounds including glycosides, alkaloids, flavonoids, arabic acid, polyphenols, hydrocarbons, phenolic acids, and fatty acids. These compounds exert antimicrobial effects through metabolic perturbations by targeting specific cellular pathways or structures within microbial cells. SME biosynthesis involves biosynthetic gene clusters (BGCs) and complex hierarchical regulatory networks, which are influenced by environmental factors and beneficial microbes associated with the plant. Studies have demonstrated the antimicrobial activity of AGs against various bacteria and fungi, highlighting their potential as natural antimicrobial agents. This chapter emphasizes the promising potential of SME extracted from AGs in improving human health and well-being, discussing their roles in treating chronic diseases, enhancing healing, and extending health. Future research directions and the potential of these metabolites in medical applications are also addressed, underscoring the significance of these AGs in various fields, including medicine, agriculture, and ecology.

A. A. M. Elnour (✉) · A. H. Nour
Faculty of Chemical and Process Engineering Technology, Universiti Malaysia Pahang Al-Sultan Abdullah (UMPSA), Pahang Gambang, Malaysia

Centre of Excellence for Advanced Research in Fluid Flow (CARIFF), University Malaysia, Pahang Gambang, Malaysia
e-mail: adamhassan@umpsa.edu.my; abrahman@umpsa.edu.my

K. E. A. Ishag
Department of Food Science and Technology, Faculty of Agriculture, University of Khartoum, Khartoum, Sudan

A. A. M. Elnour (ed.), *Gum Arabic and Breast Cancer Biology*,
https://doi.org/10.1007/978-981-97-8518-6_4

Keywords *Acacia polyacantha* gum · *Acacia senegal* gum · *Acacia seyal* gum · Antimicrobial activity · Bioactive compounds · Therapeutic properties

Highlights
1. AGs-SMEs have diverse biological applications.
2. Gums contain bioactive compounds such as glycosides, alkaloids, flavonoids, arabic acid, polyphenols, hydrocarbons, phenolic acids, and fatty acids.
3. Antimicrobial effects are exerted through metabolic perturbations that target specific cellular pathways or structures within the microbial cells.
4. The biosynthesis of AGs-SME involves BGCs and complex hierarchical regulatory networks.
5. Environmental factors and beneficial microbes associated with the plant influence the production of AGs-SME.
6. Studies have demonstrated the antimicrobial activity of AGs against various bacteria and fungi.
7. AGs-SME show promising potential as a natural antimicrobial agent for therapeutic applications.

4.1 Introduction

Acacia senegal and *Acacia seyal* are notable for their nitrogen fixation capabilities and the production of gum arabic, a substance derived from trees that is rich in soluble fibers and exhibits prebiotic properties (Bakhoum et al., 2014; Cherbut et al., 2003; Silva & Souza, 2018). Gum arabic also contains proteins that contribute to its prebiotic functionality (Calame et al., 2008; Gashua et al., 2016) and has demonstrated antioxidant and therapeutic effects (Elhasan et al., 2022; Elnour et al., 2018a). It is widely used in food, pharmaceutical, and cosmetic industries (Sethuraman & Rajendran, 2018). Furthermore, gum from *Acacia seyal* has been studied for its bioactive compounds and anti-proliferative effects, highlighting its potential in both food products and traditional medicine (Elnour et al., 2019, 2020; Ashour et al., 2022; Dauqan & Abdullah, 2013). Similarly, gums from *Acacia polyacantha* are valued for their yield and have been explored for their various properties (Adam & Fadl, 2011; Mhinzi, 2003). Thus, the multifunctional uses of gums from these AGs underscore their significant potential for diverse industrial applications, alongside their traditional medicinal roles.

AGs-SME plays essential roles in plant biological activities and has potential pharmacological benefits. These metabolites encompass a variety of compounds, including alkaloids, flavonoids, terpenes, tannins, and phenolic acids, which contribute to the antioxidant, antimicrobial, and anti-proliferative properties of plants (Batiha et al., 2022; Elhasan et al., 2022; Saheed, 2021). These metabolites not only support the traditional medicinal uses of these plants but also highlight their potential in drug discovery (Adhikari et al., 2021; Binyane & Mfengwana, 2022).

Furthermore, AGs-SME in these gums are linked to their nutritional quality, anthelmintic properties, and value in animal feed (Abraham et al., 2023; Satrija et al., 2023). The rich phytochemical composition of these AGs underscores their significance in medicine, agriculture, and ecology. Therefore, the diverse array of AGs-SME in Acacia gums underscores their traditional and pharmacological relevance and enhances their utility in nutrition and agriculture. The primary objective of this chapter is to explore the biological applications of AGs-SME, focusing on their antimicrobial, anti-inflammatory, antioxidant, and therapeutic properties. We examined the roles of these metabolites in treating chronic diseases, enhancing healing processes, and extending health spans. This chapter concludes with a discussion of future research directions and the potential of these metabolites for medical applications.

4.2 Biological Applications

4.2.1 Antimicrobial Effect of AGs-SME

It is crucial to consider the diverse bioactive compounds in AGs to assess the effectiveness of their AGs-SME in antimicrobial activities compared with conventional agents. These metabolites, including glycosides, alkaloids, flavonoids, and arabic acid, have shown significant antimicrobial potential against various pathogens. Studies have confirmed the bioactivity of AGs against a wide range of bacteria and fungi (Aryal et al., 2021; Thabet et al., 2018). Furthermore, the extraction and identification of these metabolites are essential to understand their antimicrobial properties and potential therapeutic applications (Binyane & Mfengwana, 2022). The presence of various AGs-SME underscores their potential as natural antimicrobial agents (Batiha et al., 2022). Thus, the promising antimicrobial properties of acacia metabolites underscore their potential as effective alternatives to conventional antimicrobial agents, warranting further exploration and development.

The AGs are rich in various SME- and phenolic compounds. *Acacia senegal* gum, for instance, contains a complex polysaccharide with branches linked through the 1,6-positions, including arabinose, rhamnose, and uronic acids (Fauconnier et al., 2000). *Acacia seyal* gum contains phenolic compounds such as ferulic acid, *p*-coumaric acid, caffeic acid, and sinapic acid (Elnour et al., 2019). These gums also exhibit antimicrobial activity with bioactive compounds such as polyphenols, hydrocarbons, phenolic acids, and fatty acids (Elnour et al., 2020). Additionally, the concentrations of these compounds vary among species, reflecting differences in the protein content and methoxyl levels (Elnour et al., 2019, 2020; Fauconnier et al., 2000). Therefore, the complex biochemical composition of Acacia gums contributes to their antimicrobial efficacy and highlights their potential for tailored applications in nutraceuticals and pharmaceuticals based on specific compound profiles.

The antimicrobial effects of AGs-SME are exerted through metabolic perturbations resulting from the direct or indirect action of these compounds (Sieniawska & Georgiev, 2022). These metabolites may target specific cellular pathways or structures within microbial cells, potentially affecting metabolic adaptations that are crucial for biofilm formation by gram-positive pathogens (Gélinas et al., 2020). The biosynthesis of SME, such as those found in AGs, involves BGCs and complex hierarchical regulatory networks that integrate various signals to control the expression of secondary metabolite pathways (Kallifidas et al., 2018; Yan, 2024). Additionally, SME production is influenced by environmental factors and beneficial microbes associated with the plant (Badmi, 2023). As a result, the sophisticated biosynthetic capabilities of AGs not only facilitate the production of potent antimicrobial agents, but also offer potential avenues for enhancing these properties through agricultural or genetic strategies.

The antimicrobial activities of various AGs, including *Acacia baileyana*, *Acacia dealbata*, and *Acacia melanoxylon*, have been extensively studied in a range of bacteria and fungi (Silva et al., 2016). In particular, extracts from *Acacia nilotica* are effective against carbapenem-resistant gram-negative bacteria and exhibit significant antimicrobial activity against pathogens such as *Escherichia coli*, *Bacillus cereus*, and *Candida albicans* (Gorish & Alamein, 2021; Silva et al., 2016). The potential applications of these antimicrobial agents are promising, especially in the clinical setting for treating bacterial infections. However, further research is necessary to investigate the safety profile and toxicity of these extracts to determine their suitability for in vivo use (Gorish & Alamein, 2021). Thus, while the antimicrobial properties of *A. nilotica* extracts show significant therapeutic potential, rigorous assessments are required to ensure their safety and efficacy in clinical applications. The proposed antimicrobial therapy strategies using AGs-SME are shown in Fig. 4.1.

4.2.2 Anti-inflammatory Properties of AGs-SME

AGs-SME has demonstrated promising anti-inflammatory activities against various pathogens. These gums are rich in bioactive compounds, such as alkaloids, saponins, flavonoids, and tannins, which contribute to their antimicrobial properties (Duarte & Wolf, 2005; Bhatia et al., 2023). The presence of SME enhances their effectiveness in inhibiting bacterial growth and biofilm production (Beck, 2024; Sarkisian et al., 2011). Moreover, the antimicrobial activities of these metabolites have been demonstrated both individually and in combination with other agents, demonstrating their potential as natural alternatives to conventional antimicrobial agents (Jha et al., 2022; Lim et al., 2016; Okonkwo-Uzor et al., 2022). Acacia gum exudates from AGs have also been studied for their anti-inflammatory properties and the presence of SME. These gums exhibit pharmacological activities, such as anti-inflammatory, antioxidant, and anti-proliferative

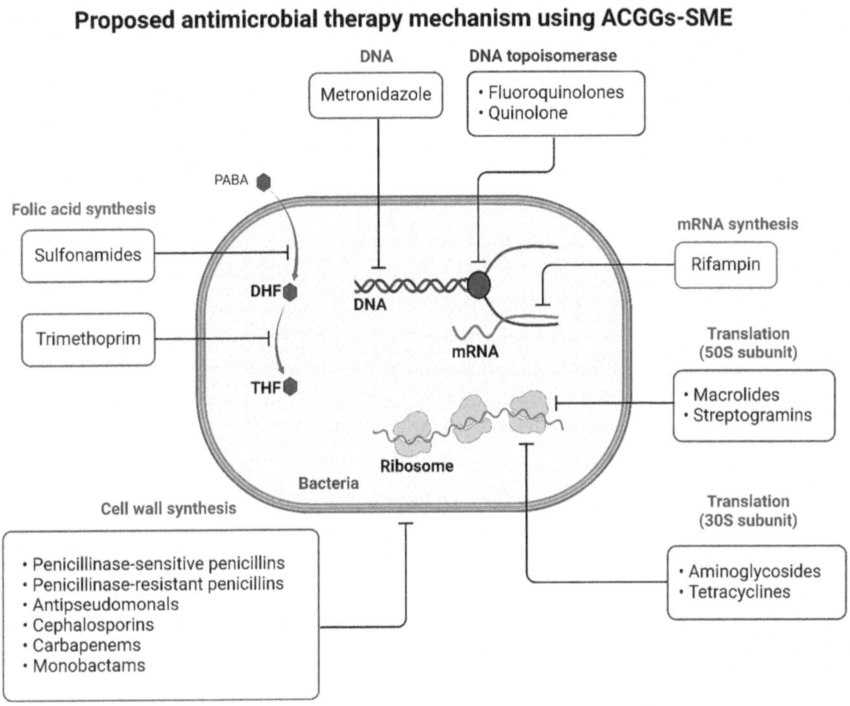

Fig. 4.1 Schematically presents the proposed antimicrobial therapy strategies using the Acacia gum secondary metabolite extraction (AGs-SME)

effects (Elnour et al., 2019; Jaafar, 2019). Gum exudates contain bioactive compounds with potential therapeutic benefits (Elnour et al., 2019, 2020). Additionally, Acacia gum has been found to modulate gut microbiota and promotes the production of short-chain fatty acids, further underscoring its health benefits (Rawi et al., 2021). The molecular composition and properties of gums from different AGs vary, which affects their functionality (Gashua et al., 2016). Thus, the diverse pharmacological properties of AGs highlight their therapeutic potential and suggest a promising future for the development of natural antimicrobial and anti-inflammatory agents.

The anti-inflammatory properties of AGs are facilitated by modulation of various cellular mechanisms. These gums contain compounds that effectively inhibit the initial stages of inflammation by suppressing the synthesis of prostaglandins and targeting inflammatory mediators, such as histamine, serotonin, and kinin (Fazmiya, 2024). Moreover, they serve as multi-target agents, influencing several signaling pathways, including NF-κB, p38/mitogen-activated protein kinase (MAPK), JNK/STAT3, and PI3K/Akt, which help downregulate inflammatory processes (Mueller et al., 2022). Additionally, these gums have been shown to reduce cellular infiltration and granulomatous inflammation by targeting cellular structures in experimentally induced arthritic rats (Ahmed & Mustafa, 2019). The overarching mechanism of their anti-inflammatory effects involves the precise targeting of cellular pathways and structures to effectively modulate the inflammatory response. Thus, the comprehensive anti-inflammatory capabilities of AGs highlight their potential as a multi-faceted therapeutic resource for managing inflammation through targeted cellular interventions.

The anti-inflammatory efficacy of AGs-SME has been corroborated by various studies, showing effects similar to those of established anti-inflammatory agents (Elhasan et al., 2022; Elnour et al., 2020; Shehu et al., 2020). Notably, *Acacia seyal* gum exhibits synergistic effects when combined with other anti-inflammatory drugs, thereby enhancing its therapeutic potential (Shehu et al., 2020). The beneficial pharmacological properties of these gums, including their antioxidant and immunomodulatory effects, significantly contribute to their anti-inflammatory actions (Elhasan et al., 2022). The bioactive compounds in these gum extracts are crucial for their anti-inflammatory and therapeutic effects (Elnour et al., 2019). Therefore, the proven anti-inflammatory performance of AGs extracts, particularly in synergy with other agents, underscores their viability as effective components in anti-inflammatory treatments, with multiple modes of action enhancing their therapeutic profiles. The anti-inflammatory properties of AGs are facilitated through the modulation of various cellular mechanisms, which result from the activation of the NF-kB/MAPK intracellular pathways, as shown in Fig. 4.2.

4.2.3 Antioxidant Properties of AGs-SME

The antioxidant properties of AGs have garnered considerable attention in recent studies. In particular, gums from *Acacia senegal* and *Acacia seyal* were noted for their antioxidant capabilities. Gum from *Acacia senegal* offers better emulsification than that from Acacia seyal (Gashua et al., 2016). Additionally, investigations of *Acacia seyal* gum revealed its ability to inhibit human breast cancer cell growth and demonstrated substantial antioxidant effects (Elnour et al., 2020). Research on *Acacia polyacantha* gum has also confirmed its rich content of phenols and flavonoids, contributing to its strong antioxidant properties (Khaled & Irani, 2021). Thus, the significant antioxidant capabilities of AGs underscore their potential as effective natural agents for cancer prevention and health promotion.

The antioxidant effects of AGs are linked to their rich composition of bioactive compounds, such as phenolic acids, alkaloids, terpenes, tannins, and flavonoids, which are known for their robust antioxidant and free radical scavenging activities (Zweil et al., 2015). Acacia gum, in particular, contains polysaccharides and glycoproteins, which are found to enhance immune functions (Mohamed, 2023). Research has not only underscored the antioxidant properties of acacia seeds but also the use of acacia gum to create antioxidant-active wound dressings (Elnour et al., 2018a; Mirghani et al., 2018). Furthermore, Acacia gum has been proven to mitigate oxidative stress, as demonstrated by studies showing reduced oxidative damage from

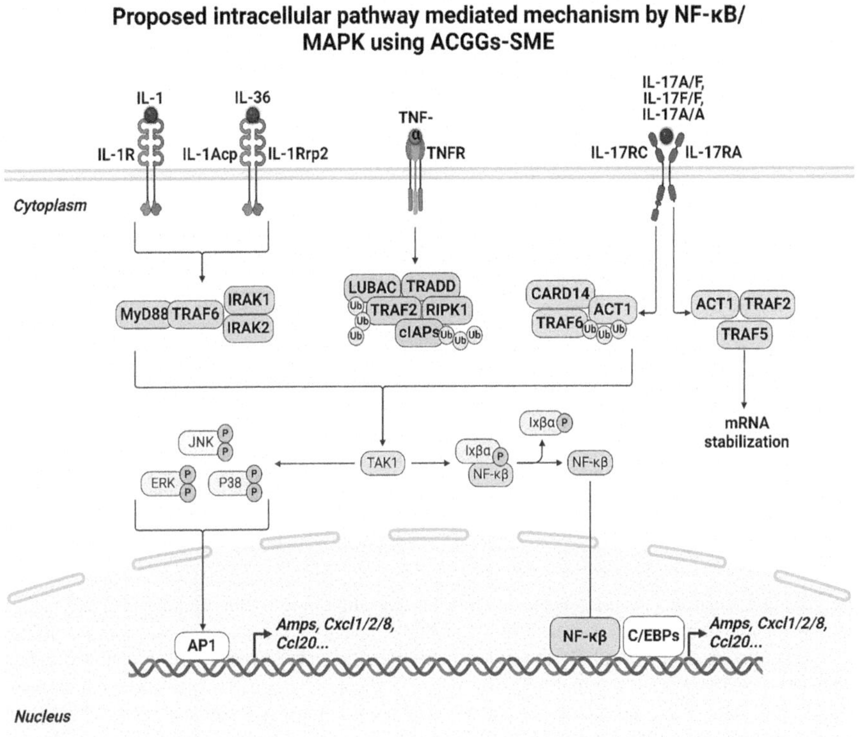

Fig. 4.2 Demonstrates how the anti-inflammatory properties of AGs-SME are mediated through the modulation of various cellular processes triggered by the activation of NF-kB/mitogen-activated protein kinase (MAPK) intracellular pathways

gentamicin with acacia gum treatment (AL-Mousawi, 2021). It has also shown promise as an antioxidant-rich hydrogel wound dressing (Sekhon et al., 2016). In addition, significant antioxidant properties and bioactive compounds have been identified in *Acacia ehrenbergiana* (Thotathil et al., 2022). Therefore, the comprehensive antioxidant activity of Acacia gums and seeds, driven by their diverse bioactive components, highlights their critical roles in oxidative stress management and therapeutic applications.

AGs-SME is recognized for its antioxidant properties because of its phenolic compounds, flavonoids, and other bioactive elements. These substances exhibit anti-inflammatory, antimicrobial, and anti-carcinogenic properties, enhancing their overall health benefits (Abu-Serie et al., 2021; Batiha et al., 2022). Specifically, *Acacia senegal* gum, noted for its prebiotic and antioxidant effects, has been studied for its ability to reduce oxidative damage and systemic toxicity, emphasizing its protective properties (Mohamed, 2023). The structure of these gums, which include phenolic rings and hydroxyl groups, lends them antioxidant and protein-binding functions, reinforcing their antioxidant role (Tchinda et al., 2020). Additionally, research on the antioxidant properties of *Acacia polyacantha* gum extracts suggests

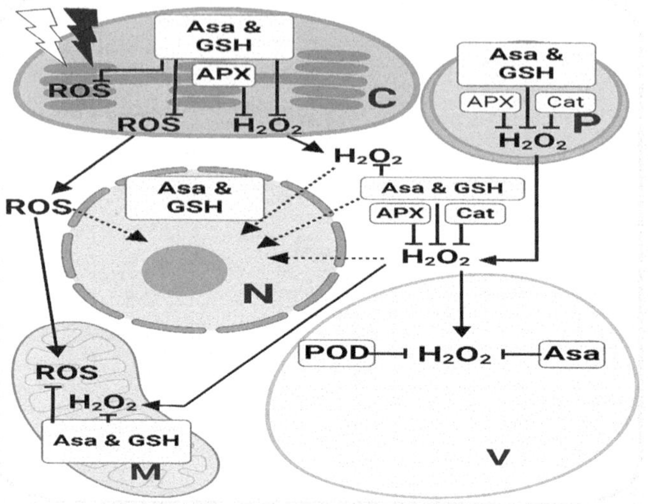

Fig. 4.3 Demonstrates a model proposing how excess white light (white thunderbolt) and far-red light (dark red thunderbolt) increase the subcellular levels of reactive oxygen species (ROS), focusing on hydrogen peroxide (H_2O_2) and its detoxification by the Acacia gums secondary metabolites (AGs-SME). Possible signaling pathways are indicated by dotted arrows. Key components include ascorbate (*AsA*), ascorbate peroxidase (*APX*), chloroplast (*C*), catalase (*Cat*), glutathione (*GSH*), mitochondria (*M*), nucleus (*N*), *P* (peroxisome), guaiacol-type peroxidase (*POD*), and vacuoles (*V*)

their potential applications in nutrition (Mwove et al., 2016). The diverse bioactive profiles of AGs-SME provide significant antioxidant benefits and offer promising avenues for nutritional and therapeutic innovations against reactive oxygen species (ROS). Cellular ROS production by light and detoxification by antioxidants in AGs-SME are presented in Fig. 4.3.

4.2.4 Neuroprotective Properties of AGs-SME

AGs-SME are known for their neuroprotective properties, primarily owing to their antioxidant and anti-inflammatory effects (Sharifi-Rad et al., 2020). These metabolites, including tannins, flavonoids, and alkaloids, have been demonstrated to alleviate oxidative stress, reduce inflammation, prevent abnormal protein accumulation, and enhance neuronal survival (Balakrishnan et al., 2022). Furthermore, these compounds have been identified as potentially effective in the treatment of neurodegenerative conditions, such as Alzheimer's and Parkinson's diseases (Reddy et al., 2020). The varied neuroprotective actions of these substances underscore their potential as treatments for neurodegenerative ailments. Thus, the potent neuroprotective capabilities of AGs-SME underscore its significant potential in developing novel therapeutic strategies for neurodegenerative diseases.

The neuroprotective effects of AGs-SME have been extensively studied, particularly the gums from *Acacia senegal* and *Acacia seyal*, which are recognized for their potential neuroprotective effects. Research indicates that *Acacia senegal* gum contains polysaccharides and glycoproteins, known for their immunomodulatory effects, which may also aid in neuroprotection (Elnour et al., 2019; Jaafar, 2019). *Acacia seyal* gum has been explored for its antioxidant and anti-inflammatory effects, which are vital for neuroprotection (Jaafar, 2019). Moreover, studies on *Acacia polyacantha* gum have highlighted its rich bioactive compounds and potential health benefits, suggesting its neuroprotective properties (Awadasseid et al., 2019). For this reason, extensive research on AGs' bioactive compounds demonstrates their significant potential in contributing to neuroprotection and managing neurodegenerative conditions.

The neuroprotective properties of AGs can be attributed to their abundant bioactive components, including amino acids, polysaccharides, and phenolic compounds, which are associated with various health benefits, particularly neuroprotection (Elnour et al., 2019, 2020). Specifically, the bioactive substances in acacia gum, notably from *Acacia senegal* and *Acacia seyal*, are essential for protecting neuronal cells against oxidative stress and inflammation, which are key factors in the progression of neurodegenerative diseases (Elnour et al., 2019, 2020). Thus, the rich bioactive makeup of AGs underscores their crucial role in neuroprotection and offers potential therapeutic benefits against neurodegenerative conditions.

The antioxidant properties of acacia gum, especially from *Acacia seyal*, are linked to its potential neuroprotective effects attributed to its capacity to neutralize free radicals and minimize oxidative damage, both of which are factors in neurodegenerative diseases (Elnour et al., 2020; Jaafar, 2019). The inclusion of phenolic compounds, flavonoids, and other antioxidants in acacia gum enhances its neuroprotective role by mitigating oxidative stress and inflammation in the brain (Elhasan et al., 2022; Elnour et al., 2020). As a result, the potent antioxidant composition of acacia gum, particularly from *Acacia seyal*, plays a crucial role in protecting brain health and offers promising avenues for treating and preventing neurodegenerative diseases.

Phenolic compounds in AGs are known for their neuroprotective properties, primarily through their antioxidant and anti-inflammatory actions. These compounds effectively scavenge free radicals, reduce oxidative stress, and mitigate inflammation, which are key processes for safeguarding neuronal cells from damage and degeneration (Elnour et al., 2018a; Elnour et al., 2020). The significant antioxidant activity of these gums, particularly noted in studies on *Acacia seyal* gum extracts, is largely due to their phenolic content (Elnour et al., 2018a). Moreover, the rich phenolic content of oils from *Acacia polyacantha* seeds also suggests their capability to provide antioxidant benefits (Tchinda et al., 2020). Therefore, the abundant phenolic compounds in AGs play an essential role in their neuroprotective and antioxidant effects, underscoring their potential for neurodegenerative disease prevention and management.

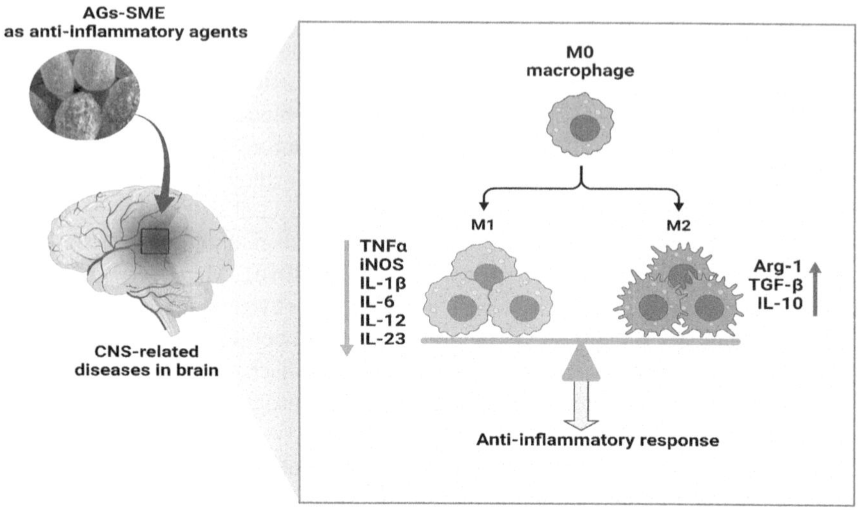

Fig. 4.4 The proposed impact of AGs-SME as a degenerative agent. *TNF-α* Tumor Necrosis Factor, *iNOS* Inducible Nitric Oxide Synthase, *IL-1β* Interleukin-1 Beta, *IL-6* Interleukin 6, *IL-12* Interleukin 12, *IL-23* Interleukin 23

The neuroprotective effects of phenolic compounds have also been reinforced by studies on the immunomodulatory attributes of gum arabic, which is a key component of AGs. Gum arabic has been recognized for its immunomodulatory capabilities that shield the nervous system from neuroinflammation and neurodegeneration (Ballal et al., 2011). Additionally, the antioxidant and anti-inflammatory properties of *Acacia seyal* gum extracts indicate their potential effectiveness in alleviating neuroinflammation, which is a critical element in the progression of various neurodegenerative disorders (Elnour et al., 2018a, 2020). Therefore, the combination of antioxidant, anti-inflammatory, and immunomodulatory effects of AGs, particularly phenolic compounds and gum arabic, offers a promising therapeutic strategy against neurodegenerative diseases. The proposed impact of AGs-SME as a degenerative agent is presented in Fig. 4.4.

4.2.5 Anti-cancer Properties of AGs-SME

AGs-SME demonstrated significant anti-cancer properties. These gums are rich in various SMEs including amines, flavonoids, alkaloids, terpenes, and tannins (Batiha et al., 2022). Research has shown that specific metabolites such as thymoquinone have cytotoxic effects, inducing apoptosis in cancer cells and inhibiting tumor angiogenesis (Raina et al., 2014). Polyphenols from *Acacia seyal* gum have been recognized for their potential as anti-cancer agents (Elnour et al., 2020). These compounds are known for their dual antioxidant and anti-cancer activities (Ghasemzadeh et al., 2015) and are crucial in modulating cellular signaling pathways associated with cell proliferation, metastasis, and apoptosis, thus playing a vital role in cancer therapy (Habli et al., 2017). In addition, AGs-SME in natural biomolecules has been shown to possess a broad-spectrum of biological activities, especially anti-cancer effects (Ullah et al., 2020). Their anti-cancer efficacy is evident from their ability to inhibit tumor cell proliferation, trigger apoptosis, control invasion and metastasis, and diminish angiogenesis. These metabolites also augment the effectiveness of chemotherapeutic drugs while reducing associated damage (Yang et al., 2022). Additionally, AGs-SME have been noted for their antibacterial and anti-cancer activities, underscoring their potential for managing diverse diseases (Damavandi, 2023). Thus, the array of AGs-SME plays a pivotal role in cancer treatment, highlighting their potential for therapeutic applications against various malignancies.

Gums from *Acacia senegal* and *Acacia seyal* feature AGPs, which are rich in sugars such as D-galactose, L-arabinose, L-rhamnose, D-glucuronic acid, and 4-*O*-methyl-D-glucuronic acid, complemented by a modest amount of proteins and minerals (Nigen et al., 2019). The gum from *Acacia senegal* primarily comprises a complex polysaccharide structure with a branched β-(1,3)-linked galactose backbone adorned with side-chains of arabinose, rhamnose, and uronic acids (Fauconnier et al., 2000). Conversely, *Acacia seyal* gum is distinguished by its bioactive constituents, including carotenoids such as astaxanthin and indolic alkaloids, and phenolic compounds such as ferulic acid, *p*-coumaric acid, caffeic acid, and sinapic acid (Elnour et al., 2019). These compositional differences underscore the rich diversity of SME found in the gums of *Acacia senegal* and *Acacia seyal*. Thus, the unique and varied compositions of *Acacia senegal* and *Acacia seyal* gums, which are rich in bioactive components, exemplify their potential for broad applications in health and industry.

AGs-SME manifests potential anti-cancer effects through multiple mechanisms. Phenolic compounds, which form a significant portion of these AGs-SME, are noted for their cytotoxic properties, antioxidant capabilities, and roles in modulating critical cellular pathways that influence cancer progression (Fathi et al., 2021). These substances are associated with pathways involved in cell proliferation, apoptosis, and metastasis (Batiha et al., 2022). Therefore, the targeted modulation of essential cellular pathways by phenolic compounds in Acacia gums underscores their potential as anti-cancer agents.

Specifically, SME, such as flavonoids, alkaloids, and tannins, in acacia gums engage with various cellular structures and signaling molecules, affecting key processes such as angiogenesis, invasion, and cell cycle regulation (Todkar et al., 2011). Additionally, these compounds exhibit antibacterial and antiradical activities, emphasizing their broad potential for addressing cancer and other health conditions (Satti, 2024). Therefore, the diverse interactions of AGs-SME with cellular mechanisms provide promising cancer therapies and a foundation for broader medicinal applications.

Extracts of *Acacia salicina*, *Acacia laeta*, *Acacia hamulosa*, and *Acacia tortilis* have demonstrated moderate cytotoxic effects against human liver carcinoma (HepG2) and breast cancer (MCF-7) cell lines, exhibiting notable anti-cancer activities (Elnour et al., 2020). Beyond their recognized anti-cancer properties, diverse AGs-SMEs have a broad-spectrum of potential applications across various sectors. In agriculture, these compounds can serve as biocontrol agents against plant pathogens, enhance plant growth, and decrease reliance on synthetic herbicides (Souza-Alonso et al., 2018). Their cytotoxic properties can also be exploited for weed control (Souza-Alonso et al., 2018). These substances can be used in industry to create nutraceuticals, pharmaceuticals, functional foods, and drug delivery systems (Oliveira et al., 2020). Additionally, their antimicrobial qualities could prove advantageous in the pharmaceutical industry, particularly in combating multidrug-resistant strains (Islam et al., 2021). Thus, the versatile and potent AGs-SME provide therapeutic benefits against cancer and offer expansive applications in agriculture and industry, highlighting their extensive utility and potential impact. To understand the main causes and factors resulted into brest cancer and how AGs-SME can be used to prevent it, as shown in Fig. 4.5.

4.2.6 Immunomodulatory Properties of AGs-SME

AGs-SME has demonstrated promising immunomodulatory effects against various cancer cells. These metabolites, including bibenzyls and phenanthrenes, are known for their ability to modulate immune responses and inhibit the growth of cancer

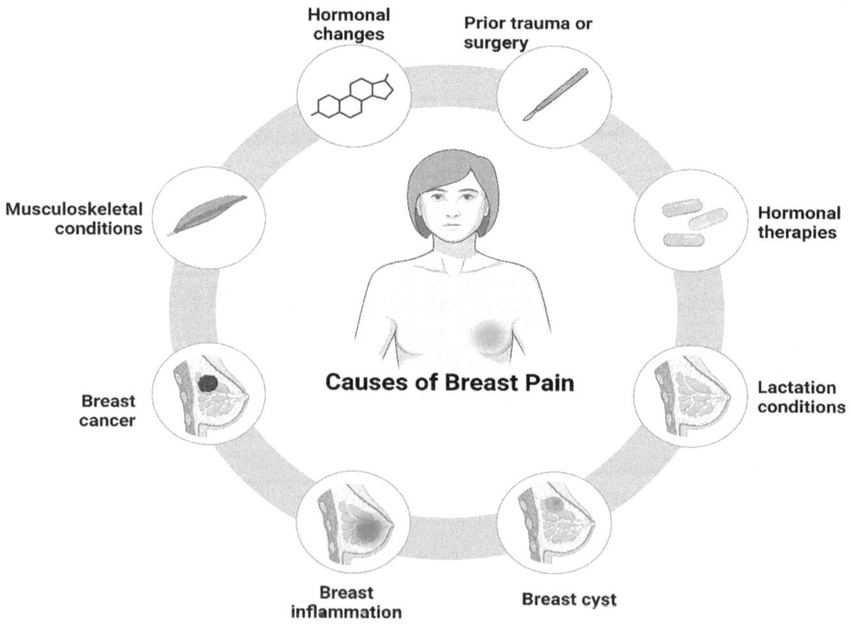

Fig. 4.5 Shows the scenarios explaining the causes of breast pain and how AGs-SME can be used to reduce or prevent the causes and symptoms

cells (Kongkatitham, 2024). Research has indicated that extracts from *Acacia seyal* are rich in phytoconstituents that possess anti-tumor capabilities, particularly against breast cancer (Zingue et al., 2018). Furthermore, *Acacia senegal* gum is recognized for its immunomodulatory, hypoglycemic, and antioxidant properties (Elhasan et al., 2022). Therefore, the diverse immunomodulatory and therapeutic properties of AGs-SME underscore its potential as an effective agent for cancer treatment and broader health applications.

AGs feature a rich array of SME, including amines, cyanogenic glycosides, flavonoids, alkaloids, diterpenes, and tannins (Batiha et al., 2022). The concentrations of these metabolites varied across the different AGs. For example, *Acacia seyal* gum contains notable bioactive compounds such as carotenoids, indolic alkaloids, and phenolic compounds (Elnour et al., 2019). Furthermore, research on *Acacia seyal* gum has highlighted its anti-tumor effects, particularly against leukemia, showing its therapeutic potential (Kabbashi, 2022). Thus, the diverse and potent AGs-SME, particularly in species such as *Acacia seyal*, offers promising therapeutic avenues for cancer treatment.

The immunomodulatory effects of AGs-SME are linked to their specific interactions with cellular pathways or structures. These metabolites, including amines, flavonoids, alkaloids, and tannins, influence the immune response (Batiha et al., 2022). *Acacia seyal* gum, for example, includes bioactive components such as carotenoids and phenolic compounds, which contribute to its immunomodulatory effects (Elnour et al., 2019). The potential mechanisms of these metabolites may involve interactions with gut microbiota, enhancement of gut probiotics, and modulation of microbial populations (Rawi et al., 2021). Additionally, proteins and minerals in gums, such as gum arabic, enhance immunomodulatory capabilities (Calame et al., 2008). Thus, the complex composition of AGs-SME is crucial for modulating immune functions, highlighting its potential as a versatile and effective immunomodulatory agent.

AGs-SME displays immunomodulatory properties primarily by interacting with specific cellular pathways. Metabolites, such as amines, flavonoids, alkaloids, and tannins, are instrumental in modulating immune responses (Batiha et al., 2022). For instance, *Acacia seyal* gum is rich in bioactive components, such as carotenoids and phenolic compounds, which enhance its immunomodulatory effects (Elnour et al., 2019). The variability in the effectiveness of extracts from different AGs likely arises from differences in the concentrations and types of SME. Therefore, the distinct immunomodulatory capacities of various AGs underscore the importance of specific AGs-SME in influencing immune system interactions and responses.

Comparative studies on the immunomodulatory properties of gum extracts from AGs-SME and other established agents have indicated that acacia gum, which is rich in polysaccharides and glycoproteins, has immunopotentiating effects (Sekhon et al., 2016). These natural extracts can enhance the efficacy of other anti-cancer agents through synergistic interactions. For example, combining *Acacia seyal* gum with *Acacia sengal* samples has shown anti-proliferative effects on human breast cancer cell lines (Elnour et al., 2020). Thus, the integration of acacia gum extract with other therapeutic agents offers a promising avenue for enhancing the efficacy of cancer treatment through synergistic immunomodulatory actions.

AGs-SMEs have demonstrated considerable promise across various sectors. These extracts display immunomodulatory properties in clinical environments, making them candidates for complementary cancer therapies via cellular pathway interactions (Batiha et al., 2022) in agriculture as natural crop protection agents or to enhance plant defense mechanisms (Selim, 2012). Industrially, they offer pharmacological benefits and could serve as novel anti-cancer agents or flavor enhancers in food production (Suneetha et al., 2015). Additionally, when combined with other anti-cancer agents, these extracts may produce synergistic effects that enhance therapeutic effectiveness (Elnour et al., 2020). Thus, the versatile applications of AGs extracts from clinical to industrial settings significantly underscore their potential to advance healthcare and agricultural practices.

AGs contain various SME, including glycosides, alkaloids, flavonoids, and arabic acid (Binyane & Mfengwana, 2022). These plants are rich in SME-like amines, cyanogenic glycosides, seed oils, and tannins (Batiha et al., 2022). Gum arabic from *Acacia senegal* and *Acacia seyal* is characterized by a heteropolysaccharide structure with a galactopyranose backbone, which branches with arabinose, rhamnose, or glucuronic acid (Gim et al., 2019). Extensive research has identified a vast spectrum of AGs-SME, including amines, cyanogenic glycosides, flavonoids, alkaloids, seed oils, cyclitols, fluoroacetate, gums, non-encompassing amino acids, compounds, diterpenes, fatty acids, terpenes, hydrolyzable tannins, and condensed tannins (Batiha et al., 2022). Thus, the rich and varied AGs-SME highlights their potential as a valuable resource for pharmaceutical, nutritional, and industrial applications. Finally, the immunomodulatory properties of AGs-SME, primarily through interactions with specific cellular pathways, are presented in Fig. 4.6.

4.2.7 Anti-diabetic Properties of AGs-SME

AGs-SME are known to contain a variety of bioactive compounds, such as tannins, flavonoids, alkaloids, and gums, showing potential anti-diabetic properties (Duarte & Wolf, 2005; Elnour et al., 2019; Elnour et al., 2020; Kabbashi, 2022; Kaddam et al., 2019). Acacia gums are well-known for their pharmacological benefits and have been traditionally utilized in various treatments (Ashour et al., 2022; Batiha et al., 2022). The physicochemical attributes of polysaccharides extracted from these species have been analyzed, revealing their stabilizing and emulsifying capabilities (Mugo et al., 2020; Palma et al., 2015; Sethuraman & Rajendran, 2018). Additionally, studies have explored the prebiotic potential of acacia gums and their ability to influence lipid profiles (Calame et al., 2008; Kaddam et al., 2019). Therefore, the extensive range of bioactive compounds in AGs and their proven pharmacological and physicochemical properties underscore their potential for therapeutic and functional food applications.

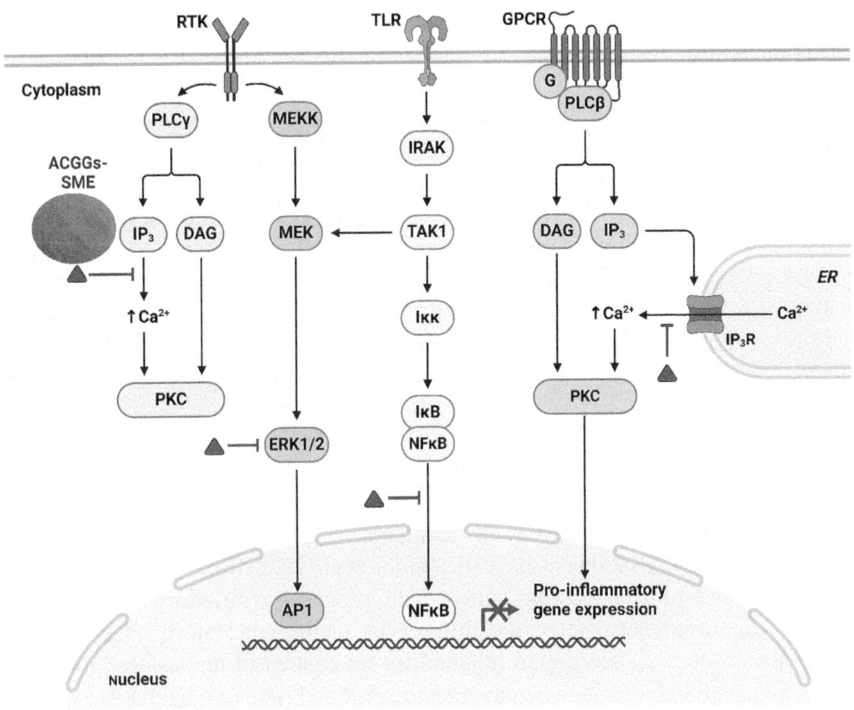

Fig. 4.6 Demonstrates the immunomodulatory properties, primarily by interacting with specific cellular pathways using AGs-SME. RTK (Receptor Tyrosine Kinase), TLR (Toll-Like Receptor), and GPCR (G-Protein-Coupled Receptor) are essential cell surface receptors that detect various molecular patterns and activate internal signal transduction pathways. PLCγ (Phospholipase C Gamma) and PLCβ (Phospholipase C Beta) are enzymes that convert phosphatidylinositol 4,5-bisphosphate into DAG (Diacylglycerol) and IP3 (Inositol Triphosphate), with DAK (Diacylglycerol Kinase) further phosphorylating DAG. PKC (Protein Kinase C) is activated by DAG, while IP3R (IP3 Receptor) binds IP3 to release calcium ions. In the MAP kinase signaling cascade, MEKK (Mitogen-Activated Protein Kinase Kinase Kinase), MEK (Mitogen-Activated Protein Kinase Kinase), and ERK1/2 (Extracellular Signal-Regulated Kinases 1 and 2) play pivotal roles in transmitting cellular signals. The NF-kB (Nuclear Factor Kappa-light-chain-enhancer of activated B cells) pathway involves Ikk (I Kappa B Kinase), IkB (Inhibitor of Kappa B), and NF-kB itself, crucial for managing inflammation and immunity. Additionally, AP1 (Activator Protein 1), TAK1 (TGF-Beta Activated Kinase 1), and Interleukin-1 Receptor-Associated Kinase (IRAK) manage the responses to stress, growth factors, and infections

AGs-SME includes alkaloids, flavonoids, terpenes, condensed tannins, and gums, with notable variability in the concentrations of these compounds among species (Batiha et al., 2022; Babiker et al., 2019; Binyane & Mfengwana, 2022; Youzbachi et al., 2019). *Acacia senegal* is particularly known for its glycoside, alkaloid, flavonoid, and Arabic contents (Binyane & Mfengwana, 2022). Meanwhile, *Acacia seyal* and *Acacia polyacantha* are characterized by their rich presence of cyanogenic glycosides, condensed tannins, terpenes, and cyclitols (Babiker et al., 2019; Batiha et al., 2022). Thus, the diverse array of SME in various AGs highlights their unique chemical profiles and their potential for targeted therapeutic and industrial applications.

The anti-diabetic effects of AGs-SME are mediated by various pharmacological mechanisms. These metabolites facilitate weight reduction and exhibit antihyperlipidemic, anti-diabetic, and anti-inflammatory effects (Fatima et al. 2019). In particular, Acacia gums have been observed to regulate lipid profiles, which can be beneficial in managing diabetes (Kaddam et al., 2019). Moreover, the prebiotic properties of Acacia gums can alter the gut microbiota and enhance the production of short-chain fatty acids, such as butyrate, which are significant in mediating anti-diabetic effects (Rawi et al., 2021). Therefore, the multi-faceted pharmacological actions of AGs-SME, particularly through lipid modulation and gut microbiota interactions, underscore their potential in comprehensive diabetes management strategies.

AGs-SME have demonstrated effectiveness in displaying anti-diabetic properties, especially in regulating lipid profiles and aiding in diabetes management (Kaddam et al., 2019; Rawi et al., 2021). These extracts are rich in bioactive compounds, such as alkaloids, flavonoids, and tannins, which are instrumental in their anti-diabetic effects (Babiker et al., 2019; Batiha et al., 2022; Binyane & Mfengwana, 2022). Although all AGs possess anti-diabetic properties, variations in their effectiveness can occur because of the differences in the concentrations of specific metabolites. *Acacia senegal* gum is noted for its content of glycosides, alkaloids, flavonoids, and arabic acid, whereas *Acacia seyal* and *Acacia polyacantha* contain high levels of cyanogenic glycosides, condensed tannins, terpenes, and cyclitols (Babiker et al., 2019; Batiha et al., 2022; Binyane & Mfengwana, 2022). Thus, the diverse and specific bioactive profiles of these AGs highlight their potential in targeted anti-diabetic treatments, contingent on the unique metabolic attributes of each species.

A systematic evaluation of glucose metabolism, insulin sensitivity, and other relevant diabetes parameters is essential to determine the anti-diabetic effectiveness of gum extracts from AGs compared with conventional anti-diabetic drugs. Although these extracts have shown promising anti-diabetic properties (Kaddam et al., 2019; Rawi et al., 2021), more research is needed to directly compare their efficacy with established anti-diabetic medications. Regarding potential synergies with anti-cancer agents, earlier studies have suggested that natural compounds might enhance the effects of chemotherapeutic agents in cancer treatments by promoting apoptosis, cell cycle arrest, and anti-angiogenesis (Eitsuka et al., 2016; Paul, 2011; Raina & Agarwal, 2007; Yao et al., 2005). However, specific investigations of the synergistic effects of acacia gum extracts and anti-cancer agents are necessary to fully understand their joint therapeutic potential. Therefore, although preliminary findings are promising, comprehensive studies are required to substantiate the anti-diabetic and anti-cancer synergies of acacia gum extracts, potentially leading to more effective and targeted treatment protocols.

Anti-diabetic agents from the AGs show great promise for multi-faceted clinical, agricultural, and industrial applications. These agents are valuable in clinical settings because of their hypoglycemic and antioxidant properties, which provide natural options for managing diabetes and its related complications (Bhateja & Singh, 2014). In the agricultural sector, bioactive compounds from acacia can be harnessed to develop functional foods or supplements with anti-diabetic and antioxidant properties (Batiha et al., 2022). Additionally, in industrial settings, these compounds can be utilized to produce high-value bioactive extracts suitable for use in the pharmaceutical, nutraceutical, cosmetic, and food industries (Correia et al., 2020). Thus, the diverse and potent properties of anti-diabetic agents from AGs underscore their potential for integration into various sectors, enhancing health outcomes and product value across multiple industries.

Extracts from *Acacia crassicarpa* have demonstrated high antioxidant activity and are rich in phenolic and hydrolyzable tannins, while also showing low toxicity (Prayogo et al., 2021). Gum arabic, derived from acacia, has been confirmed to be safe for various animal species at certain dosages (Bampidis et al., 2022). A specific study investigated the toxic potential of the *A. nilotica* extract in rats to assess its safety profile (El-Hadiyah et al., 2011). Furthermore, research on *Acacia senegal* gum has demonstrated its ability to reduce systemic toxicity in rats, underlining its protective effects (Abu-Serie et al., 2021). Acacia gum, particularly from *Acacia senegal* and *Acacia seyal*, is known for its medicinal properties and health benefits (Abu-Serie et al., 2021). Thus, Acacia extracts demonstrated safety and efficacy in reducing toxicity and enhancing health, underscoring their significant potential in medicinal and nutritional applications. A diagram of AGs-SME as an anti-diabetic agent for neuropathy pathogenesis is illustrated in Fig. 4.7.

4.2.8 Cardiovascular Protective Properties of AGs-SME

AGs-SME are abundant in alkaloids, cyanogenic glycosides, cyclitols, terpenes, flavonoids, and condensed tannins (Youzbachi et al., 2019; Gorish & Alamein, 2021; Babiker et al., 2019; Hannachi, 2011; Youzbachi et al., 2012). These metabolites have a range of therapeutic properties including antioxidant, antimicrobial, and anti-inflammatory effects (Fatimah, 2019; Fazmiya, 2024; Nurlila, 2024; Satrija et al., 2023). Polyphenols, another group of SME, are recognized for their health benefits, potentially helping prevent diseases such as cardiovascular disorders (Nugraheni et al., 2018; Olas, 2022). Additionally, these plant-derived compounds have been utilized in anticorrosion applications and are being investigated for their pharmaceutical and nutraceutical potential (Giri et al., 2023; Wang et al., 2021; Zaman et al., 2022). Thus, the extensive range of bioactive properties of AGs-SME from Acacia gums offers therapeutic benefits and holds potential for cardiovascular protection and other industrial applications.

AGs-SME are abundant in alkaloids, cyanogenic glycosides, cyclitols, terpenes, flavonoids, and condensed tannins (Youzbachi et al., 2019; Gorish & Alamein, 2021; Babiker et al., 2019; Hannachi, 2011; Youzbachi et al., 2012). These metabolites have a range of therapeutic properties including antioxidant, antimicrobial, and anti-inflammatory effects (Fatimah, 2019; Fazmiya, 2024; Nurlila, 2024; Satrija

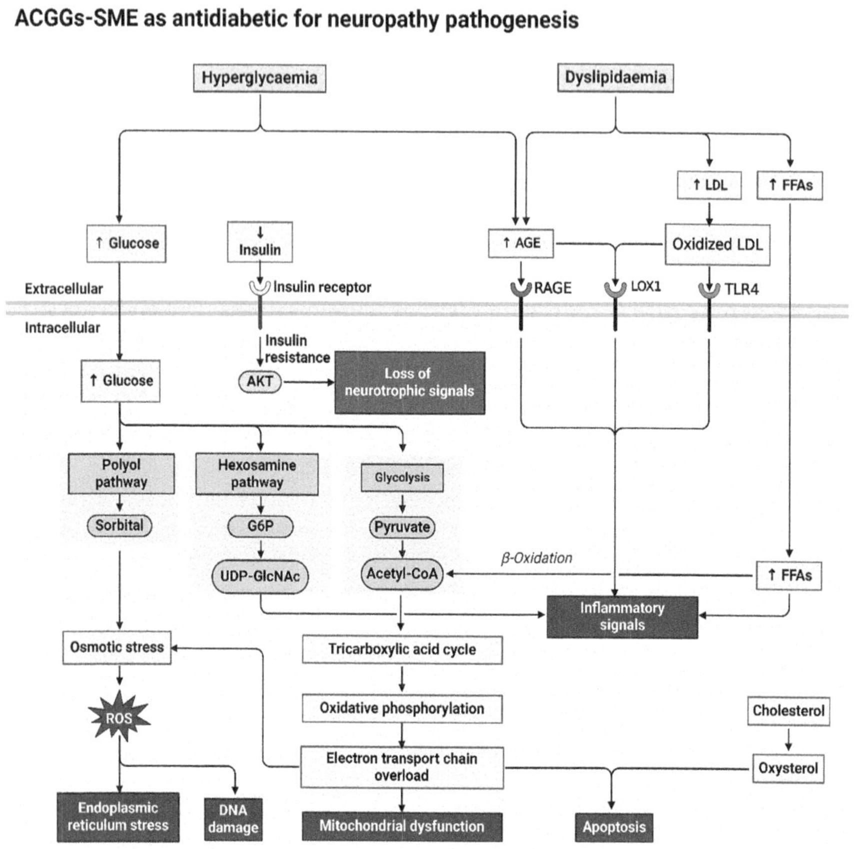

Fig. 4.7 Illustrates the mechanism of action of AGs-SME as an anti-diabetic for neuropathy pathogenesis. Protein Kinase B (AKT) is a serine/threonine-specific protein kinase essential for numerous cellular functions such as glucose metabolism and cell proliferation. Glucose 6-phosphate (G6P) plays a pivotal role in the metabolic pathways of glycolysis and gluconeogenesis. Advanced glycation end-products (AGEs) are proteins or lipids that become glycated due to sugar exposure and contribute to various degenerative diseases. The receptor for these AGEs, known as the Receptor for Advanced Glycation End-products (RAGE), activates inflammatory pathways. Lectin-type Oxidized LDL receptor 1 (LOX1) is crucial in atherosclerosis as it binds and degrades oxidized low-density lipoprotein (LDL). LDL, often termed "bad" cholesterol, transports cholesterol from the liver to cells and can accumulate in arterial walls, leading to atherosclerosis. Free Fatty Acids (FFAs) circulate in the blood and are associated with conditions such as diabetes and obesity when elevated. Toll-Like Receptor 4 (TLR4) recognizes pathogens, mediates immune responses, and plays a critical role in the body's defense mechanisms

et al., 2023). Polyphenols, another group of SME, are recognized for their health benefits, potentially helping prevent diseases such as cardiovascular disorders (Nugraheni et al., 2018; Olas, 2022). Additionally, these plant-derived compounds have been utilized in anticorrosion applications and are being investigated for their pharmaceutical and nutraceutical potential (Giri et al., 2023; Wang et al., 2021; Zaman et al., 2022). Thus, the extensive range of bioactive properties in AGs-SME not only offers therapeutic benefits, but also holds potential for cardiovascular protection and other industrial applications.

Research has extensively explored the potential synergistic effects of blending gum extracts from acacia with known cardiovascular and anti-cancer agents. For instance, one study highlighted the cytotoxic synergism achieved when a specific gum extract was combined with other compounds in human cell lines (Ramachandran et al., 2021). Furthermore, integrating herbal extracts with chemotherapeutic drugs has been shown to amplify anti-cancer effects, suggesting potent synergistic action (Cheon, 2021). Additionally, the anti-cancer efficacy of natural substances appears to increase when used alongside other chemotherapeutic agents (Eitsuka et al., 2016). Therefore, the promising outcomes of these studies suggest that combining gum extracts from acacia with conventional therapeutic agents could enhance treatment effectiveness and offer new avenues for advanced therapeutic strategies.

The widespread use of cardiovascular agents is apparent across multiple sectors. In clinical settings, aptamers are increasingly being recognized as effective ligands for both diagnostic and therapeutic applications in cardiovascular diseases (Wang et al., 2021). In the agricultural and industrial fields, biosurfactants have demonstrated potential in petroleum bio-technology and other industrial applications (Almeida et al., 2016; Dini, 2024). Furthermore, the distinctive physicochemical characteristics of β-glucans have established their value in diverse sectors including agronomy, cosmetics, pharmaceuticals, and the food industry (Caseiro et al., 2022). Thus, the versatility of cardiovascular agents in various applications highlights their potential to contribute significantly to the healthcare, agricultural, industrial, and consumer-product sectors.

Various studies have assessed the safety profiles and potential toxicity of AGs, such as those within AGs. *Acacia senegal* is widely utilized in food products (Ireland et al., 2004), and *Acacia polyacantha* has both antistaphylococcal and antibacterial properties (Ashu et al., 2020; Mambe et al., 2019). Additionally, extracts of acacia gum have been shown to have antioxidant (Elnour et al., 2018a) and antimicrobial activities (Bakshi et al., 2022). Research has further underscored the health benefits of AGs owing to their bioactive compounds (Tiwari et al., 2023). Acacia gums, including gum arabic, have been thoroughly evaluated for safety as food additives and have been confirmed to be safe for consumption (Anderson & Eastwood, 1989; Bampidis et al., 2022; Mortensen et al., 2017; Younes et al., 2019). Thus, extensive evaluation of the safety and health-promoting properties of AGs supports their continued use in food and medicinal products, reinforcing their beneficial impact on health and safety standards. Finally, AGs-SMEs can be used to reduce the impact of drug-induced cardiotoxicity, as shown in Fig. 4.8.

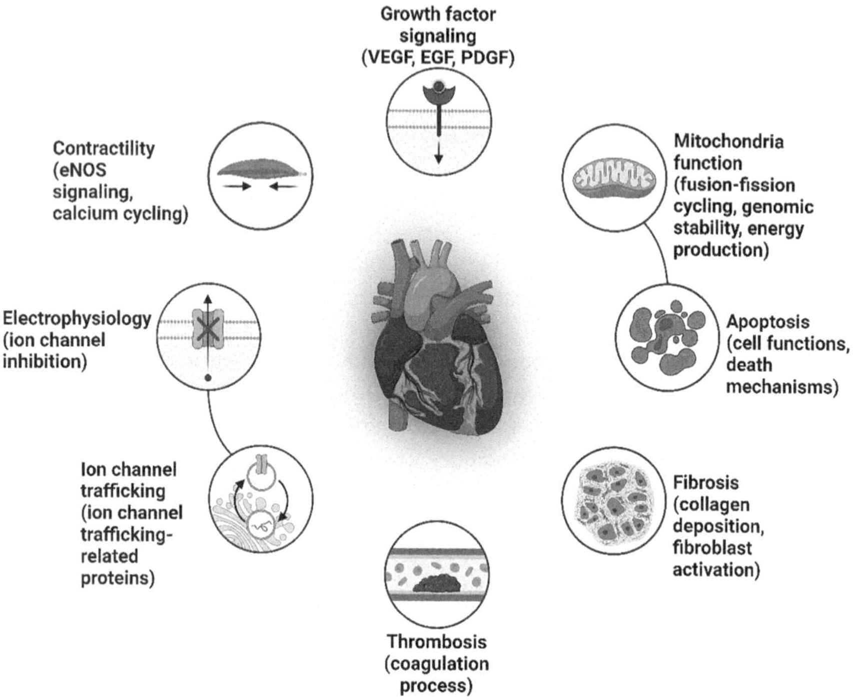

Fig. 4.8 Illustrates the proposed mechanism by which AGs-SMEs mitigate the adverse effects of drug-induced cardiotoxicity. Drug-induced cardiotoxicity is characterized by harmful effects on the cardiovascular system caused by certain pharmaceuticals. This condition alters myocardial contractility, molecular function, and vascular structure, leading to impaired cardiac performance. Cardiotoxicity may occur due to interactions between drugs and cardiac receptors, channels, or proteins, potentially inhibiting enzymatic functions, affecting vesicle transport, or disrupting signaling pathways. Understanding these mechanisms is crucial for the development of safer pharmaceuticals and enhancing drug pharmacokinetics and pharmacodynamics

4.2.9 Hepatoprotective Properties of AGs-SME

AGs-SME demonstrated hepatoprotective effects in various liver cells. These AGs-SME are rich in bioactive compounds, such as polyphenols, tannins, flavonoids, alkaloids, and other phytochemicals, which contribute to their hepatoprotective potential (Batiha et al., 2022; Palma et al., 2015). The presence of these AGs-SMEs underscores their capability as effective liver-protecting agents (Dauqan & Abdullah, 2013). Furthermore, their unique composition may provide new mechanisms of hepatoprotection compared with traditional liver health agents (Batiha et al., 2022). Therefore, the diverse and potent bioactive components of AGs offer promising alternatives for improving liver health, potentially enhancing hepatoprotective strategies with novel therapeutic mechanisms.

AGs-SME encompasses a wide range of compounds, including amines, cyanogenic glycosides, flavonoids, alkaloids, seed oils, cyclitols, fluoroacetate, non-protein amino acids, diterpenes, fatty acids, terpenes, hydrolyzable tannins, and condensed tannins (Batiha et al., 2022). The concentrations of these metabolites vary among AGs, notably in *Acacia senegal* and *Acacia seyal*, which are particularly rich in bioactive compounds, such as tannins, flavonoids, alkaloids, and gums (Duarte & Wolf, 2005). These variations in metabolite concentration, influenced by species and environmental conditions, play a crucial role in the differing hepatoprotective properties of gums (Elnour et al., 2018a). Thus, the rich and varied AGs-SME highlights their significant potential in providing targeted hepatoprotective benefits influenced by their specific composition and environmental adaptations.

The hepatoprotective properties of AGs-SME are mediated by various mechanisms. Metabolites, such as tannins, flavonoids, alkaloids, and gums, exhibit antioxidant and anti-inflammatory effects that are crucial for shielding liver cells from damage (Batiha et al., 2022). Specifically, flavonoids and phenolic compounds within these gums have been shown to target cyclooxygenase-mediated inflammation, thereby enhancing their liver-protective functions (Fazmiya, 2024). Moreover, the complex array of SME facilitates interactions with specific cellular pathways and structures in hepatocytes, influencing signaling pathways critical for liver health (Batiha et al., 2022). Therefore, diverse and potent AGs-SME play a crucial role in liver protection, offering promising therapeutic avenues for managing liver diseases through natural interventions.

AGs-SME has proven to be effective in exhibiting hepatoprotective properties under various conditions. Notably, *Acacia seyal* extract showed anti-tumor activity against leukemia cell lines (Kabbashi, 2022). The variance in the effectiveness of extracts among these AGs can be attributed to differences in their SME profiles and concentrations. *Acacia senegal* and *Acacia seyal* are noted for their high levels of bioactive compounds such as tannins, flavonoids, and alkaloids, which contribute to their liver-protective effects (Bakhoum et al., 2014). Furthermore, variations in the concentrations of these metabolites across species affect their ability to influence cellular pathways and structures crucial for hepatoprotection (Elnour et al., 2018a). Thus, the distinctive and varied secondary metabolite profiles of AGs underscore their potential to provide targeted hepatoprotective benefits, highlighting the importance of further exploration of these species for therapeutic applications.

The comparative hepatoprotective efficacy of AGs-SME against established hepatoprotective agents requires additional research. Existing studies have indicated that these extracts are rich in bioactive compounds that may confer liver protection (Elnour et al., 2020). Additionally, concerning the potential synergies with anticancer treatments, research has revealed that these gum extracts can achieve cytotoxic synergism when combined with other compounds in cancer cell lines (Ramachandran et al., 2021). Although further studies are necessary to fully understand their hepatoprotective potential, preliminary evidence supports the promising role of AGs extracts in both liver protection and cancer treatment through synergistic interactions.

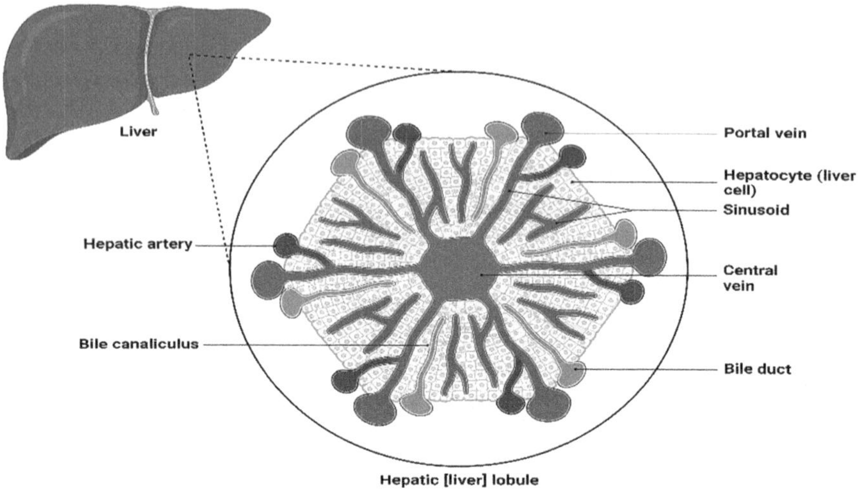

Fig. 4.9 Shows the proposed mechanism by which AGs-SMEs are promising candidates as hepatoprotective agents

Hepatoprotective agent-derived AGs-SME has demonstrated potential for diverse applications. Clinically, these agents have been explored for their ability to prevent and treat liver damage, such as that induced by acetaminophen, suggesting new avenues for managing acute liver injuries (Abdelgadir, 2023). In agriculture, these agents can be used to enhance hepatic metabolism in livestock such as lactating cows, which could help prevent metabolic disorders (Pinheiro et al., 2023). Industrially, these agents can be incorporated into formulations to safeguard against chemical-induced liver injuries, aiding the development of safer and more effective products (Rofiee et al., 2011). Additionally, Chundakkattumalayil et al. (2019) investigated the effects of probiotic Lactobacillus plantarum MBTU-HK1 and prebiotic acacia gum, used either separately or together as a synbiotic, on various health parameters in male Balb/c mice, including growth performance and reduction of procarcinogen enzyme activities. Thus, the versatile applications of hepatoprotective agents derived from AGs in clinical, agricultural, and industrial settings highlight their significant potential to improve liver health and overall well-being across different sectors. Overall, AGs-SMEs are promising hepatoprotective agents, as described in Fig. 4.9.

4.2.10 Wound-Healing Properties of AGs-SME

The wound-healing capabilities of AGs-SME include tannins, terpenes, flavonoids, saponins, alkaloids, and phenolic compounds (Cedillo-Cortezano, 2024; Tangeti et al., 2018; Tyavambiza et al., 2021). These metabolites support wound-healing through various mechanisms, including anti-inflammatory, antimicrobial, and antioxidant effects (Tyavambiza et al., 2021). The presence of these compounds enhances wound-healing by promoting wound contraction and epithelialization, and providing antibacterial, antioxidant, and anti-inflammatory benefits (Tangeti et al., 2018). Compared with traditional wound-healing agents, natural AGs-SME represent a viable alternative, offering a range of pharmacological activities that facilitate the healing process (Ahmed et al., 2022; Sheidu et al., 2021). Thus, the rich bioactive compounds in AGs position them as beneficial alternatives to conventional treatments, promising enhanced wound-healing outcomes through natural pharmacological actions.

AGs-SME, such as tannins, flavonoids, alkaloids, saponins, and phenolic compounds, are present in varying concentrations across different species (Batiha et al., 2022; Duarte & Wolf, 2005; Fazmiya, 2024). Notably, gums from *Acacia senegal* and *Acacia seyal* differ in the composition and structure of their hyperbranched exudates (Lopez-Torrez et al., 2015). Furthermore, *Acacia polyacantha* gum samples from various locations in Sudan have shown differences in their cationic compositions, with sodium being the most prevalent, followed by iron, zinc, and other trace elements (Elhenawy et al., 2014). Therefore, the diversity in the biochemical profiles of AGs, influenced by species and environmental factors, highlights their complex nature and potential for tailored applications based on specific metabolite compositions.

The wound-healing properties of AGs-SME are mediated through several mechanisms. Compounds such as tannins, flavonoids, alkaloids, saponins, and phenolic compounds enhance wound-healing by exerting anti-inflammatory, antimicrobial, and antioxidant effects (Duarte & Wolf, 2005). In particular, tannins in these gums are recognized for their astringent and antimicrobial properties that facilitate the wound-healing process (Batiha et al., 2022). These metabolites affect specific cellular pathways critical for wound-healing, including wound contraction, epithelialization, reduction of bacterial presence, oxidative stress, and inflammation (Anderson & McDougall, 1987). Additionally, varying concentrations and compositions of these metabolites across different AGs can affect their healing efficacy (Anderson & McDougall, 1987; Duarte & Wolf, 2005). Thus, the complex and varied pharmacological actions of AGs'-SME highlight their significant potential in enhancing wound-healing processes, and their effectiveness is influenced by specific botanical and regional characteristics.

AGs-SME are highly effective in promoting wound-healing due to the presence of SME, such as tannins, flavonoids, alkaloids, saponins, and phenolic compounds (Batiha et al., 2022; Duarte & Wolf, 2005). These compounds are beneficial as they target specific cellular pathways crucial for wound-healing, facilitate wound contraction and epithelialization, and provide antibacterial, antioxidant, and anti-inflammatory effects (Anderson & McDougall, 1987). Although the range of activity varies among extracts from different AGs, the underlying mechanisms of anti-inflammatory, antimicrobial, and antioxidant properties remain consistent, enhancing their overall effectiveness in wound-healing scenarios (Batiha et al., 2022; Duarte & Wolf, 2005). Therefore, diverse and potent AGs-SME play a critical role in enhancing wound-healing, with their varied actions offering comprehensive benefits across different types of wound environments.

Compared with traditional wound-healing agents, the healing efficacy of AGs-SME is garnering interest. These extracts are rich in AGs including tannins, flavonoids, alkaloids, saponins, and phenolic compounds, which are key to their healing properties (Batiha et al., 2022; Duarte & Wolf, 2005). Research indicates that acacia gum extracts facilitate wound-healing by exerting anti-inflammatory, antimicrobial, and antioxidant effects, thus promoting crucial healing processes, such as wound contraction and epithelialization (Anderson & McDougall, 1987). Additionally, when used in conjunction with other therapeutic agents, such as anti-cancer agents, these extracts might produce synergistic effects, thereby enhancing their overall therapeutic impact (Elhasan et al., 2022; Sène et al., 2023). Therefore, the multi-faceted pharmacological activities of AGs' extracts not only match but potentially surpass those of conventional healing agents, offering a potent synergistic potential that could revolutionize wound care strategies.

AGs-SME is gaining recognition for its potential as a wound-healing agent in various settings. These metabolites, including alkaloids, flavonoids, tannins, saponins, terpenoids, and anthraquinones, effectively promote wound-healing (Vijayaraghavan et al., 2017). The presence of flavonoids, terpenoids, saponins, and coumarins further enhances the wound-healing capabilities of these extracts (Ibrahim et al., 2021). Specific phenolic compounds, such as quercetin and kaempferol, are also noted for their roles in improving wound-healing processes (Adli, 2023). The diverse array of SME, including phenolic acids, flavonoids, alkaloids, and terpenoids in these plant extracts, are instrumental in triggering beneficial reactions for wound repair (Escárcega-González et al., 2018). Therefore, the rich and diverse AGs-SME support wound-healing and position these extracts as versatile and effective options for advanced wound care management.

The safety profile and potential toxicity of AGs-SME were examined both in vitro and in vivo. Research indicates that these extracts, rich in SME, facilitate wound-healing and maintain safety at the cellular and genomic levels (Vijayaraghavan et al., 2017). Despite these positive findings, additional studies are necessary to thoroughly evaluate their antimicrobial efficacy, cytotoxicity, and overall safety profile before they can be widely adopted (Ibrahim et al., 2021). In vivo toxicity assessments are essential to confirm the safety of these plant extracts, as some studies have reported potential signs of toxicity, including alterations in hematological parameters and cytotoxic effects (Adli, 2023). Thus, while AGs-SME shows promising therapeutic potential, comprehensive safety evaluations are imperative to ensure their viability and safety for clinical and commercial applications. The possible applications of AGs-SME for treating wound-healing capabilities can be demonstrated in Fig. 4.10.

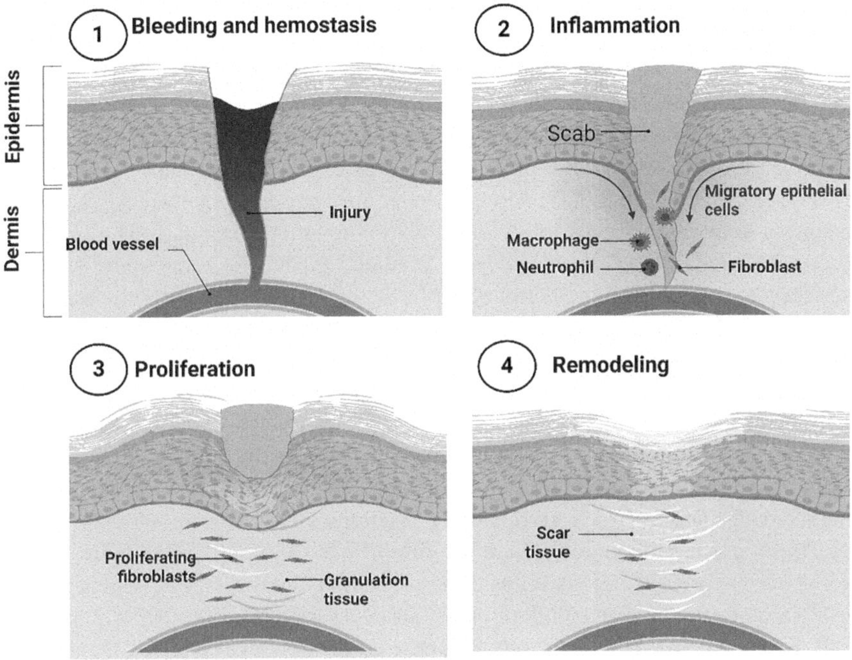

Fig. 4.10 Illustrates the potential applications of AGs-SME in enhancing wound-healing capabilities. The proposed mechanism is rooted in the understanding that wound-healing is a complex and highly coordinated process that unfolds into four distinct yet interconnected phases. Initially, the process begins with bleeding and hemostasis, during which injury-induced clot formation and platelet activation stop the blood loss. This is followed by the inflammation phase, which helps combat infection and clear debris. Subsequently, the proliferation phase takes place, which is characterized by the activation of fibroblasts, angiogenesis, and the formation of granulation tissue. Finally, during the remodeling phase, the extracellular matrix is reorganized and the tissue gradually regains its tensile strength by forming scar tissue. A thorough understanding of these biological events is crucial for developing effective wound-management strategies

4.2.11 Pain-Relief Properties of AGs-SME

AGs-SME encompasses a variety of bioactive compounds, including glycosides, alkaloids, flavonoids, and arabic acid (Binyane & Mfengwana, 2022). AGs are rich in a diverse array of SME, including amines, cyanogenic glycosides, flavonoids, alkaloids, seed oils, cyclitols, fluoroacetate, gums, non-protein amino acids, diterpenes, fatty acids, terpenes, hydrolyzable tannins, and condensed tannins, which demonstrate anti-diabetic, antioxidant, and antibacterial properties (Aryal et al., 2021; Batiha et al., 2022). The presence of tannins, flavonoids, alkaloids, and gum further underscores the therapeutic potential of these species (Duarte & Wolf, 2005). The varied chemical structures and properties of these metabolites indicate their potential for pain-relief (Todkar et al., 2011). Thus, the extensive and varied

AGs-SME highlight their broad potential for therapeutic applications ranging from pain management to combating chronic diseases.

AGs-SME are composed of diverse SME, including arabinogalactan, AGPs, glycoproteins, and various minerals (Gashua et al., 2016; Imeson, 1992; Nigen et al., 2019). *Acacia senegal* gum features a complex polysaccharide backbone with 1,6-linked branches adorned with side-chains of arabinose, rhamnose, and uronic acids (Fauconnier et al., 2000). Moreover, the nitrogen, iron, manganese, and zinc content in *Acacia senegal* varieties aligns with international standards (Lelon et al., 2013). Additionally, *Acacia seyal* gum contains bioactive compounds, such as carotenoids, indolic alkaloids, and phenolic compounds, such as ferulic acid and coumaric acid (Elnour et al., 2019). Thus, the rich composition of AGs-SME not only underscores its complex biochemical structure but also highlights its potential for various applications in food, pharmaceuticals, and nutraceuticals owing to its bioactive properties.

AGs-SME, including arabinogalactan, AGPs, and glycoproteins, potentially offers pain-relief through several mechanisms. The antioxidant properties of the phenolic rings and hydroxyl groups in these compounds may help alleviate oxidative stress and inflammation, which are often linked to pain (Abdelgadir, 2023). Moreover, the presence of alkaloids, flavonoids, and tannins in AGs indicates their possible analgesic and anti-inflammatory effects (Batiha et al., 2022; Duarte & Wolf, 2005). These metabolites may influence cellular pathways related to pain perception and inflammation, affecting neurotransmitter release, cytokine production, and oxidative stress responses (Batiha et al., 2022; Duarte & Wolf, 2005). Additionally, bioactive components such as carotenoids and indolic alkaloids in acacia gums could further enhance pain-relief properties through their impact on cellular signaling pathways (Batiha et al., 2022). Therefore, the diverse array of bioactive compounds in Acacia gums underscores their potential as natural analgesics, highlighting their therapeutic applications in managing pain and inflammation effectively.

AGs-SME effectively provides pain relief, primarily through its antioxidant, anti-inflammatory, and potentially analgesic properties (Elnour et al., 2018a, b). These extracts are rich in SME, including phenolic compounds, alkaloids, flavonoids, and tannins, which contribute to their therapeutic benefits (Batiha et al., 2022; Duarte & Wolf, 2005). Although the three species of Acacia gum share similar chemical constituents, variations in the activity spectrum might arise from differences in the concentrations and ratios of these metabolites (Elnour et al., 2018a). For example, *Acacia seyal* gum contains bioactive components, such as carotenoids, indolic alkaloids, and phenolic compounds, which may offer distinct pain-relief properties compared to *Acacia senegal* and *Acacia polyacantha* gums (Elnour et al., 2019). Thus, the diverse and potent AGs-SME contributes to its pain-relieving capabilities and suggests variability in efficacy across different species, underscoring the importance of targeted research to optimize its therapeutic use.

A study that evaluated the pain-relief properties of AGs-SME gum extracts from AGs compared the effectiveness of chewing gum and ibuprofen for orthodontic pain-relief (Delavarian et al., 2020). In a separate investigation, the pain-relieving potential of Ginkgo biloba extract in neuropathic pain was demonstrated (Taliyan & Sharma, 2012). These studies collectively underline the efficacy of natural extracts in pain management. Additionally, Boccella (2023) explored the combined effects of Boswellia serrata and *Acmella oleracea* extracts on neuropathic pain, revealing the potential synergistic benefits of these natural compounds. These findings highlight the potential of natural extracts, both individually and in combination, as effective alternatives for pain management, encouraging further exploration of their mechanisms and broader applications.

The pain-relief properties of agents derived from acacia gums have diverse potential applications in clinical, agricultural, and industrial settings. In clinical contexts, these agents could serve as natural alternatives or complementary therapies for pain management, benefiting from their antioxidant, anti-inflammatory, and analgesic properties (Elnour et al., 2018a). In agriculture, the antimicrobial and anti-inflammatory properties of these extracts can aid in plant protection and growth enhancement (Gomes et al., 2017). Industrially, these agents may be incorporated into pharmaceuticals, nutraceuticals, or cosmeceuticals, leveraging their bioactive compounds and pain-relieving potential (Gomes et al., 2017). Additionally, the synergistic effects of these agents combined with established pain-relief medications or other natural extracts could improve their effectiveness and expand their use across various sectors (Boccella, 2023). Therefore, the versatile and beneficial properties of acacia gum derivatives not only offer promising pain-relief solutions, but also encourage innovative applications in multiple fields, underscoring their potential as valuable therapeutic agents.

The safety of *Acacia senegal* gum and *Acacia senegal* gum extract in cosmetic formulations is well-established, supported by extensive safety data on gum arabic (Johnson, 2005). *Acacia senegal* gum, also known as gum arabic, is recognized by the Food and Drug Administration (FDA) as a safe dietary fiber. It has been shown to reduce systemic toxicity in rats treated with $CCl4$ by modulating the ROS/NF-κB signaling pathway (Abu-Serie et al., 2021). Additionally, gum arabic from *Acacia senegal* and *Acacia seyal* trees has been used to extract hyperbranched acidic polysaccharides (Jin et al., 2012). Moreover, research on *Acacia seyal* gum has demonstrated its anti-proliferative effects in human breast cancer cell lines, indicating its potential therapeutic benefits (Elnour et al., 2020). Thus, the extensive validation of *Acacia senegal* gum safety and its promising therapeutic effects underscores its potential for broader applications in both cosmetic and medical products.

4.2.12 Anti-aging Properties of AGs-SME

AGs-SME demonstrated potential anti-aging properties. These gums are composed of complex biopolymers enriched with glycosylated proteins and sugars as well as various SME, including alkaloids, flavonoids, and tannins (Duarte & Wolf, 2005). Research has highlighted the antioxidant, anti-inflammatory, and anti-aging activities of these compounds (Fathi et al., 2021; Usman et al., 2020). Furthermore,

Acacia gums have shown anti-diarrheal effects, underscoring their broad therapeutic potential (Mengesha et al., 2022; Panjaitan et al., 2022). The presence of bioactive compounds such as phenolic compounds in these gums is vital for enhancing human health and contributing to their anti-aging effects (Fathi et al., 2021). These findings suggest that AGs-SME is a promising alternative to conventional anti-aging agents. Thus, the diverse bioactive components of Acacia gums provide significant health benefits and position these natural extracts as effective candidates in the anti-aging sector, offering a natural alternative to synthetic treatments.

AGs-SME encompass a wide range, including amines, cyanogenic glycosides, flavonoids, alkaloids, seed oils, cyclitols, fluoroacetate, gums, non-protein amino acids, diterpenes, fatty acids, terpenes, hydrolyzable tannins, and condensed tannins (Batiha et al., 2022). These metabolites differ in concentration across species, with *Acacia senegal* and *Acacia seyal* noted for their rich content of soluble and low-viscosity fibers (Silva & Souza, 2018). Additionally, gum samples from *Acacia polyacantha* collected across various locations in Sudan displayed variations in cationic composition, with sodium being predominant, followed by Fe, Zn, Ni, Cu, Mn, Ca, and trace amounts of Pb, K, and Mg (Elhenawy et al., 2014). Variability has also been observed in the optical rotations of gum exudates from *Acacia senegal* and *Acacia seyal* (Mhinzi, 2003). Therefore, the diverse and variable composition of AGs-SME highlights the complexity of these natural gums and suggests their broad potential for various industrial and medicinal applications driven by their unique chemical properties.

AGs-SME, including amines, cyanogenic glycosides, flavonoids, alkaloids, and tannins, are known for their anti-aging properties through several mechanisms. These compounds exhibit potent antioxidant properties, attributed to phenolic rings and hydroxyl groups that effectively scavenge free radicals and mitigate oxidative stress, which is a principal contributor to aging (Abdelgadir, 2023). Furthermore, the bioactive compounds in Acacia gums interact with cellular pathways that influence inflammation and cellular damage, thereby enhancing their anti-aging effects (Usman et al., 2020). The complex chemical structures of these metabolites enable them to specifically target cellular structures and pathways related to aging, such as protein-binding and modulation of the gut microbiota, which are crucial for maintaining cellular health and function (Abdelgadir, 2023). Thus, the multifunctional and specific actions of AGs-SME underscore their potential as effective natural solutions for combating the physiological effects of aging, highlighting their therapeutic value in age-related health strategies.

AGs-SME has been proven to be effective against various age-related cellular mechanisms. *Acacia senegal* and *Acacia seyal* extracts are particularly notable for upregulating antioxidant enzymes in cells, demonstrating their capacity to combat oxidative stress, a major factor in cellular aging (Sowndhararajan et al., 2015). In contrast, *Acacia polyacantha* extracts have shown distinct antibacterial and anti-helminthic activities, indicating a varied spectrum of action compared with other AGs (Tchinda et al., 2020). Additionally, *Acacia senegal* extracts have demonstrated anti-inflammatory effects that may help mitigate age-related inflammatory conditions (Osman et al., 2022). Therefore, the diverse bioactive properties of Acacia extracts not only address multiple facets of aging but also highlight their potential as multifunctional agents in age-related health and disease management.

The effectiveness of gum extracts from AGs in demonstrating anti-aging properties can vary from that of established anti-aging agents. Although Acacia gum extracts have demonstrated antioxidant, anti-inflammatory, and anti-diabetic activities (Fathi et al., 2021; Tchinda et al., 2020; Usman et al., 2020), their performance may not always align with that of conventional anti-aging treatments. However, the synergistic effects have notable potential when combined with other anti-aging compounds. For example, berberine encapsulated in guar-acacia gum nanocomplexes showed enhanced biological efficacy against diabetes and pathogenic microorganisms (Bakshi et al., 2022). Similarly, combining *Acacia mearnsii* extract with antibiotics resulted in synergistic interactions with clinically relevant bacteria (Olajuyigbe & Afolayan, 2012). Thus, although acacia gum extracts offer beneficial anti-aging properties, their potential is significantly amplified when used with other therapeutic agents, suggesting a promising avenue for enhanced and multi-faceted anti-aging treatments.

Arabic gum acacia, sourced from *Acacia senegal* and *Acacia seyal* trees in Africa, has been recognized for its therapeutic benefits. Almohaimeed et al. (2018) found that arabic gum acacia effectively improves diabetic peripheral neuropathy in rats, providing biochemical and histopathological evidence of its efficacy. Furthermore, Abu-Serie et al. (2021) indicated that Acacia senegal gum reduced systemic toxicity in rats intoxicated with CCl_4 by modulating the ROS/NF-κB signaling pathway. The FDA recognizes gum arabic as a safe dietary fiber, highlighting its safety and origin from the branches and trunk of acacia trees (Abu-Serie et al., 2021). Therefore, gum arabic has demonstrated health benefits and regulatory recognition affirms its therapeutic potential and safety, making it a valuable natural product for addressing various health issues. The anti-aging properties of AGs-SME can be attributed to their ability to reduce ROS by controlling the 12 hallmarks of aging, which are grouped into three categories: primary, antagonistic, and integrative, as presented in Fig. 4.11.

4.2.13 Antiviral Properties of AGs-SME

AGs-SME exhibited potential antiviral properties. These metabolites, including glycosides, alkaloids, flavonoids, and arabic acid, have been associated with antiviral effects (Binyane & Mfengwana, 2022; Duarte & Wolf, 2005). Research has underscored the antiviral activities of these compounds, particularly flavonoids, polyphenols, terpenoids, and saponins, which can interfere with viral replication and cellular function (Musarra-Pizzo et al., 2021; Poy & Tohidfar, 2023; Yi et al., 2020). Moreover, specific compounds, such as tannins, flavonoids, and terpenoids, present in these gums have demonstrated antiviral capabilities (El-Darier et al., 2021; Satrija et al., 2023). Thus, the diverse array of AGs-SME not only provides

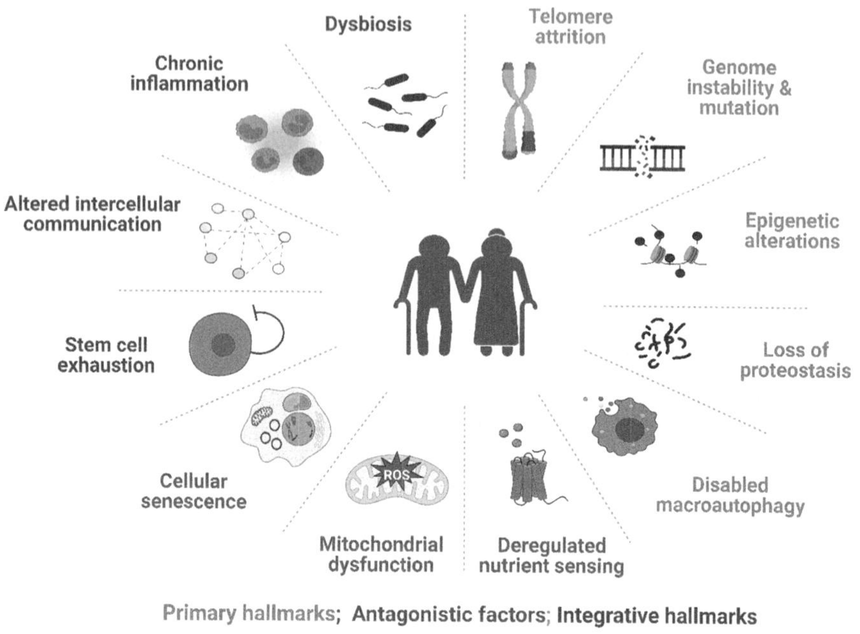

Fig. 4.11 Illustrates that the anti-aging properties of AGs-SME can be attributed to their ability to reduce reactive oxygen species (ROS) by regulating the 12 hallmarks of aging, which are grouped into three categories: primary, antagonistic, and integrative

broad-spectrum antiviral effects, but also presents a promising basis for the development of novel antiviral therapies.

AGs contain a variety of SME, such as arabinogalactan (AG), AGP, glycoproteins, and essential minerals, including potassium, calcium, magnesium, and sodium (Gashua et al., 2016; Mugo et al., 2020; Satti, 2024; Silva & Souza, 2018). The concentrations of these metabolites vary among species, with notable differences in nitrogen, Fe, Mn, and Zn concentrations (Lelon et al., 2013). *Acacia senegal* and *Acacia seyal* are rich in these bioactive compounds, making them significant sources of valuable elements (Ashour et al., 2022; Bakhoum et al., 2014). Therefore, the rich composition of SME and essential minerals in *Acacia senegal* and *Acacia seyal* underscores their nutritional and therapeutic value, highlighting their potential in diverse applications, ranging from health supplements to therapeutic agents.

AGs-SME, including amines, flavonoids, alkaloids, and tannins, possesses antiviral properties through various mechanisms. These compounds disrupt viral replication by targeting specific cellular pathways and structures. Flavonoids and alkaloids, for instance, have been shown to inhibit viral enzymes and proteins, effectively halting viral replication (Batiha et al., 2022). Tannins also play a significant role as they can bind to viral proteins and prevent their attachment to host cells, thus blocking infection (Todkar et al., 2011). Furthermore, the glycoproteins present in Acacia gums may enhance their antiviral capabilities by either modulating immune responses or obstructing viral entry into host cells (Renard et al., 2006). Thus, the diverse array of bioactive compounds in Acacia gums not only offers potential for direct antiviral action, but also suggests broader implications for enhancing host resistance against viral infections.

AGs-SME, especially *Acacia seyal*, has demonstrated effectiveness against viral cells. These extracts are rich in SMEs, such as tannins, flavonoids, and alkaloids, which are pivotal for their antiviral properties (Elmi et al., 2020; Zingue et al., 2018). These compounds target specific cellular pathways or structures within the viral cells, effectively inhibiting viral replication and entry (Zingue et al., 2018). *Acacia seyal* has been shown to have potent cytotoxic effects against various cell lines, showing broad antiviral potential (Zingue et al., 2018). To further assess the antiviral efficacy of gum extracts from AGs relative to established antiviral agents, studies such as those by Elnour et al. (2020) and Hunt et al. (2021) are crucial, as they explore anti-proliferative and virucidal activities. Additionally, investigating the synergistic effects with other antiviral agents, as demonstrated by Ramachandran et al. (2021), can elucidate the combinatory benefits that may enhance overall antiviral efficacy. Therefore, the diverse and potent antiviral properties of *Acacia seyal* extracts highlight their potential as natural alternatives or adjunctive treatments against viral infections, meriting further investigation into their therapeutic applications.

Antiviral agents derived from AGs have shown significant potential across multiple sectors. In the pharmaceutical industry, these agents could serve as anti-tumor, anti-inflammatory, and antiviral treatments, showcasing their broad therapeutic capabilities (Sitthiyotha et al., 2018). In the agricultural sector, the antiviral properties of these agents could play a crucial role in combating plant viral diseases, thereby enhancing crop health and increasing yields (Dutta et al., 2022). In industrial applications, these compounds may be used to develop biosurfactants with versatile functions ranging from emulsifying agents to biocides in the petrochemical industry (Almeida et al., 2016). Therefore, the diverse applications of antiviral agents from AGs underscore their value in healthcare and agriculture and highlight their potential to innovate product development in various industrial domains.

Several studies have highlighted that AGs are rich sources of diverse SME. According to Youzbachi et al. (2019), these include alkaloids, cyanogenic glycosides, cyclitols, gums, terpenes, flavonoids, and condensed tannins. Maslin et al. reported similar findings, noting the presence of amines, alkaloids, cyanogenic glycosides, cyclitols, gums, non-protein amino acids, terpenes, flavonoids, and condensed tannins within various AGs (Youzbachi et al., 2012). Furthermore, Duarte and Wolf (2005) reinforced that the Acacia genus is characterized by a wealth of SME, such as tannins, flavonoids, alkaloids, and gums, which showcase the biochemical diversity and ecological adaptability of these plants. Thus, the extensive array of SME identified in the Acacia genus not only highlights its rich chemical diversity, but also underscores the potential for these compounds in pharmaceutical and ecological applications.

4.2.14 Anti-obesity Properties of AGs-SME

AGs-SME demonstrated potential anti-obesity properties. These metabolites effectively inhibit lipid- and carbohydrate-metabolizing enzymes, reduce adipogenesis and appetite, and boost energy metabolism (Bhardwaj et al., 2021). In particular, phenolic compounds such as flavonoids influence the pathways associated with fat accumulation and obesity (Olaokun et al., 2017). Flavonoids, a significant group of plant phenolic SME, exhibit anti-obesogenic, anti-inflammatory, and immunomodulatory activities (Sudhakaran & Doseff, 2020). The mechanisms contributing to the anti-obesity effects of these SME include inhibition of lipid absorption and metabolism, impact on satiety signals, and suppression of adipocyte differentiation (Chu & Phang, 2016). Therefore, the diverse and potent mechanisms of action of AGs-SME make them promising candidates for natural, multi-faceted approaches to managing and preventing obesity.

AGs-SME demonstrated significant anti-obesity properties. These metabolites function by inhibiting lipid and carbohydrate-metabolizing enzymes, reducing adipogenesis and appetite, and boosting energy metabolism (Bhardwaj et al., 2021). Notably, phenolic compounds, such as flavonoids, in these extracts are known to regulate pathways associated with adiposity and obesity (Olaokun et al., 2017). Additionally, flavonoids, a prominent class of plant phenolic SME, exhibit anti-obesogenic, anti-inflammatory, and immunomodulatory effects (Sudhakaran & Doseff, 2020). The mechanisms underlying the anti-obesity effects of SME include inhibition of lipid absorption and metabolism, modulation of satiety signals, and suppression of adipocyte differentiation (Chu & Phang, 2016). Thus, the multifunctional properties of AGs-SME not only offer a holistic approach to managing obesity, but also highlight the potential for developing natural, plant-based therapies that target multiple aspects of this complex condition.

AGs contain a variety of specific SME, including arabinogalactan, AGP, glycoproteins, and complex polysaccharides with 1,6-linked branches featuring side-chains of arabinose, rhamnose, and uronic acids (Fauconnier et al., 2000; Gashua et al., 2016). The concentrations of these metabolites vary among species, with *Acacia senegal* and *Acacia seyal* being particularly notable for their rich content of soluble and low-viscosity fibers (Lopez-Torrez et al., 2015). Moreover, these gums also contain proteins, minerals, and sugars, such as D-galactose, L-arabinose, L-rhamnose, D-glucuronic acid, and 4-*O*-methyl-D-glucuronic acid, which contribute to their diverse functional properties (Elnour et al., 2019; Nigen et al., 2019). Therefore, the complex biochemical composition of AGs highlights their nutritional and functional diversity, and underscores their potential for various applications in the food, pharmaceutical, and nutraceutical industries.

AGs-SME exhibit anti-obesity properties through diverse mechanisms. These metabolites, including flavonoids and tannins, target specific cellular pathways and structures that are involved in obesity management. They are known to inhibit lipid- and carbohydrate-metabolizing enzymes, suppress adipogenesis, and enhance energy metabolism (Bhardwaj et al., 2021). Moreover, these compounds modulate critical pathways related to lipid metabolism, such as fatty acid synthase and acetyl-CoA carboxylase, which play significant roles in lipid accumulation (Son et al., 2020). Additionally, some SME function as insulin sensitizers, influence lipid metabolism pathways, and provide anti-inflammatory effects, further contributing to their anti-obesity capabilities (Gopal et al., 2016). Thus, the broad-spectrum of actions by which SME from natural sources, such as AGs, affect obesity underscores their potential as comprehensive therapeutic agents for weight management and metabolic health.

To compare the anti-obesity efficacy of gum extracts from AGs with that of established anti-obesity agents, it is essential to examine their mechanisms of action and effectiveness. Gum extracts have demonstrated potential in inhibiting lipid- and carbohydrate-metabolizing enzymes, suppressing adipogenesis, and enhancing energy metabolism (Bhardwaj et al., 2021). These effects are comparable to those of conventional anti-obesity agents, which often target pathways involved in lipid metabolism and adipogenesis (Huang et al., 2015). Furthermore, combining gum extracts with other anti-obesity agents, such as polyphenols or flavonoids, could lead to synergistic effects by simultaneously modulating multiple pathways (Zhu et al., 2021). Therefore, gum extracts from AGs hold significant promise as effective natural anti-obesity agents, potentially enhancing the efficacy of established treatments through synergistic interactions.

AGs-SME, including glycosides, alkaloids, flavonoids, and arabic acid, have demonstrated potential anti-obesity effects. Notably, metabolites from *Acacia senegal* can function as prebiotics, boosting the production of propionate and acetate, which may aid in the control of obesity (Ahallil et al., 2020). These gums also find applications in various industries, such as food, pharmaceuticals, cosmetics, and lithography, because of their biopolymer properties (Sethuraman & Rajendran, 2018). Additionally, bioactive compounds in these gums, such as polyphenols, have shown anti-proliferative effects against human breast cancer cell lines, suggesting their potential use in clinical settings (Elnour et al., 2020). Thus, the diverse applications and health benefits of AGs-SME underscore its significant potential in both industrial and clinical settings, particularly in managing obesity and other health conditions.

AGs-SME has shown potential in reducing toxicity in vivo, as demonstrated by *Acacia senegal* gum, which mitigates systemic toxicity in rats (Abu-Serie et al., 2021). Additionally, these gums have been safely used as stabilizers in food products such as low-fat yogurt, indicating their favorable safety profiles (Mugo et al., 2020). Furthermore, the utilization of these gums, particularly from *Acacia senegal* and *Acacia seyal*, supports the restoration of food systems with nutritious native plants in African drylands, emphasizing their role in promoting health and well-being (Sacande, 2023). Thus, the safety and efficacy of AGs-SME not only enhances food product stability and health benefits, but also contributes significantly to sustainable food systems and nutritional security in arid regions by reducing body weight and serving as an anti-obesity agent by reducing several cancer types, as presented in Fig. 4.12.

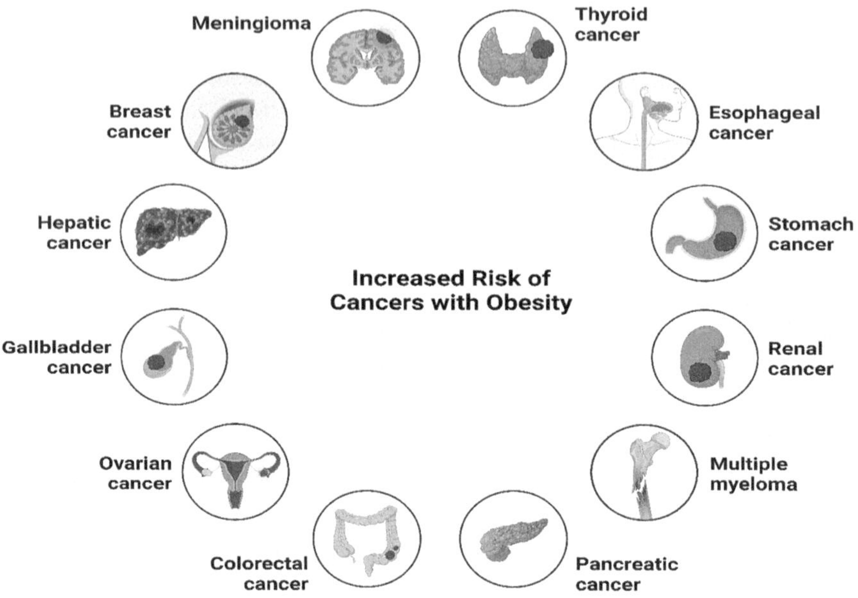

Fig. 4.12 Demonstrates the mechanism of the anti-obesity agents in AGs-SME, highlighting their role in reducing several types of cancer

4.2.15 Antifungal Properties of AGs-SME

AGs-SME showed promising antifungal properties. These gums, particularly *Acacia senegal* and *Acacia seyal*, are rich in bioactive compounds, such as polyphenols, which exhibit significant antifungal and antioxidant effects (Elnour et al., 2019, 2020). The molecular composition and properties of these gums differ, with *Acacia senegal* gum being more effective in stabilizing emulsions because of its higher adsorption at interfaces than *Acacia seyal* gum (Gashua et al., 2016). Traditionally, Acacia gum has been used for its various health benefits, highlighting its potential as a natural antifungal agent (Calame et al., 2008). The diverse phytochemical compositions, including tannins and other bioactive compounds, contribute to their antifungal properties (Abdelgadir, 2023). Overall, AGs have demonstrated significant potential as natural antifungal agents compared with conventional antifungal treatments. Additionally, polyphenols from *Acacia senegal* gum may be potential anti-cancer agents (Elnour et al., 2020). Thus, the multi-faceted benefits of AGs-SME, particularly their antifungal and potential anti-cancer properties, underscore their value as natural therapeutic agents, offering a promising alternative to synthetic compounds in medical and agricultural applications.

AGs contain specific SME such as arabinogalactan (AG), AGP, and glycoproteins (Gashua et al., 2016). These metabolites vary in concentration among different AGs, with *Acacia senegal* and *Acacia seyal* being particularly rich in bioactive compounds, such as polyphenols (Elnour et al., 2019, 2020). The composition and structure of hyperbranched plant exudates from *Acacia senegal* and *Acacia seyal* exhibit similarities and variations in specific optical rotations (Lopez-Torrez et al., 2015; Mhinzi, 2003). Additionally, the bioactive compounds in *Acacia seyal* gum include carotenoids, indolic alkaloids, and phenolic compounds such as ferulic acid and coumaric acid (Elnour et al., 2019). Therefore, the diverse and rich composition of AGs-SME underscores its potential for various therapeutic and industrial applications, making them valuable resources in medicine, nutrition, and beyond.

The antifungal properties of AGs-SME are exerted through various mechanisms. These metabolites target critical microbial processes such as cell wall formation, protein and nucleic acid synthesis, and metabolic pathways, leading to antimicrobial effects (Agustha, 2024). Specific pathways responsible for secondary metabolite production, including type I and III polyketide synthases and non-ribosomal peptide synthase, play a crucial role in the antifungal activity of these compounds (Adra, 2023). SMEs with antifungal activity disrupt fungal growth by affecting the cellular metabolism, oxidative phosphorylation, and ergosterol biosynthesis pathways (Bautista-Crescencio, 2023; Liu et al., 2020; Morales et al., 2017). Additionally, these metabolites may inhibit fungal viability by disrupting mitochondrial iron-sulfur cluster biosynthesis and interfering with essential cellular processes (Tripathi et al., 2017). Thus, the multi-faceted antifungal mechanisms of AGs-SME highlight its potential as a powerful natural alternative to synthetic antifungal agents, offering promising applications in medicine, agriculture, and industry.

AGs-SME have demonstrated antifungal efficacy against various fungal species, with some extracts showing more pronounced activity against specific fungi. For instance, extracts of *E. torquata* and *C. atlantica* exhibit strong antifungal effects by inhibiting mycelial growth, sporulation, and spore germination in a dose-dependent manner (Bouhlali et al., 2020). Additionally, ethyl acetate and aqueous extracts of *A. catechu* have shown significant inhibition of *Staphylococcus aureus* (Aryal et al., 2021). Thus, the potent antifungal properties of AGs extracts highlight their potential as effective natural antifungal agents, paving the way for their use in medical, agricultural, and industrial applications.

The antifungal properties of AGs have demonstrated effectiveness comparable to that of established antifungal agents (Kanimozhi & Rose, 2023; Tharakan, 2023). Combined with conventional antifungal agents, these extracts have shown synergistic effects, enhanced antifungal potential and potentially reducing the side-effects caused by oxidative stress in mammalian cells (Kanimozhi & Rose, 2023; Yassin et al., 2023). Synergistic interactions between plant extracts and known antifungal agents have been reported to increase antifungal activity, particularly against resistant strains, indicating that combination therapies can improve the treatment efficacy (Kim et al., 2018; Santoso et al., 2022; Yassin, 2023). Furthermore, the use of medicinal plants in conjunction with conventional antifungal drugs has been suggested as a strategy to combat fungal infections, especially in regions with high drug resistance and limited access to expensive antifungal agents (Haji et al., 2022). Thus, integrating AGs-SME with established antifungal treatments offers a promising approach to enhance antifungal efficacy and address drug resistance, providing a valuable resource for both developed and resource-limited regions.

SME extracted from *Acacia senegal* and *Acacia seyal* gums has diverse potential applications in both clinical and industrial settings. In clinical applications, these antifungal agents can be used to treat fungal infections, providing natural and effective alternatives to conventional antifungal medications (Kanimozhi & Rose, 2023; Tharakan, 2023). Furthermore, the synergistic effects observed when combining these gum extracts with established antifungal agents could enhance the treatment efficacy and reduce side-effects (Santoso et al., 2022; Yassin et al., 2023). In industrial settings, these gum extracts have shown promise in various applications, such as food packaging, pharmaceuticals, cosmetics, and lithography, because of their unique emulsifying and stabilizing properties (Palma et al., 2015; Sethuraman & Rajendran, 2018). Additionally, their biocompatibility and functional characteristics make them suitable for use in bio-technology, nanotechnology, and materials science, thereby offering a wide range of possibilities for innovative product development (Gashua et al., 2016). Therefore, the versatile applications of AGs-SME underscore their significant potential for advancing medical treatments and industrial technologies.

SME reported from various AGs widely distributed worldwide include polysaccharides (gums) and several polyphenols, such as condensed and hydrolyzable tannins and flavonoids (Taboada et al., 2019). Specific SME identified in extracts from *Acacia senegal* plants include glycosides, alkaloids, flavonoids, and arabic acid (Binyane & Mfengwana, 2022). Additionally, ultrahigh-performance liquid chromatography–high-resolution mass spectrometry analysis identified 95 SME in *Echinops ritro* L. extracts for the first time, including acylquinic acids, flavonoids, and coumarins (Zheleva-Dimitrova et al., 2023). The efficacy of the secondary metabolite crude extract of *Serratia nematodiphila* GCSR38 against Xoo was tested using the detached leaf technique, demonstrating disease control with 24.45% and 15.56% disease severity when applied before and after inoculation, respectively (Sutthisa, 2022). Compounds isolated from various classes of plant SME, including flavonoids, terpenoids, tannins, alkaloids, phytosterols, and coumarins, have shown significant anti-inflammatory activity both in vitro and in vivo (Anokwah et al., 2022). Thus, the diverse range of AGs-SME and other plants underscores their potential in various therapeutic and agricultural applications, highlighting the importance of continued research on their bioactive properties. The antifungal properties of AGs-SME are exerted through various mechanisms, as shown in Fig. 4.13.

4.3 Case Studies and Research Findings

Case studies and research findings on AGs-SME highlight these compounds' diverse biological applications and potential therapeutic properties. AGs-SME exhibits a wide range of beneficial effects, including antimicrobial, anti-inflammatory, antioxidant, neuroprotective, anti-cancer, immunomodulatory, anti-diabetic, cardiovascular protective, hepatoprotective, wound-healing, pain-relief, anti-aging, antiviral, anti-obesity, and antifungal properties. These bioactive compounds, such as

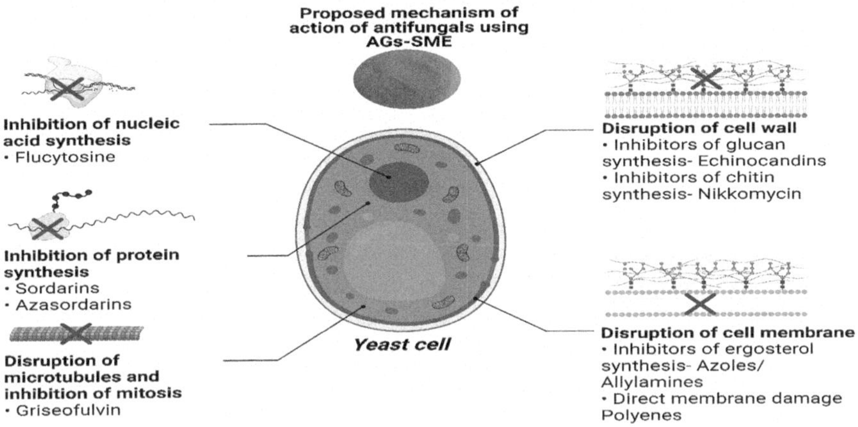

Fig. 4.13 Explains the antifungal mechanism exerted by AGs-SME, detailing the various mechanisms involved

glycosides, alkaloids, flavonoids, arabic acid, polyphenols, hydrocarbons, phenolic acids, and fatty acids, play crucial roles in acacia gum's biological activities and potential pharmacological benefits. Studies have demonstrated the antimicrobial activity of *Acacia spp*. against various bacteria and fungi, highlighting their potential as natural antimicrobial agents for therapeutic applications. Furthermore, AGs-SME have been linked to their nutritional quality, anthelmintic properties, and impact on animal feed value, underscoring their significance in medicine, agriculture, and ecological interactions. The rich phytochemical composition of these AGs emphasizes their potential for different applications and importance in traditional medicinal uses and drug discovery. In conclusion, case studies and research findings on AGs-SME provide valuable insights into the diverse biological applications and therapeutic properties of these compounds. Further research is essential to fully understand the mechanisms of action, optimize their applications, and unlock their full potential in medicine, agriculture, and industry.

4.4 Challenges and Limitations

Challenges and limitations in the study of AGs-SMEs

1. Challenges in Extraction and Purification
   - (i) Extraction and purification of AGs-SME are challenging due to their complex nature and the need for specific techniques to isolate them effectively.
   - (ii) Variability in the chemical composition of Acacia gums can make the extraction and purification processes challenging, requiring tailored methods to isolate specific bioactive compounds.

2. Variability in Chemical Composition and Its Impact on Biological Activity
   (i) The variability in the chemical composition of Acacia gums among different species can affect the biological activity of AGs-SME.
   (ii) Differences in the concentrations and types of SME among AGs may lead to variations in their pharmacological properties and health benefits, thereby affecting their efficacy in different applications.
3. Limitations of Current Research and Potential Areas for Improvement
   (i) Current research on AGs-SME may have limitations in understanding the full spectrum of their biological activities and mechanisms of action.
   (ii) Further exploration and research are needed to uncover these SME's complete potential, identify their specific mechanisms of action, and optimize their applications in medicine, agriculture, and industry.

4.5 Future Directions and Applications

1. *Emerging Technologies in Extraction and Analysis*
   (i) Advancements in extraction and analysis technologies can enhance the efficiency and accuracy of isolating and characterizing SME from acacia gum.
   (ii) Techniques such as chromatography, mass spectrometry, and bioinformatics can improve the extraction, purification, and identification of bioactive compounds for various applications.
2. *Potential New Applications in Medicine and Industry.*
   (i) The diverse AGs-SMEs offer potential medical, agriculture, and industry applications beyond their known properties.
   (ii) These bioactive compounds could be explored for their therapeutic effects in treating various diseases, for agricultural applications such as biocontrol agents, and in industrial settings for the development of nutraceuticals and pharmaceuticals.
3. *Role of Sustainable Sourcing and Production*
   (i) Sustainable practices for sourcing and producing acacia gums can ensure these valuable natural resources' long-term availability and environmental impact.
   (ii) Implementing sustainable sourcing methods, such as ethical harvesting and cultivation practices, can help preserve biodiversity and ecosystem balance while meeting the increasing demand for acacia gum products.

4.6 Conclusion

The chapter findings on AGs-SME highlight their diverse biological applications and significant potential across various fields. The bioactive compounds present in AGs-SME exhibit a range of beneficial properties, including antimicrobial, anti-inflammatory, antioxidant, neuroprotective, anti-cancer, immunomodulatory, anti-diabetic, cardiovascular protective, hepatoprotective, wound-healing, pain-relief, anti-aging, antiviral, anti-obesity, and antifungal effects. These AGs SME interact with specific cellular pathways and structures, influencing processes, such as angiogenesis, invasion, and cell cycle regulation. The potential applications of these compounds extend beyond their known properties, with promising uses in medicine, agriculture, and industry. The need to explore their full potential, identify specific mechanisms of action, and optimize their applications in various fields underscores the importance of further research on AGs-SME. The diverse chemical structures of these metabolites enable them to target specific cellular structures and pathways, thereby offering a wide array of potential applications in clinical, agricultural, and industrial settings. In conclusion, the promising biological applications of AGs-SME underscore their versatility and potential impact across different sectors. Further research is essential to unlock the full potential of these bioactive compounds, paving the way for advancements in science, medicine, agriculture, and industry.

Acknowledgments The authors thank the Chemical and Process Engineering Technology at the University of Malaysia, Pahang-UMPSA, Malaysia, for supporting this research.

Fund This research was funded by Eastern Unity Technology under Grant No. UIC 190806.

References

Abdelgadir, M. I. (2023). Expected benefits of administration of tannins extracted from Acacia seyal plants: A biochemical and traditional medicine overview. *International Journal of Scientific World, 9*(1), 12–15. https://doi.org/10.14419/5ksxnn82

Abraham, G., Kechero, Y., & Andualem, D. (2023). Nutritional quality of indigenous legume browse in the southern Ethiopia: farmers' preference and correlation of local valuation of feed value with scientific indicators. *Frontiers in Veterinary Science, 10*, 1198212. https://doi.org/10.20944/preprints202302.0034.v2

Abu-Serie, M., Hamouda, A., & Habashy, N. (2021). Acacia senegal gum attenuates systemic toxicity in ccl4-intoxicated rats via regulation of the ros/nf-κb signaling pathway. *Scientific Reports, 11*(1), 20316. https://doi.org/10.1038/s41598-021-99953-y

Adam, I., & Fadl, K. (2011). Effect of tapping tools and date of tapping on Acacia polyacantha gum yield in southern Kordofan state, Sudan. *Journal of Forestry Research, 22*(1), 117–122. https://doi.org/10.1007/s11676-011-0136-1

Adhikari, B., Aryal, B., & Bhattarai, B. (2021). A comprehensive review of the chemical composition and pharmacological activities of A. catechu (l.f.). *Journal of Chemistry, 2021*, 1–11. https://doi.org/10.1155/2021/2575598

Adli, M. (2023). Phytochemical assessment, antioxidant activity, and in-vitro wound-healing potential of *P. minus* huds. *Journal of Current Science and Technology, 14*(1), 18. https://doi.org/10.59796/jcst.v14n1.2024.18

Adra, C. (2023). Untargeted ms-based metabolomic analysis of termite gut-associated Streptomycetes with antifungal activity against Pyrrhoderma noxium. *Antibiotics, 12*(9), 1373. https://doi.org/10.3390/antibiotics12091373

Agustha, A. (2024). Comparative analysis of antifungal properties in diverse extracts of Arcangelisia flava (L.). *Pharmacy Education, 24*(2), 99–103. https://doi.org/10.46542/pe.2024.242.99103

Ahallil, H., Maskat, M., Abdullah, A., & Sarbini, S. (2020). The effect of Acacia senegal as potential prebiotic on obese gut microbiota. *Food Research, 4*(3), 814–822. https://doi.org/10.26656/fr.2017.4(3).350

Ahmed, R., & Mustafa, D. (2019). Green synthesis of silver nanoparticles mediated by traditionally used medicinal plants in Sudan. *International Nano Letters, 10*(1), 1–14. https://doi.org/10.1007/s40089-019-00291-9

Ahmed, M., Ameyaw, E., Armah, F., Acheampong, D., Gathumbi, P., Adinortey, M., et al. (2022). In-vitro and in-vivo toxicological evaluation of Avicennia africana P: Beauv. (Avicenniaceae) leaf extract in a rat model. *Journal of Toxicology, 2022*, 1–11. https://doi.org/10.1155/2022/3434383

Almeida, D., Silva, R., Luna, J., Rufino, R., Santos, V., Banat, I., et al. (2016). Biosurfactants: Promising molecules for petroleum bio-technology advances. *Frontiers in Microbiology, 7*, 1718. https://doi.org/10.3389/fmicb.2016.01718

Almohaimeed, H., Amin, H., El-Aziz, G., & Saleh, H. (2018). Arabic gum acacia improves diabetic peripheral neuropathy in rats: A biochemical and histopathological evidence. *International Journal of Basic & Clinical Pharmacology, 7*(6), 1065. https://doi.org/10.18203/2319-2003.ijbcp20181965

AL-Mousawi, Z. (2021). Some physiological effect of acacia gum on acute renal failure induced by gentamicin in male rabbits. *Indian Journal of Forensic Medicine & Toxicology, 15*(3), 3850–3855. https://doi.org/10.37506/ijfmt.v15i3.15895

Anderson, D., & Eastwood, M. (1989). The safety of gum arabic as a food additive and its energy value as an ingredient: A brief review. *Journal of Human Nutrition and Dietetics, 2*(3), 137–144. https://doi.org/10.1111/j.1365-277x.1989.tb00017.x

Anderson, D. M. W., & McDougall, F. J. (1987). Degradative studies of gum arabic (Acacia senegal (L.) Willd.) with special reference to the fate of the amino acids present.

Anokwah, D., Kwatia, E., Amponsah, I., Jibira, Y., Harley, B., Ameyaw, E., et al. (2022). Evaluation of the anti-inflammatory and antioxidant potential of the stem-bark extract and some constituents of Aidia genipiflora (DC.) dandy (rubiaceae). *Heliyon, 8*(8), e10082. https://doi.org/10.1016/j.heliyon.2022.e10082

Aryal, B., Adhikari, B., Aryal, N., Bhattarai, B., Khadayat, K., & Parajuli, N. (2021). Lc-hrms profiling and anti-diabetic, antioxidant, and antibacterial activities of Acacia catechu (L.F.) Willd. *BioMed Research International, 2021*, 1–16. https://doi.org/10.1155/2021/7588711

Ashour, M., Fatima, W., Imran, M., Ghoneim, M., Alshehri, S., & Shakeel, F. (2022). A review on the main phytoconstituents, traditional uses, inventions, and patent literature of gum arabic emphasizing Acacia seyal. *Molecules, 27*(4), 1171. https://doi.org/10.3390/molecules27041171

Ashu, F., Na-Iya, J., Wamba, B., Kamga, J., Nayim, P., Ngameni, B., et al. (2020). Antistaphylococcal activity of extracts, fractions, and compounds of Acacia polyacantha Wild (Fabaceae). *Evidence-based Complementary and Alternative Medicine, 2020*, 1–10. https://doi.org/10.1155/2020/2654247

Awadasseid, A., Adam, I., Jamal, M., & Teia, F. (2019). Physicochemical properties of Kakamut gum (Acacia polyacantha) and Hashab gum (Acacia senegal): a comparative analysis. *International Journal of Applied Agricultural Sciences, 5*(6), 138. https://doi.org/10.11648/j.ijaas.20190506.12

Babiker, E., Al-Juhaimi, F., Alqah, H., Adisa, A., Adiamo, O., Ahmed, I., et al. (2019). The effect of Acacia nilotica seed extract on the physicochemical, microbiological and oxidative stability of chicken patties. *Journal of Food Science and Technology, 56*(8), 3910–3920. https://doi.org/10.1007/s13197-019-03862-y

Badmi, R. (2023). SME and their role in strawberry defense. *Planning Theory, 12*(18), 3240. https://doi.org/10.3390/plants12183240

Bakhoum, N., Galiana, A., Roux, C., Kane, A., Duponnois, R., Ndoye, F., et al. (2014). Phylogeny of nodulation genes and symbiotic diversity of Acacia senegal (L.) Willd. and A. seyal (Del.) Mesorhizobium strains from different regions of Senegal. *Microbial Ecology, 69*(3), 641–651. https://doi.org/10.1007/s00248-014-0507-1

Bakshi, J., Mehra, M., Grewal, S., Dhingra, D., & Kumari, S. (2022). Synthesis, characterization and evaluation of in-vitro antimicrobial and anti-diabetic activity of berberine encapsulated in guar-acacia gum nanocomplexes. *Journal of Bioactive and Compatible Polymers, 37*(4), 233–251. https://doi.org/10.1177/08839115221106700

Balakrishnan, R., Cho, D., Kim, I., Seol, S., & Choi, D. (2022). Molecular mechanisms and therapeutic potential of α- and β-asarone in the treatment of neurological disorders. *Antioxidants, 11*(2), 281. https://doi.org/10.3390/antiox11020281

Ballal, A., Bobbala, D., Qadri, S., Föller, M., Kempe, D., Nasir, O., et al. (2011). Anti-malarial effect of gum arabic. *Malaria Journal, 10*(1), 139. https://doi.org/10.1186/1475-2875-10-139

Bampidis, V., Azimonti, G., Bastos, M., Dusemund, B., Durjava, M., Kouba, M., et al. (2022). Safety and efficacy of a feed additive consisting of acacia gum (gum Arabic) for all animal species (A.I.P.G. Association for International Promotion of Gums). *EFSA Journal, 20*(4), e07252. https://doi.org/10.2903/j.efsa.2022.7252

Batiha, G., Akhtar, N., Alsayegh, A., Abusudah, W., Almohmadi, N., Shaheen, H., et al. (2022). Bioactive compounds, pharmacological actions, and pharmacokinetics of genus Acacia. *Molecules, 27*(21), 7340. https://doi.org/10.3390/molecules27217340

Bautista-Crescencio, C. (2023). Streptomyces albidoflavus Q antifungal metabolites inhibit the ergosterol biosynthesis pathway and yeast growth in fluconazole-resistant Candida glabrata: Phylogenomic and metabolomic analyses. *Microbiology Spectrum, 11*(5), e0127123. https://doi.org/10.1128/spectrum.01271-23

Beck, F. (2024). Fermented rapeseed and soybean alone and in combination with macro algae inhibit human and pig pathogenic bacteria in-vitro. *Microorganisms, 12*(5), 891. https://doi.org/10.3390/microorganisms12050891

Bhardwaj, M., Yadav, P., Vashishth, D., Sharma, K., Kumar, A., Chahal, J., et al. (2021). A review on obesity management through natural compounds and a green nanomedicine-based approach. *Molecules, 26*(11), 3278. https://doi.org/10.3390/molecules26113278

Bhateja, P., & Singh, R. (2014). Anti-diabetic activity of Acacia tortilis (Forsk.) Hayne ssp. raddiana polysaccharide on streptozotocin-nicotinamide induced diabetic rats. *BioMed Research International, 2014*, 1–9. https://doi.org/10.1155/2014/572013

Bhatia, S., Al-Harrasi, A., Shah, Y., Altoubi, H., Sharma, P., Anwer, M., et al. (2023). Fabrication, characterization, and antioxidant potential of sodium alginate/acacia gum hydrogel-based films loaded with cinnamon essential oil. *Gels, 9*(4), 337. https://doi.org/10.3390/gels9040337

Binyane, M., & Mfengwana, P. (2022). Traditional medicinal plants as the potential adjuvant, prophylactic and treatment therapy for covid-19 disease: A review. In *Medicinal plants*. IntechOpen. https://doi.org/10.5772/intechopen.104491

Boccella, S. (2023). Synergistic effects of Boswellia serrata and Acmella oleracea extract combination for treating neuropathic pain in a preclinical model of spared nerve injury. *Phytotherapy Research, 38*(4), 1731–1734. https://doi.org/10.1002/ptr.8001

Bouhlali, E., Derouich, M., Ben-Amar, H., Meziani, R., & Essarioui, A. (2020). Exploring the potential of using bioactive plant products in the management of Fusarium oxysporum f.sp. albedinis: The causal agent of Bayoud disease on date palm (Phoenix dactylifera l.). *Beni-Suef University Journal of Basic and Applied Sciences, 9*(1), 46. https://doi.org/10.1186/s43088-020-00071-x

Calame, W., Weseler, A., Viebke, C., Flynn, C., & Siemensma, A. (2008). Gum arabic establishes prebiotic functionality in healthy human volunteers in a dose-dependent manner. *The British Journal of Nutrition, 100*(6), 1269–1275. https://doi.org/10.1017/s0007114508981447

Caseiro, C., Dias, J., Fontes, C., & Bule, P. (2022). From cancer therapy to winemaking: The molecular structure and applications of β-glucans and β-1, 3-glucanases. *International Journal of Molecular Sciences, 23*(6), 3156. https://doi.org/10.3390/ijms23063156

Cedillo-Cortezano, M. (2024). Use of medicinal plants in the process of wound-healing: A literature review. *Pharmaceuticals, 17*(3), 303. https://doi.org/10.3390/ph17030303

Cheon, C. (2021). Synergistic effects of herbal medicines and anti-cancer drugs. *Medicine, 100*(46), e27918. https://doi.org/10.1097/md.0000000000027918

Cherbut, C., Michel, C., Raison, V., Kravtchenko, T., & Meance, S. (2003). Acacia gum is a bifidogenic dietary fiber with high digestive tolerance in healthy humans. *Microbial Ecology in Health and Disease, 15*(1), 43–50. https://doi.org/10.1080/08910600310014377

Chu, W., & Phang, S. (2016). Marine algae as a potential source for anti-obesity agents. *Marine Drugs, 14*(12), 222. https://doi.org/10.3390/md14120222

Chundakkattumalayil, H., Kumar, S., Narayanan, R., & Raghavan, K. (2019). Role of L. plantarum KX519413 as probiotic and acacia gum as prebiotic in gastrointestinal tract strengthening. *Microorganisms, 7*(12), 659. https://doi.org/10.3390/microorganisms7120659

Correia, R., Quintela, J., Duarte, M., & Goncalves, M. (2020). Insights for the valorization of biomass from Portuguese invasive Acacia spp. in a biorefinery perspective. *Forests, 11*(12), 1342. https://doi.org/10.3390/f11121342

Damavandi, M. (2023). The anti-cancer and antibacterial potential of bioactive SME derived from bacterial endophytes in association with Artemisia absinthium. *Scientific Reports, 13*(1), 18473. https://doi.org/10.1038/s41598-023-45910-w

Dauqan, E., & Abdullah, A. (2013). Utilization of gum arabic for industries and human health. *American Journal of Applied Sciences, 10*(10), 1270–1279. https://doi.org/10.3844/ajassp.2013.1270.1279

Delavarian, M., Imani, M., Delavarian, F., & Bayani, S. (2020). Comparison of chewing gum and ibuprofen in alleviating orthodontic pain: A single center, randomized clinical trial. *Australasian Orthodontic Journal, 36*(1), 38–44. https://doi.org/10.21307/aoj-2020-005

Dini, S. (2024). The physicochemical and functional properties of biosurfactants: A review. *Molecules, 29*(11), 2544. https://doi.org/10.3390/molecules29112544

Duarte, M., & Wolf, S. (2005). Anatomical characters of the phyllode and stem of Acacia podalyriifolia A. cunn. ex G. Don (Fabaceae). *Revista Brasileira De Farmacognosia, 15*(1), 71–76. https://doi.org/10.1590/s0102-695x2005000100015

Dutta, P., Kumari, A., Mahanta, M., Biswas, K., Dudkiewicz, A., Thakuria, D., et al. (2022). Advances in nanotechnology as a potential alternative for plant viral disease management. *Frontiers in Microbiology, 13*, 935193. https://doi.org/10.3389/fmicb.2022.935193

Eitsuka, T., Tatewaki, N., Nishida, H., Nakagawa, K., & Miyazawa, T. (2016). Synergistic anticancer effect of tocotrienol combined with chemotherapeutic agents or dietary components: A review. *International Journal of Molecular Sciences, 17*(10), 1605. https://doi.org/10.3390/ijms17101605

El-Darier, S., Rashed, S., Fayez, A., Abdel-Mawgood, A., Sharaby, M., Tawfik, N., et al. (2021). Medicinal plant-derived compounds as potential phytotherapy forcovid-19: Future perspectives. *Journal of Pharmacognosy and Phytotherapy, 13*(3), 68–81. https://doi.org/10.5897/jpp2021.0603

El-Hadiyah, T., Abdulhadi, N., Badico, E., & Mohammed, E. (2011). Toxic potential of ethanolic extract of *Acacia nilotica* (Garad) in rats. *Sudan Journal of Medical Sciences, 6*(1), 67269. https://doi.org/10.4314/sjms.v6i1.67269

Elhasan, L., Ahmed, B., Abdelgadir, T., Dafaallah, S., & Othman, O. (2022). The effect of ethanolic extract of Acacia seyal bark on induced diabetic rats. *bioRxiv*, 2022.01.21.476925. https://doi.org/10.1101/2022.01.21.476925

Elhenawy, A., Omer, E., Alomari, A., & Osman, M. (2014). A comparative analytical studies on Acaciapolyacantha gum samples collected from three different locations in Sudan. *Journal of Advances in Natural Sciences, 1*(1), 1–9. https://doi.org/10.24297/jns.v1i1.379

Elmi, A., Spina, R., Risler, A., Philippot, S., Merito, A., Duval, R., et al. (2020). Evaluation of antioxidant and antibacterial activities, cytotoxicity of Acacia seyal del bark extracts and isolated compounds. *Molecules, 25*(10), 2392. https://doi.org/10.3390/molecules25102392

Elnour, A., Mirghani, M., Kabbashi, N., Alam, Z., & Musa, K. (2018a). Gum arabic: An optimization of ultrasonic-assisted extraction of antioxidant activity. *Studia Universitatis Babeş-Bolyai Chemia, 63*(3), 95–116. https://doi.org/10.24193/subbchem.2018.3.08

Elnour, A., Mirghani, M., Na, K., Alam, Z., & Musa, K. (2018b). Study of antioxidant and anti-inflammatory crude methanol extract and fractions of Acacia seyal gum. *American Journal of Pharmacology and Pharmacotherapeutics, 5*(1), 3. https://doi.org/10.21767/2393-8862.100013

Elnour, A., Mirghani, M., & Kabbashi, N. (2019). Significant bioactive compounds in crude methanol extracts and their fractions of Acacia seyal gum. *International Journal of Sudan Research, 8*(2), 99–112. https://doi.org/10.47556/j.ijsr.8.2.2018.3

Elnour, A., Mirghani, M., Kabbashi, N., Daddiouaissa, D., Musa, K., Alam, Z., et al. (2020). Active fractions of methanol crude obtained from Acacia seyal gum and their anti-proliferative effects against human breast cancer cell lines. *Global Journal of Science Frontier Research, 20*(C3), 51–64. https://doi.org/10.34257/gjsfrcvol20is3pg51

Escárcega-González, C., Garza-Cervantes, J., Vázquez-Rodriguez, A., Montelongo-Peralta, L., Treviño-González, M., Barriga, C., et al. (2018). In-vivo antimicrobial activity of silver nanoparticles produced via a green chemistry synthesis using *Acacia rigidula* as a reducing and capping agent. *International Journal of Nanomedicine, 13*, 2349–2363. https://doi.org/10.2147/ijn.s160605

Fathi, F., Ebrahimi, S., Pereira, D., Estevinho, B., & Rocha, F. (2021). Microencapsulation of enriched extracts of two Satureja species by spray-drying, evaluation of the controlled-release mechanism and cytotoxicity. *Pharmaceutical Sciences, 28*(1), 145–155. https://doi.org/10.34172/ps.2021.54

Fatima, Khan, M. S., Anwar, M., Rahman, S., & Sajad, M. A. (2019). Antifungal activity of crude extracts of stem-bark of Acacia ampliceps Maslin (Family Leguminosae). *Pure and Applied Biology, 8*(2), 1690–1697. https://doi.org/10.19045/bspab.2019.80112

Fatimah, A. (2019). Antifungal activity and Fourier-transform infrared spectrometric characterization of aqueous extracts of Acacia senegal and Acacia tortilis on phytopathogenic fungi. *Journal of Pharmaceutical Research International, 31*(2), 1–11. https://doi.org/10.9734/jpri/2019/v31i230295

Fauconnier, M., Blecker, C., Groin, J., Razafindralambo, H., Vanzeveren, E., Marlier, M., et al. (2000). Characterization of two acacia gums and their fractions using a Langmuir film balance. *Journal of Agricultural and Food Chemistry, 48*(7), 2709–2712. https://doi.org/10.1021/jf990749x

Fazmiya, M. (2024). Efficacy of a vaginal suppository formulation prepared with Acacia arabica (Lam.) Willd. gum and Cinnamomum camphora (L.) J. Presl. in heavy menstrual bleeding analyzed using a machine-learning technique. *Frontiers in Pharmacology, 15*, 1331622. https://doi.org/10.3389/fphar.2024.1331622

Gashua, I., Williams, P., & Baldwin, T. (2016). Molecular characteristics, association and interfacial properties of gum arabic harvested from both Acacia senegal and Acacia seyal. *Food Hydrocolloids, 61*, 514–522. https://doi.org/10.1016/j.foodhyd.2016.06.005

Gélinas, M., Museau, L., Milot, A., & Beauregard, P. (2020). Cellular adaptation and the importance of the purine biosynthesis pathway during biofilm formation in gram-positive pathogens. *bioRxiv*, 2020.12.11.422287. https://doi.org/10.1101/2020.12.11.422287

Ghasemzadeh, A., Jaafar, H., Rahmat, A., & Ashkani, S. (2015). SME constituents and antioxidant, anti-cancer and antibacterial activities of Etlingera elatior (Jack) R.M.Sm grown in different Malaysia locations. *BMC Complementary and Alternative Medicine, 15*(1), 4361–4370. https://doi.org/10.1186/s12906-015-0838-6

Gim, S., Zhu, Y., Seeberger, P., & Delbianco, M. (2019). Carbohydrate-based nanomaterials for biomedical applications. *Wiley Interdisciplinary Reviews. Nanomedicine and Nanobiotechnology, 11*(5), e1558. https://doi.org/10.1002/wnan.1558

Giri, A., Gautam, M., Roka, A., Bhattarai, N., & Bhattarai, J. (2023). Performance of anticorrosive measures of steel in concrete infrastructure by plant-based extracts. *Macromolecular Symposia, 410*(1), 2200115. https://doi.org/10.1002/masy.202200115

Gomes, K., Duarte, R., & Bastos, M. (2017). Lantibiotics produced by actinobacteria and their potential applications (a review). *Microbiology, 163*(2), 109–121. https://doi.org/10.1099/mic.0.000397

Gopal, T., Muthukumaran, P., Kumar, B., Muthusamy, V., & Lakshmi, B. (2016). Selective inhibition of ptp1b by Vitalboside A from Syzygium cumini enhances insulin sensitivity and attenuates lipid accumulation via partial agonism to PPARγ: In-vitro and in silico investigation. *Chemical Biology & Drug Design, 88*(2), 302–312. https://doi.org/10.1111/cbdd.12757

Gorish, B., & Alamein, U. (2021). Antimicrobial activity of Acacia nilotica pods against carbapenem-resistant gram-negative bacteria isolated from different hospitals in Khartoum state. *Journal of Pharmaceutical Research & Reports, 2*(3), 1–6. https://doi.org/10.47363/jprsr/2021(2)120

Habli, Z., Toumieh, G., Fatfat, M., Rahal, O., & Gali-Muhtasib, H. (2017). Emerging cytotoxic alkaloids in the battle against cancer: Overview of molecular mechanisms. *Molecules, 22*(2), 250. https://doi.org/10.3390/molecules22020250

Haji, M., Mohamed, A., Albashir, A., & Mirghani, M. (2022). A review of antifungal activity of combined plant extracts or plant exudates from medicinal plants either together or with known antifungal agents. *European Journal of Medicinal Plants, 33*(8), 16–47. https://doi.org/10.9734/ejmp/2022/v33i830483

Hannachi, H. (2011). Chemicals profiling and antioxidants activities of acacia seeds. *Journal of Medicinal Plants Research, 5*(31), 6869–6875. https://doi.org/10.5897/jmpr11.1390

Huang, L., Chen, J., Cao, P., Pan, H., Ding, C., Xiao, T., et al. (2015). Anti-obese effect of glucosamine and chitosan oligosaccharide in high-fat diet-induced obese rats. *Marine Drugs, 13*(5), 2732–2756. https://doi.org/10.3390/md13052732

Hunt, N., Suleman, L., Josling, P., & Popov, T. (2021). Virucidal activity of Nasaleze® Cold & Flu Blocker and Nasaleze® Travel in cell cultures infected with human pathogenic Coronavirus 229-e. *bioRxiv*, 2021.09.23.461483. https://doi.org/10.1101/2021.09.23.461483

Ibrahim, H., Omosa, L., Nchiozem-Ngnitedem, V., Onyari, J., Maru, S., Guefack, M., et al. (2021). Antibacterial activities and phytochemical screening of crude extracts from Kenyan macaranga species toward MDR phenotypes expressing efflux pumps. *Pharmacognosy Communications, 11*(2), 119–126. https://doi.org/10.5530/pc.2021.2.22

Imeson, A. P. (1992). Exudate gums. In *Thickening and gelling agents for food* (pp. 66–97). Springer. https://doi.org/10.1007/978-1-4615-3552-2_4

Ireland, H., Clutterbuck, A., Cloquet, J., Thurston, M., Williams, P., Cronk, Q., et al. (2004). The development of immunoassays to identify and quantify species source of gum arabic. *Journal of Agricultural and Food Chemistry, 52*(26), 7804–7808. https://doi.org/10.1021/jf048864h

Islam, R., Sun, L., & Zhang, L. (2021). Biomedical applications of Chinese herb-synthesized silver nanoparticles by phytonanotechnology. *Nanomaterials, 11*(10), 2757. https://doi.org/10.3390/nano11102757

Jaafar, N. (2019). Clinical effects of arabic gum (acacia): A mini review. *Iraqi Journal of Pharmaceutical Sciences*. (P-Issn 1683–3597 E-Issn 2521–3512), *28*(2), 9–16. https://doi.org/10.31351/vol28iss2pp9-16

Jha, V., Jain, T., Nikumb, D., Gharat, Y., Koli, J., Jadhav, N., et al. (2022). Streptomyces peucetius M1 and Streptomyces lavendulae M3 soil isolates as a promising source for antimicrobials discovery. *Journal of Pharmaceutical Research International, 34*(50B), 7–19. https://doi.org/10.9734/jpri/2022/v34i50b36438

Jin, H., Huang, W., Zhu, X., Zhou, Y., & Yan, D. (2012). Biocompatible or biodegradable hyperbranched polymers: From self-assembly to cytomimetic applications. *Chemical Society Reviews, 41*(18), 5986. https://doi.org/10.1039/c2cs35130g

Johnson, W. (2005). Final report of the safety assessment of Acacia catechu gum, Acacia concinna fruit extract, Acacia dealbata leaf extract, Acacia dealbata leaf wax, Acacia decurrens extract, Acacia farnesiana extract, Acacia farnesiana flower wax, Acacia farnesiana gum, AGs gum extract. *International Journal of Toxicology, 24*, 75–118.

Kabbashi, N. (2022). Extraction of bioactive compound from Acacia seyal gum, in-vitro evaluation of anti-tumor activity of its crude extract against leukemia. *International Journal of Plant-Based Pharmaceuticals, 2*(1), 82–88. https://doi.org/10.62313/ijpbp.2022.18

Kaddam, L., Fadl-Elmula, I., Eisawi, O., Abdelrazig, H., & Saeed, A. (2019). Acacia senegal (gum arabic) supplementation modulate lipid profile and ameliorated dyslipidemia among sickle cell anemia patients. *Journal of Lipids, 2019*, 1–5. https://doi.org/10.1155/2019/3129461

Kallifidas, D., Jiang, G., Ding, Y., & Luesch, H. (2018). Rational engineering of Streptomyces albus J1074 for the overexpression of secondary metabolite gene clusters. *Microbial Cell Factories, 17*(1), 25. https://doi.org/10.1186/s12934-018-0874-2

Kanimozhi, A., & Rose, J. (2023). Screening and evaluation of potential antifungal plant extracts against the skin-infecting fungus Trichophyton rubrum. *Pharmaceutical Research, 15*(2), 328–337. https://doi.org/10.5530/pres.15.2.035

Khaled, K., & Irani, R. (2021). Antioxidant activity and anti-nutritional factors in Acacia nilotica gum. *International Journal of Pharma and Bio Sciences, 11*(6), 42–51. https://doi.org/10.22376/ijpbs/lpr.2021.11.3.l42-51

Kim, S., Woo, E., & Lee, D. (2018). Synergistic antifungal activity of isoquercitrin: Apoptosis and membrane permeabilization related to reactive oxygen species in Candida albicans. *IUBMB Life, 71*(2), 283–292. https://doi.org/10.1002/iub.1973

Kongkatitham, V. (2024). Diverse modulatory effects of bibenzyls from dendrobium species on human immune cell responses under inflammatory conditions. *PLoS One, 19*(2), e0292366. https://doi.org/10.1371/journal.pone.0292366

Lelon, J., Jumba, I., Keter, J., & Oduor, F. (2013). Influence of Acacia Ssenegal varieties on quality of gum arabic in Baringo district, Kenya. *African Journal of Plant Science, 7*(6), 190–200. https://doi.org/10.5897/ajps10.185

Lim, A., Subhan, N., Jazayeri, J., John, G., Vanniasinkam, T., & Obied, H. (2016). Plant phenols as antibiotic boosters: In-vitro interaction of olive leaf phenols with ampicillin. *Phytotherapy Research, 30*(3), 503–509. https://doi.org/10.1002/ptr.5562

Liu, H., Wang, Y., Yang, Q., Zhao, W., Cui, L., Wang, B., et al. (2020). Genomics and lc-ms reveal diverse active SME in Bacillus amyloliquefaciens WS-8. *Journal of Microbiology and Biotechnology, 30*(3), 417–426. https://doi.org/10.4014/jmb.1906.06055

Lopez-Torrez, L., Nigen, M., Williams, P., Doco, T., & Sánchez, C. (2015). Acacia senegal vs. Acacia seyal gums—Part 1: Composition and structure of hyperbranched plant exudates. *Food Hydrocolloids, 51*, 41–53. https://doi.org/10.1016/j.foodhyd.2015.04.019

Mambe, F., Na-Iya, J., Fotso, G., Ashu, F., Ngameni, B., Ngadjui, B., et al. (2019). Antibacterial and antibiotic-modifying potential of crude extracts, fractions, and compounds from Acacia polyacantha willd. Against MDR gram-negative bacteria. *Evidence-based Complementary and Alternative Medicine, 2019*, 1–13. https://doi.org/10.1155/2019/7507549

Mengesha, A., Birru, E., & Adugna, M. (2022). Anti-diarrheal activities of hydromethanolic crude extract and solvent fractions of Acacia seyal (Fabaceae) roots in mice. *Clinical Pharmacology Advances and Applications, 14*, 99–110. https://doi.org/10.2147/cpaa.s383896

Mhinzi, G. (2003). Intra-species variation of the properties of gum exudates from *Acacia senegal* var. *Senegal* and *Acacia seyal* var. *fistula* from Tanzania. *Bulletin of the Chemical Society of Ethiopia, 17*(1), 67–74. https://doi.org/10.4314/bcse.v17i1.61734

Mirghani, M., Elnour, A., Kabbashi, N., Alam, Z., Musa, K., & Abdullah, A. (2018). Determination of antioxidant activity of gum arabic: An exudation from two different locations. *ScienceAsia, 44*(3), 179. https://doi.org/10.2306/scienceasia1513-1874.2018.44.179

Mohamed, R. (2023). Effect of prebiotic dietary supplement Acacia senegal on hormonal and metabolic markers in polycystic ovary syndrome patients: A pilot study. *Cureus, 15*(9), e45480. https://doi.org/10.7759/cureus.45480

Morales, J., Mendoza, L., & Cotoras, M. (2017). Alteration of oxidative phosphorylation as a possible mechanism of the antifungal action of p-coumaric acid against Botrytis cinerea. *Journal of Applied Microbiology, 123*(4), 969–976. https://doi.org/10.1111/jam.13540

Mortensen, A., Aguilar, F., Crebelli, R., Domenico, A., Frutos, M., Galtier, P., et al. (2017). Re-evaluation of acacia gum (E 414) as a food additive. *EFSA Journal, 15*(4), e04741. https://doi.org/10.2903/j.efsa.2017.4741

Mueller, A., Brockmueller, A., Kunnumakkara, A., & Shakibaei, M. (2022). Modulation of inflammation by plant-derived nutraceuticals in tendinitis. *Nutrients, 14*(10), 2030. https://doi.org/10.3390/nu14102030

Mugo, E., Mahungu, S., Chikamai, B., & Mwove, J. (2020). Evaluation of gum arabic from Acacia senegal var kerensis and Acacia senegal var senegal as a stabilizer in low-fat yogurt. *International Journal of Food Studies, 9*, 210952003. https://doi.org/10.7455/ijfs/9.si.2020.a9

Musarra-Pizzo, M., Pennisi, R., Ben-Amor, I., Mandalari, G., & Sciortino, M. (2021). Antiviral activity exerted by natural products against human viruses. *Viruses, 13*(5), 828. https://doi.org/10.3390/v13050828

Mwove, J., Gogo, L., Chikamai, B., Omwamba, M., & Mahungu, S. (2016). Preparation and quality evaluation of extended beef rounds containing gum Arabic from *Acacia senegal* var. *kerensis*. *Food and Nutrition Sciences, 7*(11), 977–988. https://doi.org/10.4236/fns.2016.711096

Nigen, M., Apolinar-Valiente, R., Iturmendi, N., Williams, P., Doco, T., Moine, V., et al. (2019). The colloidal stabilization of young red wine by Acacia senegal gum: The involvement of the protein backbone from the protein-rich arabinogalactan-proteins. *Food Hydrocolloids, 97*, 105176. https://doi.org/10.1016/j.foodhyd.2019.105176

Nugraheni, K., Hutchinson, S., Thorne, J., & Boesch, C. (2018). Modulation of nuclear receptor liver x receptor alpha by polyphenols. *The Proceedings of the Nutrition Society, 77*(OCE4), 10–12. https://doi.org/10.1017/s0029665118002422

Nurlila, R. (2024). Test of antihyperuricemia activity of methanol extract of plantain peel (Musa paradisiaca var. sapientum) against caffeine- induced mice (Mus musculus). *The Pharmaceutical Journal, 16*(1), 226–230. https://doi.org/10.5530/pj.2024.16.32

Okonkwo-Uzor, N., Okezie, U., Eze, P., Ikegbune, C., Okoye, F., & Esimone, C. (2022). In-vitro antimicrobial activity of SME of an endophytic aspergillus brunneoviolaceus isolated from Bryophyllum pinnatum (Lam.) oken. *Journal of Current Biomedical Research, 2*(4), 249–254. https://doi.org/10.54117/jcbr.v2i4.1

Olajuyigbe, O., & Afolayan, A. (2012). Synergistic interactions of methanolic extract of Acacia mearnsii De Wild. with antibiotics against bacteria of clinical relevance. *International Journal of Molecular Sciences, 13*(7), 8915–8932. https://doi.org/10.3390/ijms13078915

Olaokun, O., Mkolo, N., Mogale, M., & King, P. (2017). Phytochemical screening, antioxidant, anti-inflammatory, and glucose utilization activities of three South African plants used traditionally to treat diseases. *Biologie et Médecine, 09*(05), 89695651. https://doi.org/10.4172/0974-8369.1000412

Olas, B. (2022). New perspectives on the effect of dandelion, its food products and other preparations on the cardiovascular system and its diseases. *Nutrients, 14*(7), 1350. https://doi.org/10.3390/nu14071350

Oliveira, C., Moreira, P., Resende, J., Cruz, M., Pereira, C., Silva, A., et al. (2020). Characterization and cytotoxicity assessment of the lipophilic fractions of different morphological parts of Acacia dealbata. *International Journal of Molecular Sciences, 21*(5), 1814. https://doi.org/10.3390/ijms21051814

Osman, W., Mostafa-He, G., Osman, A., Rashed, L., & Labib, D. (2022). Effect of arabic gum (Acacia senegal) on paracetamol-induced chronic nephrotoxicity in albino rats. *International Journal of Pharmacology, 18*(8), 1593–1604. https://doi.org/10.3923/ijp.2022.1593.1604

Palma, S., Carvalho, A., Silva, J., Martins, P., Marciello, M., Fernandes, A., et al. (2015). Covalent coupling of gum arabic onto superparamagnetic iron oxide nanoparticles for MRI cell labeling: Physicochemical and in-vitro characterization. *Contrast Media & Molecular Imaging, 10*(4), 320–328. https://doi.org/10.1002/cmmi.1635

Panjaitan, N., Wang, L., & Handayani, S. (2022). A response to article "anti-diarrheal activity of hydromethanolic crude extract and solvent fractions of Acacia seyal (Fabaceae) roots in mice" [letter]. *Clinical Pharmacology Advances and Applications, 14*, 111–112. https://doi.org/10.2147/cpaa.s398217

Paul, W. (2011). Tocotrienol combination therapy results in synergistic anti-cancer response. *Frontiers in Bio-science-Elite, 16*(1), 3183. https://doi.org/10.2741/3905

Pinheiro, J., Neto, S., Signoretti, R., Neto, J., Henriques, L., & Rufino, M. (2023). Effects of a hepatoprotectant and different feed energy sources on milk composition and nutrient partitioning in lactating cows. *Revista Brasileira de Saúde e Produção Animal, 24*(1), 20220028. https://doi.org/10.1590/s1519-994020220028

Poy, D., & Tohidfar, M. (2023). An in silico investigation and introduction of some SME of medicinal plants with antiviral activity and effective vitamins for the treatment of MPox. *Research Square*. https://doi.org/10.21203/rs.3.rs-2381680/v1

Prayogo, Y., Syafii, W., Sari, R., & Batubara, I. (2021). Pharmacological activity and phytochemical profile of acacia heartwood extracts. *Scientia Pharmaceutica, 89*(3), 37. https://doi.org/10.3390/scipharm89030037

Raina, K., & Agarwal, R. (2007). Combinatorial strategies for cancer eradication by silibinin and cytotoxic agents: Efficacy and mechanisms. *Acta Pharmacologica Sinica, 28*(9), 1466–1475. https://doi.org/10.1111/j.1745-7254.2007.00691.x

Raina, H., Soni, G., Jauhari, N., Sharma, N., & Bharadvaja, N. (2014). Phytochemical importance of medicinal plants as potential sources of anti-cancer agents. *Turkish Journal of Botany, 38*, 1027–1035. https://doi.org/10.3906/bot-1405-93

Ramachandran, C., Quirin, K., & Melnick, S. (2021). Cytotoxic synergism between a proprietary Commiphora mukul gum extract gu-mct810, 2-deoxy-d-glucose, and metformin in human alveolar rhabdomyosarcoma and hepatoma cell lines in-vitro. *Journal of Complementary and Alternative Medical Research, 16*(4), 185–195. https://doi.org/10.9734/jocamr/2021/v16i430307

Rawi, M., Abdullah, A., Ismail, A., & Sarbini, S. (2021). Manipulation of gut microbiota using acacia gum polysaccharide. *ACS Omega, 6*(28), 17782–17797. https://doi.org/10.1021/acsomega.1c00302

Reddy, V., Aryal, P., Robinson, S., Rafiu, R., Obrenovich, M., & Perry, G. (2020). Polyphenols in Alzheimer's disease and in the gut–brain axis. *Microorganisms, 8*(2), 199. https://doi.org/10.3390/microorganisms8020199

Renard, D., Lavenant-Gourgeon, L., Ralet, M., & Sánchez, C. (2006). Acacia senegal gum: Continuum of molecular species differing by their protein to sugar ratio, molecular weight, and charges. *Biomacromolecules, 7*(9), 2637–2649. https://doi.org/10.1021/bm060145j

Rofiee, M., Zakaria, Z., Somchit, M., Zuraini, A., Arifah, A., Teh, L., et al. (2011). Involvement of cyp450 system in hepatoprotective activity of Malaysian agricultural research and development institute (MARDI)-produced virgin coconut oils. *African Journal of Pharmacy and Pharmacology, 5*(23), 2526–2536. https://doi.org/10.5897/ajpp11.021

Sacande, M. (2023). Restoring food systems with nutritious native plants: Experiences from the African drylands. *Food and Nutrition Bulletin, 44*(2_suppl), S58–S68. https://doi.org/10.1177/03795721231190779

Saheed, Y. (2021). Antifungal potential of silver nanoparticles from Acacia nilotica pod against dermatophytes. *Journal of Drug Delivery and Therapeutics, 11*(5-S), 85–95. https://doi.org/10.22270/jddt.v11i5-s.5024

Santoso, P., Minamihata, K., Ishimine, Y., Taniguchi, H., Komada, T., Sato, R., et al. (2022). Enhancement of the antifungal activity of chitinase by palmitoylation and the synergy of palmitoylated chitinase with amphotericin b. *ACS Infectious Diseases, 8*(5), 1051–1061. https://doi.org/10.1021/acsinfecdis.2c00052

Sarkisian, S., Janssen, M., Matta, H., Henry, G., LaPlante, K., & Rowley, D. (2011). Inhibition of bacterial growth and biofilm production by constituents from Hypericum spp. *Phytotherapy Research, 26*(7), 1012–1016. https://doi.org/10.1002/ptr.3675

Satrija, F., Nurhidayah, N., Astuti, D., Retnani, E., & Murtini, S. (2023). The diversity and quality of forages and their potency as herbal anthelmintic for swamp buffalo in Brebes district, central java. *Veterinary World, 2023*, 1496–1504. https://doi.org/10.14202/vetworld.2023.1496-1504

Satti, A. (2024). Analytical data of Acacia nilotica var. Nilotica gum. *Bulletin of the Chemical Society of Ethiopia, 38*(4), 839–852. https://doi.org/10.4314/bcse.v38i4.2

Sekhon, B., Roubin, R., Li, Y., Devi, P., Nammi, S., & Fan, K. (2016). Evaluation of selected immunomodulatory glycoproteins as an adjunct to cancer immunotherapy. *PLoS One, 11*(1), e0146881. https://doi.org/10.1371/journal.pone.0146881

Selim, K. (2012). Biology of endophytic fungi. *Current Research in Environmental & Applied Mycology, 2*(1), 31–82. https://doi.org/10.5943/cream/2/1/3

Sène, M., Diop, N., Diallo, M., Sarr, A., Barboza, F., Ndiaye, M., et al. (2023). Healing, anti-inflammatory and analgesic activities of the hydro-methanolic extract of Acacia nilotica pods (Mimosaceae). *International Journal of Pharmacology, 19*(1), 40–51. https://doi.org/10.3923/ijp.2023.40.51

Sethuraman, S., & Rajendran, K. (2018). Multicharacteristic behavior of tyrosine present in the microdomains of the macromolecule gum arabic at various PH conditions. *ACS Omega, 3*(12), 17602–17609. https://doi.org/10.1021/acsomega.8b02928

Sharifi-Rad, M., Lankatillake, C., Dias, D., Docea, A., Mahomoodally, M., Lobine, D., et al. (2020). Impact of natural compounds on neurodegenerative disorders: From preclinical to pharmacotherapeutics. *Journal of Clinical Medicine, 9*(4), 1061. https://doi.org/10.3390/jcm9041061

Shehu, A., Anyip, B., & Magaji, M. (2020). Antidepressant effect of methanol root bark extract of Acacia seyal Del. (Fabaceae): Possible involvement of the inflammatory pathway. *Tropical Journal of Pharmaceutical Research, 19*(7), 1459–1464. https://doi.org/10.4314/tjpr.v19i7.18

Sheidu, A., Zezi, A., Ahmed, A., Chindo, B., & Magaji, G. (2021). Antimicrobial and wound-healing properties of the methanol extract of Ficus platyphylla Del. (Moraceae) stem-bark. *Journal of Pharmacy & Bioresources, 17*(2), 121–130. https://doi.org/10.4314/jpb.v17i2.6

Sieniawska, E., & Georgiev, M. I. (2022). Metabolomics: towards acceleration of antibacterial plant-based leads discovery. *Phytochemistry Reviews, 21*(3), 765–781.

Silva, E., Fernandes, S., Bacelar, E., & Sampaio, A. (2016). Antimicrobial activity of aqueous, ethanolic and methanolic leaf extracts from Acacia spp. and Eucalyptus nicholii. *African journal of traditional, complementary and alternative medicines, 13*(6), 130–134.

Silva, T., & Souza, J. (2018). The influence of environmental factors in the production of gum arabic (Acacia senegal, Fabaceae)—A meta-analysis. *Revista Brasileira De Climatologia, 23*(Jul/Dez), 33–44. https://doi.org/10.5380/abclima.v23i0.58884

Sitthiyotha, T., Pichyangkura, R., & Chunsrivirot, S. (2018). Molecular dynamics provides insight into how n251a and n251y mutations in the active site of Bacillus licheniformis RN-01 levansucrase disrupt production of long-chain levan. *PLoS One, 13*(10), e0204915. https://doi.org/10.1371/journal.pone.0204915

Son, H., Shin, H., Jang, E., Moon, B., Lee, C., & Lee, J. (2020). Gochujang prepared using rice and wheat koji partially alleviates high-fat diet-induced obesity in rats. *Food Science & Nutrition, 8*(3), 1562–1574. https://doi.org/10.1002/fsn3.1443

Souza-Alonso, P., Puig, C., Pedrol, N., Freitas, H., Rodriguez-Echeverría, S., & Lorenzo, P. (2018). Exploring the use of residues from the invasive Acacia sp. for weed control. *Renewable Agriculture and Food Systems, 35*(1), 26–37. https://doi.org/10.1017/s1742170518000170

Sowndhararajan, K., Hong, S., Jhoo, J., Kim, S., & Chin, N. (2015). Effect of acetone extract from stem-bark of AGs (A. dealbata, A. ferruginea and A. leucophloea) on antioxidant enzymes status in hydrogen peroxide-induced HepG2 cells. *Saudi Journal of Biological Sciences, 22*(6), 685–691. https://doi.org/10.1016/j.sjbs.2015.03.010

Sudhakaran, M., & Doseff, A. (2020). The targeted impact of flavones on obesity-induced inflammation and the potential synergistic role in cancer and the gut microbiota. *Molecules, 25*(11), 2477. https://doi.org/10.3390/molecules25112477

Suneetha, S., Lakshminarasu, M., Prabhakar, T., & Rao, P. (2015). Screening and isolation of l-glutaminase producing marine bacterial cultures. *International Journal of Advances in Pharmacy and Bio-technology, 1*(1), 28–32. https://doi.org/10.38111/ijapb.20150101003

Sutthisa, W. (2022). Comparison of the antagonistic potential of the entomopathogenic bacterium Serratia nematodiphila GCSR38 with other effective microorganisms for the control of rice bacterial leaf blight. *Journal of Pure and Applied Microbiology, 16*(1), 557–566. https://doi.org/10.22207/jpam.16.1.54

Taboada, F., Habib, N., & Genta, S. (2019). Gastroprotective effect of Acacia aroma Gill. ex Hook. and Arn. leaves extracts in gastric ulcer models in rats. *International Journal of Pharmaceutical Investigation, 9*(4), 187–194. https://doi.org/10.5530/ijpi.2019.4.35

Taliyan, R., & Sharma, P. (2012). Protective effect and potential mechanism of Ginkgo biloba extract EGb 761 on STZ-induced neuropathic pain in rats. *Phytotherapy Research, 26*(12), 1823–1829. https://doi.org/10.1002/ptr.4648

Tangeti, S., Gabbita, P., Ponnaluri, R., & Kolasani, B. (2018). Comparative study of wound-healing effect of topical Acacia catechu extract and silver sulfadiazine on excisional wound model in Guinea pigs. *International Journal of Basic & Clinical Pharmacology, 7*(12), 2347. https://doi.org/10.18203/2319-2003.ijbcp20184844

Tchinda, J., Tchebe, T., Tchoukoua, A., Yona, A., Fauconnier, M., Kor, M., et al. (2020). Fatty acid profiles, antioxidant, and phenolic contents of oils extracted from Acacia polyacantha and Azadirachta indica (neem) seeds using green solvents. *Journal of Food Processing and Preservation, 45*(2), e15115. https://doi.org/10.1111/jfpp.15115

Thabet, A., Youssef, F., El-Shazly, M., & Singab, A. (2018). Sterculia and Brachychiton: A comprehensive overview on their ethnopharmacology, biological activities, phytochemistry and the role of their gummy exudates in drug delivery. *The Journal of Pharmacy and Pharmacology, 70*(4), 450–474. https://doi.org/10.1111/jphp.12876

Tharakan, M. (2023). The antifungal potential of cardamom-based Denteez™ mouthwash against opportunistic fungal infections in the oral cavity. *Journal of Advanced Oral Research, 14*(2), 169–176. https://doi.org/10.1177/23202068231194895

Thotathil, V., Rizk, H. H., Fakrooh, A., & Sreerama, L. (2022). Phytochemical Analysis of Acacia ehrenbergiana (Hayne) Grown in Qatar: Identification of Active Ingredients and Their Biological Activities. *Molecules, 27*(19), 6400.

Tiwari, M., Panghal, A., Mittal, V., & Gupta, R. (2023). Bioactive compounds of acacia, health benefits and its utilization in food processing industry: A critical review. *Nutrition & Food Science, 53*(7), 1125–1146. https://doi.org/10.1108/nfs-08-2022-0274

Todkar, S., Todkar, R., Kavathekar, V., Kulkarni, S., & Kulkarni, A. (2011). Secondary metabolite profiling of Acacia concinna. *Biosciences, Biotechnology Research Asia, 8*(2), 653–660. https://doi.org/10.13005/bbra/916

Tripathi, S. K., Xu, T., Feng, Q., Avula, B., Shi, X., Pan, X., et al. (2017). Two plant-derived aporphinoid alkaloids exert their antifungal activity by disrupting mitochondrial iron-sulfur cluster biosynthesis. *Journal of Biological Chemistry, 292*(40), 16578–16593.

Tyavambiza, C., Dube, P., Goboza, M., Meyer, S., Madiehe, A., & Meyer, M. (2021). Wound-healing activities and potential of selected African medicinal plants and their synthesized biogenic nanoparticles. *Planning Theory, 10*(12), 2635. https://doi.org/10.3390/plants10122635

Ullah, A., Munir, S., Badshah, S., Khan, N., Ghani, L., Poulson, B., et al. (2020). Important flavonoids and their role as a therapeutic agent. *Molecules, 25*(22), 5243. https://doi.org/10.3390/molecules25225243

Usman, H., Ullah, M., Jan, H., Siddiquah, A., Drouet, S., Anjum, S., et al. (2020). Interactive effects of wide-spectrum monochromatic lights on phytochemical production, antioxidant and biological activities of Solanum xanthocarpum callus cultures. *Molecules, 25*(9), 2201. https://doi.org/10.3390/molecules25092201

Vijayaraghavan, K., Rajkumar, J., & Ali-Seyed, M. (2017). Efficacy of Chromolaena odorata leaf extracts for the healing of rat excision wounds. *Veterinary Medicine, 62*(10), 565–578. https://doi.org/10.17221/161/2016-vetmed

Wang, J., Yang, K., Yao, L., Ma, Z., Li, C., Si, E., et al. (2021). Metabolomics analyses provide insights into nutritional value and abiotic stress tolerance in halophyte Halogeton glomeratus. *Frontiers in Plant Science, 12*, 703255. https://doi.org/10.3389/fpls.2021.703255

Yan, Y. (2024). The roles of SARP family regulators involved in secondary metabolism in Streptomyces. *Frontiers in Microbiology, 15*, 1368809. https://doi.org/10.3389/fmicb.2024.1368809

Yang, Z., Li, X., Luo, W., Zhu, J., Zhao, J., Wang, M., et al. (2022). Allicin in digestive system cancer: From biological effects to clinical treatment. *Frontiers in Pharmacology, 13*, 903259. https://doi.org/10.3389/fphar.2022.903259

Yao, C., Lai, G., Chan, C., Cheng, A., Yang, Y., & Chuang, S. (2005). Dramatic synergistic anticancer effect of clinically achievable doses of lovastatin and troglitazone. *International Journal of Cancer, 118*(3), 773–779. https://doi.org/10.1002/ijc.21361

Yassin, M. (2023). Synergistic anticandidal effectiveness of greenly synthesized zinc oxide nanoparticles with antifungal agents against nosocomial candidal pathogens. *Microorganisms, 11*(8), 1957. https://doi.org/10.3390/microorganisms11081957

Yassin, M., Elgorban, A., Al-Askar, A., Sholkamy, E., Ameen, F., & Maniah, K. (2023). Synergistic anticandidal activities of greenly synthesized ZnO nanomaterials with commercial antifungal agents against candidal infections. *Micromachines, 14*(1), 209. https://doi.org/10.3390/mi14010209

Yi, M., Lin, S., Zhang, B., Jin, H., & Ding, L. (2020). Antiviral potential of natural products from marine microbes. *European Journal of Medicinal Chemistry, 207*, 112790. https://doi.org/10.1016/j.ejmech.2020.112790

Younes, M., Aquilina, G., Castle, L., Engel, K., Fowler, P., Fürst, P., et al. (2019). Opinion on the re-evaluation of acacia gum (e 414) as a food additive in foods for infants below 16 weeks of age and the follow-up of its re-evaluation as a food additive for uses in foods for all population groups. *EFSA Journal, 17*(12), e0592. https://doi.org/10.2903/j.efsa.2019.5922

Youzbachi, N., Elfalleh, W., Tlili, N., Grégoire, S., Berdeaux, O., Salles, C., et al. (2012). Unexploited Acacia cyanophylla seeds: Potential food sources of ω6 fatty acids and antioxidants? *Journal of the Science of Food and Agriculture, 92*(7), 1526–1532. https://doi.org/10.1002/jsfa.4737

Youzbachi, N., Trabelsi, H., Elfalleh, W., Khaldi, A., Nasri, N., & Tlili, N. (2019). Fatty acids and triacylglycerols composition from Tunisian Acacia species seed oil. *Arabian Journal of Chemistry, 12*(8), 3302–3308.

Zaman, G., Alshahrani, M., Barua, P., Aladel, A., Zaman, F., Ahmad, I., et al. (2022). Screening of the antioxidant and antibacterial effects of extracted essential oils from Thunbergia coccinea, Acacia polyacantha, Polygonum micrpcephallum, Abies spectabilis and Clerodendrum colebrookianum. *Cellular and Molecular Biology, 67*(4), 56–67. https://doi.org/10.14715/cmb/2021.67.4.7

Zheleva-Dimitrova, D., Simeonova, R., Kondeva-Burdina, M., Savov, Y., Balabanova, V., Zengin, G., & Gevrenova, R. (2023). Antioxidant and hepatoprotective potential of Echinops ritro L. extracts on induced oxidative stress in-vitro/in-vivo. *International Journal of Molecular Sciences, 24*(12), 9999. https://doi.org/10.3390/ijms24129999

Zhu, M., Zhou, F., Ouyang, J., Wang, Q., Li, Y., Wu, J., et al. (2021). Combined use of epigallocatechin-3-gallate (EGCG) and caffeine in low doses exhibits marked anti-obesity synergy through regulation of gut microbiota and bile acid metabolism. *Food & Function, 12*(9), 4105–4116. https://doi.org/10.1039/d0fo01768j

Zingue, S., Njuh, A., Tueche, A., Tamsa, J., Tchoupang, E., Kakene, S., et al. (2018). In vitro cytotoxicity and in vivo antimammary tumor effects of the hydroethanolic extract of Acacia seyal (Mimosaceae) stem-bark. *BioMed Research International, 2018*, 1–13. https://doi.org/10.1155/2018/2024602

Zweil, H., Kamel, K., Zahran, S., & AbdELRahman, M. (2015). Effects of Acacia nilotica leaf extract supplementation on physiological parameters and antioxidant activity in growing rabbits. *Journal of the Advances in Agricultural Researches, 20*(2), 216–227. https://doi.org/10.21608/jalexu.2015.161389

Chapter 5
Breast Cancer-the-State-of-the Art in the World and Islamic Development Banks Members Countries (IsDB-MCs)

Ahmed A. M. Elnour

Abstract

Background

Breast cancer (BC) is a significant health concern and a leading cause of death among women in many Islamic Development Bank Member Countries (IsDB-MCs) within the Organization of Islamic Cooperation (OIC). Despite extensive global research efforts, BC survival rates remain low in several IsDB-MCs due to various barriers hindering the implementation of effective programs, including religious, social, and scientific challenges.

Aims

This chapter aims to provide an in-depth assessment of the current state of BC research in IsDB-MCs, focusing on OIC-member countries. The main objectives include evaluating research output and performance, examining research patterns and themes, assessing scientific strengths and weaknesses, and analyzing international collaborations in BC research within these countries.

Results

The analysis reveals that while IsDB-MCs have been extensively published in BC research, their output AGs behind that of developed nations from non-OIC groups. The leading areas of BC research within OIC-MCs aim to improve the health quality for affected or at-risk women by developing comprehensive analyses that provide insights to support and promote BC research within IsDB-MCs.

A. A. M. Elnour (✉)
Faculty of Chemical and Process Engineering Technology, Universiti Malaysia Pahang Al-Sultan Abdullah (UMPSA), Pahang Gambang, Malaysia

Centre of Excellence for Advanced Research in Fluid Flow (CARIFF), University Malaysia, Pahang Gambang, Malaysia
e-mail: adamhassan@umps.edu.my

A. A. M. Elnour (ed.), *Gum Arabic and Breast Cancer Biology*,
https://doi.org/10.1007/978-981-97-8518-6_5

Recommendations
To improve BC outcomes and reduce poverty in IsDB-MCs, it is recommended to increase research funding and establish research priorities, support international collaborations among researchers, enhance research capacity and infrastructure, and improve data collection and analysis on the burden of BC. Additionally, it is crucial to encourage greater involvement from researchers and institutions with strong publication records and address the cultural, social, and economic challenges related to BC. This c underscores the importance of collaborative efforts, financial support, and strategic research directions to advance BC research and improve outcomes in IsDB-MCs within the OIC.

Keywords Breast cancer · Islamic Development Bank (IsDB) · Member Countries (MCs) · Organization of Islamic Cooperation (OIC) · Survival rate

Highlights
1. BC is a leading cause of death and is the most frequently diagnosed cancer among many IsDB-MCs.
2. Despite global research efforts, survival rates remain low in several IsDB-MCS owing to various barriers.
3. This chapter assessed the current state of BC research in IsDB-MCs, focusing on OIC countries.
4. The study reveals that while OIC-MCS have been published extensively on breast cancer, they lag developed non-OIC nations.
5. The leading areas of BC research within OIC-MCS are highlighted to improve health and quality of life.
6. The analysis provides insights and recommendations to support and promote BC research on IsDB-MCS.
7. Collaboration and investment in research infrastructure are key to achieving quality output and progress in the IsDB-MCS.

5.1 Introduction

BC is a leading cause of death among women and is the most frequently diagnosed cancer among many IsDB-MCS (Lewison et al., 2020). The OIC has MCS, in which the incidence of and mortality from BC is particularly high (GhaviDel & Mostafavi, 2021). Although much cancer research has been conducted globally, research outcomes in OIC-Member Countries (OIC-MCS) have not yet been well documented (Fatma & Siddique, 2023). Hence, this chapter, funded by the Islamic Development Bank, presents an in-depth assessment of the current state of BC research on IsDB-MCS.

Despite extensive research on BC conducted worldwide, several countries belonging to the IsDB have low survival rates due to factors such as insufficient research contribution and religious, social, and scientific barriers that impede the implementation of BC programs (Lewison et al., 2020). The financial resources allocated by IsDB-MCS to BC research are proportional to the disease burden (Hricak et al., 2021). Over the past 5 years, Iran, Saudi Arabia, Turkey, Malaysia, Egypt, Pakistan, and Jordan have significantly contributed to BC research in this region (Al-Mousa et al., 2020). Collaboration among institutions in IsDB-MCS is crucial for achieving research progress. To enhance the quality of research output, it is recommended that these countries invest in research infrastructure (ESCAP, U. N., 2024). Promoting scientific research and supporting BC research throughout the research process using the various mechanisms available within the IsDB can help address the current societal challenges related to BC (Veljkovikj et al., 2022). In conclusion, utilizing IsDB resources and fostering collaboration are crucial for overcoming barriers and improving BC survival rates in IsDB-MCS.

IsDB-MCS, historically called OIC-MCS, encompass 57 nations across four continents with varying levels of gross national income (Kamarudin, 2021). The IsDB aims to promote social and economic development to reduce poverty and enhance the living conditions of its inhabitants (Ali et al., 2021). Scientific research serves as a key means of achieving this ambitious objective (Gottschalk & Poon, 2020). BC is among numerous most frequently diagnosed cancers among numerous IsDB-MCS (Lewison et al., 2020). Research on BC is crucial for improving survival rates (Wilkinson & Gathani, 2022). This chapter evaluates the current state of BC research in IsDB-MCS, which are part of the OIC (Hussain et al., 2023). Hence, enhancing BC research in IsDB-MCS is vital for improving survival rates and achieving broader social and economic development goals.

Although the research output of OIC-MCS has been extensive in the BC field. AGs behind other developed nations outside the OIC. This chapter aims to evaluate the research performance of 10 major OIC-MCS, which are also G20 members, and to assess the patterns and concentrations of research in terms of major research themes within BC research. Additionally, the study aimed to evaluate the scientific strengths and weaknesses of BC research within OIC-MCS, as well as international collaborations. Finally, the study provides an in-depth analysis model and report template for further assessment.

BC is a significant topic in women's health and cancer research, as highlighted by Chlebowski et al. (2020). Despite this, the current state of BC research on IsDB-MCS has not been extensively explored, as noted by Lewison et al. (2020). The OIC encompasses a substantial portion of the world and 57 IsDB-MCS, as indicated by Elzahi Saaid Ali (2022). In this chapter, we provide an overview of the state of BC research in OIC-MCS as a group and in ten large or scientifically advanced OIC-MCS, which are also members of the G20, using the large-scale data and visualization techniques available in the in-house developed science, Technology, and Innovation Research Excellence Mapping Online (S, T, and IR-Map) tools, as described by Abu Huson et al. (2024) and Gignac et al. (2024). Therefore, a comprehensive assessment of BC research in OIC-MCS can help identify gaps and opportunities for collaboration, ultimately leading to improved outcomes in women affected by this disease.

5.2 Epidemiology of BC in IsDB-MCS

BC is a crucial health concern involving research efforts worldwide (Lei et al., 2021). In countries belonging to the IsDB, both in-country and international collaborative research endeavors are pursued to address this issue (Chankseliani, 2023). These collaborative efforts have led to accelerated advancements in BC research and the associated challenges (Pashayan et al., 2020). This chapter presents the main areas of BC research conducted in the MCS of the OIC (Lewison et al., 2020). The shared objective of these research initiatives is to discover solutions that will enhance the overall health and quality of life of women with BC and those at-risk of developing the disease (Konieczny et al., 2020). Therefore, enhancing BC research through collaboration and innovation in OIC-MCS is essential to improve women's health and quality of life.

BC is a significant public health issue and a focal point of healthcare research in almost all countries (Duggan et al., 2021). It is the most prevalent type of cancer among women and the primary cause of cancer-related fatalities in women. Approximately 1.7 million women are diagnosed with this disease annually, and more than half a million women die from it (Samaroo et al., 2021). The development of BC is influenced by reproductive, hormonal, and genetic factors that interact with environmental variables (Calaf et al., 2020). A considerable and preventable portion of the current global BC burden is attributable to conventional reproductive factors (Basu et al., 2020). This disease burden affects not only women, but also their families, communities, healthcare systems, and economies (Trapani et al., 2022). Therefore, targeted research and preventive strategies are crucial to reduce the global burden of BC and its far-reaching impacts.

5.2.1 Global Burden of BC

The rationale for the widespread adoption of mammographic screening programs is clear. Setting up these programs is relatively straightforward, and commercial interest has been invested in promoting screening (Ressler, 2022). BC treatment is a highly profitable venture (Shah et al., 2024). Additionally, patient advocacy groups lobby for access to the latest and often costly therapeutic agents (Cardoso et al., 2021). In contrast, implementing health education programs to change dietary habits, promote physical activity, reduce exposure to environmental carcinogens, and offer chemo-preventive agents is far more daunting (Rock et al., 2020). This is because such preventive measures target BC and address a range of other chronic diseases that women in the target age group may suffer from (Beau et al., 2020). However, in many countries, adopting a prevention-focused strategy is the most realistic long-term approach to mitigating the growing burden of BC (Basu et al., 2020). Therefore, while mammographic screening and advanced treatments are critical, a prevention-focused strategy offers the most sustainable solution for reducing BC burden globally.

BC is the most prevalent form of cancer among women worldwide and in most high-income countries (Francies et al., 2020). Its incidence is rising rapidly in low- and middle-income countries, where it accounts for the highest number of cancer-related deaths in women (Canfell et al., 2020). The global burden of this disease is significant, and in many cases, it can be prevented by implementing appropriate measures (Arnold et al., 2022). Without suitable prevention strategies, the trend of increasing incidence is likely to continue, placing considerable strain on the healthcare systems in many countries (Sulis et al., 2022). The term "prevention" is often associated with BC. Still, in many countries, national BC control programs emphasize screening and early detection rather than primary prevention strategies that could have a much more significant impact (Ginsburg et al., 2020). Therefore, shifting the focus from screening and early detection to primary prevention is essential for effectively reducing the global burden of BC. For more information about BC's global burden, see Chap. 6, Sect. 6.2.2, Fig. 6.2.

5.2.2 *Regional Differences in Incidence and Mortality*

BC presents significant disparities in incidence, prevalence, and mortality rates between developed and less-developed countries and within specific regions and ethnic groups (Huang et al., 2021). For instance, the absolute risk of developing BC by age 75 is approximately 10–14% in most developed countries and only 1–3% in most less-developed countries (Groll, 2021). These disparities are illustrated by GLOBOCAN, which provides the estimated incidence rates for 184 countries (Fan et al., 2022). Its successor, the International Agency for Research on Cancer (IARC), offers detailed statistics, including time trends on a global scale. At the same time, the World Health Organization (WHO) regularly issues fact sheets on cancer, with a particular focus on less-developed countries (Sung et al., 2021). The 2009 report on BC in the Muslim World highlights the significance of these regional differences and their religious and ethnic subgroups (Alatrash, 2020). Additionally, it discusses the major risk factors and recent developments in the Middle East (Memish et al., 2020). Therefore, addressing the BC incidence and outcomes disparities between regions and ethnic groups is crucial for global cancer control.

Sung et al. (2021) reported the incidence of BC is rising globally, with a disproportionately high increase observed in less-developed countries. Reproductive risk factors were significant contributors to this increase. Ding et al. (2022) note large regional variations in incidence and mortality rates. In high- and middle-income countries, mortality rates have declined due to early detection through mammographic screening and improved treatment, whereas in low-income countries, mortality rates have increased, as reported by Francies et al. (2020). Early detection

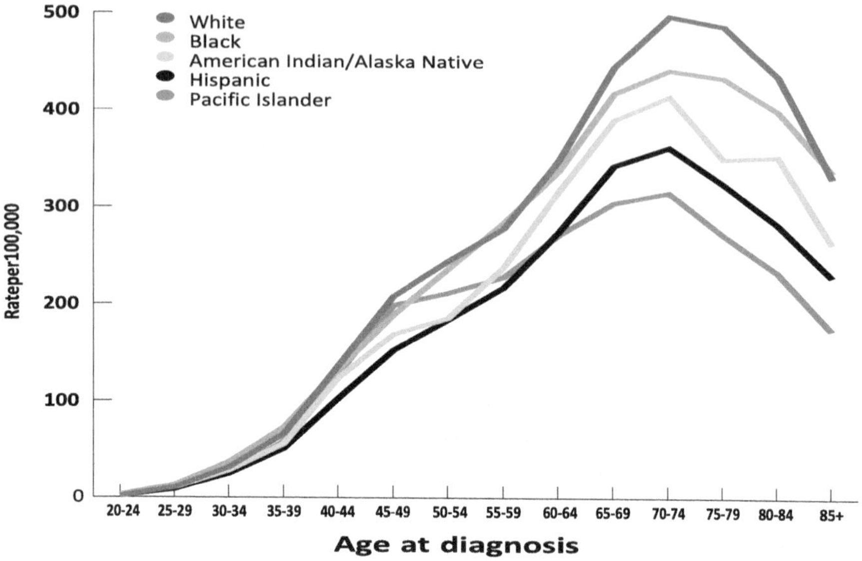

Fig. 5.1 Regional differences in incidence and mortality. The rates are per 100,000 people, adjusted for age based on the 2000 US standard population, and account for reporting delays. The racial data exclude individuals of Hispanic origin. Data are adjusted from previously published data by Desantis et al. (2024)

through screening can reduce mortality by identifying the disease at an early, curable stage, and improved treatment can reduce mortality by curing patients or prolonging their life, as stated by Nardin et al. (2020). Treatment success is heavily dependent on the stage at diagnosis, with a 5-year relative survival of almost 100% for localized stage diagnosis and approximately 25% for distant metastatic stage diagnosis, as noted by Hellman et al. (2020). In low-income countries, only 10–30% of patients diagnosed at a late-stage have a chance of long-term survival, as stated by Mishra, Chung et al. (2021) and Mishra, Pimple et al. (2021). Enhancing early detection and access to effective treatments are crucial for improving BC survival rates, particularly in low-income countries. Figure 5.1 demonstrates the variations in incidence and mortality for the USA region as an example due to BC.

5.3 Risk Factors and Protective Factors

Post-menopausal women can benefit from using agents that either block the effects of estrogen on the breast or reduce endogenous estrogen production, such as selective estrogen receptor modulators (SERMs) and aromatase inhibitors (Kharb et al., 2020). In addition, protective factors against BC may include factors that reduce mental depression and secretion of cortisol, melatonin, and other hormones that suppress the immune system or enhance the growth of BC cells (Ahmad et al.,

2023). Other protective factors may involve the effects of diet and nutrition on breast tissue (Buja et al., 2020). These may include the consumption of specific macro- and micronutrients that affect the risk of BC through their interactions with sex hormones, nervous and immune systems, and intrinsic biological rhythms (Kudamba et al., 2021). Furthermore, genetic susceptibility and gene–environment interactions may be involved in the etiology of some BCs (Mbemi et al., 2020). Exploiting such gene–environment interactions could potentially lead to developing more effective and novel therapeutic and preventive strategies for BC (Mbemi et al., 2020). Therefore, leveraging hormonal and dietary interventions and understanding gene–environment interactions can enhance preventive and therapeutic approaches to BC, particularly in post-menopausal women.

The primary risk factors for BC that are associated with sex hormones, as reported by Daly et al. (2021), include early menarche, late menopause, late age at first full-term pregnancy, nulliparity, post-menopausal use of hormone replacement therapy or combined estrogen and progesterone preparations, high body weight, especially after menopause, and physical inactivity. Other well-established risk factors for BC, as identified by Justiz-Vaillant et al. (2021), include family history, long-term oral contraceptive use, various reproductive and gynecological abnormalities, and tobacco or alcohol use. Recent research has also recognized exposure to ionizing radiation, especially before age 30, as an additional risk factor for BC (Seidler & Huber, 2020). Protective factors against BC are generally the opposite of risk factors reported by Britt et al. (2020). These protective factors include factors that lower estrogen levels in premenopausal or post-menopausal women, such as late age at menarche, early age at first full-term pregnancy, short or long breastfeeding duration, earlier age at menopause, and the use of SERMs or aromatase inhibitors, as described by Mishra, Chung et al. (2021) and Mishra, Pimple et al. (2021). Therefore, understanding and mitigating risk factors while enhancing protective factors are crucial for reducing the incidence of BC (Table 5.1).

5.3.1 Genetic Factors Recognized

Furthermore, several genes with low-to-moderate penetrance have been reported in BC. Women WHO possess the WHO risk allele in any of these genes have a 20–50% chance of developing the disease (Peleg Hasson et al., 2020). Among these genes, ESR1, TGFB1, FGFR2, MAP 3K1, LSP1, and MRPS30 were the most noteworthy (Özgöz et al., 2020). These genes influence estrogen hormone and growth factor signaling, intracellular signal transduction, cell growth regulation, and mRNA biogenesis (Liu et al., 2021). In addition to these well-established BC genes, new genes have been discovered using next-generation sequencing techniques (Hu et al., 2022). Genetic polymorphisms in various novel genes, such as the 8q24.21 region of chromosome 8 (Myc), 2q35, 3p, 10p15, 10q, and 16q, and KLHDC9, TERT, ZNF365, and PPP2R2A, have been associated with BC (Bahreini et al., 2021). Given the varying frequencies of allele risk and the differences in the effect size of

Table 5.1 Factors that increase the relative risk of invasive breast cancer in women

Relative risk	Factors
>4.0	Age (65+ compared to under 65 years, with risk increasing across all ages up to 80)
	Atypical hyperplasia confirmed by biopsy
	Specific inherited genetic mutations linked to breast cancer (e.g., BRCA1, BRCA2, PALB2, TP53)
	Lobular neoplasia
	Personal history of breast cancer diagnosed before age 40
2.1–4.0	Ductal carcinoma confined to the ducts
	Elevated endogenous estrogen or testosterone levels (post-menopausal)
	High-dose radiation exposure to the chest (e.g., treatment for Hodgkin lymphoma)
	Breasts with 26% or greater mammographic density (compared to 11–25% density).
	Personal history of breast cancer diagnosed at age 40 or older
	Two or more first-degree relatives diagnosed with breast cancer
1.1–2.0	Consumption of alcohol.
	Early onset of menstruation (before age 11).
	Height (tall).
	First full-term pregnancy at an older age (after 30 years).
	Onset of menopause at age 55 or later.
	Never having breastfed a child.
	Absence of full-term pregnancies. A first-degree relative diagnosed with breast cancer.
	Post-menopausal obesity
	Personal history of endometrial or ovarian cancer
	Lack of physical activity
	Proliferative breast disease without atypia (such as usual ductal hyperplasia or fibroadenoma)
	Recent and prolonged use of menopausal hormone therapy containing both estrogen and progestin
	Recent use of hormonal contraceptives
	Type 2 diabetes mellitus
	Increased body weight acquired in adulthood

established genetic variants in diverse ethnic populations, it is essential to identify the genetic profile of these populations (Lewison et al., 2020). This will facilitate accurate risk assessment and the development of appropriate prevention strategies for affected populations (Budreviciute et al., 2020). Therefore, identifying and understanding genetic variations in diverse populations is crucial for accurate risk assessment and development of effective BC prevention strategies.

Genetic factors play a highly significant role in the development of BC, accounting for 2–85% of all BC incidences (Francies et al., 2020). Rare genetic mutations and common genetic polymorphisms contribute to this high heritability (Kierczak et al., 2022). More than 142 genes and 171 gene polymorphisms have been reported to be associated with BC (Daniyal et al., 2021). The most well-known group of BC susceptibility genes are the high-penetrance genes, including BRCA1, BRCA2, P53, PTEN, STK11/LKB1, CDH1, CHEK2, ATM, BRIP1, PALB2, RAD51C, and RAD51D (Felbor et al., 2022). The WHO carries the risk allele in one of these high-penetrance genes and has an approximately 70% chance of developing BC during their lifetime (Torabi Dalivandan et al., 2021). These genes are involved in DNA damage repair, cell cycle checkpoint control, and transcriptional regulation of tumor suppressors (Matthews et al., 2022). Thus, the identification of high-penetrance genes and their roles in BC underscores the importance of genetic screening and personalized prevention strategies for high-risk individuals.

5.3.2 Lifestyle Factors

The utilization of hormone replacement therapy has been identified as a potential risk factor for the development of BC, according to various sources (Goh, 2020). Additionally, excessive consumption of refined sugar, which is commonly available in many Muslim countries, has been linked to an increased risk of BC (Aljamali et al., 2022). Some Islamic scholars have expressed their views on the permissibility of consuming sugar (Bouchareb et al., 2022). Similarly, increased consumption of fast food and lack of physical activity have also been reported as risk factors for BC (Rock et al., 2020). In their fatwas, scholars addressed the issue of sedentary eating habits and recommended engaging in physical activity (Ibrahim et al., 2024). Furthermore, recent research trends in lifestyle, genetics, and preventive measures are likely to be influenced by growing BC awareness, which is becoming increasingly recognized as a global concern (Filippou et al., 2024). Hence, recognizing lifestyle risk factors and promoting healthy habits, supported by religious and cultural guidelines, are crucial for reducing BC risk in Muslim countries and globally.

Numerous modifiable lifestyle-related factors contribute to the high BC prevalence. One such factor is delaying the first pregnancy or having no children at all as well as having fewer children than the average (Akdeniz et al., 2020). Women choose not to have children and may be at a higher risk for BC (Marklund et al., 2021). Recent studies have shown an inconsistent trend in the age-at-marriage and childbirth, with a declining trend in the age-at-marriage and an increasing trend in the age at childbirth (Batyra & Pesando, 2021). These changes have contributed to the implementation of family planning programs (Ding et al., 2020). Family planning services are available for women who wish to space out or limit their number of children (Götmark & Andersson, 2020). Family planning empowers women by allowing them to make decisions about when and how many children they have, resulting in improved physical and mental health outcomes (Kc et al., 2021). Overall, addressing modifiable lifestyle factors, including reproductive choices and family planning, is essential to reduce the prevalence of BC and promote overall health in women.

5.3.3 *Hormonal Factors*

Numerous studies have investigated the relationship between anthropometric factors and BC. However, these studies have been inconsistent, and more research is needed in this area (Hayati et al., 2022). This chapter examines current research on hormonal factors, focusing on findings from studies conducted in IsDB-MCS within the OIC (Rodriguez-García et al., 2020). A total of 89 studies have explored the association between various hormonal factors and BC risk (Roheel et al., 2023). Fifteen studies investigated the relationship between peptide sex hormones and BC risk (Tin Tin et al., 2021). Among the peptide hormones, insulin-like growth factors have been the most studied (LeRoith et al., 2021). Eight IsDB-MCS evaluated the inconsistent association between peptide hormones and BC risk, with insulin and insulin-like growth factors being the most studied (Katsimperis et al., 2024). This comprehensive evaluation aims to contribute to the ongoing discussion on the potential link between hormonal factors and BC within OIC-MCS (Hayati et al., 2022; World Health Organization, 2023). Therefore, further research on hormonal factors and their relationship with BC is essential, particularly for IsDB-MCS, to clarify inconsistencies and guide effective prevention and treatment strategies.

Hormonal factors are widely recognized for their significant roles in BC etiology (Smolarz et al., 2022). Several studies have investigated these associations, resulting in a substantial body of evidence (Cave et al., 2020). Researchers have explored various aspects of the relationship between hormonal factors and BC, including the use of exogenous hormones (e.g., post-menopausal hormone therapy and oral contraceptives) and endogenous hormones (e.g., sex steroids and circulating peptide hormones) (Tin Tin et al., 2021). It is well-known that reproductive and anthropometric factors are correlated with circulating peptide hormone levels and, to some extent, sex steroid hormone levels (Dhiman et al., 2023). Therefore, understanding the intricate relationship between hormonal factors and BC is crucial for developing targeted prevention and treatment strategies, highlighting the need for ongoing research in this field.

5.4 Screening and Early Detection

According to IARC, mammography screening in certain populations can lead to a 35% decrease in BC mortality (Canelo-Aybar et al., 2021). Despite these benefits, mammography is resource-intensive, expensive, and unavailable to most of the global population (Schousboe et al., 2022). In Muslim countries, cultural and religious factors, such as the purdah (female seclusion) and the mahram (close male relative to whom a woman can be seen unveiled by), impact the acceptability and feasibility of mass mammographic screening (Salem & Daher-Nashif, 2020). Observant women must balance the loss of modesty with the potential threat of BC, which often presents without symptoms in the early stages (Yi, 2023). Given these challenges, providing universal access to high-quality clinical breast examination

(CBE) and breast self-examination (BSE) instructions may be a more practical solution for these women, as it would likely identify symptomatic and some asymptomatic cancers (Duggan, Dvaladze, Scheel et al., 2020). It may also be beneficial to educate religious leaders about the importance of CBE so that they can endorse and promote it within their communities (Chen et al., 2022). Subsequently, while mammography screening has proven benefits, alternative strategies such as CBE and BSE, supported by community and religious leaders, may provide a more feasible and culturally sensitive approach to BC detection in Muslim countries.

Effective methods for reducing BC deaths across diverse religious and ethnic groups involve early detection, ideally before cancer progresses beyond the breast (Yi, 2023). Various screening techniques have been employed to identify BC at an early-stage (Ginsburg et al., 2020). The most recognized and utilized screening methods include CBE, mammography, BSE, and more recently, breast ultrasound and magnetic resonance imaging (MRI) (Luo et al., 2022). Among these methods, mammography provides the strongest supporting evidence. The United States Preventive Services Task Force and American Cancer Society have recommended regular mammographic screening since the 1980s (Henderson et al., 2024), and various other professional organizations have made similar recommendations.

5.4.1 Mammography

In several countries belonging to the OIC, wealthy individuals have established exclusive private diagnostic centers equipped with cutting-edge technologies for diagnosing and treating BC (Moiyadi, 2021). However, a significant proportion of women in IsDB-MCS, particularly those residing in rural areas, face difficulties accessing these advanced BC mammography services (Hricak et al., 2021). Apart from the scarcity of resources, there is a shortage of trained personnel, including radiologists and technologists, a lack of awareness among women about their susceptibility to developing BC, and the importance of undergoing screening and early detection for timely treatment (Mehrotra & Yadav, 2022). As a result, in many OIC-MCS, the incidence of BC is increasing, while the inadequate availability of mammographic services and underutilization of existing services continue to pose significant public health challenges (Norfjord van Zyl et al., 2020). Therefore, addressing the obstacles confronting current mammographic screening programs is crucial for advancing BC research in OIC-MCS (Omisore et al., 2023). Therefore, overcoming barriers to mammographic screening and improving accessibility, especially in rural areas, are essential for advancing BC research and reducing the incidence rates of OIC-MCS.

Mammography is widely regarded as the most effective method for early BC detection (Khan & Arora, 2021). Numerous studies have confirmed its efficacy in reducing mortality rates by detecting cancer at an early-stage and enabling timely treatment (Lim et al., 2022). In addition to its role in the early detection of BC, mammography can also identify preinvasive lesions and other non-invasive and invasive malignant consequences (Khan & Arora, 2021). According to the American Cancer Society, mammography conducted every 2 years in women aged between 50 and 74 years can decrease BC mortality by approximately 33% (and potentially up to 46% when performed more frequently) (García-Albéniz et al., 2020). Despite its demonstrated value in early cancer detection, a substantial number of women in both high- and low-income countries do not undergo mammographic screening (Omisore et al., 2023). Thus, enhancing the awareness and accessibility of mammographic screening is critical to leveraging its proven benefits in reducing BC mortality rates worldwide.

5.4.2 Clinical Breast Examination

This chapter provides an overview of the available research and data on clinical breast examinations of selected OIC-MCS (Zaluzec & Sempere, 2024). The studies reviewed in this section highlight the importance of CBE as part of the early detection process for BC, regardless of whether it is conducted in the general population or in specific groups of women with unique characteristics, such as age, marital status, or type of healthcare service attended (Mittra et al., 2021). It should be noted that this chapter focuses specifically on clinical breast examinations conducted as part of routine screenings as well as those performed in response to abnormal complaints or findings from other diagnostic tests (McKinney et al., 2020). Thus, the integration of clinical breast examinations into routine screening and diagnostic protocols is essential to enhance early detection efforts and improve BC outcomes in OIC-MCS.

Early detection of BC is crucial for its successful treatment. CBE is one of the methods used for detecting BC, as demonstrated in several studies (Ginsburg et al., 2020; McKinney et al., 2020). In some cultures and regions, healthcare providers play a vital role in community education, motivation, and performance of breast examinations (Sangwan et al., 2023). This examination can lead to the early detection of cancer in symptomatic women (Whitaker, 2020), although screening CBE has not been proven to affect the stage distribution of BC at diagnosis (Mittra et al., 2021). Moreover, it is a relatively low-cost procedure (Huang et al., 2022). Thus, promoting CBE as a routine practice can enhance early detection efforts, particularly in resource-limited settings, thereby contributing to better BC outcomes.

5.5 Diagnosis and Staging

This chapter explores current research initiatives and advancements in the field of BC in IsDB-MCS within the OIC. Given the prevalence of BC and the need for comprehensive research that addresses various aspects of diagnosis, treatment, outcomes, and quality of life of affected individuals (Invernizzi et al., 2022), the collaborative research network "Breast Health Global Initiative" emphasizes the importance of multidisciplinary research teams that address the different priority areas identified within each country or region (Duggan, Dvaladze, Rositch, et al., 2020). Commonly identified priority areas of research, guided by limited resources for BC control, include prevention, early detection, pathology and diagnostic imaging, surgery and radiotherapy, medical oncology, supportive care, and palliative care (Lewison et al., 2020). A definitive diagnosis of BC is made through careful clinical evaluation, radiographic studies, and histopathological confirmation (Fox et al., 2023). Accordingly, fostering multidisciplinary and collaborative research efforts is essential for advancing BC control and improving outcomes in patients with IsDB-MCS.

5.5.1 Imaging Technique

Mammography is a highly effective low-dose X-ray imaging technique commonly used to visualize the internal structure of the breast. This technique is particularly useful for screening asymptomatic women and diagnostic purposes in symptomatic patients (Hendrick, 2020). Numerous studies have demonstrated that mammography can significantly reduce BC-related mortality by detecting cancer at an early-stage and providing an opportunity for cure (Katalinic et al., 2020). However, the sensitivity and specificity of mammography can vary depending on the presence of an abnormal mass or calcification as well as the skill of the radiologist (Nicosia et al., 2023). Additionally, it is important to consider the potential risks associated with exposure to ionizing radiation, particularly in young women with sensitive breast tissues (Hendrick, 2020). As such, there is growing interest in exploring alternative imaging techniques that do not involve radiation exposure (Hendrick, 2020). This is especially relevant in countries such as Saudi Arabia, where a significant proportion of the population is young (Al-Khraif et al., 2020). Moreover, it is important for women with dense breasts to have access to information regarding alternative breast imaging techniques (Mann et al., 2022). Overall, while mammography remains a cornerstone of BC screening and early detection, exploring and providing access to alternative imaging techniques, particularly for young women and those with dense breast tissue, is essential for comprehensive breast healthcare.

Breast imaging is a non-invasive method for diagnosing BC at the earliest stages (Li, Guan et al., 2020). Mammography, MRI, and ultrasonography are the most used imaging techniques (Pereira et al., 2020). Mammography and MRI are the most sensitive and specific modalities for BC detection (Aristokli et al., 2022). Mammography has long been used for screening and diagnosis, and numerous studies have confirmed its sensitivity and specificity (Liu et al., 2023). Other imaging modalities include scintimammography, positron emission tomography, optical imaging, thermography, and contrast-enhanced digital mammography (Hussain et al., 2022). Thus, leveraging a variety of breast imaging techniques is crucial for the accurate early detection and diagnosis of BC, enabling timely and effective treatment.

5.5.2 *Biopsy Procedures*

Several specialized types of BC biopsy are available (Rakha et al., 2022). One commonly used technique is sentinel lymph node biopsy (Giammarile et al., 2022). This procedure can be performed outpatient under local anesthesia (Gu et al., 2022). The primary advantage of BC biopsy is its ability to avoid axillary lymph node dissection in patients without affected lymph nodes (Chung et al., 2021). However, in certain cases, surgical excisional biopsy may be necessary to remove the entire tumor in patients who have already been diagnosed with BC and are currently undergoing neoadjuvant therapy (Heil et al., 2022). Overall, selecting an appropriate biopsy technique is essential for accurate diagnosis and effective treatment planning for BC management.

Diagnosis typically commences with physical examination and patient history, as per Jarvis (2023). Subsequently, imaging studies such as mammography, ultrasonography, and MRI were performed, as per Aristokli et al. (2022). A biopsy is required when an abnormality is detected because not all discovered masses are malignant, as by Barba et al. (2021) reported. Biopsy is the standard diagnostic tool that Hendriks et al. (2021) reported. It is typically performed through one of the following procedures: fine-needle aspiration, core biopsy, or surgical biopsy, as described by Nadda et al. (2022). Each can be performed in different ways. Additional tests, such as blood, genetic, or molecular tests, may also be necessary, as reported by Chen et al. (2020). The results of these tests are then used to determine the most effective course of BC treatment for the patient, which may include surgery, radiation, chemotherapy, or other targeted therapies (Trayes & Cokenakes, 2021). The patient's medical team works closely with them throughout the biopsy procedure to provide support and guidance, as reported by Neuberger et al. (2020). Hence, a comprehensive diagnostic approach, including physical examination, imaging, biopsy, and additional tests, is essential to diagnose BC and plan effective treatment strategies accurately.

5.6 Treatment Modalities

There is growing concern about the increasing number of publications reporting the inappropriate use of anti-cancer medications and herbal remedies, which can negatively impact treatment outcomes (Choudhury et al., 2023). Factors contributing to this issue include a lack of education for healthcare professionals and patients' misguided preferences for alternative therapies (Szmaglinska et al., 2024). Access to palliative care for patients with advanced BC is limited (Giap et al., 2023) and it is essential to implement measures to improve BC treatment and outcomes (Mutebi et al., 2020). BC can be treated using various modalities, either alone or in combination, including surgery, radiation therapy, hormone therapy, and chemotherapy (Correia et al., 2021; Masfarré et al., 2021; Wang & Tepper, 2021). For IsDB-MCS, available treatment options for BC include surgery, chemotherapy, and radiation therapy (Hricak et al., 2021). For advanced disease, systemic therapies such as immunotherapy, targeted therapy, hormone therapy, and chemotherapy are used (Lau et al., 2022). Therefore, addressing the inappropriate use of anti-cancer medications and enhancing education for both healthcare professionals and patients are crucial steps toward improving BC treatment and outcomes, especially in IsDB-MCS.

Further research is being conducted on targeted therapies and immunotherapies for BC treatment (Ye et al., 2023). Research on BC has led to significant advancements in developing new treatment modalities, such as targeted therapies and immunotherapies (Bou-Dargham et al., 2021). These modalities have demonstrated promising results in improving patient outcomes and reducing disease burden. Additionally, the use of precision medicine and personalized treatment approaches has gained traction in the management of BC by tailoring therapies to individual patients based on their genetic and molecular profiles. Overall, the evolving landscape of BC treatment modalities in IsDB-MCS within the OIC reflects continuous efforts to improve patient care and outcomes in the region. However, treatment modalities are limited, and surgery is often performed in the late stages of the disease (Riis, 2020). BC radiation therapy services are also inadequate (Mehrotra & Yadav, 2022). Moreover, the limited availability and affordability of all essential medicines, especially those included in the WHO Model List of Essential Medicines for BC, is a serious concern (Jenei et al., 2022). Therefore, advancing research on targeted therapies and immunotherapies and improving access to essential medicines and comprehensive treatment services is critical for enhancing BC care in IsDB-MCS. Figure 5.2 demonstrates the proposed mechanism and medical treatment of BC using imaging modalities.

5.6.1 Surgery

Large-scale prospective studies or meta-analyses of smaller studies may offer insights into various aspects of BC research that require only minor adjustments for citizens of IsDB-MCS (Ricci-Cabello et al., 2020). Establishing a specialized registry for BC would streamline compiling a national BC report for each member country (Mirbagheri et al., 2022). Preparing a country-specific report is a logical step toward creating a broader geographical report, such as a regional report for groups of IsDB-MCS in the same geographical area (World Health Organization, 2021). To fulfill the need for migration studies, a dedicated registry, designed as an extension of existing registries, could be employed (Axelsson et al., 2022). It

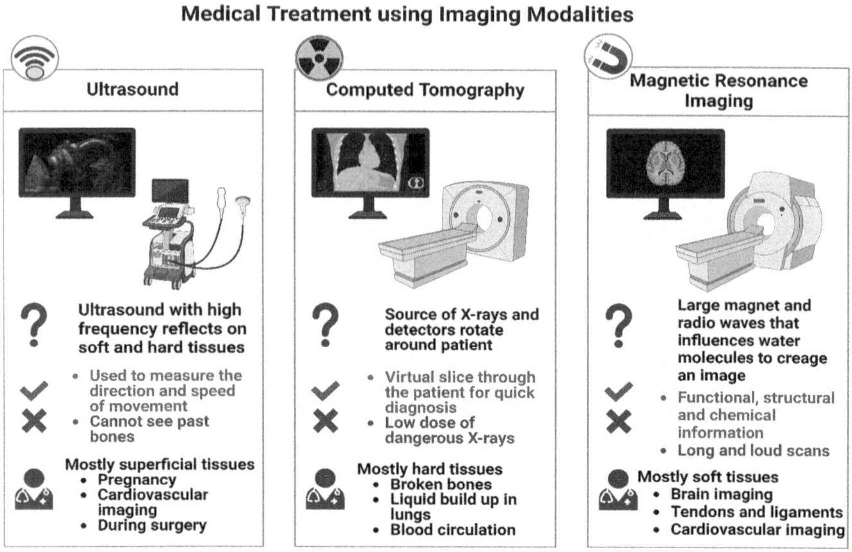

Fig. 5.2 Provides an overview of some of the most used medical imaging modalities in clinical treatment, including those used for breast cancer

is essential to apply guidelines with flexibility, ensuring that the general principles, quality, and standards of BC research are maintained while allowing for more variations in design and execution to accommodate the unique circumstances of IsDB-MCS (El-Halees & Tafish, 2020). Hence, establishing specialized BC registries and adapting research guidelines to the unique context of IsDB-MCS are critical steps for advancing BC research and improving outcomes across the region.

Surgery is typically the initial treatment option for early-stage BC, as recommended by Brackstone et al. (2021). Breast-conserving surgery (BCS), when combined with radiation therapy, is the preferred treatment for most women with stage I and II BC. This approach is considered safe and effective in an oncological context (Kim et al., 2021). Compared to mastectomy, women who undergo BCS have demonstrated a lower adverse impact from surgery (Christiansen et al., 2022). However, factors such as family history and fear of cancer may exacerbate the psychological stress experienced by patients with BC (Ban et al., 2021). Notably, the Islamic perspective on BC screening among asymptomatic women plays a significant and encouraging role in OIC-MC (Al-Shamsi, 2022). In conclusion, while BCS combined with radiation therapy remains a highly effective treatment for early-stage BC, addressing psychological stress and leveraging supportive cultural perspectives are essential for comprehensive patient care in OIC-MCS.

5.6.2 Chemotherapy

The mysteries of natural and therapy-induced resistance and the relationship between breast tumors and their microenvironment remain largely unknown (Mehraj et al., 2021). Achieving control or eradication of BC depends on the balance between pro-inflammatory and suppressive effectors produced by the microenvironment (Li, Yu, et al., 2020). This information will pave the way for the development of novel therapeutic strategies (Islam et al., 2022). Capitalizing on this balance or the shift in the direction of neutrophils, macrophages, mast cells, myeloid-derived suppressor cells, cytokines, chemokines, growth factors, and adaptive immunity T-cells, together with immunotherapies, radiotherapy, and targeted therapy, may prove to be a promising approach (Diana et al., 2020; Zhong et al., 2021). Although chemotherapy remains one of the mainstays of BC treatment, recent developments have identified more precisely targeted molecules and improved strategies for overcoming resistance and applying newer, more potent agents (Dutta et al., 2021). However, much work remains on chemo-modulation and exploring new targets (Zhong et al., 2021). Recent advances in developing anti-cancer immune responses have opened a novel and innovative approach for treating BC and other solid tumors by modulating their resistance (Burguin et al., 2021). This discovery adds immunotherapy to the treatment principles of preventing disease and controlling advanced disease (Dhar et al., 2021). Thus, understanding the interactions between breast tumors and their microenvironment, along with advancements in immunotherapy and targeted treatments, holds the key to overcoming resistance and improving outcomes in BC patients. Figure 5.3 illustrates the proposed mechanism and basic methods of chemotherapy for BC treatment.

5.6.3 Radiation Therapy

BC is the most prevalent form of cancer in women and ranks second in cancer-related fatalities (Afifi et al., 2020). Early detection programs have played a significant role in identifying many BC cases (Ginsburg et al., 2020). These tumors are typically small and do not spread to lymph nodes (Min et al., 2021). Many BC patients present with more advanced diseases (Testa et al., 2020). Locally advanced BC is characterized by large tumors with extensive skin or chest wall involvement and/or regional lymph node metastasis (Yadav et al., 2022). Metastatic BC, on the other hand, refers to a more advanced-stage of the disease, spreading beyond the breast, chest wall, and regional lymph nodes (DiPaola et al., 2022). Despite this, many BC cases are locally advanced and metastatic (Cuthrell & Tzenios, 2023). Therefore, while early detection programs are crucial for identifying BC at an early, more treatable stage, addressing the challenges of locally advanced and metastatic BC remains essential for improving patient outcomes and reducing cancer-related fatalities.

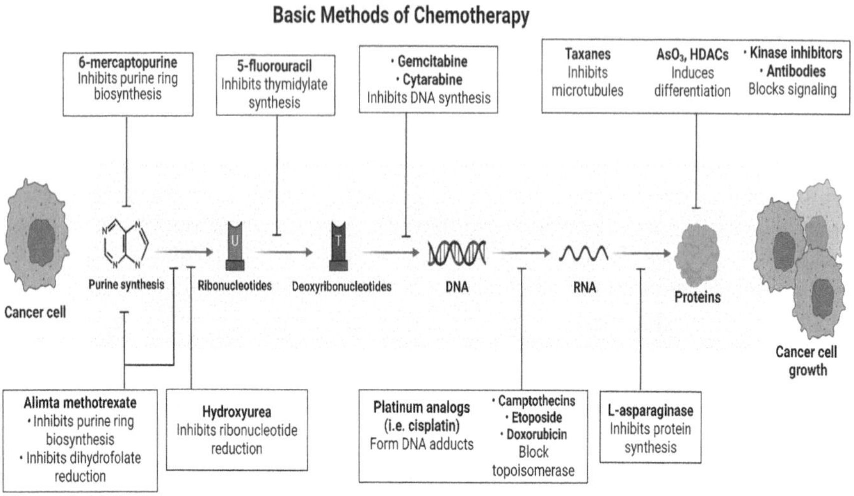

Fig. 5.3 Depicts the suggested mechanism and fundamental techniques of chemotherapy for treating breast cancer

Radiation therapy employs high-energy rays to combat diseases, particularly cancers (Ronga et al., 2021). Approximately 50% of all cancer patients receive radiation therapy, which is typically administered with a curative intent (Grégoire et al., 2020). Locally advanced BC often requires radiation therapy after surgery to minimize the risk of local recurrence as well as in the metastatic setting to alleviate symptoms (Aebi et al., 2022). Radiation therapy may be used alone or in conjunction with surgery, chemotherapy, and/or hormonal therapy, depending on the clinical situation (Wang & Tepper, 2021). Technological advances have enhanced the precision of radiation therapy and reduced its side-effects (Grégoire et al., 2020). Accelerated partial breast irradiation is a widely used treatment, but caution should be exercised when selecting suitable candidates for this approach (Meattini et al., 2023). This article explains the current recommendations, recent advances, common practices, and available resources for radiation therapy of BC in IsDB-MCS (Mangla et al., 2022). Overall, radiation therapy remains a cornerstone of BC treatment, with advancements in technology and precision improving outcomes and reducing side-effects, particularly in IsDB-MCS, where access to resources and tailored treatment approaches are crucial. Figure 5.4 depicts the suggested mechanism and fundamental techniques of radiotherapy resistant during BC treatment.

5.7 Targeted Therapies and Immunotherapy

It is important to recognize that most current knowledge on BC and its treatment has been derived from research conducted in developed countries with advanced infrastructure for high-quality, multidisciplinary research (Lei et al., 2021). As a result,

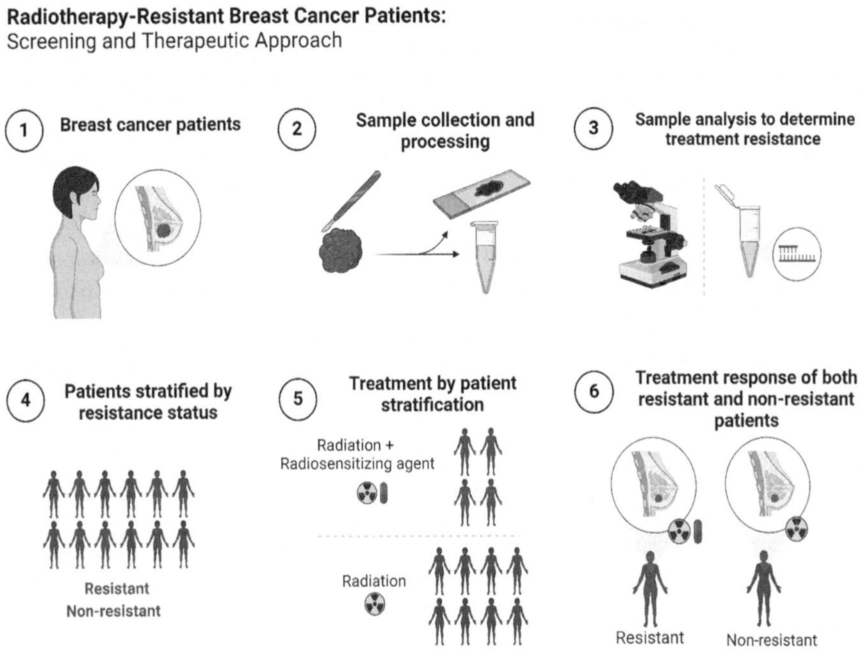

Fig. 5.4 Shows the proposed mechanism and essential methods for managing radiotherapy resistance in breast cancer treatment

much of the research conducted on IsDB-MCS has focused on translational and collaborative research and the outcomes of clinical trials conducted primarily in developed countries (Micoli et al., 2021). Collaborative network groups, such as the Breast Health Global Initiative (BHGI), the Union for International Cancer Control, and European Society for Medical Oncology, play a significant role in facilitating such collaborations (Schachter et al., 2021). Despite the challenges faced in the research and development of new therapies for BC in IsDB-MCS, significant progress has been made in some countries (Veljkovikj et al., 2022). Targeted therapies, including monoclonal antibodies, tyrosine kinase inhibitors, mTOR inhibitors, and PARP inhibitors, have revolutionized the therapeutic landscape of BC (Subhan et al., 2023). However, the high-cost of these treatments, as well as the intrinsic and acquired resistance and adverse effects of these drugs, pose major challenges in the management of BC (Luque-Bolivar et al., 2020). Combination therapy and drug scheduling are important strategies for addressing these issues (Jin et al., 2023). Immunotherapy, a novel strategy, has shown promise in treating BC with the recent development of immune checkpoint inhibitors, such as anti-PD-1 and anti-PDL1 antibodies (Debien et al., 2023). Hence, addressing costs, resistance, and side-effects is crucial despite advancements. Collaborative research and tailored strategies are key to improving BC outcomes in the IsDB countries.

5.7.1 HER2-Targeted Therapies for BC

Developing targeted BC therapies represents one of the most significant achievements in biomedical research (Buocikova et al., 2020). Since the identification of HER2 oncogene (Hamilton et al., 2021), numerous HER2-targeted therapies have been developed. The first FDA-approved HER2-targeted therapy was trastuzumab, a humanized monoclonal antibody targeting HER2 (Kunte et al., 2020). Trastuzumab has revolutionized the treatment of HER2-positive BC and dramatically improved the outcome of patients with this aggressive sub-type (Patel et al., 2020). Trastuzumab is included in the World Health Organization's "Model List of Essential Medicines" and is considered one of the most important advances in oncology (Zhou et al., 2023). Due to its significant clinical benefit, trastuzumab is the "gold standard" among HER2-targeted therapies (Ariga, 2023). However, many patients resist trastuzumab, and some do not respond to treatment with trastuzumab (Derakhshani et al., 2020). To address these issues, other HER2-targeted therapies are either approved or in various clinical development stages (Mercogliano et al., 2023). In conclusion, while trastuzumab has set the "gold standard" for HER2-targeted therapies, developing new treatments is essential to overcome resistance and improve patient outcomes. The proposed mechanism and essential methods for managing HER2-targeted therapies in breast cancer treatment are explained in Fig. 5.5.

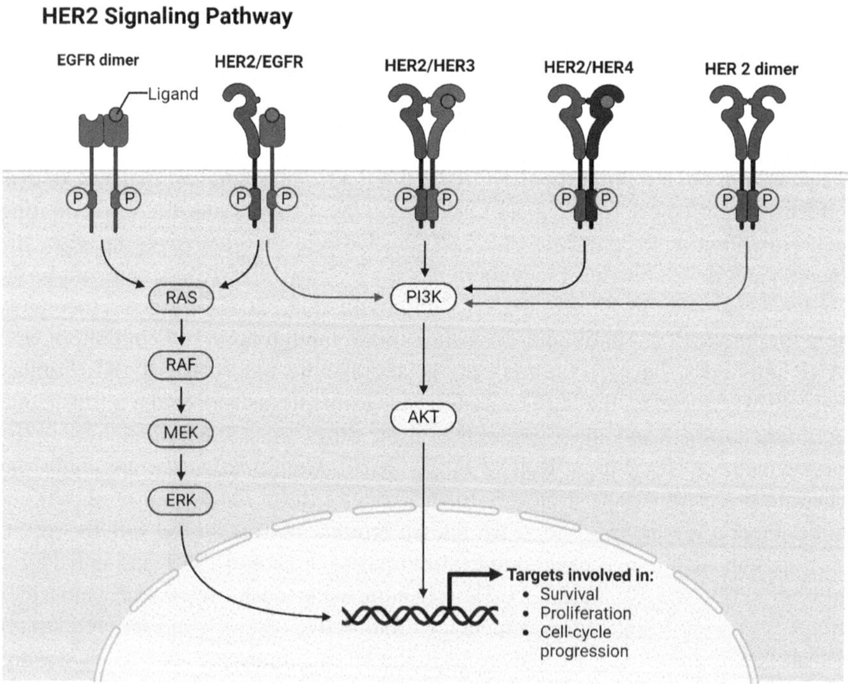

Fig. 5.5 Illustrates the proposed mechanism and key approaches for managing HER2-targeted therapies in breast cancer treatment

5.7.2 *Hormone-Receptor-Targeted Therapies*

Many BCs rely on estrogen for growth (Yu et al., 2021). This is called estrogen-positive (ER+) BC (Reinert et al., 2022). SERMs prevent estrogen from attaching to the ER+ cancer cells (Rej et al., 2023). Tamoxifen is an illustration of SERM (Liu, 2020). Fulvestrant hinders estrogen activity by degrading the estrogen receptor and is classified as an estrogen receptor downregulator (Chen et al., 2022). It is used for the treatment of advanced hormone-receptor-positive BC in post-menopausal women and the progression of advanced disease following anti-estrogen therapy (Lei et al., 2020). In 2017, the WHO listed "Tamoxifen and Fulvestrant" among the essential medicines (Cheng et al., 2020). It is essential to ensure that the production capabilities for these crucial drugs are present and available in IsDB-MCS (Qin et al., 2023). Thus, securing the availability and production of essential drugs, such as tamoxifen and fulvestrant, is vital for effective BC treatment in IsDB-MCS.

A recent study by Mansouri and Attar (2022) has brought attention to the research and development efforts made in OIC-MCS to discover drugs for the modern pharmacological management of BC. Currently, the standard of care for hormone-receptor-positive BC includes Tamoxifen, Fulvestrant, and aromatase inhibitors, such as Anastrozole, Letrozole, and Exemestane (Andrahennadi et al., 2021). Despite the use of these drugs, resistance to tamoxifen is common, leading to the need for new drug discovery (Afzal et al., 2022). In many cases, patients with OIC-MCS are diagnosed with advanced-stage BC (Lewison et al., 2020), making the development and deployment of new drugs for BC therapy an urgent necessity (Buocikova et al., 2020). It is crucial that these modern targeted drugs be made accessible to patients to improve their outcomes (Anand et al., 2023). Therefore, enhancing drug development and ensuring access to modern targeted therapies are essential for improving BC treatment outcomes in OIC-MCS.

5.8 Supportive Care and Survivorship

Complementary and alternative medicine (CAM) is often used by patients to improve their overall well-being, manage symptoms, and mitigate treatment side-effects (Tangkiatkumjai et al., 2020). However, the use of CAM can be detrimental when patients rely on unproven treatments as substitutes for conventional medical therapy (Knecht et al., 2020). On the other hand, CAM can be beneficial when used in conjunction with conventional treatment for BC (Hammersen et al., 2020). According to numerous users, the use of CAM can reduce the adverse effects of conventional BC therapy (Hammersen et al., 2020). Some of the most popular CAM modalities used for BC include dietary supplements, mind-body techniques, healing, energy, touch, messaging, and physical activities (Orticelli, 2022). In rural areas and war-torn countries, the daily diet consists of fresh organic vegetables, mushrooms, tomatoes, garlic, and fruits (Aldrich, 2023). Guided imagery, a mind-body technique, has been used to improve the quality of life (Silva et al., 2021). Herbal medicine can stimulate the immune system and protect the liver (Xu et al., 2020). As a result, while CAM can enhance quality of life and support conventional BC treatments, it is crucial to ensure that its use does not replace proven medical therapies. Integrating safe and evidence-based CAM practices can be beneficial for BC patients.

BC and its treatment often engender a variety of physical, emotional, social, and financial repercussions for patients and their families (Ibid). Palliative care, which encompasses supportive care, aims to enhance the quality of life of BC patients and their families, beginning at the time of diagnosis and continuing through survivorship or end-of-life care (Chou et al., 2020). Survivorship care, a facet of supportive care, addresses the physical and emotional ramifications of cancer and its treatment, enabling patients to cope with these consequences and lead a healthy life as far as possible (Chan et al., 2021). This section examines several aspects of supportive care and survivorship for patients with BC (Nardin et al., 2020), including the application of CAM in supportive care (Bajwah et al., 2020), physical therapy management of upper quadrant dysfunction and related concerns (Paolucci, Saggino et al., 2021), and psychological issues and potential interventions for patients with BC (Dinapoli et al., 2021). Comprehensive supportive care and survivorship strategies, including the integration of complementary medicine, physical therapy, and psychological support, are crucial for enhancing the quality of life and overall outcomes of BC patients and their families.

5.8.1 Palliative Care

BC patients in numerous countries who are members of the IsDB in the OIC often present at an advanced-stage, as reported by Lewison et al. (2020). Without surgical intervention, invasive BC typically results in miserable, fungating breasts with nerve involvement, as indicated by Dyer (2023). Uncontrolled pain is a common feature of end-of-life care in these situations (Hagarty et al., 2020). In some Muslim communities, there is a strong desire for a spiritual healing approach that embraces the ultimate destiny of God's will (Qadr), along with measures to ensure the patient's acceptance and spiritual well-being (Qureshi et al., 2020). The issues of permissible pain-relief and sedation at end-of-life within Islamic Fiqh (jurisprudence) have generated considerable controversy and confusion, as highlighted by Rashid (2022). This area, which has become increasingly important in BC management, necessitates a comprehensive Islamic perspective on palliative and supportive care, including its spiritual dimension, as emphasized by Sharif et al. (2021). Given that a cure for BC is yet to be discovered, research on BC in OIC-MCS should concentrate on palliative and supportive care, including pain management and the patient's spiritual well-being, as suggested by Wang et al. (2023). Therefore, research on OIC-MCS should prioritize palliative and supportive care, integrating pain management and spiritual well-being, to enhance the quality of life of patients with BC.

According to a study by Huang et al., BC is the most frequently diagnosed cancer and the leading cause of cancer-related deaths in women worldwide. (2021). In many OIC-MCS, BC is not only the most common malignancy affecting women, but also the primary cause of cancer deaths, with high mortality rates attributed to late-stage presentation and inadequate treatment facilities, as noted by Lewison et al. (2020). Research has shown that patients with advanced BC experience economic transition (Gharzai et al., 2021). Patients' experience needs related to work and financial disruptions throughout their illness and treatment must be addressed to optimize patient outcomes, including early and continued work involvement and treatment adherence (Ayub et al., 2024). Patients express significant concerns about financial toxicity, which may impede their decision-making regarding treatment options, as they worry about the affordability of any treatment (Abrams et al., 2021). Consequently, addressing the financial and economic impacts of BC is essential for improving patient outcomes in OIC-MCS, as financial toxicity can significantly hinder treatment adherence and decision-making.

5.8.2 Rehabilitation Services

Rehabilitation services for women with BC are virtually nonexistent in many Muslim countries (World Health Organization, 2022b). However, there has been increasing interest from private institutions in the Muslim sector to provide BC rehabilitation services (Sabgul et al., 2021). Ideally, BC rehabilitation services should be provided within the multidisciplinary framework of specialized breast clinics (Horton et al., 2020). Currently, the role of physiotherapists in BC rehabilitation is better defined and more widely accepted than that of other specialists (Rafn et al., 2020). Physiotherapy plays an important restorative, educational, and preventive role in managing BC patients. Physiotherapists can also provide valuable input for vocational rehabilitation of women with BC (Devoogdt & De Groef, 2024). The role of occupational therapy in BC rehabilitation is crucial (Lai et al., 2021). Women with BC who are left with disabilities require assistance in activities of daily living (Blackwood et al., 2020). Their ability to perform these tasks must be assessed, and their families must be helped to care for them.

Rehabilitation for BC survivors is a key aspect of comprehensive BC treatment, from the time of diagnosis to end-of-life (Invernizzi et al., 2022). This rehabilitation should be provided by a multidisciplinary team that focuses on the physical, functional, emotional, and vocational well-being of survivors (Paltrinieri et al., 2022). The objective of physical rehabilitation is to restore patients to the highest possible level of function, which may involve addressing impairments such as lymphedema, upper extremity weakness or pain, and shoulder joint restriction caused by surgery, radiation therapy, or systemic treatment (Cheville et al., 2021). In addition to physical rehabilitation, BC survivors may also benefit from emotional rehabilitation to address the many psychological issues that often accompany a diagnosis of BC and its treatment (Paolucci, Bernetti, et al., 2021) as well as vocational rehabilitation to help them return to work or find new employment that is fulfilling and appropriate, given their physical and emotional health (Hiltrop, 2022). Functional rehabilitation aims to assist survivors in re-engaging in their activities of daily living, work, and leisure (Stevens et al., 2020) and may involve appropriate exercise programs that enhance strength, flexibility, and overall physical health (Ortega et al., 2020). Therefore, a comprehensive rehabilitation program for BC survivors that addresses physical, emotional, and vocational needs is essential to restore function and improve quality of life.

5.9 Research and Innovation in BC

Developing national research and innovation systems is crucial for addressing the specific environmental, genetic, and cultural factors that contribute to cancer development and progression within OIC-MCS (Lewison et al., 2020). Investing in scientific research and innovation is essential to confront the growing cancer burden and ensure the ongoing improvement of cancer outcomes (Pramesh et al., 2022). Moreover, research on cancer is a significant area for exploring the potential beneficial effects of convergence and integrating art and science (STEAM) activities, which can help unlock human creativity and develop innovative ideas that can have a profound impact on society and improve human health (Lewison et al., 2021). Thus, enhancing research and innovation systems, particularly through STEAM integration, is vital for addressing cancer burden and improving health outcomes in OIC-MCS.

BC is a major contributor to morbidity and mortality worldwide. According to recent statistics, it is one of the leading causes of new cancer cases (Abdullah et al., 2020). In addition, women in low- and middle-income countries face several challenges in addressing this health issue, such as lack of awareness, inadequate screening programs, limited access to treatment, shortage of medicine, and high costs of therapy (Arnold et al., 2022). BC is the most prevalent cancer and the leading cause of cancer-related deaths among women (Frankland & Brown, 2022). Each year, more than 1.6 million new cases of BC are diagnosed, and more than 500,000 women lose their lives to the disease (Frankland & Brown, 2022). Notably, half of the cases and 60% of BC-related deaths occur in high-income countries (Francies et al., 2020). Therefore, there is an urgent need to establish cutting-edge research programs to address emerging BC issues and foster effective collaborations at national, regional, and international levels to create conducive environments for innovation in BC research (Carriero et al., 2023). Therefore, addressing BC effectively requires the establishment of advanced research programs and fostering collaboration to innovate and improve outcomes, especially in low- and middle-income countries.

5.9.1 Genomic Studies

The growing importance of collaborative studies has led to the discovery of new knowledge and enhancement of research policies, best practices, and research governance frameworks within regional research centers (Aassouli et al., 2023). The results of collaborative studies and data reported to the SIBCC have contributed to the development of recommendations for clinical practice and public health initiatives as well as the establishment of regional training programs for shared decision-making and quality assurance that consider the unique features of regional and ethnic variations (Gordon et al., 2020; Pereira et al., 2022). In addition to Islamic and some OIC and Arab quality of life initiatives, several countries, including Bahrain, KSA, UAE, and Jordan, have established BC research and registry systems in collaboration with the BC Research Group at KFSH&RC, which has led to the development of collaborative molecular-driven mapping in several Arab countries (Lewison et al., 2020). Therefore, collaborative studies and regional partnerships are crucial for advancing BC research, improving clinical practices, and tailoring public health initiatives to regional and ethnic variations.

In the field of genomics research of BC in IsDB-MCS of the OIC in the Middle East and North Africa (MENA) region (Al-Shamsi et al., 2023), various studies have reported unique characteristics and associations, such as increased levels of consanguinity, worse survival rates for specific gene expression subtypes, and unique germline mutations among young individuals (Nassir & Esheba, 2022). Additionally, there are associations between germline and somatic mutations, population-specific variations in BC predisposition genes, PAM50 signature gene expression patterns, somatic mutations, and viral genomic integration in BC (Hamid, 2022). Hence, understanding the unique genomic characteristics of BC in the MENA region is crucial for developing targeted interventions and improving outcomes in patients with OIC-MCS.

Although genomic research has contributed to a better understanding of various aspects of BC, there is a significant gap in research in genomics and its current state in IsDB-MCS (Campos et al., 2022). To address this gap, we recently extensively reviewed genomics and epigenomic studies in the broader OIC, highlighting some associations in non-communicable disease research and leveraging precision medicine initiatives. Despite the relatively low contribution of genomic research output, our review underscores the importance of further exploring the potential of genomics in the region (GhaviDel & Mostafavi, 2021). Thus, bridging the gap in genomics research on IsDB-MCS is crucial for advancing BC understanding and treatment, emphasizing the need for increased focus and investment in this field.

5.9.2 *Clinical Trials*

Fulfilling the unmet need for advanced research necessitates proven data collection activities, including the establishment of a cancer registry that should initially be based on national or regional cooperative efforts, focusing on simple yet critical data elements related to the study of patterns of care (Rollison et al., 2022). The specific area where research can most feasibly be conducted is the subgroup of women diagnosed with operable BC (Weber et al., 2021). The questions surrounding stage distribution, presence of clinical guidelines, and adherence to guidelines are considered priorities to aid in planning further interventions in the participating IsDB-MCS (Pericoli, 2020). It is worth noting that the low ratio of clinical trials reflects the general status of cancer research due to a lack of available funding or facilities for specific IsDB-MCS (Lewison et al., 2020). This low ratio also partly reflects the high proportion of preventable or treatable cancers that is carried out through the implementation of evidence-based primary prevention and early detection programs (Duggan et al., 2021). The availability and utilization of research evidence linking exploratory, molecular, and epidemiological work to incorporate new treatment modalities is still required in participating countries (Campbell et al., 2021). Collaboration in multi-center trials and registry studies can help establish feasible frameworks for increasing research activity (Mikita et al., 2021). Meeting these challenges would contribute to the design and conduct of regional and international collaborative studies within scientific and ethical standards that are considered acceptable (WHO, 2022a). Hence, establishing robust cancer registries and fostering collaboration in multi-center trials are essential for advancing BC research and improving the outcomes of IsDB-MCS.

Clinical trials for BC are formal research programs that evaluate novel approaches for preventing and treating BC (Britt et al., 2020). Different types of BC clinical trials are typically classified into various categories, such as surgery, radiotherapy, chemotherapy, endocrine therapy, biological therapy, CAM, and prevention (Korde et al., 2021). However, clinical trials on BC have not been conducted for several IsDB-MCS within the OIC (Lewison et al., 2020). Despite the importance of cancer and biomedical research, the level of support from both public and private sectors in OIC countries is extremely low (Al-Shamsi, 2022). Therefore, increasing support and conducting clinical trials on IsDB-MCS are crucial for advancing BC research and improving treatment outcomes.

5.10 Challenges and Opportunities in BC Management

BC management presents several challenges for OIC-MCS, as outlined by Lewison et al. (2020). These challenges include social beliefs, cultural practices, and low awareness, contributing to late diagnosis and poor survival rates (Mehrotra & Yadav, 2022). Furthermore, many OIC-MCS lack national cancer control plans and registries (Al-Shamsi, 2022). However, there has been a growing trend in scientific publications on BC from OIC-MCS (Lewison et al., 2020). Collaboration between OIC-MCS and within these countries has also increased (Haq & Tanveer, 2020). Funding opportunities abound. The development of IOMS Oncology by the Islamic Organization for Medical Sciences is a positive step toward improving cancer care in OIC-MCS (Obafemi & Ada, 2023). In conclusion, efforts should focus on enhancing BC research output, developing national cancer plans and registries, and improving research governance and infrastructure (Trapani et al., 2022). Moreover, research collaborations should be fostered, and funding for BC research in OIC-MCS should be increased (GhaviDel & Mostafavi, 2021). Therefore, addressing these challenges requires boosting BC research, developing cancer plans and registries, improving infrastructure, fostering collaboration, and increasing funding in the OIC countries.

5.10.1 Resource Allocation

Limited financial and human resources allocated to BC research in low-income countries, compared to high-income countries, may reflect inadequate resource allocation research (Bourgeois et al., 2024). The findings from the study on resource allocation and intervention in BC research may be applicable to other cancer sites and non-communicable diseases (NCDs) (Veljkovikj et al., 2022). Recommendations from resource allocation research should guide stakeholders involved in NCDs research in IsDB-MCS, which are part of the OIC countries that are members of the IsDB (Agbo, 2024). Despite the substantial burden of BC in OIC countries compared to high-income countries, the financial and human resources allocated to BC research are disproportionately low (Jing et al., 2020). Furthermore, there are significant global disparities in research output related to low-income and middle-income populations (Pramesh et al., 2022). Thus, addressing these disparities requires increased resource allocation and focused research efforts to improve BC outcomes in low-income and middle-income countries.

5.10.2 Healthcare Infrastructure

Islamic teachings place significant importance on the preservation of good health and provision of medical care when necessary (Ahmed, 2021). The need to support healthcare research is recognized in the culture of OIC-MCS as well as in other regions of the world (Orjiakor et al., 2023). In addition, religious authorities have issued fatwas to encourage and support research and development in health sciences (Yusli et al., 2021). These countries have made considerable efforts to increase healthcare expenditures and improve their healthcare systems by enhancing healthcare delivery and access and utilizing advancements in scientific knowledge

(Canton, 2021). This chapter discusses the distinctive patterns of BC observed in IsDB countries and examines the current state of various research initiatives in diagnosis, treatment, and healthcare delivery (Xiao et al., 2024). Hence, addressing these challenges requires boosting BC research, developing cancer plans and registries, improving infrastructure, fostering collaboration, and increasing funding in the OIC countries.

The healthcare infrastructure in countries that are members of the IsDB varies significantly, ranging from highly advanced facilities in some of the wealthiest nations to extremely limited primary healthcare centers in some of the poorest countries (Daurenova et al., 2021). Investment in healthcare infrastructure within OIC-MCS has been a topic of interest for many policymakers concerned with healthcare reforms (Steinacker et al., 2023). Based on income as the primary determinant, four categories of countries with distinct healthcare financing and delivery methods were identified (WAGstaff, A., & Neelsen, S., 2020). Several studies have provided evidence of a connection between the healthcare infrastructure and patient outcomes (Jimenez et al., 2021). Relying solely on primary healthcare centers to treat patients with serious illnesses or complex surgeries is insufficient (Bäuerle et al., 2022). A hierarchical system, with well-developed referral systems, should be implemented (Tao et al., 2020). Thus, enhancing the healthcare infrastructure and establishing robust referral systems are essential for improving patient outcomes in IsDB-MCS.

5.11 Conclusion and Future Directions

BC is diagnosed at relatively high survival rates in many countries that are members of the IsDB (Gunn et al., 2024). However, most research on this topic has been conducted in high-income countries. To improve the quality and relevance of research, it is important to investigate survival rates after diagnosis, access to diagnostic and treatment services, and the quality of treatment for BC in low- and middle-income countries (Nardin et al., 2020). Supporting surviving women can also increase research benefits from a broader community perspective, encourage local researchers to collaborate internationally, and enhance their capacity for high-quality research (Bi et al., 2021). The BHGI program, led by the BHGI Foundation, aims to promote collaboration and partnership in research in a strategic direction (Duggan, Dvaladze, Rositch, et al., 2020). The IsDB can assist OIC-MCS in participating in this program. In summary, significant advancements have been made in BC research within the OIC; however, there are still numerous opportunities for further exploration (Rej et al., 2023). In particular, obtaining greater financial support from local and national/international public and private sector stakeholders will be instrumental in facilitating BC research in (OIC-MCS) as reported by El Damanhoury (2023). This, in turn, will benefit communities across the region from north to south and west to east. Furthermore, implementing knowledge translation and quality care initiatives in OIC-MCS is crucial for substantially impacting healthcare and outcomes within the next 10 years (Song et al., 2024).

5.11.1 Key Findings and Recommendations

Several recommendations have been proposed to increase the research productivity and participation of OIC-MCS in the BC field. These include: (1) increasing research funding and establishing priorities for research; (2) encouraging and supporting international collaborations among researchers; (3) improving research capacity and investing in research infrastructure; (4) enhancing data collection and analysis on the burden of BC; (5) encouraging more involvement from researchers and institutions with a strong publication record; (6) addressing cultural, social, and economic challenges related to BC prevention and control; and (7) involvement of women and their representative organizations in research governance and setting the research agenda. Hopefully, the findings and recommendations presented in this chapter will motivate further research activities on BC to reduce the burden in both developed and developing countries.

This chapter examines state-of-the-art BC research conducted in the MCS of the IsDB within the OIC. Twenty countries were considered, representing 68.6% of the OIC's population (Chaudhary et al., 2024). The researchers analyzed 10 years using various bibliometric indicators (Yu, 2022). It was observed that scientific productivity related to BC in the region is noteworthy (Lei et al., 2021). Nine of the top ten most prolific countries are from the Middle East (Wilson, 2021). When adjusted for population size and research funding, Saudi Arabia emerged as the leading country in terms of productivity (Shehri et al., 2023). King Saud University has been identified as the most prolific institution in this region (Hefner & Zaman). The analysis of collaborations revealed that the most common type of partnership was national, followed by international and interinstitutional collaborations (Vogel et al., 2022).

5.11.2 Future Research Directions

The field of BC research and treatment has witnessed significant advancements owing to the application of nanotechnology, as demonstrated in recent studies (Tran et al., 2020). Nanotechnology is a promising avenue for the development of innovative diagnostic and therapeutic strategies for BC. Islamic member states within the OIC have actively contributed to academic research (Mahmud et al., 2021). In addition to therapeutic breakthroughs, nanotechnology has led to the creation of novel contrast agents that enable BC diagnosis at extremely early stages (Liu & Grodzinski, 2021).

This chapter aims to provide a comprehensive overview of the current state-of-the-art in BC research, focusing on the application of nanotechnology in IsDB-MCS (Mangla et al., 2022). Additionally, we discuss potential future research directions and the challenges that must be addressed for IsDB-MCS to play a more significant role in international BC research efforts (Sun et al., 2020).

Recently, the focus has shifted toward conducting extensive statistical analyses on more diverse populations, such as genome-wide association studies (Tibbs Cortes et al., 2021). Despite the high-cost and requirement of advanced technological resources, which are currently lacking in most Muslim countries, Islamic teachings emphasize ethical and spiritual obligations for promoting health and healing, thereby facilitating medical research within Islamic communities (Brockopp & Eich, 2021). BC research must transition toward developing practical applications that effectively address this issue (Britt et al., 2020). Nanotechnology is a promising area of research that has gained considerable attention, particularly for treating various cancers, including BC (Jain et al., 2020). There are numerous potential applications for the diagnosis and treatment of BC, ranging from the development of intelligent drugs and vaccines to innovative gene-targeted therapies, as well as the use of novel imaging and therapeutic agents, which may also be utilized by the larger Muslim community (Gadag et al., 2020).

Acknowledgment The authors thank the Chemical and Process Engineering Technology at the University of Malaysia, Pahang-UMPSA, Malaysia, for supporting this research.

Fund This research was funded by Eastern Unity Technology under Grant No. UIC 190806.

References

Aassouli, D., Akande, A., & Jureidini, R. (2023). Comparative analysis of sustainable food governance and the alignment of food security policies to sustainable development: A case study of OIC countries. *Sustainability, 15*(22), 15789.

Abdullah, R., Guo, P., & Harding, R. (2020). Preferences and experiences of Muslim patients and their families in Muslim-majority countries for end-of-life care: A systematic review and thematic analysis. *Journal of Pain and Symptom Management, 60*(6), 1223–1238.

Abrams, H. R., Durbin, S., Huang, C. X., Johnson, S. F., Nayak, R. K., Zahner, G. J., & Peppercorn, J. (2021). Financial toxicity in cancer care: Origins, impact, and solutions. *Translational Behavioral Medicine, 11*(11), 2043–2054.

Abu Huson, Y., Sierra-García, L., & García-Benau, M. A. (2024). A bibliometric review of information technology, artificial intelligence, and blockchain on auditing. *Total Quality Management & Business Excellence, 35*(1–2), 91–113.

Aebi, S., Karlsson, P., & Wapnir, I. L. (2022). Locally advanced breast cancer. *Breast, 62*(Suppl 1), S58–S62.

Afifi, A. M., Saad, A. M., Al-Husseini, M. J., Elmehrath, A. O., Northfelt, D. W., & Sonbol, M. B. (2020). Causes of death after breast cancer diagnosis: A US population-based analysis. *Cancer, 126*(7), 1559–1567.

Afzal, S., Hassan, M., Ullah, S., Abbas, H., Tawakkal, F., & Khan, M. A. (2022). Breast cancer; discovery of novel diagnostic biomarkers, drug resistance, and therapeutic implications. *Frontiers in Molecular Biosciences, 9*, 783450.

Agbo, S. (2024). *Improving health budgeting and management practices in low-middle income countries (LMIC): Using three country case studies—Nigeria, Kenya and Tanzania*. Intechopen. https://doi.org/10.5772/intechopen.1004996

Ahmad, S. B., Ali, A., Bilal, M., Rashid, S. M., Wani, A. B., Bhat, R. R., & Rehman, M. U. (2023). Melatonin and health: Insights of melatonin action, biological functions, and associated disorders. *Cellular and Molecular Neurobiology, 43*(6), 2437–2458.

Ahmed, A. A. (2021). Health and disease: An Islamic framework. In *Caring for Muslim patients* (pp. 35–43). Routledge.

Akdeniz, D., Klaver, M. M., Smith, C. Z., Koppert, L. B., & Hooning, M. J. (2020). The impact of lifestyle and reproductive factors on the risk of a second new primary cancer in the contralateral breast: A systematic review and meta-analysis. *Cancer Causes & Control, 31*, 403–416.

Alatrash, M. (2020). Prevalence, perceived benefits, and perceived barriers regarding breast cancer screening among three Arab American women subgroups. *Journal of Transcultural Nursing, 31*(3), 242–249.

Aldrich, E. (2023). *Casseroles, can openers, and jell-O: American food and the cold war, 1947–1959*. State University of new York press.

Ali, M., Hashmi, S. H., Nazir, M. R., Bilal, A., & Nazir, M. I. (2021). Does financial inclusion enhance economic growth? Empirical evidence from the IsDB member countries. *International Journal of Finance & Economics, 26*(4), 5235–5258.

Aljamali, N. M., Mujjed, A. N., & Jasim, D. A. W. (2022). Relationship between food type and increased risk of cancer. *Biomedical Research and Clinical Reviews, 6*(5), 107. https://doi.org/10.31579/2690-4861/107

Al-Khraif, R., Abdul Salam, A., & Abdul Rashid, M. F. (2020). Family demographic transition in Saudi Arabia: Emerging issues and concerns. *SAGE Open, 10*(1), 215824402091455.

Al-Mousa, D. S., Alakhras, M., Hossain, S. Z., Al-Sa'di, A. G., Al Hasan, M., Al-Hayek, Y., & Brennan, P. C. (2020). Knowledge, attitude and practice around breast cancer and mammography screening among Jordanian women. *Breast Cancer: Targets and Therapy, 12*, 231–242.

Al-Shamsi, H. O. (2022). The state of cancer care in The United Arab Emirates in 2022. *Clinics and Practice, 12*(6), 955–985.

Al-Shamsi, H. O., Abdelwahed, N., Al-Awadhi, A., Albashir, M., Abyad, A. M., Rafii, S., et al. (2023). Breast cancer in The United Arab Emirates. *JCO Global Oncology, 9*, e2200247.

Anand, U., Dey, A., Chandel, A. K. S., Sanyal, R., Mishra, A., Pandey, D. K., & de la Lastra, J. M. P. (2023). Cancer chemotherapy and beyond: Current status, drug candidates, associated risks and progress in targeted therapeutics. *Genes & Diseases, 10*(4), 1367–1401.

Andrahennadi, S., Sami, A., Manna, M., Pauls, M., & Ahmed, S. (2021). Current landscape of targeted therapy in hormone-receptor-positive and HER2-negative breast cancer. *Current Oncology, 28*(3), 1803–1822.

Ariga, S. (2023). History and future of HER2-targeted therapy for advanced gastric cancer. *Journal of Clinical Medicine, 12*(10), 3391.

Aristokli, N., Polycarpou, I., Themistocleous, S. C., Sophocleous, D., & Mamais, I. (2022). Comparison of the diagnostic performance of magnetic resonance imaging (MRI), ultrasound and mammography for detection of breast cancer based on tumor type, breast density and patient's history: A review. *Radiography, 28*(3), 848–856.

Arnold, M., Morgan, E., Rumgay, H., Mafra, A., Singh, D., Laversanne, M., et al. (2022). Current and future burden of breast cancer: Global statistics for 2020 and 2040. *The Breast, 66*, 15–23.

Axelsson, L., Hedberg, C., Pettersson, N., & Zhang, Q. (2022). Re-visiting the 'black box'of migration: State-intermediary co-production of regulatory spaces of labor migration. *Journal of Ethnic and Migration Studies, 48*(3), 594–612.

Ayub, M., Khan, K., Khan, M., & Ismail, M. (2024). Waqf for accelerating socio-economic development: A proposed model with focus on Pakistan. *Qualitative Research in Financial Markets*. https://doi.org/10.1108/QRFM-07-2023-0161

Bahreini, F., Rayzan, E., & Rezaei, N. (2021). microRNA-related single-nucleotide polymorphisms and breast cancer. *Journal of Cellular Physiology, 236*(3), 1593–1605.

Bajwah, S., Wilcock, A., Towers, R., Costantini, M., Bausewein, C., Simon, S. T., et al. (2020). Managing the supportive care needs of those affected by COVID-19. *European Respiratory Journal, 55*(4), 2000815.

Ban, Y., Li, M., Yu, M., & Wu, H. (2021). The effect of fear of progression on quality of life among breast cancer patients: The mediating role of social support. *Health and Quality of Life Outcomes, 19*(1), 178.

Barba, D., León-Sosa, A., Lugo, P., Suquillo, D., Torres, F., Surre, F., et al. (2021). Breast cancer, screening and diagnostic tools: All you need to know. *Critical Reviews in Oncology/Hematology, 157*, 103174.

Basu, P., Zhang, L., Hariprasad, R., Carvalho, A. L., & Barchuk, A. (2020). A pragmatic approach to tackle the rising burden of breast cancer through prevention & early detection in countries 'in transition'. *Indian Journal of Medical Research, 152*(4), 343–355.

Batyra, E., & Pesando, L. M. (2021). Trends in child marriage and new evidence on the selective impact of changes in age at marriage laws on early marriage. *SSM-Population Health, 14*, 100811.

Bäuerle, P., Schneider, U., Holtkamp, M., Gloveli, T., & Dugladze, T. (2022). Outlines to initiate epilepsy surgery in low-and middle-income countries. *Journal of Integrative Neuroscience, 21*(5), 134.

Beau, A. B., Napolitano, G. M., Ewertz, M., Vejborg, I., Schwartz, W., Andersen, P. K., & Lynge, E. (2020). Impact of chronic diseases on effect of breast cancer screening. *Cancer Medicine, 9*(11), 3995–4003.

Bi, W., Wang, H., Yang, G., & Zhu, C. (2021). A longitudinal cohort study on benefit finding evolution in Chinese women breast cancer survivals. *Scientific Reports, 11*(1), 20640.

Blackwood, J., Karczewski, H., Huang, M. H., & Pfalzer, L. (2020). Katz activities of daily living disability in older cancer survivors by age, stage, and cancer type. *Journal of Cancer Survivorship, 14*, 769–778.

Bouchareb, S., Chrifou, R., Bourik, Z., Nijpels, G., Hassanein, M., Westerman, M. J., & Elders, P. J. (2022). "I am my own doctor": A qualitative study of the perspectives and decision-making process of Muslims with diabetes on Ramadan fasting. *PLoS One, 17*(3), e0263088.

Bou-Dargham, M. J., Draughon, S., Cantrell, V., Khamis, Z. I., & Sang, Q. X. A. (2021). Advancements in human breast cancer targeted therapy and immunotherapy. *Journal of Cancer, 12*(23), 6949.

Bourgeois, A., Horrill, T., Mollison, A., Stringer, E., Lambert, L. K., & Stajduhar, K. (2024). Barriers to cancer treatment for people experiencing socio-economic disadvantage in high-income countries: A scoping review. *BMC Health Services Research, 24*(1), 670.

Brackstone, M., Baldassarre, F. G., Perera, F. E., Cil, T., Chavez Mac Gregor, M., Dayes, I. S., & Eisen, A. F. (2021). Management of the axilla in early-stage breast cancer: Ontario health (Cancer Care Ontario) and ASCO guideline. *Journal of Clinical Oncology, 39*(27), 3056–3082. ascopubs.org.

Britt, K. L., Cuzick, J., & Phillips, K. A. (2020). Key steps for effective breast cancer prevention. *Nature Reviews Cancer, 20*(8), 417–436.

Brockopp, J. E., & Eich, T. (2021). *Muslim medical ethics: From theory to practice*. University of South Carolina Press.

Budreviciute, A., Damiati, S., Sabir, D. K., & Kodzius, R. (2020). Management and prevention strategies for non-communicable diseases (NCDs) and their risk factors. *Frontiers in Public Health, 8*, 574111.

Buja, A., Pierbon, M., Lago, L., Grotto, G., & Baldo, V. (2020). Breast cancer primary prevention and diet: An umbrella review. *International Journal of Environmental Research and Public Health, 17*(13), 4731.

Buocikova, V., Rios-Mondragon, I., Pilalis, E., Chatziioannou, A., Miklikova, S., Mego, M., et al. (2020). Epigenetics in breast cancer therapy—New strategies and future nanomedicine perspectives. *Cancers, 12*(12), 3622.

Burguin, A., Diorio, C., & Durocher, F. (2021). Breast cancer treatments: Updates and new challenges. *Journal of Personalized Medicine, 11*(8), 808.

Calaf, G. M., Ponce-Cusi, R., Aguayo, F., Muñoz, J. P., & Bleak, T. C. (2020). Endocrine disruptors from the environment affecting breast cancer. *Oncology Letters, 20*(1), 19–32.

Campbell, E. M., Boyles, A., Shankar, A., Kim, J., Knyazev, S., Cintron, R., & Switzer, W. M. (2021). MicrobeTrace: Retooling molecular epidemiology for rapid public health response. *PLoS Computational Biology, 17*(9), e1009300. plos.org.

Campos, B., Pickering, A. C., Rocha, L. S., Aguilar, A. P., Fabres-Klein, M. H., deOliveira Mendes, T. A., et al. (2022). Diversity and pathogenesis of Staphylococcus aureus from bovine mastitis: Current understanding and future perspectives. *BMC Veterinary Research, 18*(1), 115.

Canelo-Aybar, C., Ferreira, D. S., Ballesteros, M., Posso, M., Montero, N., Solà, I., et al. (2021). Benefits and harms of breast cancer mammography screening for women at average risk of breast cancer: A systematic review for the European Commission initiative on breast cancer. *Journal of Medical Screening, 28*(4), 389–404.

Canfell, K., Kim, J. J., Brisson, M., Keane, A., Simms, K. T., Caruana, M., et al. (2020). Mortality impact of achieving WHO cervical cancer elimination targets: A comparative modeling analysis in 78 low-income and lower-middle-income countries. *The Lancet, 395*(10224), 591–603.

Canton, H. (2021). Islamic Development Bank—IsDB. In *In the Europa directory of international organizations 2021* (pp. 649–651). Routledge.

Cardoso, F., MacNeill, F., Penault-Llorca, F., Eniu, A., Sardanelli, F., Nordström, E. B., & Poortmans, P. (2021). Why is appropriate healthcare inaccessible for many European breast cancer patients?–the EBCC twelve manifesto. *The Breast, 55*, 128–135.

Carriero, S., Lanza, C., Pellegrino, G., Ascenti, V., Sattin, C., Pizzi, C., et al. (2023). Ablative therapies for breast cancer: State of art. *Technology in Cancer Research & Treatment, 22*, 15330338231157193.

Cave, L., Cooper, M. N., Zubrick, S. R., & Shepherd, C. C. (2020). A systematic review of racial discrimination and child and adolescent health in longitudinal studies. *Social Science & Medicine, 250*, 112864.

Chan, R. J., Nekhlyudov, L., Duijts, S. F., Hudson, S. V., Jones, J. M., Keogh, J., et al. (2021). Future research in cancer survivorship. *Journal of Cancer Survivorship, 15*(5), 659–667.

Chankseliani, M. (2023). Who funds the production of globally visible research in the global south? *Scientometrics, 128*(1), 783–801.

Chaudhary, F. A., Ahmad, B., Arjumand, B., & Alharkan, H. M. (2024). The association between economic status and religious identity with Oral health disparities and inequalities around the world. *Cureus, 16*(1), e51917.

Chen, X., Gole, J., Gore, A., He, Q., Lu, M., Min, J., et al. (2020). Non-invasive early detection of cancer four years before conventional diagnosis using a blood test. *Nature Communications, 11*(1), 1–10.

Chen, Y. C., Yu, J., Metcalfe, C., DeBruyn, T., Gelzleichter, T., Malhi, V., et al. (2022). Latest generation estrogen receptor degraders for the treatment of hormone-receptor-positive breast cancer. *Expert Opinion on Investigational Drugs, 31*(6), 515–529.

Cheng, R., Qi, L., Kong, X., Wang, Z., Fang, Y., & Wang, J. (2020). Identification of the significant genes regulated by estrogen receptor in estrogen receptor-positive breast cancer and their expression pattern changes when tamoxifen or fulvestrant resistance occurs. *Frontiers in Genetics, 11*, 538734.

Cheville, A., Smith, S., Barksdale, T., & Asher, A. (2021). Cancer rehabilitation. In *Braddom's physical medicine and rehabilitation* (pp. 568–593). https://doi.org/10.1016/B978-0-323-62539-5.00029-1

Chlebowski, R. T., Aragaki, A. K., Anderson, G. L., Pan, K., Neuhouser, M. L., Manson, J. E., et al. (2020). Dietary modification and breast cancer mortality: Long-term follow-up of the Women's health initiative randomized trial. *Journal of Clinical Oncology, 38*(13), 1419–1428.

Chou, Y. H., Hsieh, V. C. R., Chen, X., Huang, T. Y., & Shieh, S. H. (2020). Unmet supportive care needs of survival patients with breast cancer in different cancer stages and treatment phases. *Taiwanese Journal of Obstetrics and Gynecology, 59*(2), 231–236.

Choudhury, A., Singh, P. A., Bajwa, N., Dash, S., & Bisht, P. (2023). Pharmacovigilance of herbal medicines: Concerns and future prospects. *Journal of Ethnopharmacology, 309*, 116383.

Christiansen, P., Mele, M., Bodilsen, A., Rocco, N., & Zachariae, R. (2022). Breast-conserving surgery or mastectomy?: Impact on survival. *Annals of Surgery Open, 3*(4), e205.
Chung, H. L., Le-Petross, H. T., & Leung, J. W. T. (2021). Imaging updates to breast cancer lymph node management. *Radiographics, 41*(5), 1283–1299.
Correia, A. S., Gärtner, F., & Vale, N. (2021). Drug combination and repurposing for cancer therapy: The example of breast cancer. *Heliyon, 7*(1), e05948.
Cuthrell, K. M., & Tzenios, N. (2023). Breast cancer: Updated and deep insights. *International Research Journal of Oncology, 6*(1), 104–118.
Daly, A. A., Rolph, R., Cutress, R. I., & Copson, E. R. (2021). A review of modifiable risk factors in young women for the prevention of breast cancer. *Breast Cancer: Targets and Therapy, 13*, 241–257.
Daniyal, A., Santoso, I., Gunawan, N. H. P., Barliana, M. I., & Abdulah, R. (2021). Genetic influences in breast cancer drug resistance. *Breast Cancer: Targets and Therapy, 13*, 59–85. tandfonline.com.
Daurenova, A., Ospanova, A., Kilybaeva, P., & Sergazin, Y. (2021). The ISLAMIC development BANK and its role in socio-economic reforms in the CENTRAL ASIAN countries. *Central Asia & the Caucasus, 21*(2), 14046091.
Debien, V., De Caluwé, A., Wang, X., Piccart-Gebhart, M., Tuohy, V. K., Romano, E., & Buisseret, L. (2023). Immunotherapy in breast cancer: An overview of current strategies and perspectives. *NPJ Breast Cancer, 9*(1), 7.
Derakhshani, A., Rezaei, Z., Safarpour, H., Sabri, M., Mir, A., Sanati, M. A., et al. (2020). Overcoming trastuzumab resistance in HER2-positive breast cancer using combination therapy. *Journal of Cellular Physiology, 235*(4), 3142–3156.
Desantis, C., Siegel, R., & Jemal, A. (2024). *Breast cancer facts and figures 2022-2024* (pp. 1–48). American Cancer Society.
Devoogdt, N., & De Groef, A. (2024). Physiotherapy management of breast cancer treatment-related sequelae. *Journal of Physiotherapy, 70*(2), 90–105.
Dhar, R., Seethy, A., Singh, S., Pethusamy, K., Srivastava, T., Talukdar, J., et al. (2021). Cancer immunotherapy: Recent advances and challenges. *Journal of Cancer Research and Therapeutics, 17*(4), 834–844.
Dhiman, D., Kumar, A., & Shukla, S. (2023). Association of preoperative serum adipokines, insulin, and sex steroid hormones with breast cancer risk in the Indian women. *Indian Journal of Cancer, 60*(4), 548–555.
Diana, A., Carlino, F., Franzese, E., Oikonomidou, O., Criscitiello, C., De Vita, F., et al. (2020). Early triple-negative breast cancer: Conventional treatment and emerging therapeutic landscapes. *Cancers, 12*(4), 819.
Dinapoli, L., Colloca, G., Di Capua, B., & Valentini, V. (2021). Psychological aspects to consider in breast cancer diagnosis and treatment. *Current Oncology Reports, 23*(3), 38.
Ding, C., Wu, Y., Chen, X., Chen, Y., Wu, Z., Lin, Z., & Chen, F. (2022). Global, regional, and national burden and attributable risk factors of neurological disorders: The global burden of disease study 1990–2019. *Frontiers in Public Health, 10*, 952161.
Ding, Y., Chen, X., Zhang, Q., & Liu, Q. (2020). Historical trends in breast cancer among women in China from age-period-cohort modeling of the 1990–2015 breast cancer mortality data. *BMC Public Health, 20*(1), 1280.
DiPaola, V., Mazzotta, G., Pignatelli, V., Bufi, E., D'Angelo, A., Conti, M., et al. (2022). Beyond N staging in breast cancer: Importance of MRI and ultrasound-based imaging. *Cancers, 14*(17), 4270.
Duggan, C., Dvaladze, A., Rositch, A. F., Ginsburg, O., Yip, C. H., Horton, S., et al. (2020). The breast health global initiative 2018 global summit on improving breast healthcare through resource-stratified phased implementation: Methods and overview. *Cancer, 126*, 2339–2352.
Duggan, C., Dvaladze, A., Scheel, J. R., Stevens, L. M., & Anderson, B. O. (2020). Situational analysis of breast health care systems: Why context matters. *Cancer, 126*, 2405–2415.

Duggan, C., Trapani, D., Ilbawi, A. M., Fidarova, E., Laversanne, M., Curigliano, G., et al. (2021). National health system characteristics, breast cancer stage at diagnosis, and breast cancer mortality: A population-based analysis. *The Lancet Oncology, 22*(11), 1632–1642.

Dutta, J., Singh, S., Jaiswal, A., Ray, A., Das, P., & Mabalirajan, U. (2021). Role of oxidative stress and DNA damage/repair in lung cancer. In *Handbook of oxidative stress in cancer: Mechanistic aspects* (pp. 1–21). Springer Singapore.

Dyer, K. (2023). *The phenomenon of coping for women with primary breast cancer*. DCPsych thesis Middlesex University. https://repository.mdx.ac.uk/item/8q39y

El Damanhoury, K. (2023). Fighting religious extremism with faith-based entertainment-education: The portrayal of is in Arab Drama. *Journal of Religion, 11*(2), 171–197.

El-Halees, A., & Tafish, M. (2020). Breast cancer severity predication using deep learning techniques. *Jordanian Journal of Computers and Information Technology, 6*(1), 94–102.

Elzahi Saaid Ali, A. (2022). The challenges of Islamic trade finance in promoting SMEs in IsDB member countries. In *Empowering the poor through financial and social inclusion in Africa: An Islamic perspective* (pp. 103–117). Springer International Publishing.

ESCAP, U. N. (2024). *Digital and sustainable trade facilitation in the Organization of Islamic Cooporation (OIC) Asian Countries*.

Fan, K. M., Rimal, J., Zhang, P., & Johnson, N. W. (2022). Stark differences in cancer epidemiological data between GLOBOCAN and GBD: Emphasis on oral cancer and wider implications. *EClinicalMedicine, 54*, 101673.

Fatma, H., & Siddique, H. R. (2023). Research and patents status of selected phytochemicals against cancer: How close and how far? *Recent Patents on Anti-Cancer Drug Discovery, 18*(4), 428–447. [HTML].

Felbor, U., Bülow, R., Schmutzler, R. K., & Rath, M. (2022). Hereditary breast and ovarian cancer Service in Sparsely Populated Western Pomerania. *Healthcare, 10*(10), 2021. mdpi.com.

Filippou, C., Themistocleous, S. C., Marangos, G., Panayiotou, Y., Fyrilla, M., Kousparou, C. A., et al. (2024). Microbial therapy and breast cancer management: Exploring mechanisms, clinical efficacy, and integration within the one health approach. *International Journal of Molecular Sciences, 25*(2), 1110.

Fox, A. H., Nishino, M., Osarogiagbon, R. U., Rivera, M. P., Rosenthal, L. S., Smith, R. A., et al. (2023). Acquiring tissue for advanced lung cancer diagnosis and comprehensive biomarker testing: A National Lung Cancer Roundtable best-practice guide. *CA: a Cancer Journal for Clinicians, 73*(4), 358–375.

Francies, F. Z., Hull, R., Khanyile, R., & Dlamini, Z. (2020). Breast cancer in low-middle income countries: Abnormality in splicing and lack of targeted treatment options. *American Journal of Cancer Research, 10*(5), 1568.

Frankland, M., & Brown, T. (2022). For nurse practitioners, advanced practice nurses and allied health professionals. In *Breast health. Each Woman's menopause: An evidence-based resource* (pp. 347–403). Springer.

Gadag, S., Sinha, S., Nayak, Y., Garg, S., & Nayak, U. Y. (2020). Combination therapy and nanoparticulate systems: Smart approaches for effectively treating breast cancer. *Pharmaceutics, 12*(6), 524.

García-Albéniz, X., Hernán, M. A., Logan, R. W., Price, M., Armstrong, K., & Hsu, J. (2020). Continuation of annual screening mammography and breast cancer mortality in women older than 70 years. *Annals of Internal Medicine, 172*(6), 381–389.

Gharzai, L. A., Ryan, K. A., Szczygiel, L., Goold, S., Smith, G., Hawley, S., et al. (2021). Financial toxicity during breast cancer treatment: A qualitative analysis to inform strategies for mitigation. *JCO Oncology Practice, 17*(10), e1413–e1423.

GhaviDel, S., & Mostafavi, I. (2021). Comparative evaluation of top papers outputs of OIC-member countries in the essential science indicators database. *Malaysian Journal of Library and Information Science, 26*(2), 89–106.

Giammarile, F., Vidal-Sicart, S., Paez, D., Pellet, O., Enrique, E. L., Mikhail-Lette, M., & Mariani, G. (2022). Sentinel lymph node methods in breast cancer. *Seminars in Nuclear Medicine, 52*(5), 551–560.

Giap, F., Ma, S. J., Oladeru, O. T., Hong, Y. R., Yu, B., Mailhot Vega, R. B., et al. (2023). Palliative care utilization and racial and ethnic disparities among women with de novo metastatic breast cancer in the United States. *Breast Cancer Research and Treatment, 200*(3), 347–354.

Gignac, P. M., Aceves, V., Baker, S., Barnes, J. J., Bell, J., Boyer, D., et al. (2024). The role of networks to overcome large-scale challenges in tomography: The non-clinical tomography users research network. *Tomography of Materials and Structures, 5*(1852), 100031.

Ginsburg, O., Yip, C. H., Brooks, A., Cabanes, A., Caleffi, M., Dunstan Yataco, J. A., et al. (2020). Breast cancer early detection: A phased approach to implementation. *Cancer, 126*, 2379–2393.

Goh, J. N. (2020). *Becoming a Malaysian trans-man: Gender, society, body and faith*. Palgrave Macmillan Singapore. https://doi.org/10.1007/978-981-15-4534-4

Gordon, W. J., Landman, A., Zhang, H., & Bates, D. W. (2020). Beyond validation: Getting health apps into clinical practice. *npj Digital Medicine, 3*, 14.

Götmark, F., & Andersson, M. (2020). Human fertility in relation to education, economy, religion, contraception, and family planning programs. *BMC Public Health, 20*(1), 265.

Gottschalk, R., & Poon, D. (2020). Scaling up finance for the sustainable development goals. In *Southern-led development finance: Solutions from the global south*. Routledge.

Grégoire, V., Guckenberger, M., Haustermans, K., Lagendijk, J. J., Ménard, C., Pötter, R., et al. (2020). Image guidance in radiation therapy for better cure of cancer. *Molecular Oncology, 14*(7), 1470–1491.

Groll, D. (2021). *Conceiving people: Genetic knowledge and the ethics of sperm and egg donation*. Oxford University Press.

Gu, C., Wang, L., He, Y., Ouyang, T., Li, J., Wang, T., et al. (2022). Effect of local versus general anesthesia in breast-conserving surgery on cancer recurrence and cost. *Cancer Control, 29*, 10732748221083078.

Gunn, A. H., Warraich, H. J., & Mentz, R. J. (2024). Costs of care and financial hardship among patients with heart failure. *American Heart Journal, 269*, 94–107.

Hagarty, A. M., Bush, S. H., Talarico, R., Lapenskie, J., & Tanuseputro, P. (2020). Severe pain at the end-of-life: A population-level observational study. *BMC Palliative Care, 19*, 1–13.

Hamid, G. A. (2022). Breast cancer care in Yemen. *Age, 9*(3), 24–29.

Hamilton, E., Shastry, M., Shiller, S. M., & Ren, R. (2021). Targeting HER2 heterogeneity in breast cancer. *Cancer Treatment Reviews, 100*, 102286.

Hammersen, F., Pursche, T., Fischer, D., Katalinic, A., & Waldmann, A. (2020). Use of complementary and alternative medicine among young patients with breast cancer. *Breast Care, 15*(2), 163–170.

Haq, I. U., & Tanveer, M. (2020). Status of research productivity and higher education in the members of OIC. *Library Philosophy and Practice* (e-journal). https://digitalcommons.unl.edu/libphilprac/3845,researchgate.net

Hayati, Z., Jafarabadi, M. A., & Pirouzpanah, S. (2022). Dietary inflammatory index and breast cancer risk: An updated meta-analysis of observational studies. *European Journal of Clinical Nutrition, 76*(8), 1073–1087.

Heil, J., Pfob, A., Sinn, H. P., Rauch, G., Bach, P., Thomas, B., et al. (2022). Diagnosing pathologic complete response in the breast after neoadjuvant systemic treatment of breast cancer patients by minimal invasive biopsy: Oral presentation at the San Antonio breast cancer symposium on Friday, December 13, 2019, program number GS5-03. *Annals of Surgery, 275*(3), 576–581.

Hellman, K., Holmberg, E., Bjurberg, M., Borgfeldt, C., Dahm-Kähler, P., Rådestad, A. F., et al. (2020). Primary treatment and relative survival by stage and age in vulvar squamous cell carcinoma: A population-based SweGCG study. *Gynecologic Oncology, 159*(3), 663–671.

Henderson, J. T., Webber, E. M., Weyrich, M. S., Miller, M., & Melnikow, J. (2024). Screening for breast cancer: Evidence report and systematic review for the US preventive services task force. *Journal of the American Medical Association, 331*(22), 1931–1946.

Hendrick, R. E. (2020). Radiation doses and risks in breast screening. *Journal of Breast Imaging, 2*(3), 188–200. [HTML].

Hendriks, R. J., Vander Leest, M. M., Israel, B., Hannink, G., YantiSetiasti, A., Cornel, E. B., et al. (2021). Clinical use of the SelectMDx urinary-biomarker test with or without mpMRI in prostate cancer diagnosis: A prospective, multi-center study in biopsy-naïve men. *Prostate Cancer and Prostatic Diseases, 24*(4), 1110–1119.

Hiltrop, K. S. (2022). *The role of job changes in sustainable return to work for breast cancer survivors–patterns of interpretation and coping strategies*. Doctoral thesis.

Horton, S., Camacho Rodriguez, R., Anderson, B. O., Aung, S., Awuah, B., Delgado Pebe, L., et al. (2020). Health system strengthening: Integration of breast cancer care for improved outcomes. *Cancer, 126*, 2353–2364.

Hricak, H., Abdel-Wahab, M., Atun, R., Lette, M. M., Paez, D., Brink, J. A., et al. (2021). Medical imaging and nuclear medicine: A lancet oncology commission. *The Lancet Oncology, 22*(4), e136–e172.

Hu, T., Li, J., Long, M., Wu, J., Zhang, Z., Xie, F., et al. (2022). Detection of structural variations and fusion genes in breast cancer samples using third-generation sequencing. *Frontiers in Cell and Developmental Biology, 10*, 854640.

Huang, J., Chan, P. S., Lok, V., Chen, X., Ding, H., Jin, Y., et al. (2021). Global incidence and mortality of breast cancer: A trend analysis. *Aging (Albany NY), 13*(4), 5748.

Huang, Y., Wang, H., Lyu, Z., Dai, H., Liu, P., Zhu, Y., et al. (2022). Development and evaluation of the screening performance of a low-cost high-risk screening strategy for breast cancer. *Cancer Biology & Medicine, 19*(9), 1375.

Hussain, H., Siddiqui, H., & Gerothanassis, I. P. (2023). Prof. Dr. M. Iqbal Choudhary-A lifetime career dedicated to remarkable services in "natural products sciences". *Frontiers in Pharmacology, 14*, 1119419.

Hussain, S., Mubeen, I., Ullah, N., Shah, S. S. U. D., Khan, B. A., Zahoor, M., & Sultan, M. A. (2022). Modern diagnostic imaging technique applications and risk factors in the medical field: A review. *BioMed Research International, 2022*, 5164970.

Ibrahim, M., Ba-Essa, E., Alarouj, M., Annabi, F., Armstrong, D. G., Bennakhi, A., & Umpierrez, G. E. (2024). Recommendations for management of diabetes and its complications during hajj (Muslim pilgrimage)–2024 update. *Diabetes Research and Clinical Practice, 212*, 111647.

Invernizzi, M., DeSire, A., Venetis, K., Cigna, E., Carda, S., Borg, M., et al. (2022). Quality of life interventions in breast cancer survivors: State-of-the-art in targeted rehabilitation strategies. *Anti-Cancer Agents in Medicinal Chemistry, 22*(4), 801–810.

Islam, W., Niidome, T., & Sawa, T. (2022). Enhanced permeability and retention effect as a ubiquitous and epoch-making phenomenon for the selective drug targeting of solid tumors. *Journal of Personalized Medicine, 12*(12), 1964.

Jain, V., Kumar, H., Anod, H. V., Chand, P., Gupta, N. V., Dey, S., & Kesharwani, S. S. (2020). A review of nanotechnology-based approaches for breast cancer and triple-negative breast cancer. *Journal of Controlled Release, 326*, 628–647.

Jarvis, C. (2023). *Physical examination and health assessment-Canadian E-book*. Elsevier Health Sciences.

Jenei, K., Aziz, Z., Booth, C., Cappello, B., Ceppi, F., de Vries, E. G., & Moja, L. (2022). Cancer medicines on the WHO model list of essential medicines: Processes, challenges, and a way forward. *The Lancet Global Health, 10*(12), e1860–e1866.

Jimenez, M. P., DeVille, N. V., Elliott, E. G., Schiff, J. E., Wilt, G. E., Hart, J. E., & James, P. (2021). Associations between nature exposure and health: A review of the evidence. *International Journal of Environmental Research and Public Health, 18*(9), 4790.

Jin, H., Wang, L., & Bernards, R. (2023). Rational combinations of targeted cancer therapies: Background, advances and challenges. *Nature Reviews Drug Discovery, 22*(3), 213–234.

Jing, J., Feng, R., Zhang, X., Li, M., & Gao, J. (2020). Financial toxicity and its associated patient and cancer factors among women with breast cancer: A single-center analysis of low-middle income region in China. *Breast Cancer Research and Treatment, 181*, 435–443.

Justiz-Vaillant, A., Gardiner, L., Mohammed, M., Surajbally, M., Maharaj, L., Ramsingh, L., et al. (2021). Narrative literature review on risk factors involved in breast cancer. *Brain Cancer, Colon Rectal Cancer, Gynecological Malignancy, Lung Cancer, and Prostate Cancer, 11*(4), 11–28.

Kamarudin, M. M. I. (2021). Poverty and income inequality: Evidence from the OIC countries, based on Islamic values. In *The human development link between financial development*. PQDT-Global.

Katalinic, A., Eisemann, N., Kraywinkel, K., Noftz, M. R., & Hübner, J. (2020). Breast cancer incidence and mortality before and after implementation of the German mammography screening program. *International Journal of Cancer, 147*(3), 709–718.

Katsimperis, S., Kapriniotis, K., Manolitsis, I., Bellos, T., Angelopoulos, P., Juliebø-Jones, P., & Tzelves, L. (2024). Early investigational agents for the treatment of benign prostatic hyperplasia. *Expert Opinion on Investigational Drugs, 33*(4), 359–370. (just-accepted).

Kc, H., Shrestha, M., Pokharel, N., Niraula, S. R., Pyakurel, P., & Parajuli, S. B. (2021). Women's empowerment for abortion and family planning decision-making among marginalized women in Nepal: A mixed method study. *Reproductive Health, 18*, 1–11.

Khan, A. A., & Arora, A. S. (2021). Thermography as an economical alternative modality to mammography for early detection of breast cancer. *Journal of Healthcare Engineering, 2021*(6), 5543101.

Kharb, R., Haider, K., Neha, K., & Yar, M. S. (2020). Aromatase inhibitors: Role in postmenopausal breast cancer. *Archiv der Pharmazie, 353*(8), e2000081.

Kierczak, M., Rafati, N., Höglund, J., Gourlé, H., Lo Faro, V., Schmitz, D., et al. (2022). Contribution of rare whole-genome sequencing variants to plasma protein levels and the missing heritability. *Nature Communications, 13*(1), 2532.

Kim, H., Lee, S. B., Nam, S. J., Lee, E. S., Park, B. W., Park, H. Y., et al. (2021). Survival of breast-conserving surgery plus radiotherapy versus total mastectomy in early breast cancer. *Annals of Surgical Oncology, 28*(9), 5039–5047.

Knecht, K., Kinder, D., & Stockert, A. (2020). Biologically-based CAM use in cancer patients: The good, the bad, the misunderstood. *Frontiers in Nutrition, 6*, 196.

Konieczny, M., Cipora, E., Sygit, K., & Fal, A. (2020). Quality of life of women with breast cancer and socio-demographic factors. *Asian Pacific Journal of Cancer Prevention: APJCP, 21*(1), 185. nih.gov.

Korde, L. A., Somerfield, M. R., Carey, L. A., Crews, J. R., Denduluri, N., Hwang, E. S., et al. (2021). Neoadjuvant chemotherapy, endocrine therapy, and targeted therapy for breast cancer: ASCO guideline. *Journal of Clinical Oncology, 39*(13), 1485.

Kudamba, A., Jamil, E. S., Okurut, S. A., Kafero, H. M., Nsubuga, H., & Walusansa, A. (2021). Assessment of knowledge, perception, attitude, risk factors prevention and treatment options of cancer among natives in Elgon Sub-Region, Uganda. *Fortune Journal of Health Sciences, 5*, 472–487.

Kunte, S., Abraham, J., & Montero, A. J. (2020). Novel HER2–targeted therapies for HER2–positive metastatic breast cancer. *Cancer, 126*(19), 4278–4288.

Lai, L. L., Player, H., Hite, S., Satyananda, V., Stacey, J., Sun, V., et al. (2021). Feasibility of remote occupational therapy services via telemedicine in a breast cancer recovery program. *The American Journal of Occupational Therapy, 75*(2), 7502205030p1–7502205030p9.

Lau, K. H., Tan, A. M., & Shi, Y. (2022). New and emerging targeted therapies for advanced breast cancer. *International Journal of Molecular Sciences, 23*(4), 2288.

Lei, S., Zheng, R., Zhang, S., Wang, S., Chen, R., Sun, K., et al. (2021). Global patterns of breast cancer incidence and mortality: A population-based cancer registry data analysis from 2000 to 2020. *Cancer Communications, 41*(11), 1183–1194.

Lei, W., Li, H., Song, G., Zhang, R., Ran, R., Yan, Y., et al. (2020). Efficacy and safety of fulvestrant 500mg in hormone-receptor positive human epidermal receptor two negative advanced breast cancer: A real-world study in China. *Journal of Cancer, 11*(22), 6612.

LeRoith, D., Holly, J. M. P., & Forbes, B. E. (2021). Insulin-like growth factors: Ligands, binding proteins, and receptors. *Molecular Metabolism, 52*, 101245.

Lewison, G., Hussain, S. F., Guo, P., Harding, R., Mukherji, D., Sittah, G. A., & Sullivan, R. (2020). Cancer research in the 57 OIC countries, 2008–17. *Ecancermedicalscience, 14*, 1094.

Lewison, G., Sullivan, R., & Kiliç, C. (2021). Mental health disorders research in the countries of the Organization of Islamic Cooperation (OIC), 2008–17, and the disease burden: Bibliometric study. *PLoS One, 16*(4), e0250414.

Li, J., Guan, X., Fan, Z., Ching, L. M., Li, Y., Wang, X., et al. (2020). Non-invasive biomarkers for early detection of breast cancer. *Cancers, 12*(10), 2767.

Li, L., Yu, R., Cai, T., Chen, Z., Lan, M., Zou, T., et al. (2020). Effects of immune cells and cytokines on inflammation and immunosuppression in the tumor microenvironment. *International Immunopharmacology, 88*, 106939.

Lim, Y. X., Lim, Z. L., Ho, P. J., & Li, J. (2022). Breast cancer in Asia: Incidence, mortality, early detection, mammography programs, and risk-based screening initiatives. *Cancers, 14*(17), 4218.

Liu, C. H., & Grodzinski, P. (2021). Nanotechnology for cancer imaging: Advances, challenges, and clinical opportunities. *Radiology: Imaging Cancer, 3*(3), e200052.

Liu, J., Lei, J., Ou, Y., Zhao, Y., Tuo, X., Zhang, B., & Shen, M. (2023). Mammography diagnosis of breast cancer screening through machine-learning: A systematic review and meta-analysis. *Clinical and Experimental Medicine, 23*(6), 2341–2356.

Liu, J. H. (2020). Selective estrogen receptor modulators (SERMS): Keys to understanding their function. *Menopause, 27*(10), 1171–1176.

Liu, T., Huang, Y., & Lin, H. (2021). Estrogen disorders: Interpreting the abnormal regulation of aromatase in granulosa cells. *International Journal of Molecular Medicine, 47*(5), 1–12.

Luo, C., Wang, L., Zhang, Y., Lu, M., Lu, B., Cai, J., et al. (2022). Advances in breast cancer screening modalities and status of global screening programs. *Chronic Diseases and Translational Medicine, 8*(02), 112–123.

Luque-Bolivar, A., Pérez-Mora, E., Villegas, V. E., & Rondón-Lagos, M. (2020). Resistance and overcoming resistance in breast cancer. *Breast Cancer, 12*, 211–229.

Mahmud, K. M., Niloy, M. S., Shakil, M. S., & Islam, M. A. (2021). Ruthenium complexes: An alternative to platinum drugs in colorectal cancer treatment. *Pharmaceutics, 13*(8), 1295.

Mangla, M., Shinde, S. K., Mehta, V., Sharma, N., & Mohanty, S. N. (Eds.). (2022). *Handbook of research on machine-learning: Foundations and applications*. CRC Press.

Mann, R. M., Athanasiou, A., Baltzer, P. A., Camps-Herrero, J., Clauser, P., Fallenberg, E. M., & European Society of Breast Imaging (EUSOBI). (2022). Breast cancer screening in women with extremely dense breasts recommendations of the European Society of Breast Imaging (EUSOBI). *European Radiology, 32*(6), 4036–4045.

Mansouri, Y., & Attar, A. (2022). *The role of walkability in creating sustainable cities and communities Case of the city of Bejaia in Algeria* (pp. 274–277). OIC Youth Scientific Congress. uliege.be.

Marklund, A., Lundberg, F. E., Eloranta, S., Hedayati, E., Pettersson, K., & Rodriguez-Wallberg, K. A. (2021). Reproductive outcomes after breast cancer in women with vs without fertility preservation. *JAMA Oncology, 7*(1), 86–91.

Masfarré, L., Vidal, J., Fernandez-Rodriguez, C., & Montagut, C. (2021). ctDNA to guide adjuvant therapy in localized CRC. *Cancers, 13*(12), 2869.

Matthews, H. K., Bertoli, C., & de Bruin, R. A. (2022). Cell cycle control in cancer. *Nature Reviews Molecular Cell Biology, 23*(1), 74–88.

Mbemi, A., Khanna, S., Njiki, S., Yedjou, C. G., & Tchounwou, P. B. (2020). Impact of gene-environment interactions on cancer development. *International Journal of Environmental Research and Public Health, 17*(21), 8089.

McKinney, S. M., Sieniek, M., Godbole, V., Godwin, J., Antropova, N., Ashrafian, H., et al. (2020). International evaluation of an artificial intelligence (AI) system for breast cancer screening. *Nature, 577*(7788), 89–94.

Meattini, I., de Oliveira Franco, R., Salvestrini, V., & Hijal, T. (2023). Partial breast irradiation. *The Breast, 69*, 401–409. sciencedirect.com.

Mehraj, U., Dar, A. H., Wani, N. A., & Mir, M. A. (2021). Tumor microenvironment promotes breast cancer chemoresistance. *Cancer Chemotherapy and Pharmacology, 87*(2), 147–158.

Mehrotra, R., & Yadav, K. (2022). Breast cancer in India: Present scenario and the challenges ahead. *World Journal of Clinical Oncology, 13*(3), 209.

Memish, Z. A., Perlman, S., Van Kerkhove, M. D., & Zumla, A. (2020). Middle East respiratory syndrome. *The Lancet, 395*(10229), 1063–1077.

Mercogliano, M. F., Bruni, S., Mauro, F. L., & Schillaci, R. (2023). Emerging targeted therapies for HER2-positive breast cancer. *Cancers, 15*(7), 1987.

Micoli, F., Bagnoli, F., Rappuoli, R., & Serruto, D. (2021). The role of vaccines in combating antimicrobial resistance. *Nature Reviews Microbiology, 19*(5), 287–302.

Mikita, J. S., Mitchel, J., Gatto, N. M., Laschinger, J., Tcheng, J. E., Zeitler, E. P., et al. (2021). Determining the suitability of registries for embedding clinical trials in the United States: A project of the clinical trials transformation initiative. *Therapeutic Innovation & Regulatory Science, 55*, 6–18.

Min, S. K., Lee, S. K., Woo, J., Jung, S. M., Ryu, J. M., Yu, J., et al. (2021). Relation between tumor size and lymph node metastasis according to subtypes of breast cancer. *Journal of Breast Cancer, 24*(1), 75.

Mirbagheri, E., Shafiee, M., Shanbezadeh, M., & Kazemi-Arpanahi, H. (2022). Developing the required data set for the integration of breast cancer registry systems in Iran. *Informatics in Medicine Unlocked, 32*, 101011.

Mishra, G. A., Pimple, S. A., Mittra, I., & Badwe, R. A. (2021a). Screening for breast cancer: Cost-effective solutions for low-& middle-income countries. *Indian Journal of Medical Research, 154*(2), 229–236.

Mishra, S. R., Chung, H. F., Waller, M., & Mishra, G. D. (2021b). Duration of estrogen exposure during reproductive years, age at menarche and age at menopause, and risk of cardiovascular disease events, all-cause and cardiovascular mortality: A systematic review and meta-analysis. *BJOG: An International Journal of Obstetrics & Gynecology, 128*(5), 809–821.

Mittra, I., Mishra, G. A., Dikshit, R. P., Gupta, S., Kulkarni, V. Y., Shaikh, H. K. A., & Badwe, R. A. (2021). Effect of screening by clinical breast examination on breast cancer incidence and mortality after 20 years: Prospective, cluster randomized controlled trial in Mumbai. *BMJ, 372*, 256.

Moiyadi, A. V. (2021). *XIX Annual Conference on Evidence-Based Management of Cancers in India "Technology and Cancer Care–Promise and Reality of the Brave New World"*.

Mutebi, M., Anderson, B. O., Duggan, C., Adebamowo, C., Agarwal, G., Ali, Z., & Eniu, A. (2020). Breast cancer treatment: A phased approach to implementation. *Cancer, 126*, 2365–2378. wiley.com.

Nadda, R., Sahani, A. K., & Repaka, R. (2022). A systematic review of real-time fine-needle aspiration biopsy methods for soft tissues. *IETE Technical Review, 39*(5), 1011–1026.

Nardin, S., Mora, E., Varughese, F. M., D'Avanzo, F., Vachanaram, A. R., Rossi, V., et al. (2020). Breast cancer survivorship, quality of life, and late toxicities. *Frontiers in Oncology, 10*, 864.

Nassir, R., & Esheba, G. (2022). Implementation of next-generation sequencing in Saudi Arabia for HER2-positive breast cancer. *Saudi Journal of Biological Sciences, 29*(3), 1808–1812.

Neuberger, J., Patel, J., Caldwell, H., Davies, S., Hebditch, V., Hollywood, C., et al. (2020). Guidelines on the use of liver biopsy in clinical practice from the British Society of Gastroenterology, the Royal College of Radiologists and the Royal College of pathology. *Gut, 69*(8), 1382–1403.

Nicosia, L., Gnocchi, G., Gorini, I., Venturini, M., Fontana, F., Pesapane, F., & Cassano, E. (2023). History of mammography: Analysis of breast imaging diagnostic achievements over the last century. *Healthcare, 11*(11), 1596. MDPI.

Norfjord van Zyl, M., Akhavan, S., Tillgren, P., & Asp, M. (2020). Non-participation in mammographic screening–experiences of women from a region in Sweden. *BMC Public Health, 20*, 219. springer.com.

Obafemi, F. A., & Ada, A. I. (2023). A review on Ficus Benjamina as a potential viable candidate in the struggle against cancer. *Journal of Global Biosciences, 12*(3), 9674–9729.

Omisore, A. D., Olasehinde, O., Wuraola, F. O., Sutton, E. J., Sevilimedu, V., Omoyiola, O. Z., et al. (2023). Improving access to breast cancer screening and treatment in Nigeria: The triple mobile assessment and patient navigation model (NCT05321823): A study protocol. *PLoS One, 18*(6), e0284341.

Orjiakor, C. T., Onwujekwe, O., McKee, M., Hutchison, E., Agwu, P., & Balabanova, D. (2023). "I can't kill myself": Local narratives and meanings that foster absenteeism in Nigerian primary health centers. *Journal of Global Health, 13*, 04129.

Ortega, M. A., Fraile-Martínez, O., García-Montero, C., Pekarek, L., Guijarro, L. G., Castellanos, A. J., et al. (2020). Physical activity as an imperative support in breast cancer management. *Cancers, 13*(1), 55.

Orticelli, B. (2022). *Integrating prayer and faith in complementary alternative medicine into treating breast cancer: Lived experiences*. A Dissertation

Özgöz, A., Mutlu İçduygu, F., Yükseltürk, A., ŞamlI, H., Hekİmler Ozturk, K., & Başkan, Z. (2020). Low-penetrance susceptibility variants and post-menopausal estrogen receptor-positive breast cancer. *Journal of Genetics, 99*, 1–10.

Paltrinieri, S., Costi, S., Pellegrini, M., Díaz Crescitelli, M. E., Vicentini, M., Mancuso, P., et al. (2022). Adaptation of the core set for vocational rehabilitation for cancer survivors: A qualitative consensus-based study. *Journal of Occupational Rehabilitation, 32*(4), 718–730.

Paolucci, T., Bernetti, A., Bai, A. V., Segatori, L., Monti, M., Maggi, G., et al. (2021). The sequelae of mastectomy and quadrantectomy to the reaching movement in breast cancer survivors: Evidence for an integrated rehabilitation protocol during oncological care. *Supportive Care in Cancer, 29*, 899–908.

Paolucci, T., Saggino, A., Agostini, F., Paoloni, M., Bernetti, A., Mangone, M., & Tommasi, M. (2021). The influence of rehabilitation on quality of life in breast cancer survivors: A clinical study. *International Journal of Environmental Research and Public Health, 18*(16), 8585. mdpi.com.

Pashayan, N., Antoniou, A. C., Ivanus, U., Esserman, L. J., Easton, D. F., French, D., et al. (2020). Personalized early detection and prevention of breast cancer: ENVISION consensus statement. *Nature Reviews Clinical Oncology, 17*(11), 687–705.

Patel, A., Unni, N., & Peng, Y. (2020). The changing paradigm for the treatment of HER2-positive breast cancer. *Cancers, 12*(8), 2081.

Peleg Hasson, S., Menes, T., & Sonnenblick, A. (2020). Comparison of patient susceptibility genes across breast cancer: Implications for prognosis and therapeutic outcomes. *Pharmacogenomics and Personalized Medicine, 13*, 227–238.

Pereira, R. D. O., Luz, L. A. D., Chagas, D. C., Amorim, J. R., Nery-Júnior, E. D. J., Alves, A. C. B. R., & Silva, B. B. D. (2020). Evaluation of the accuracy of mammography, ultrasound and magnetic resonance imaging in suspect breast lesions. *Clinics, 75*, e1805. scielo.br.

Pereira, V. C., Silva, S. N., Carvalho, V. K., Zanghelini, F., & Barreto, J. O. (2022). Strategies for the implementation of clinical practice guidelines in public health: An overview of systematic reviews. *Health Research Policy and Systems, 20*(1), 13.

Pericoli, A. (2020). Islamic finance and charity in the Muslim world. The role of the Islamic Development Bank in financing aid. *Journal of economics and business Aseanomics, 5*(2), 113–133. yarsi.ac.id.

Pramesh, C. S., Badwe, R. A., Bhoo-Pathy, N., Booth, C. M., Chinnaswamy, G., Dare, A. J., et al. (2022). Priorities for cancer research in low-and middle-income countries: A global perspective. *Nature Medicine, 28*(4), 649–657.

Qin, R. X., Yoon, S., Fowler, Z. G., Jayaram, A., Stankey, M., Samad, L., & Park, K. B. (2023). Financing surgical, obstetric, anesthesia, and trauma care in the Asia Pacific region: Proceedings. *BMC Proceedings, 17*(5), 10.

Qureshi, N. A., Khalil, A. A., & Alsanad, S. M. (2020). Spiritual and religious healing practices: Some reflections from Saudi National Center for Complementary and Alternative Medicine, Riyadh. *Journal of Religion and Health, 59*(2), 845–869.

Rafn, B. S., Midtgaard, J., Camp, P. G., & Campbell, K. L. (2020). Shared concern with current breast cancer rehabilitation services: A focus group study of survivors' and professionals' experiences and preferences for rehabilitation. *BMJ Open, 10*(7), e037280. bmj.com.

Rakha, E., Toss, M., & Quinn, C. (2022). Specific cell differentiation in breast cancer: A basis for histological classification. *Journal of Clinical Pathology, 75*(2), 76–84.

Rashid, R. (2022). *An Islamic bioethical framework for withholding and withdrawing life-sustaining treatment. In end-of-life care, dying and death in the Islamic moral tradition* (pp. 282–329). Brill.

Reinert, T., Cascelli, F., Resende, C. A. A. D., Goncalves, A. C., Godo, V. S. P., & Barrios, C. H. (2022). Clinical implication of low estrogen receptor (ER-low) expression in breast cancer. *Frontiers in Endocrinology, 13*, 1015388.

Rej, R. K., Thomas, J. E., Acharyya, R. K., Rae, J. M., & Wang, S. (2023). Targeting the estrogen receptor for breast cancer treatment: Recent advances and challenges. *Journal of Medicinal Chemistry, 66*(13), 8339–8381.

Ressler, A. M. (2022). *Addressing barriers through A telephone intervention to promote screening mammogram adherence*. An evidence-based doctoral project

Ricci-Cabello, I., Vásquez-Mejía, A., Canelo-Aybar, C., Niño de Guzman, E., Pérez-Bracchiglione, J., Rabassa, M., et al. (2020). Adherence to breast cancer guidelines is associated with better survival outcomes: A systematic review and meta-analysis of observational studies in EU countries. *BMC Health Services Research, 20*, 1–12.

Riis, M. (2020). Modern surgical treatment of breast cancer. *Annals of Medicine and Surgery, 56*, 95–107.

Rock, C. L., Thomson, C., Gansler, T., Gapstur, S. M., McCullough, M. L., Patel, A. V., et al. (2020). American Cancer Society guideline for diet and physical activity for cancer prevention. *CA: a Cancer Journal for Clinicians, 70*(4), 245–271.

Rodriguez-García, A., Oliva-Ramirez, J., Bautista-Flores, C., & Hosseini, S. (2020). 3D in-vitro human organ mimicry devices for drug discovery, development, and assessment. *Advances in Polymer Technology, 2020*, 1–41.

Roheel, A., Khan, A., Anwar, F., Akbar, Z., Akhtar, M. F., Imran Khan, M., et al. (2023). Global epidemiology of breast cancer based on risk factors: A systematic review. *Frontiers in Oncology, 13*, 1240098.

Rollison, D. E., Levin, G. M., Warner, J. L., Pinder, R., Havener, L. A., Behera, M., et al. (2022). Current and emerging informatics initiatives impactful to cancer registries. *Journal of Registry Management, 49*(4), 153.

Ronga, M. G., Cavallone, M., Patriarca, A., Leite, A. M., Loap, P., Favaudon, V., et al. (2021). Back to the future: Very high-energy electrons (VHEEs) and their potential application in radiation therapy. *Cancers, 13*(19), 4942.

Sabgul, A. A., Qattan, A. M., Hashmi, R., & Al-Hanawi, M. K. (2021). Husbands' knowledge of breast cancer and their wives' attitudes and practices related to breast cancer screening in Saudi Arabia: Cross-sectional online survey. *Journal of Medical Internet Research, 23*(2), e25404.

Salem, H., & Daher-Nashif, S. (2020). Psychosocial aspects of female breast cancer in the Middle East and North Africa. *International Journal of Environmental Research and Public Health, 17*(18), 6802.

Samaroo, K., Hosein, A., Olivier, L. K., & Ali, J. (2021). Breast cancer in the Caribbean. *Cureus, 13*(8), e17042.

Sangwan, R. K., Huda, R. K., Panigrahi, A., Toteja, G. S., Sharma, A. K., Thakor, M., & Kumar, P. (2023). Strengthening breast cancer screening program through health education of women and capacity-building of primary healthcare providers. *Frontiers in Public Health, 11*, 1276853.

Schachter, N. F., Adams, J. R., Skowron, P., Kozma, K. J., Lee, C. A., Raghuram, N., et al. (2021). Single allele loss-of-function mutations select and sculpt conditional cooperative networks in breast cancer. *Nature Communications, 12*(1), 5238.

Schousboe, J. T., Sprague, B. L., Abraham, L., O'Meara, E. S., Onega, T., Advani, S., et al. (2022). Cost-effectiveness of screening mammography beyond age 75 years: A cost-effectiveness analysis. *Annals of Internal Medicine, 175*(1), 11–19.

Seidler, S. J., & Huber, D. E. (2020). Overview of diagnosis and treatment of breast cancer in young women. *EC Gynaecology, 2*, 18–25.

Shah, S. A., Zhang, Y., Correa, A. M., Hijaz, B. A., Yang, A. Z., Fayanju, O. M., & Cerullo, M. (2024). Rates of price disclosure associated with the surgical treatment of early-stage breast cancer one year after implementation of federal regulations. *Breast Cancer Research and Treatment, 203*(2), 397–406.

Sharif, S. P., Lehto, R. H., Amiri, M., Ahadzadeh, A. S., Nia, H. S., & Haghdoost, A. A.,& Goudarzian, A. H. (2021). Spirituality and quality of life in women with breast cancer: The role of hope and educational attainment. *Palliative & Supportive Care, 19*(1), 55–61.

Shehri, T. A., Braun, J. F., Howarth, N., Lanza, A., & Luomi, M. (2023). Saudi Arabia's climate change policy and the circular carbon economy approach. *Climate Policy, 23*(2), 151–167.

Silva, L. A. A. D., Santana, E. D. O., Marcondes, L., Silva, M. N. D., Scherer, A. B., Sawada, N. O., & Kalinke, L. P. (2021). Does guided imagery improve health-related quality of life in cancer patients? Integrative review. *Cogitare Enfermagem, 26*, e75334.

Smolarz, B., Nowak, A. Z., & Romanowicz, H. (2022). Breast cancer—Epidemiology, classification, pathogenesis and treatment (review of literature). *Cancers, 14*(10), 2569.

Song, M., Anees, A., Rahman, S. U., & Ali, M. S. E. (2024). Technology transfer for green investments: Exploring how technology transfer through foreign direct investments can contribute to sustainable practices and reduced environmental impact in OIC economies. *Environmental Science and Pollution Research, 31*(6), 8812–8827.

Steinacker, J. M., Van Mechelen, W., Bloch, W., Börjesson, M., Casasco, M., Wolfarth, B., & Pitsiladis, Y. P. (2023). Global alliance for the promotion of physical activity: The Hamburg declaration. *BMJ Open Sport & Exercise Medicine, 9*(3), e001626.

Stevens, C. J., Hegel, M. T., Bakitas, M. A., Bruce, M., Azuero, A., Pisu, M., et al. (2020). Study protocol for a multisite randomized controlled trial of a rehabilitation intervention to reduce participation restrictions among female breast cancer survivors. *BMJ Open, 10*(2), e036864.

Subhan, M. A., Parveen, F., Shah, H., Yalamarty, S. S. K., Ataide, J. A., & Torchilin, V. P. (2023). Recent advances with precision medicine treatment for breast cancer including triple-negative sub-type. *Cancers, 15*(8), 2204.

Sulis, G., Sayood, S., & Gandra, S. (2022). Antimicrobial resistance in low-and middle-income countries: Current status and future directions. *Expert Review of Anti-Infective Therapy, 20*(2), 147–160.

Sun, W., Starly, B., Daly, A. C., Burdick, J. A., Groll, J., Skeldon, G., et al. (2020). The bioprinting roadmap. *Biofabrication, 12*(2), 022002.

Sung, H., Ferlay, J., Siegel, R. L., Laversanne, M., Soerjomataram, I., Jemal, A., & Bray, F. (2021). Global cancer statistics 2020: GLOBOCAN estimates of incidence and mortality worldwide for 36 cancers in 185 countries. *CA: a Cancer Journal for Clinicians, 71*(3), 209–249.

Szmaglinska, M., Andrew, L., Massey, D., & Kirk, D. (2024). Exploring the underutilized potential of clinical hypnosis: A scoping review of healthcare professionals' perceptions, knowledge, and attitudes. *International Journal of Clinical and Experimental Hypnosis, 72*(2), 109–138.

Tangkiatkumjai, M., Boardman, H., & Walker, D. M. (2020). Potential factors that influence usage of complementary and alternative medicine worldwide: A systematic review. *BMC Complementary Medicine and Therapies, 20*, 1–15.

Tao, Z., Cheng, Y., & Liu, J. (2020). Hierarchical two-step floating catchment area (2SFCA) method: Measuring the spatial accessibility to hierarchical healthcare facilities in Shenzhen, China. *International Journal for Equity in Health, 19*, 164.

Testa, U., Castelli, G., & Pelosi, E. (2020). Breast cancer: A molecularly heterogenous disease needing sub-type-specific treatments. *Medical Science, 8*(1), 18.

Tibbs Cortes, L., Zhang, Z., & Yu, J. (2021). Status and prospects of genome-wide association studies in plants. *The plant genome, 14*(1), e20077.

Tin Tin, S., Reeves, G. K., & Key, T. J. (2021). Endogenous hormones and risk of invasive breast cancer in pre-and post-menopausal women: Findings from the UK biobank. *British Journal of Cancer, 125*(1), 126–134.

Torabi Dalivandan, S., Plummer, J., & Gayther, S. A. (2021). Risks and function of breast cancer susceptibility alleles. *Cancers, 13*(16), 3953.

Tran, P., Lee, S. E., Kim, D. H., Pyo, Y. C., & Park, J. S. (2020). Recent advances of nanotechnology for the delivery of anti-cancer drugs for breast cancer treatment. *Journal of Pharmaceutical Investigation, 50*, 261–270.

Trapani, D., Ginsburg, O., Fadelu, T., Lin, N. U., Hassett, M., Ilbawi, A. M., et al. (2022). Global challenges and policy solutions in breast cancer control. *Cancer Treatment Reviews, 104*, 102339.

Trayes, K. P., & Cokenakes, S. E. H. (2021). Breast cancer treatment. *American Family Physician, 104*(2), 171–178.

Veljkovikj, I., Ilbawi, A. M., Roitberg, F., Luciani, S., Barango, P., Corbex, M., et al. (2022). Evolution of the joint International Atomic Energy Agency (IAEA), IARC, and WHO cancer control assessments (imPACT reviews). *The Lancet Oncology, 23*(10), e459–e468.

Vogel, R., Göbel, M., Grewe-Salfeld, M., Herbert, B., Matsuo, Y., & Weber, C. (2022). Cross-sector partnerships: Mapping the field and advancing an institutional approach. *International Journal of Management Reviews, 24*(3), 394–414.

WAGstaff, A. & Neelsen, S. (2020). A comprehensive assessment of universal health coverage in 111 countries: A retrospective observational study. *The Lancet Global Health, 8*(1), e39–e49.

Wang, J., Liu, R., Chen, H., Chen, A., & Chen, L. (2023). *Ent*-16*β*, 17-dihydroxy-kauran-19-oic acid (DKA), a kaurane diterpenoid from *Sigesbeckia pubescens (Makino) Makino*, inhibits the migration of MDA-MB-231 breast cancer. *Natural Product Research*, 1–6. https://doi.org/10.1080/14786419.2023.2287177

Wang, K., & Tepper, J. E. (2021). Radiation therapy-associated toxicity: Etiology, management, and prevention. *CA: A Cancer Journal for Clinicians, 71*(5), 437–454.

Weber, W. P., Matrai, Z., Hayoz, S., Tausch, C., Henke, G., Zwahlen, D. R., et al. (2021). Tailored axillary surgery in patients with clinically node-positive breast cancer: Pre-planned feasibility substudy of TAXIS (OPBC-03, SAKK 23/16, IBCSG 57-18, ABCSG-53, GBG 101). *The Breast, 60*, 98–110.

Whitaker, K. (2020). Earlier diagnosis: The importance of cancer symptoms. *The Lancet Oncology, 21*(1), 6–8.

Wilkinson, L., & Gathani, T. (2022). Understanding breast cancer as a global health concern. *The British Journal of Radiology, 95*(1130), 20211033.

Wilson, R. (2021). *Economic development in the Middle East.*

Word Health Organization. (2021). *Global status report on the public health response to dementia.* fiu.edu.

Word Health Organization. (2022a). *Framework for the responsible use of life sciences: Summary report of consultations on the principles, gaps and challenges of biorisk management*, May 2022. who.int.

Word Health Organization. (2022b). *Regional review: Role of the private health sector in the provision of essential health services in the Eastern Mediterranean Region.*

Word Health Organization. (2023). *Universal health coverage partnership annual report 2021: Health systems strengthening and health emergencies beyond COVID-19.*

Xiao, T., Kong, S., Zhang, Z., Hua, D., & Liu, F. (2024). A review of big data technology and its application in cancer care. *Computers in Biology and Medicine, 10*, 108577.

Xu, Y., Guo, W., Zhang, C., Chen, F., Tan, H. Y., Li, S., et al. (2020). Herbal medicine in the treatment of non-alcoholic fatty liver diseases-efficacy, action mechanism, and clinical application. *Frontiers in Pharmacology, 11*, 601.

Yadav, D., Shukla, N. K., & Mishra, M. C. (2022). Management of locally advanced breast cancer. In *Breast cancer: Comprehensive management* (pp. 305–337). Springer Nature Singapore.

Ye, F., Dewanjee, S., Li, Y., Jha, N. K., Chen, Z. S., Kumar, A., et al. (2023). Advancements in clinical aspects of targeted therapy and immunotherapy in breast cancer. *Molecular Cancer, 22*(1), 105.

Yi, H. S. (2023). *Understanding social, cultural, and religious factors influencing medical decision-making on breast cancer genetic testing in the orthodox Jewish*. Columbia University.

Yu, K. D., Cai, Y. W., Wu, S. Y., Shui, R. H., & Shao, Z. M. (2021). Estrogen receptor-low breast cancer: Biology chaos and treatment paradox. *Cancer Communications, 41*(10), 968–980. wiley.com.

Yu, Z. (2022). Development of public administration research with a bibliometric analysis. *Scientific Programming, 1*, 4959930.

Yusli, A. Y., Asni, F., & Umar, B. A. I. (2021). The role of Fiqh method in strengthening fatwa related to vaccine injection for prevent Covid-19 by Perlis state mufti. *Journal Of Academic Research In Business And Social Sciences, 11*(8), 1439–1449. researchgate.net.

Zaluzec, E. K., & Sempere, L. F. (2024). Systemic and local strategies for primary prevention of breast cancer. *Cancers, 16*(2), 248.

Zhong, L., Li, Y., Xiong, L., Wang, W., Wu, M., Yuan, T., et al. (2021). Small molecules in targeted cancer therapy: Advances, challenges, and future perspectives. *Signal Transduction and Targeted Therapy, 6*(1), 1–48.

Zhou, Y., Naci, H., Chen, D., Bai, L., Shi, L., Guan, X., & Wagner, A. K. (2023). Overall survival benefits of cancer drugs in the WHO model list of essential medicines, 2015–2021. *BMJ Global Health, 8*(9), ee012899.

Chapter 6
The Impact of Breast Cancer on Sustainable Development Goals (SDGs)

Ibrahim Bani, Essam M. Hamad, and Ahmed A. M. Elnour

Abstract Breast cancer is a significant global health issue, with over one million cases expected by 2025. It disproportionately affects women in developing countries, where access to affordable treatment and early detection are limited. The increasing prevalence of breast cancer compromises the achievement of the SDGs, particularly SDG 3 (ensuring healthy lives), SDG 1 (ending poverty), and SDG 10 (reducing inequality). Breast cancer survivors face long-term physical, emotional, and psychological effects that impact their quality of life and ability to contribute to society and the economy. The risk factors for breast cancer include age, genetic mutations, family history, and hormonal factors. Early detection through breast screening is crucial for successful treatment of breast cancer. The 2030 Agenda recognizes the negative impact of chronic diseases, including breast cancer, on most SDGs.

Keywords Breast cancer · Sustainable development goals (SDGs) · Environmental determinants · Income inequality · Health policy

I. Bani
Department of Pathological Sciences, College of Medicine, Ajman University, Ajman, UAE

Department of Chronic Disease Epidemiology, Yale School of Public Health, Yale University, New Haven, CT, USA
e-mail: i.bani@ajman.ac.ae

E. M. Hamad
Department of Food Science and Human Nutrition, College of Agriculture and Veterinary Medicine, Qassim University, Buraidah, Saudi Arabia
e-mail: e.hamad@qu.edu.sa

A. A. M. Elnour (✉)
Faculty of Chemical and Process Engineering Technology, Universiti Malaysia Pahang Al-Sultan Abdullah (UMPSA), Pahang Gambang, Malaysia

Centre of Excellence for Advanced Research in Fluid Flow (CARIFF), University Malaysia, Pahang Gambang, Malaysia
e-mail: adamhassan@umpsa.edu.my

A. A. M. Elnour (ed.), *Gum Arabic and Breast Cancer Biology*,
https://doi.org/10.1007/978-981-97-8518-6_6

Highlights

1. Breast cancer is a major global health issue, with over one million cases projected by 2025.
2. It disproportionately affects women in developing countries, with limited access to affordable treatment.
3. The rising prevalence of breast cancer has hampered the development of the SDGs.
4. Survivors have long-term physical, emotional, and psychological effects that affect their quality of life.
5. Risk factors include age, genetic mutations, family history, and hormonal factors.
6. Early detection through screening is essential for a successful treatment.
7. The 2030 Agenda acknowledges the detrimental impact of chronic diseases, including breast cancer, on achieving the SDGs.

6.1 Introduction

Of the estimated 19 million cancer cases by 2025, over one million will be breast cancer, which is the most diagnosed cancer in women (Wojtyla et al., 2021). Eighty percent of the women who survive breast cancer live in developing countries. However, the cost of drugs remains high in many countries, and access to affordable medicines is scarce (Mutebi et al., 2020). Even when free or subsidized cancer medicines are available, treatment is not always accessible owing to long waiting times or travel costs (Cortes et al., 2020). According to the American Cancer Society, the death rate from cancer is increasing in developing countries due to lifestyle changes, which result in late-stage detection of cancer, making it harder and less successful to treat. In addition, many countries have outdated radiology equipment, which increases the time for diagnosis (Lin et al., 2021). Thus, addressing these multi-faceted challenges is crucial to improve cancer outcomes in developing countries.

It has been projected that by 2030, over 13 million people will be living with cancer, many of whom will need access to safe and effective cancer medicines (Perera et al., 2021). This increase makes it urgent to ensure the availability, accessibility, and affordability of these medicines without compromising SDG3 (Sachs & Sachs, 2021). The fight to prevent and treat cancer is critical not only because it is a growing NCD in low- and middle-income countries (LMICs), where resources are frequently already stretched (Azadnajafabad et al., 2023), but also because it affects many families and communities that SDGs 1 and 10—to end poverty and reduce inequality—in the 2030 Agenda are impaired (Anderson et al., 2021). Therefore, ensuring comprehensive cancer care is essential for achieving broader SDGs.

Women with BC must deal with complex adverse events throughout the course of their disease, which can result in long-term negative physical, emotional, and psychological effects for survivors and their families (Tran et al., 2022). BC significantly affects the health-related quality of life of survivors; consequently, the WHO has a high-level of stress and a lack of self-confidence (Shen et al., 2020). Many roles are interconnected in families and social bonds, and are restricted in many ways (Settersten Jr et al., 2020). Since then, BC has had a significant impact on the ability of those affected to contribute to society and the economy (Adam & Koranteng, 2020). Currently, gender equality is one of the SDGs and is associated with many social and economic benefits (Castro Núñez et al., 2020). BC in women is an important factor to consider in intervening processes (Bowen et al., 2021). Health directly and indirectly influences most goals and targets (Scharlemann et al., 2020). The perception of the political focus on this process brings some controversy regarding the universe of BC survivors (Tarantino et al., 2023). Therefore, a comprehensive approach to BC care is vital to achieve gender equality and overall societal progress.

Cancer is considered a disease in developed countries, and many countries initially ignored its existence because they considered it an additional health problem (Reid et al., 2021). However, the prevalence and incidence of cancer in the world are high, and this burden results in high costs to finance the recovery process and care (Chen et al., 2023). Irrational behaviors, the lack of interest in people's health, the excessive consumption of food with high trans-fat content, sweetened drinks (Doran, 2020), and food additives associated with a reduction in nutritional quality directly affect the quality of life and prevalence of many diseases listed in the SDGs (Platta & Radzymińska, 2023). The negative impacts of chronic diseases on 16 of the 17 SDGs are known and discussed, as is their relevance to the economic development of countries, with a focus on infectious and non-infectious diseases, tobacco consumption, alcohol consumption, obesity, and physical inactivity (World Health Organization, 2020). Globally, many elements that directly affect the incidence and prevalence of NCDs are within the scope of the 2030 Agenda (Allen et al., 2021). Therefore, addressing lifestyle and dietary factors is crucial for reducing the global burden of cancer and achieving the SDGs.

6.2 Understanding BC

The cause of BC is unknown; however, the risk of developing BC increases with age. Other factors also increase the risk of BC (Daly et al. 2021a, 2021b). These risk factors may include being a woman, age, inheriting certain genetic mutations from the family, having a family history of BC, genes, ovarian hormones (Daly et al. 2021a, 2021b), having a first period before age 12, entering menopause after age 55, no history of pregnancy or first full-term pregnancy after the age of 30, and having a child WHO started BC (Kashyap et al., 2022). Breast screening involves checking the breasts for BC in women. The WHO does not have any symptoms of BC, as it can be detected at its earliest stage (Ginsburg et al., 2020). Although BC can be invasive, some types, such as in situ BC and lobular carcinoma in situ, remain in the milk ducts and do not develop further (Tabár et al., 2022). Therefore, understanding and monitoring risk factors are essential for the early detection and prevention of BC.

BC is the uncontrolled growth of cells that form breast tissue (Bahcecioglu et al., 2020). These cells form a tumor often seen on an X-ray or felt as a lump (Shaikh et al., 2021). BC primarily occurs in women but can also occur in men (Yao et al., 2022). This invasive cancer invades the surrounding breast tissue over time (Risom et al., 2022). Cancer cells can also break away from the primary tumor and enter the bloodstream or the lymphatic system. BC can form tumors called metastases in other body parts (Kim, 2021). These cells grow faster than the surrounding normal tissue, continue growing without stopping, or do not die entirely. This results in the accumulation of cells and a lump or tumor (Yuan et al., 2023). They may also be spread throughout the body (Yu et al., 2022). Early detection and treatment are crucial to prevent the spread and complications of BC.

6.2.1 Definition and Types

BC is the most common malignancy among women due to its late presentation and treatment failure (Foroozani et al., 2020). Only a few patients are diagnosed early, and their diseases are cured in LMICs (Haque et al., 2020). Women with metastatic and locally advanced BC in LMICs and high-income countries tend to live between 40 and 60 months (Horton et al., 2020). The illness is expensive and may result in catastrophic health expenditures, especially in LMICs (Murphy et al., 2020). During BC treatment, socio-economic and physical consequences can cause the dismantling of family structures (Bowen et al., 2021). The illness may also favor an increase in unemployment and the number of people living below the poverty threshold (Ngubane et al., 2023). The illness is accountable not only for its impact on the mortality and morbidity of people but also for a bond with human rights (D'Cruz & Banerjee, 2020). Therefore, improving early diagnosis and affordable treatment are critical for mitigating the widespread impact of BC.

BC is the most prevalent malignancy among women and its incidence is gradually increasing in developing countries (Mubarik et al., 2022). This illness is associated with a high financial burden and substantial morbidity and frequently has a major effect on how family's function (Bello et al., 2022). BC appears to have direct and indirect effects on seven of the eight SDGs; thus, its current treatment is not cost-effective (Pashayan et al., 2020). The Women WHO relied on their partners for financial support and care, while the ill are particularly vulnerable to the indirect impact of BC on SDGs (Schantz et al., 2024). Accordingly, addressing the multifaceted impact of breast cancer is essential for achieving the SDGs. The global distribution of BC cases is shown in Fig. 6.1.

Fig. 6.1 Global incidence of breast cancer. (The authors adapted from Ferlay et al. (2020)'s source, which is available from https://gco.iarc.who.int/today, accessed [01-06-2024])

6.2.2 *Epidemiology and Global Burden*

The development of the United Nations 2030 Agenda for SDGs has enabled governments, industries, private organizations, and academic institutions to promote human rights, environmental integrity, economic prosperity, and social welfare worldwide (Ricardo, 2021). A separate target of SDG Indicator 3.4.1 is to reduce premature mortality caused by NCDs by one-third through prevention and treatment by 2030 (Frieden et al., 2020). A global response plan for the prevention of BC is an essential part of efforts to achieve 'healthier lives and promote well-being for all ages' (Kaminsky et al., 2022). Globally, BC is the most frequently diagnosed cancer among women (Luo et al., 2022). In 2018, 2.1 million women were diagnosed with BC and 630,000 women died (Gnanamuttupulle et al., 2021). BC may occur at any age, although it has an increasing incidence in patients approximately 35 years of age (van Der Meer et al., 2021). The prevalence of BC has increased significantly in recent decades due in part to demographic transition, the worldwide spread of the Western lifestyle, and the cumulative exposure of women to recognized and potential risk factors throughout their lives (Huang et al., 2021). Consequently, a concerted global effort to address BC is crucial for meeting the SDGs. The BC cases and deaths by region are shown in Fig. 6.2.

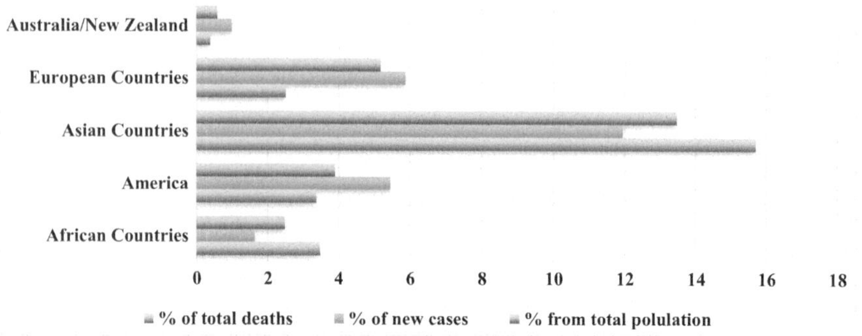

Fig. 6.2 BC Cases and Deaths by Region. (Data are adjusted from previously published data from Arnold et al. (2022)'s source, which is available from https://gco.iarc.who.int/today, accessed [01-06-2024])

6.3 The Intersection of BC and SDGs

The WHO timeline for BC highlights a range of policies aligned with a specific focus on increasing BC awareness, detecting disease early, and providing effective and affordable treatment (Ginsburg et al., 2020). A holistic approach to cancer control increases the likelihood of diagnostic and therapeutic success, allowing those affected to have a good-quality of life (Mao et al., 2022). A substantial increase in the disease will force nations to invest in prevention measures to lessen the burden of BC (Britt et al., 2020). BC care can influence the achievement of multiple SDGs (Panda et al., 2024). The first mapping highlighting these intersections shows that BC should not simply be viewed as an SDG three concern, but as a major public health issue necessary for the success of disease prevention and care's economic, social, and environmental impact on society (Duggan et al., 2020). Facilitators of positive engagement hope to stimulate interest in obtaining a comprehensive overview of the sustainable development dimensions of BC management that demand an informed connection, education, and a collective organizational goal (Anderson et al., 2021). Thus, integrating BC care into a broader sustainable development framework is essential for comprehensive progress in global health.

Sustainable development goals: The SDGs are 17 global goals created to achieve a better and more sustainable future (Halkos & Gkampoura, 2021). SDGs were first proposed by the Open Working Group of the United Nations Assembly in 2012. The SDGs are integrated and indivisible, and balance the three core elements of sustainable development: economic growth, social inclusion, and environmental protection (Coscieme et al., 2020). The goals are broad-based and interdependent, requiring coordinated actions, such as eradicating poverty, addressing inequalities, promoting peace, ensuring access to clean water and education, and providing good-quality jobs and life-changing innovations (Cernev & Fenner, 2020). An intentional focus was placed on the health goal (SDG 3) to ensure healthy lives and promote

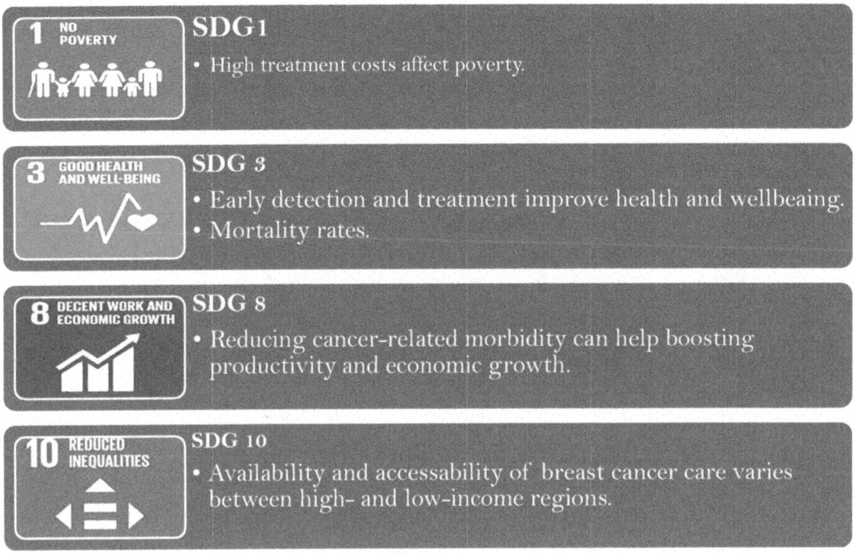

Fig. 6.3 BC and sustainable development goal (SDGs) intersections

well-being at all ages (Macassa, 2021). Hence, comprehensive implementation of SDGs is essential for fostering sustainability and equality. The intersections between BC and SDGs are illustrated in Fig. 6.3.

6.3.1 Overview of SDGs

To reduce or eliminate the incidence of the most relevant BC conditions, the implementation of the 17 SDGs should achieve the following results: SDG 1–End poverty in all forms everywhere (Dhahri & Omri, 2020); SDG 2–End hunger, achieving food security, improving nutrition, and promoting sustainable agriculture (Anderson et al., 2021). BC prevention measures are effective when implemented throughout an individual's life (Pashayan et al., 2020). Due to the hereditary nature of these lifestyles, it is important to invest in physical education, good nutrition, and healthy lifestyles, especially during the school years, to build long-term behaviors that reduce cancer risk (Bull et al., 2020; Rock et al., 2020). Working with families for early prevention and providing proper healthcare and social support is also important (Nyashanu et al., 2020). Therefore, comprehensive and lifelong approaches to education and health are vital for effective BC prevention.

The Millennium Development Goals (MDGs) were a 15-year action plan adopted by all United Nations Member States in 2000 to address specific global problems and improve the quality of life of humans by 2015 (Kc et al., 2020). It was the first global effort to tackle reproductive and sexual health issues and aimed to end

Fig. 6.4 United Nations' the millennium development goals MDGs

extreme poverty (Chandra-Mouli et al., 2020). Approximately, 500 million people live in extreme poverty (Rentschler et al., 2022). The 15-year timeframe established by the MDGs ended in 2015 and was considered successful in eradicating extreme poverty and improving living conditions for families (Bhukta, 2020). The United Nations replaced the MDGs with a strategic roadmap for the future, proposing 17 SDGs to protect people and the planet (Blissman, 2021). In addition to combating inequality and promoting sustainable development, it is important to recognize that ending suffering requires collective effort (Leal Filho et al., 2020). Thus, transitioning from MDGs to SDGs marks a continued commitment to address global challenges through unified action. The United Nations MDGs are shown in Fig. 6.4.

6.3.2 *Specific Goals Impacted by BC*

These groups are also more likely to receive emergency room visits or delayed care (Nourazari et al., 2021). They are also more likely to be uninsured, with five times as many people lacking health insurance as their white counterparts (Artiga et al., 2020). They are less likely to receive an early diagnosis and treatment. Better data on BC are needed, but what is known suggests that women of color receive less effective treatment than other women (Poteat et al., 2021). Disparities in cancer treatment are often influenced by a combination of systemic and socioeconomic factors. These include overcrowded or underfunded healthcare facilities, shortages of specialized oncologists, limited access to health insurance, and financial constraints. Additionally, low health literacy, cultural and linguistic barriers, and inadequate transportation options exacerbate the inequities faced by patients in accessing timely and effective care (Mutebi et al., 2020). Women of color WHO may also face other marginalized barriers in general, including those who have been diagnosed with disabilities, sexual orientation, or HIV (Azhar & Gunn, 2021). Within the framework of health-related Sustainable Development Goals (SDGs), Goal 3 emphasizes the importance of implementing strategies to prevent deaths from treatable or preventable diseases. This includes strengthening healthcare systems, enhancing access to medical resources, promoting public health education, and advancing research and development to combat both communicable and non-communicable diseases. Achieving this goal requires a collaborative global effort to reduce health inequities and ensure universal health coverage (Nabukalu et al., 2020). The higher rates of morbidity from BC among women of color suggest that these populations face barriers to early detection, diagnosis, access to care, and poor treatment outcomes (Crown et al., 2022). Women of color and low-income women, in general, have high rates of poor health, associated risk behaviors, and late-stage diagnosis, underscoring the need for more comprehensive healthcare (Ansari et al., 2023). Therefore, addressing disparities in healthcare access and quality is essential for improving outcomes for women of color and for achieving health equity.

6.4 Health Systems and BC Care

On January 1, 2016, the United Nations' 17 new SDGs officially took effect (Rasche, 2020). The SDGs are based on the principle that ending poverty must go hand-in-hand with strategies that build economic growth and address a range of social needs, while simultaneously tackling climate change and environmental protection (Belaïd, 2022). Toward the same end, good health is one of the foundations for promoting the healthy and productive lives described by SDG 3: 'ensure healthy lives and promote well-being for all at all ages' (Fernandez, 2020). It is recognized as a target of numerous other goals and an essential outcome of development efforts (Alongi & Anese, 2021). Progress in addressing global health challenges has been suboptimal, and without renewed global commitment, SDGs will certainly not be achieved (Elomrani et al., 2024). To measure and manage the real impact of BC on the SDGs, health professionals and decision-makers need to adopt a new mindset that connects multidisciplinary training and research with mutual accountability to develop new high-impact interventions (Cankurtaran & Beverland, 2020). Consequently, a holistic and collaborative approach is crucial for advancing global health and achieving the SDGs.

Health systems and programs are critical to achieving global health goals (Lal et al., 2021). They include all public and private organizations, institutions, and people that ensure that people are healthy, such as hospitals, families, clinics, and local health workers (Sethi et al., 2020). At the state level, well-functioning health systems are necessary for basic issues, such as reaching people with health services and promoting essential public health actions (Malakoane et al., 2020). At the household level, health systems also play a central role in determining whether lives can be saved, and prospects for living long and healthy lives are possible (Timmis & Ramos, 2021). Hence, robust health systems are fundamental for improving public health and achieving the SDGs. The suggested policies and programs for BC prevention and control are shown in Fig. 6.5.

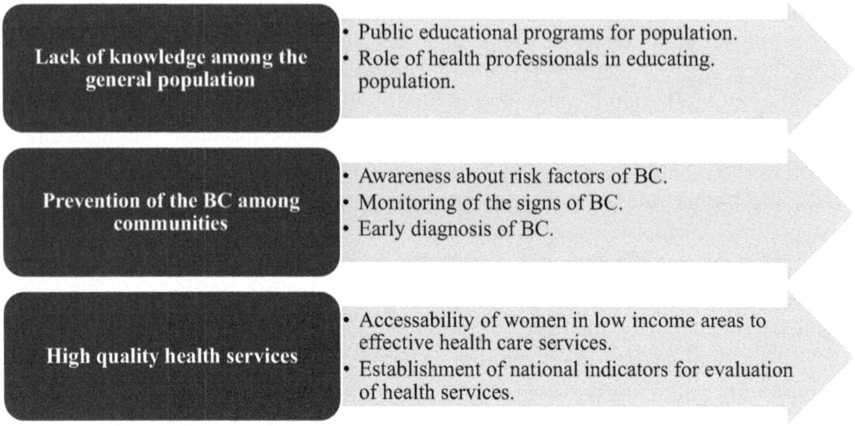

Fig. 6.5 Proposed policies and programs for BC prevention and control

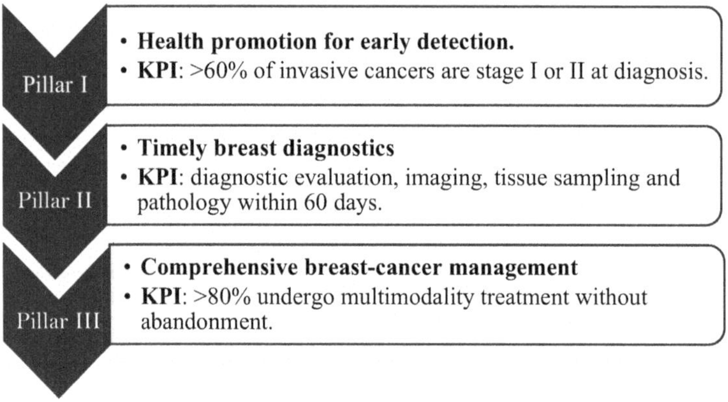

Fig. 6.6 Pillars and KPIs of WHO—GBCI

6.4.1 The Global BC Initiative (GBCI)

As mentioned earlier, BC is the most diagnosed type of cancer globally, and the leading cause of death in women. It is a common global cancer found in 86% of countries. Moreover, BC is considered the most prevalent malignancy, with the highest number of living patients diagnosed within the previous 5 years. It is the leading cause of cancer-related deaths among women in 58% of countries (Ferlay et al., 2020). The WHO established an evidence-based framework called GBCI in 2021 to reduce global BC deaths. This initiative aims to improve early detection, effective diagnosis, high-quality treatment, and supportive services for women with BC. This ultimate goal will be achieved through the following three pillars and key performance indicators (KPI): 1) Pillar I: Health promotion for early detection

(KPI: >60% of invasive cancers are stage I or II at diagnosis), 2) Pillar II: Timely breast diagnostics (KPI: diagnostic evaluation, imaging, tissue sampling, and pathology within 60 days), and 3) Pillar III: Comprehensive BC management (KPI:>80% undergo multimodality treatment without abandonment) (WHO, 2023). Therefore, GBCI aims to reduce BC mortality through structured and measurable strategies. The pillars and KPIs of WHO—GBCI are shown in Fig. 6.6.

6.4.2 Challenges and Opportunities

It is important to realize regional disparities and closely monitor the behavior of governments (Liu et al., 2022). One of the best solutions to this problem is to invest in human resources to grow, moving from a short-term to a long-term model (Ridha & Parwanto, 2020). Consequently, it is equally important for the next-generation (Neumann et al., 2022). BC also directly impacts education because 90% of women diagnosed with BC are in the process of losing the income that supports their families (Jerome-D'Emilia et al., 2021). Then, from this point of view, SDG 4, 'Quality Education,' and SDG 5, 'Gender Equality,' are directly affected (Leal Filho et al., 2020). People who become ill cannot work, and their family members are heavily burdened by the costs of health and other needs (Zakayo et al., 2020). Why do women overwhelmingly bear the impact of BC over others? To obtain a concrete answer to this question, we need to investigate what STI activities are taking place (Dittrich, 2020). Therefore, addressing the broader socio-economic impact of BC is essential for promoting gender equality and sustainable development.

Considerable developments have been made in technology and healthcare worldwide. However, it is facing the challenge of focusing on the health infrastructure (Mbunge et al., 2021). The enormous amount of energy, time, and cost required to combat health problems such as the BC epidemic directly impacts human development (Awada et al., 2021; Dlamini et al., 2024). The United Nations has made a joint statement with the WHO to address the four health challenges: (i) critical actions at universality, (ii) affordable health, (iii) opening the horizon of the future of humanitarian action, and (iv) aligning with the WHO to ensure the effectiveness of the action (Canton, 2021). These challenges critically focus on the first three SDGs: 'No poverty, zero hunger, and good health' (Okello et al., 2021). Hence, strengthening the health infrastructure is imperative to overcome these challenges and promote sustainable human development.

6.5 Gender Perspectives in BC

Femininity, the symbol of being female, not only carries biological significance but also possesses a type of gender and a symbol representing the attribute (Bordo, 2020). Each individual embraces different types of behavior and identities of femininity beliefs while displaying these behaviors in different ways and with varying intensities (Zemenick et al., 2022). Furthermore, the impact of feminine beliefs varies among different groups of people (Abuhashesh et al., 2021). However, it is obvious that past behaviors and femininity beliefs are no longer useful for modern educational concepts, living environments, social organizations, and traditional practices; these past practices hinder gender equality (Lee et al., 2020). Women should no longer be placed in a specific role defined by society at any cost; they have the right to explore unique aspects of their lives (Teixeira et al., 2021). Generally, patients with cancer, their families, and society mistakenly believe that BC survivors have similar gender characteristics. This opinion can be vulnerable to the eyes of others, sending an incorrect message (Yan et al., 2023). Therefore, surgery, chemotherapy, targeted therapy, hormonal therapy, and other treatments linked to female personality characteristics emphasize the need for a gender-specific component of BC treatment (Wang et al. 2020a, 2020b). Hence, modern approaches to BC treatment must incorporate sex-specific considerations to respect and support women's diverse experiences.

BC is a sex-related disease (Lopes-Ramos et al., 2020). Sex refers to the biological characteristics of living organisms, including structural and functional differences in animals and plants (Khalidi, 2020). Gender is the social role that society considers appropriate for men and women, varying between regions and cultures (Bonvillain, 2020). Sex does not equate to sex. Nevertheless, sex-and gender are highly correlated (Ballering et al., 2020). Whether from health status and outcomes, or mortality rates, cancer incidence, recurrence, and survival, the biological differences reflected in the incidence rates of males and females with the same cancer are directly related to sex, reflecting gender differences (Wang et al., 2022). Gender perspectives should be combined with female BC treatment guidelines, and sex differences in cancer incidence and outcomes should be used as a guide for health policy to ensure that the national health and wellness strategy realizes better women's health in the context of policy implementation to promote economic growth (Chao et al., 2020). It is time to consider the gender perspective and sex differences in different parts of national strategies and report the causes of gender gaps (Campbell et al., 2021). Hence, integrating sex differences into health policies and treatment guidelines is crucial for improving BC outcomes and promoting overall health in women.

6.5.1 Societal Norms and Stigma

The pain and courage (accepting vulnerability) of men and children must be acknowledged, and we believe that it is important to provide counseling and support for all family members from the moment of diagnosis (Bootsma et al., 2020). Patients with BC will not discuss treatment with family members because they think it will negatively impact the person they are discussing the issue (Daniel et al., 2022). Patients also refuse treatment for fear of losing their human aspect (hair and breast) (Iddrisu et al., 2020). This fear implicates the body part and bond of femininity, which is deeply connected with emotional well-being (Allen et al., 2022). Women's perception is that everyone views them as sexual individuals; therefore, if they lose any aspect of their human aspect, they feel a loss of their femininity and sexual relationship (Panzeri et al., 2020). This results in deep social stigma. Therefore, we believe that adding to this stigma by anyone in the healthcare field will have a negative effect (Muncan et al., 2020). Therefore, oncologists must be especially aware of how they approach patients with such issues (Mao et al., 2022). Thus, a sensitive and supportive approach from healthcare providers is crucial to address patients' emotional and psychological needs with BC.

Society has conditioned women to care for everyone else, including their families, and to put themselves last (Mayra et al., 2022). Therefore, after a BC diagnosis, the patient journeys inward to find her own strength when she is accustomed to giving it to others (Chan, 2022). When she emerges, she is the epitome of vulnerability, both psychologically and (temporarily) physically (Göttsche, 2021). The resultant anger, depression, fights with partners, and withdrawal from families and society cause even more stress and can lead to cancer recurrence or a negative health outcome (Riaz et al., 2023). They can also harm families and society. Wives can become demanding and irrational owing to their intense vulnerability (Uddin, 2021). Hence, providing psychological support and fostering resilience in BC patients is essential for their families' well-being and health. The impact of BC on gender perspectives and societal stigma is shown in Fig. 6.7.

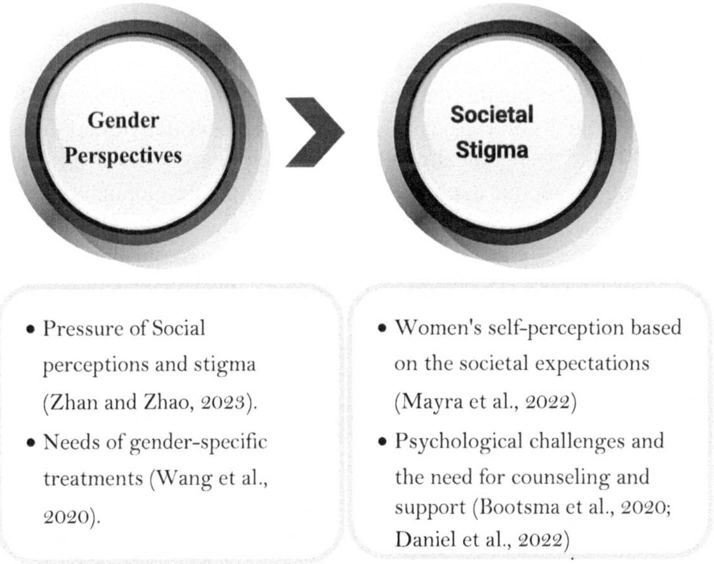

Fig. 6.7 Impacts of BC on gender perspectives and societal stigma

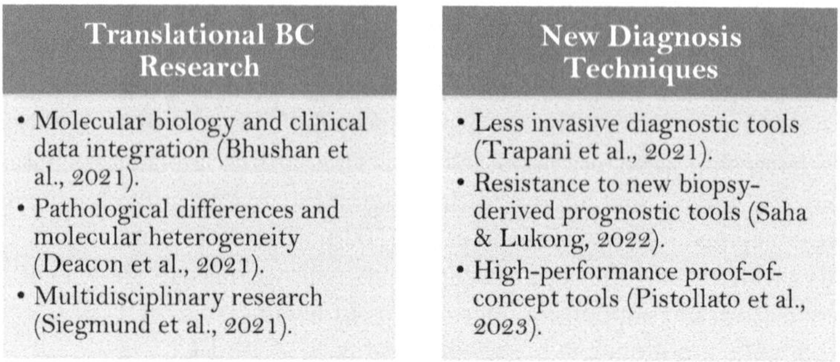

Fig. 6.8 Advancements in BC diagnosis and treatment

6.6 Innovations in BC Diagnosis and Treatment

Innovations in BC diagnosis and treatment are in this section, and the research is relatively restricted to the publications that follow this box (Bhushan et al., 2021). The focused topics include translational BC research, new diagnostic and surveillance techniques, new focal treatments, new adjuvant drug-based and physical-based strategies, and failure-driven steps to potential breakthrough cancer research (Duggan et al., 2020). BC requires new diagnostic and prognostic tools that offer less restricted and lower resource setting detection, longitudinal monitoring, and response to therapy (Trapani et al., 2021). Pathological differences, molecular heterogeneity, and limited predictiveness of classical biomarkers in clinical decisions create a gap (Deacon et al., 2021); despite being very user-friendly, the barrier of resistance to new biopsy-derived prognostic and predictive tools is significant among the treating community because of fragmented and uncertain method applications and costs, ultimately compromising the total cost of BC surgery (Saha & Lukong, 2022). Smaller health innovation companies rely mostly on public parties to pay for added benefits and struggle to maintain the continuity of applied research (Allers et al., 2023). Discrepant aims and multidisciplinary focus within the investigation team during academic applied research hinder the development of a complete solution from the same research (Siegmund et al., 2021). The initial high-performance proof-of-concept tools passed through a screening and validation process to be considered COTS, unable to overcome investment breakdowns in the transition into clinical use (Pistollato et al., 2023). Innovative pathways should be designed to maintain continuity in developing these tools (Xu et al., 2020). Therefore, addressing these challenges is crucial for advancing BC diagnostic and treatment innovations. The advantages of BC diagnosis and treatment are summarized in Fig. 6.8.

6.7 Research Gaps and Future Directions

Findings from these research gaps can provide useful contributions to health and economic research (AlOmari, 2021). In areas where women require vaccinations or check-ups, they can be bundled with education on BC detection (Lubi et al., 2022). Special health camps can be designed for areas with fewer health facilities, which may help identify the situation of BC among women in these areas (Horton et al., 2020). The government should also consider specific programs to increase awareness of BC (Ginsburg et al., 2020). These measures will help early detection and reduce treatment complexity and costs (Ginsburg et al., 2020). Treatment costs are sometimes as high as the cost of living, resulting in a greater depletion of available resources (Scanlon et al., 2023). These measures may help reduce the multidimensional poverty headcount ratio (Vollmer & Alkire, 2022). Therefore, by identifying social inequalities that prevent women from receiving necessary treatment, the government can design, create, and direct appropriate programs (Shaw et al., 2021). Such programs will increase health insurance coverage and social protection and reduce poverty (Abdoul-Azize & El Gamil, 2021). Thus, implementing these initiatives is essential to improve women's health and economic stability. We are aware of several research gaps and suggest avenues for advancing our understanding of BC in the context of SDGs (Horton et al., 2020). Research on health-related costs due to BC is scarce in LMICs (Martinez-Cannon et al., 2021). There is also no existing literature on the relationship between the unmet need for family planning and breastfeeding, and delays in the detection of BC (Callaway & Dos Santos, 2023). There is only one study in a low-income country on how women displaced from living in conflict regions face delays in diagnosis and treatment (Hanafi et al., 2023). Research investigating the effects of these health conditions on other SDGs is also lacking (Pizzi et al., 2020). As a result, addressing these research gaps is crucial for a comprehensive understanding and policy formulation regarding BC in the context of the SDGs.

6.8 Conclusion and Recommendations

The spread of NCD is wider among young and old age groups. The last segment contributes heavily to the cost of national healthcare spending, which can lead to unfounded public and private financial burdens. In addition, BC affects not only the health budgets of all the Global Partnership for Education health systems but also creates indirect non-health budget losses for states, societies, and countries' economies. A highly controversial issue has shifted to those suffering from inequality in social security and healthcare funding. The epidemic readiness at all levels is crucial in providing truly governed healthcare governance. We should be ready to meet the challenge and reduce the pain and loss associated with the recession of the national economy, negative social and demographic processes, and disability of age shifts.

The ability to work in a uniform is vital for combating BC epidemics. To gain the positive impact of the age shift in BC, life after the epidemic should be carefully planned at the state level. All previous actions should be synchronized to obtain the state goals. We should be ready to meet this challenge and reduce the pain and loss associated with BC. Maintaining a healthy lifestyle should be the main strategy for reducing the risks associated with poor condition and disability of age shifts. Furthermore, technological innovations are essential to provide easy access to fast and precise early diagnostic tests; effective, humane, and less expensive treatments and medications; and continuous quality surveillance to minimize errors and negligence.

Acknowledgments The authors thank the Chemical and Process Engineering Technology at the University of Malaysia, Pahang-UMPSA, Malaysia, for supporting this research.

Fund This research was funded by Eastern Unity Technology under Grant No. UIC 190806.

References

Abdoul-Azize, H. T., & El Gamil, R. (2021). Social protection as a key tool in crisis management: Lessons learned from the COVID-19 pandemic. *Global Social Welfare, 8*, 107.

Abuhashesh, M. Y., Alshurideh, M. T., Ahmed, A. E., Sumadi, M., & Masa'deh, R. E. (2021). The effect of culture on customers' attitudes toward Facebook advertising: The moderating role of gender. *Review of International Business and Strategy, 31*(3), 416–437.

Adam, A., & Koranteng, F. (2020). Availability, accessibility, and impact of social support on breast cancer treatment among breast cancer patients in Kumasi, Ghana: A qualitative study. *PLoS One, 15*(4), e0231691.

Allen, L. N., Wigley, S., & Holmer, H. (2021). Implementation of non-communicable disease policies from 2015 to 2020: A geopolitical analysis of 194 countries. *Lancet Global Health, 9*, e1528.

Allen, K. A., Gray, D. L., Baumeister, R. F., & Leary, M. R. (2022). The need to belong: A deep dive into a foundational construct's origins, implications, and future. *Educational Psychology Review, 34*(2), 1133–1156.

Allers, S., Eijkenaar, F., van Raaij, E. M., & Schut, F. T. (2023). … road toward payment for healthcare innovation with high societal value but limited commercial value: A comparative case study of devices and health …. *Technology in Society, 75*, 102405.

AlOmari, F. (2021). Measuring gaps in healthcare quality using SERVQUAL model: Challenges and opportunities in developing countries. *Measuring Business Excellence, 25*, 407.

Alongi, M., & Anese, M. (2021). Re-thinking functional food development through a holistic approach. *Journal of Functional Foods, 81*, 104466.

Anderson, B. O., Ilbawi, A. M., Fidarova, E., Weiderpass, E., Stevens, L., Abdel-Wahab, M., & Mikkelsen, B. (2021). The Global Breast Cancer Initiative: A strategic collaboration to strengthen health care for non-communicable diseases. *The Lancet Oncology, 22*(5), 578–581.

Ansari, N., Jehangir, F., Ansari, R., Ali, H. S., Aslam, S., & Salman, B. (2023). Visual inspection with acetic acid, a low-cost screening method for cervical cancer in low-resource settings, and low socio-economic strata: Primary health care physicians and general practitioners—An appropriate entry point for screening. *Journal of the Society of Obstetricians and Gynecologists of Pakistan, 13*(3), 313–318.

Arnold, M., Morgan, E., Rumgay, H., Mafra, A., Singh, D., Laversanne, M., et al. (2022). Current and future burden of breast cancer: Global statistics for 2020 and 2040. *The Breast, 66*, 15–23.

Artiga, S., Orgera, K., & Pham, O. (2020). *Disparities in health and health care: Five key questions and answers*. Kaiser Family Foundation.

Awada, M., Becerik-Gerber, B., Hoque, S., O'Neill, Z., Pedrielli, G., Wen, J., & Wu, T. (2021). Ten questions concerning occupant health in buildings during normal operations and extreme events including the COVID-19 pandemic. *Building and Environment, 188*, 107480.

Azadnajafabad, S., Saeedi Moghaddam, S., Mohammadi, E., Rezaei, N., Rashidi, M. M., Rezaei, N., et al. (2023). Burden of breast cancer and attributable risk factors in the North Africa and Middle East region, 1990–2019: A systematic analysis for the Global Burden of Disease Study 2019. *Frontiers in Oncology, 13*, 1132816.

Azhar, S., & Gunn, A. J. (2021). Navigating intersectional stigma: Strategies for coping among cisgender women of color. *Qualitative Health Research, 31*, 2194.

Bahcecioglu, G., Basara, G., Ellis, B. W., Ren, X., & Zorlutuna, P. (2020). Breast cancer models: Engineering the tumor microenvironment. *Acta Biomaterialia, 106*, 1.

Ballering, A. V., Bonvanie, I. J., Hartman, T. C. O., Monden, R., & Rosmalen, J. G. (2020). Gender and sex independently associate with common somatic symptoms and lifetime prevalence of chronic disease. *Social Science & Medicine, 253*, 112968.

Belaïd, F. (2022). Implications of poorly designed climate policy on energy poverty: Global reflections on the current surge in energy prices. *Energy Research & Social Science, 92*, 102790.

Bello, A. K., Okpechi, I. G., Osman, M. A., Cho, Y., Htay, H., Jha, V., et al. (2022). Epidemiology of hemodialysis outcomes. *Nature Reviews Nephrology, 18*(6), 378–395.

Bhukta, A. (2020). No poverty: How much, how far. In S. Hazra & A. Bhukta (Eds.), *Sustainable development goals* (pp. 1–11). Springer International Publishing.

Bhushan, A., Gonsalves, A., & Menon, J. U. (2021). Current state of breast cancer diagnosis, treatment, and theranostics. *Pharmaceutics., 13*(5), 723.

Blissman, B. (2021). The United Nations' sustainable development goals (SDGs): A strategic plan for humanity. In *World scientific encyclopedia of climate change: Case studies of climate risk, action, and opportunity* (Vol. 2, pp. 43–61).

Bonvillain, N. (2020). *Women and men: Cultural constructs of gender*. Rowman & Littlefield Publishers.

Bootsma, T. I., Duijveman, P., Pijpe, A., Scheelings, P. C., Witkamp, A. J., & Bleiker, E. M. (2020). Unmet information needs of men with breast cancer and health professionals. *Psycho-Oncology, 29*(5), 851–860.

Bordo, S. R. (2020). The body and the reproduction of femininity: A feminist appropriation of Foucault. In *The new social theory reader*. Routledge.

Bowen, D. J., Fernandez Poole, S., White, M., Lyn, R., Flores, D. A., Haile, H. G., & Williams, D. R. (2021). The role of stress in breast cancer incidence: Risk factors, interventions, and directions for the future. *International Journal of Environmental Research and Public Health, 18*(04), 1871.

Britt, K. L., Cuzick, J., & Phillips, K. A. (2020). Key steps for effective breast cancer prevention. *Nature Reviews Cancer., 20*, 417.

Bull, F. C., Al-Ansari, S. S., Biddle, S., Borodulin, K., Buman, M. P., Cardon, G., et al. (2020). World Health Organization 2020 guidelines on physical activity and sedentary behavior. *British Journal of Sports Medicine, 54*(24), 1451–1462.

Callaway, M. K., & Dos Santos, C. O. (2023). Gestational breast cancer—A review of outcomes, pathophysiology, and model systems. *Journal of Mammary Gland Biology and Neoplasia, 28*(1), 16.

Campbell, O. L. K., Bann, D., & Patalay, P. (2021). The gender gap in adolescent mental health: A cross-national investigation of 566,829 adolescents across 73 countries. *SSM-Population Health, 13*, 100742.

Cankurtaran, P., & Beverland, M. B. (2020). Using design thinking to respond to crises: B2B lessons from the 2020 COVID-19 pandemic. *Industrial Marketing Management, 88*, 255.

Canton, H. (2021). Food and agriculture organization of the United Nations—FAO. In *The Europa directory of international organizations 2021* (pp. 297–305). Routledge.

Castro Núñez, R. B., Bandeira, P., & Santero-Sánchez, R. (2020). Social economy, gender equality at work and the 2030 agenda: Theory and evidence from Spain. *Sustainability*.

Cernev, T., & Fenner, R. (2020). The importance of achieving foundational sustainable development goals in reducing global risk. *Futures, 115*, 102492.

Chan, S. (2022). Exploration of the trajectory of being a breast cancer patient through collaborative writing in imaginal dialogue.

Chandra-Mouli, V., Akwara, E., Engel, D., Plessons, M., Asnake, M., Mehra, S., et al. (2020). Progress in adolescent sexual and reproductive health and rights globally between 1990 and 2016: What progress has been made, what contributed to this, and what are the implications for the future? *Sexual and Reproductive Health Matters, 28*(1), 1741495.

Chao, C. A., Huang, L., Visvanathan, K., Mwakatobe, K., Masalu, N., & Rositch, A. F. (2020). Understanding women's perspectives on breast cancer is essential for cancer control: Knowledge, risk awareness, and care-seeking in Mwanza, Tanzania. *BMC Public Health, 20*, 1–11.

Chen, S., Cao, Z., Prettner, K., Kuhn, M., Yang, J., Jiao, L., et al. (2023). Estimates and projections of the global economic cost of 29 cancers in 204 countries and territories from 2020 to 2050. *JAMA Oncology, 9*(4), 465–472.

Cortes, J., Perez-García, J. M., Llombart-Cussac, A., Curigliano, G., El Saghir, N. S., Cardoso, F., et al. (2020). Enhancing global access to cancer medicines. *CA: A Cancer Journal for Clinicians, 70*(2), 105–124.

Coscieme, L., Mortensen, L. F., Anderson, S., Ward, J., Donohue, I., & Sutton, P. C. (2020). Going beyond gross domestic product as an indicator to bring coherence to the sustainable development goals. *Journal of Cleaner Production, 248*, 119232.

Crown, A., Ramiah, K., Siegel, B., & Joseph, K. A. (2022). The role of safety-net hospitals in reducing disparities in breast cancer care. *Annals of Surgical Oncology., 29*, 4067.

Daly, A. A., Rolph, R., Cutress, R. I., & Copson, E. R. (2021a). A review of modifiable risk factors in young women for the prevention of breast cancer. In *Breast cancer: Targets and therapy* (pp. 241–257).

Daly, M. B., Pal, T., Berry, M. P., Buys, S. S., Dickson, P., Domchek, S. M., et al. (2021b). Genetic/familial high-risk assessment: Breast, ovarian, and pancreatic, version 2.2021, NCCN clinical practice guidelines in oncology. *Journal of the National Comprehensive Cancer Network, 19*(1), 77–102.

Daniel, S., Venkateswaran, C., Singh, C., Hutchinson, A., & Johnson, M. J. (2022). "So, when a woman becomes ill, the total structure of the family is affected, they can't do anything…" Voices from the community on women with breast cancer in India: A qualitative focus group study. *Supportive Care in Cancer, 30*, 951–963.

D'Cruz, M., & Banerjee, D. (2020). 'An invisible human rights crisis': The marginalization of older adults during the COVID-19 pandemic—An advocacy review. *Psychiatry Research, 292*, 113369.

Deacon, D. C., Smith, E. A., & Judson-Torres, R. L. (2021). Molecular biomarkers for melanoma screening, diagnosis and prognosis: Current state and future prospects. *Frontiers in Medicine, 8*, 642380.

Dhahri, S., & Omri, A. (2020). Foreign capital toward SDGs 1 & 2—Ending poverty and hunger: The role of agricultural production. *Structural Change and Economic Dynamics., 53*, 208.

Dittrich, E. (2020). *Homeless with breast cancer on skid row, Los Angeles: A review of services for women*. Pepperdine University.

Dlamini, Z., Molefi, T., Khanyile, R., Mkhabele, M., Damane, B., Kokoua, A., et al. (2024). From incidence to intervention: A comprehensive look at breast cancer in South Africa. *Oncology and Therapy, 12*(1), 1–11.

Doran, K. (2020). Fat: Saturated, unsaturated. In *Encyclopedia of behavioral medicine*. Springer.
Duggan, C., Dvaladze, A., Rositch, A. F., Ginsburg, O., Yip, C. H., Horton, S., et al. (2020). The Breast Health Global Initiative 2018 global summit on improving breast healthcare through resource-stratified phased implementation: Methods and overview. *Cancer, 126*, 2339–2352.
Elomrani, S., Bezad, R., De Brouwere, V., Campbell, O. M. R., Lange, I. L., Oswald, W. E., & Assarag, B. (2024). Approaching the SDG targets with sustained political commitment: Drivers of Morocco's notable decline in maternal and neonatal mortality. *BMJ Global Health, 9*(Suppl 2), e011278.
Ferlay, J., Ervik, M., Lam, F., Colombet, M., Mery, L., Piñeros, M., et al. (2020). *Observatory: "Cancer Today"*. Lyon International Agency Res. Cancer.
Fernandez, R. M. (2020). SDG3 good health and well-being: Integration and connection with other SDGs. In *Good health and well-being*. Springer.
Foroozani, E., Ghiasvand, R., Mohammadianpanah, M., Afrashteh, S., Bastam, D., Kashefi, F., et al. (2020). Determinants of delay in diagnosis and end stage at presentation among breast cancer patients in Iran: A multi-center study. *Scientific Reports, 10*(1), 21477.
Frieden, T. R., Cobb, L. K., Leidig, R. C., Mehta, S., & Kass, D. (2020). Reducing premature mortality from cardiovascular and other non-communicable diseases by one third: Achieving sustainable development goal indicator 3.4 …. *Global Heart, 15*, 50.
Ginsburg, O., Yip, C. H., Brooks, A., Cabanes, A., Caleffi, M., Dunstan Yataco, J. A., et al. (2020). Breast cancer early detection: A phased approach to implementation. *Cancer, 126*, 2379–2393.
Göttsche, S. (2021). The accumulation of vulnerability aspects in the migrant figure: A theoretical approach. In *Power in vulnerability: A multidimensional review of migrants' vulnerabilities* (pp. 3–27). Springer.
Halkos, G., & Gkampoura, E. C. (2021). Where do we stand on the 17 sustainable development goals? An overview on progress. *Economic Analysis and Policy, 70*, 94.
Hanafi, I., Alsalkini, M., Husein, S., & Salamoon, M. (2023). The delay of breast cancer diagnosis and management during the Syrian war. *Cancer Epidemiology, 82*, 102290.
Haque, M., Islam, T., Sartelli, M., Abdullah, A., & Dhingra, S. (2020). Prospects and challenges of precision medicine in lower-and middle-income countries: A brief overview. *Bangladesh Journal of Medical Science, 19*(1), 32.
Horton, S., Camacho Rodriguez, R., Anderson, B. O., Aung, S., Awuah, B., Delgado Pebe, L., et al. (2020). Health system strengthening: Integration of breast cancer care for improved outcomes. *Cancer, 126*, 2353–2364.
Huang, J., Chan, P. S., Lok, V., Chen, X., Ding, H., Jin, Y., et al. (2021). Global incidence and mortality of breast cancer: A trend analysis. *Aging (Albany NY), 13*(4), 5748.
Iddrisu, M., Aziato, L., & Dedey, F. (2020). Psychological and physical effects of breast cancer diagnosis and treatment on young Ghanaian women: A qualitative study. *BMC Psychiatry, 20*, 353.
Jerome-D'Emilia, B., Suplee, P. D., Robles-Rodriguez, E., & D'Emilia, W. (2021). The impact of delays in low-income women's breast cancer experiences. *Cancer Nursing, 44*(1), E43–E52.
Kaminsky, L. A., German, C., Imboden, M., Ozemek, C., Peterman, J. E., & Brubaker, P. H. (2022). The importance of healthy lifestyle behaviors in the prevention of cardiovascular disease. *Progress in Cardiovascular Diseases, 70*, 8–15.
Kashyap, D., Pal, D., Sharma, R., Garg, V. K., Goel, N., Koundal, D., et al. (2022). Global increase in breast cancer incidence: Risk factors and preventive measures. *BioMed Research International, 2022*, 9605439.
Kc, A., Jha, A. K., Shrestha, M. P., Zhou, H., Gurung, A., Thapa, J., & Budhathoki, S. S. (2020). Trends for neonatal deaths in Nepal (2001–2016) to project progress toward the SDG target in 2030, and risk factor analyses to focus action. *Maternal and Child Health Journal, 24*, 5–14.
Khalidi, M. A. (2020). Are sexes natural kinds? In *Current controversies in philosophy of science*. Routledge.
Kim, M. Y. (2021). Breast cancer metastasis. In *Translational research in breast cancer*. Springer.

Leal Filho, W., Brandli, L. L., Lange Salvia, A., Rayman-Bacchus, L., & Platje, J. (2020). COVID-19 and the UN sustainable development goals: Threat to solidarity or an opportunity? *Sustainability, 12*(13), 5343.

Lee, I. C., Hu, F., & Li, W. Q. (2020). Cultural factors facilitating or inhibiting the support for traditional household gender roles. *Journal of Cross-Cultural Psychology, 51*, 333.

Lin, L., Li, Z., Yan, L., Liu, Y., Yang, H., & Li, H. (2021). Global, regional, and national cancer incidence and death for 29 cancer groups in 2019 and trends analysis of the global cancer burden, 1990–2019. *Journal of Hematology & Oncology, 14*, 197.

Liu, H., Zhou, R., Yao, P., & Zhang, J. (2022). Assessing Chinese governance low-carbon economic peer effects in local government and under sustainable environmental regulation. *Environmental Science and Pollution Research, 30*, 1–20.

Lopes-Ramos, C. M., Quackenbush, J., & DeMeo, D. L. (2020). Genome-wide sex-and gender differences in cancer. *Frontiers in Oncology, 10*, 597788.

Lubi, K., Sildver, K., Sokolova, D., Savicka, V., Nool, I., Mets-Oja, S., & Tupits, M. (2022). Existing health practices in shaping the decision to participate during breast cancer screening among 50–69 years old women in Estonia. *SAGE Open Nursing, 8*, 23779608221124293.

Luo, C., Li, N., Lu, B., Cai, J., Lu, M., Zhang, Y., et al. (2022). Global and regional trends in incidence and mortality of female breast cancer and associated factors at national level in 2000 to 2019. *Chinese Medical Journal, 135*(01), 42–51.

Macassa, G. (2021). Can sustainable health behavior contribute to ensure healthy lives and well-being for all at all ages (SDG 3)? A viewpoint. *Journal of Public Health Research, 10*, 2051.

Malakoane, B., Heunis, J. C., Chikobvu, P., Kigozi, N. G., & Kruger, W. H. (2020). Public health system challenges in the Free State, South Africa: A situation appraisal to inform health system strengthening. *BMC Health Services Research, 20*, 1–14.

Mao, J. J., Pillai, G. G., Andrade, C. J., Ligibel, J. A., Basu, P., Cohen, L., et al. (2022). Integrative oncology: Addressing the global challenges of cancer prevention and treatment. *CA: A Cancer Journal for Clinicians, 72*(2), 144–164.

Martinez-Cannon, B. A., Barragan-Carrillo, R., & Villarreal-Garza, C. (2021). Young women with breast cancer in resource-limited settings: What we know and what we need to do better. *Breast Cancer: Targets and Therapy, 13*, 641–650.

Mayra, K., Matthews, Z., & Padmadas, S. S. (2022). Why do some health care providers disrespect and abuse women during childbirth in India? *Women and Birth, 35*, e49.

Mbunge, E., Muchemwa, B., & Batani, J. (2021). Sensors and healthcare 5.0: Transformative shift in virtual care through emerging digital health technologies. *Global Health Journal, 5*, 169.

Mubarik, S., Yu, Y., Wang, F., Malik, S. S., Liu, X., Fawad, M., et al. (2022). Epidemiological and socio-demographic transitions of female breast cancer incidence, death, case fatality and DALYs in 21 world regions and globally, from 1990 to 2017: An age-period-cohort analysis. *Journal of Advanced Research, 37*, 185–196.

Muncan, B., Walters, S. M., Ezell, J., & Ompad, D. C. (2020). "They look at us like junkies": Influences of drug use stigma on the healthcare engagement of people who inject drugs in New York City. *Harm Reduction Journal, 17*, 53.

Murphy, A., Palafox, B., Walli-Attaei, M., Powell-Jackson, T., Rangarajan, S., Alhabib, K. F., et al. (2020). The household economic burden of non-communicable diseases in 18 countries. *BMJ Global Health, 5*(2), e002040.

Mutebi, M., Anderson, B. O., Duggan, C., Adebamowo, C., Agarwal, G., Ali, Z., & Eniu, A. (2020). Breast cancer treatment: A phased approach to implementation. *Cancer, 126*, 2365–2378.

Nabukalu, J. B., Asamani, J. A., & Nabyonga-Orem, J. (2020). Monitoring sustainable development goals 3: Assessing the readiness of low-and middle-income countries. *International Journal of Health Policy and Management, 9*(7), 297.

Neumann, J., Petranikova, M., Meeus, M., Gamarra, J. D., Younesi, R., Winter, M., & Nowak, S. (2022). Recycling of lithium-ion batteries—Current state-of-the art, circular economy, and next-generation recycling. *Advanced Energy Materials, 12*(17), 2102917.

Ngubane, M. Z., Mndebele, S., & Kaseeram, I. (2023). Economic growth, unemployment and poverty: Linear and non-linear evidence from South Africa. *Heliyon, 9*, e20267.

Nourazari, S., Davis, S. R., Granovsky, R., Austin, R., Straff, D. J., Joseph, J. W., & Sánchez, L. D. (2021). Decreased hospital admissions through emergency departments during the COVID-19 pandemic. *The American Journal of Emergency Medicine, 42*, 203–210.

Nyashanu, M., Pfende, F., & Ekpenyong, M. (2020). Exploring the challenges faced by front-line workers in health and social care amid the COVID-19 pandemic: Experiences of front-line workers in the English Midlands region, UK. *Journal of Interprofessional Care, 34*(5), 655–661.

Okello, M., Lamo, J., Ochwo-Ssemakula, M., & Onyilo, F. (2021). Challenges and innovations in achieving zero hunger and environmental sustainability through the lens of sub-Saharan Africa. *Outlook on Agriculture, 50*(2), 141–147.

Panzeri, M., Ferrucci, R., Cozza, A., & Fontanesi, L. (2020). Changes in sexuality and quality of couple relationship during the COVID-19 lockdown. *Frontiers in Psychology, 11*, 565823.

Pashayan, N., Antoniou, A. C., Ivanus, U., Esserman, L. J., Easton, D. F., French, D., et al. (2020). Personalized early detection and prevention of breast cancer: ENVISION consensus statement. *Nature Reviews Clinical Oncology, 17*(11), 687–705.

Perera, S. K., Jacob, S., Wilson, B. E., Ferlay, J., Bray, F., Sullivan, R., & Barton, M. (2021). Global demand for cancer surgery and an estimate of the optimal surgical and anesthesia work-force between 2018 and 2040: A population-based modeling study. *The Lancet Oncology, 22*(2), 182–189.

Pistollato, F., Campia, I., Daskalopoulos, E. P., Bernasconi, C., Desaintes, C., Di Virgilio, S., et al. (2023). Gauging innovation and health impact from biomedical research: Survey results and interviews with recipients of EU-funding in the fields of Alzheimer's disease, breast cancer and prostate cancer. *Health Research Policy and Systems, 21*(1), 66.

Pizzi, S., Caputo, A., Corvino, A., & Venturelli, A. (2020). Management research and the UN sustainable development goals (SDGs): A bibliometric investigation and systematic review. *Journal of Cleaner Production, 276*, 124033.

Platta, A., & Radzymińska, M. (2023). Dietary behavior of adolescents based on the example of fat intake as a measure of consumption of unhealthy foods in the context of setting sustainable development goals. *Scientific Journal of Gdynia Maritime University, 126*, 33–40.

Poteat, T. C., Adams, M. A., Malone, J., Geffen, S., Greene, N., Nodzenski, M., et al. (2021). Delays in breast cancer care by race and sexual orientation: Results from a national survey with diverse women in the United States. *Cancer, 127*(19), 3514–3522.

Rasche, A. (2020). The United Nations Global Compact and the sustainable development goals. In *Research handbook of responsible management*. Edward Elgar.

Reid, F., Bhatla, N., Oza, A. M., Blank, S. V., Cohen, R., Adams, T., & Jones, A. (2021). The World Ovarian Cancer Coalition Every Woman Study: Identifying challenges and opportunities to improve survival and quality of life. *International Journal of Gynecologic Cancer, 31*(2), 238.

Rentschler, J., Salhab, M., & Jafino, B. A. (2022). Flood exposure and poverty in 188 countries. *Nature Communications, 13*, 3527.

Riaz, A., Zakar, R., & Farrah, H. A. (2023). Social and psychological issues experienced by women suffering from breast cancer. *Journal of Positive School Psychology, 7*(8), 188–197.

Ricardo, M. A. (2021). *Knowledge management and the 2030 agenda for sustainable development in the United Nations context*. Ciencias Administrativas.

Ridha, M. R., & Parwanto, N. B. (2020). The effect of foreign direct investment, human development and macroeconomic condition on economic growth: Evidence from Indonesia. *Journal of Indonesian Applied Economics, 8*, 46.

Risom, T., Glass, D. R., Averbukh, I., Liu, C. C., Baranski, A., Kagel, A., & Angelo, M. (2022). Transition to invasive breast cancer is associated with progressive changes in the structure and composition of tumor stroma. *Cell, 185*(2), 299–310.

Rock, C. L., Thomson, C., Gansler, T., Gapstur, S. M., McCullough, M. L., Patel, A. V., & Doyle, C. (2020). American Cancer Society guideline for diet and physical activity for cancer prevention. *CA: A Cancer Journal for Clinicians, 70*(4), 245–271.

Sachs, L. E., & Sachs, J. D. (2021). Health priorities for sustainable development. In *Critical issues in human rights and development* (pp. 186–212). Edward Elgar Publishing.

Saha, T., & Lukong, K. E. (2022). Breast cancer stem-like cells in drug resistance: A review of mechanisms and novel therapeutic strategies to overcome drug resistance. *Frontiers in Oncology, 12*, 856974.

Scanlon, B. R., Fakhreddine, S., Rateb, A., de Graaf, I., Famiglietti, J., Gleeson, T., & Zheng, C. (2023). Global water resources and the role of groundwater in a resilient water future. *Nature Reviews Earth & Environment, 4*(2), 87–101.

Schantz, C., Coulibaly, A., Traoré, A., Traoré, B. A., Faye, K., Robin, J., et al. (2024). Access to oncology care in Mali: A qualitative study on breast cancer. *BMC Cancer, 24*(1), 81.

Scharlemann, J. P., Brock, R. C., Balfour, N., Brown, C., Burgess, N. D., Guth, M. K., et al. (2020). Toward understanding interactions between sustainable development goals: The role of environment–human linkages. *Sustainability Science, 15*, 1573–1584.

Sethi, B. A., Sethi, A., Ali, S., & Aamir, H. S. (2020). Impact of coronavirus disease (COVID-19) pandemic on health professionals. *Pakistan Journal of Medical Sciences, 36*(COVID19-S4), S6.

Settersten, R. A., Jr., Bernardi, L., Härkönen, J., Antonucci, T. C., Dykstra, P. A., Heckhausen, J., & Thomson, E. (2020). Understanding the effects of Covid-19 through a life-course lens. *Advances in Life-Course Research, 45*, 100360.

Shaikh, K., Krishnan, S., Thanki, R., Shaikh, K., Krishnan, S., & Thanki, R. (2021). An introduction to breast cancer. In *Artificial intelligence in breast cancer early detection and diagnosis* (pp. 1–20).

Shaw, J., Brewer, L. P. C., & Veinot, T. (2021). *Recommendations for health equity and virtual care arising from the COVID-19 pandemic: Narrative review* (Vol. 5, p. e23233). JMIR Formative Research.

Shen, A., Qiang, W., Wang, Y., & Chen, Y. (2020). Quality of life among breast cancer survivors with triple-negative breast cancer—Role of hope, self-efficacy and social support. *European Journal of Oncology Nursing, 46*, 101771.

Siegmund, D., Metz, S., Peinecke, V., Warner, T. E., Cremers, C., Grevé, A., & Apfel, U. P. (2021). Crossing the valley of death: From fundamental to applied research in electrolysis. *JACS Au, 1*(5), 527–535.

Tabár, L., Dean, P. B., Tucker, F. L., Yen, A. M. F., Fann, J. C. Y., Lin, A. T. Y., & Chen, T. H. H. (2022). Breast cancer originates from the terminal ductal lobular units: In situ and invasive acinar adenocarcinoma of the breast, AAB. *European Journal of Radiology, 152*, 110323.

Tarantino, P., Viale, G., Press, M. F., Hu, X., Penault-Llorca, F., Bardia, A., & Curigliano, G. (2023). ESMO expert consensus statements (ECS) on the definition, diagnosis, and management of HER2-low breast cancer. *Annals of Oncology, 34*(8), 645–659.

Teixeira, M. B. M., Galvão, L. L. D. C., Mota-Santos, C. M., & Carmo, L. J. O. (2021). Women and work: Film analysis of most beautiful thing. *Revista de Gestão, 28*(1), 66–83.

Timmis, K., & Ramos, J. L. (2021). The soil crisis: The need to treat as a global health problem and the pivotal role of microbes in prophylaxis and therapy. *Microbial Bio-technology, 14*, 769.

Tran, T. X. M., Jung, S. Y., Lee, E. G., Cho, H., Kim, N. Y., Shim, S., & Cho, H. (2022). Fear of cancer recurrence and its negative impact on health-related quality of life in long-term breast cancer survivors. *Cancer Research and Treatment: Official Journal of Korean Cancer Association, 54*(4), 1065.

Trapani, D., Douillard, J. Y., Winer, E. P., Burstein, H., Carey, L. A., Cortes, J., et al. (2021). The global landscape of treatment standards for breast cancer. *JNCI: Journal of the National Cancer Institute, 113*(9), 1143–1155.

Uddin, M. (2021). Addressing work life balance challenges of working women during COVID-19 in Bangladesh. *International Social Science Journal, 71*, 7.

van Der Meer, D. J., Kramer, I., van Maaren, M. C., van Diest, P. J., Linn, S. C., Maduro, J. H., & Voogd, A. C. (2021). Comprehensive trends in incidence, treatment, survival and mortality of first primary invasive breast cancer stratified by age, stage and receptor sub-type in the Netherlands between 1989 and 2017. *International Journal of Cancer, 148*(9), 2289–2303.

Vollmer, F., & Alkire, S. (2022). Consolidating and improving the assets indicator in the global multidimensional poverty index. *World Development, 158*, 105997.

Wang, J., Barth, J., Göttgens, I., Emchi, K., Pach, D., & Oertelt-Prigione, S. (2020a). An opportunity for patient-centered care: Results from a secondary analysis of sex-and gender-based data in mobile health trials for chronic medical conditions. *Maturitas, 138*, 1–7.

Wang, L., Geng, X., Ji, L., Lu, G., & Lu, Q. (2020b). Treatment decision-making, family influences, and cultural influences of Chinese breast cancer survivors: A qualitative study using an expressive writing method. *Supportive Care in Cancer, 28*, 3259.

Wang, B., He, F., Hu, Y., Wang, Q., Wang, D., Sha, Y., & Wu, J. (2022). Cancer incidence, mortality, and risk factors in member countries of the Belt and Road initiative. *BMC Cancer, 22*, 582.

WHO. (2023). *GBCI implementation framework: Assessing, strengthening, and scaling up services for the early detection and management of BC*. World Health Organization.

Wojtyla, C., Bertuccio, P., Wojtyla, A., & La Vecchia, C. (2021). European trends in breast cancer mortality, 1980–2017 and predictions to 2025. *European Journal of Cancer, 152*, 4–17.

World Health Organization. (2020). *Ending the neglect to attain the sustainable development goals: A road map for neglected tropical diseases 2021–2030*. World Health Organization.

Xu, H., Winnink, J., Yue, Z., Liu, Z., & Yuan, G. (2020). Topic-linked innovation paths in science and technology. *Journal of Informetrics, 14*, 101014.

Yan, R., Wang, J., & Yu, J. (2023). Association of sexual attitudes with sexual dysfunction and sexual distress among Chinese breast cancer survivors: A cross-sectional study. *Supportive Care in Cancer, 31*, 154.

Yao, N., Shi, W., Liu, T., Siyin, S. T., Wang, W., Duan, N., & Qu, J. (2022). Clinicopathologic characteristics and prognosis for male breast cancer compared to female breast cancer. *Scientific Reports, 12*(1), 220.

Yu, J., Mu, Q., Fung, M., Xu, X., Zhu, L., & Ho, R. J. Y. (2022). Challenges and opportunities in breast cancer treatments: Nano-drug combinations delivered preferentially to metastatic cells may enhance therapeutic response. *Pharmacology & Therapeutics, 236*, 108108.

Yuan, Z., Li, Y., Zhang, S., Wang, X., Dou, H., Yu, X., & Xiao, M. (2023). Extracellular matrix remodeling in tumor progression and immune escape: From mechanisms to treatments. *Molecular Cancer, 22*(1), 48.

Zakayo, S. M., Njeru, R. W., Sanga, G., Kimani, M. N., Charo, A., Muraya, K., & Molyneux, S. (2020). Vulnerability and agency across treatment-seeking journeys for acutely ill children: How family members navigate complex healthcare before, during and after hospitalization in a rural Kenyan setting. *International Journal for Equity in Health, 19*, 1–17.

Zemenick, A. T., Turney, S., Webster, A. J., Jones, S. C., & Weber, M. G. (2022). Six principles for embracing gender and sexual diversity in postsecondary biology classrooms. *Bio-Science, 72*(5), 481–492.

Chapter 7
Opportunities for Investing in Organ-on-a-Chip (OoC) Technology, Breast-on-a-Chip (BoC) as an Example Technology Transfer into Islamic Development Bank Members Countries (IsDB-MCs)

Ahmed A. M. Elnour, Nureddin Ashammakhi, and El Bashier Sallam

Abstract Cancer-on-a-chip technology holds immense potential for advancing personalized medicine and cancer research in Islamic Development Bank (IsDB) member countries. This study explores the opportunity to introduce this innovative technology in the region, considering its ability to replicate intricate cancer models, simulate tumor microenvironments, and facilitate drug development. Despite the availability of relevant infrastructure and human resources, no established mechanism currently exists to implement cancer-on-a-chip technology in IsDB member countries. The proposed strategy involves initiating a coordination platform among universities and governmental organizations, eventually spinning it off into a registered international company. This company would provide continuous quality services, offering a one-stop solution for cancer diagnosis and individualized treatment concepts. The growing organ-on-a-chip market, projected to reach $331 million by 2023, further highlights the potential of this technology. However, addressing the gap between developed and developing countries in accessing cancer technologies

A. A. M. Elnour (✉)
Faculty of Chemical and Process Engineering Technology, Universiti Malaysia Pahang Al-Sultan Abdullah (UMPSA), Pahang Gambang, Malaysia

Centre of Excellence for Advanced Research in Fluid Flow (CARIFF), University Malaysia, Pahang Gambang, Malaysia
e-mail: adamhassan@umpsa.edu.my

N. Ashammakhi
Department of Biomedical Engineering, Institute for Quantitative Health Science and Engineering, College of Engineering, and College of Human Medicine, Michigan State University, East Lansing, MI, USA
e-mail: ashammak@msu.edu

E. B. Sallam
Freelance Global and Public Health Expert (FGPHE), Khartoum, Sudan

A. A. M. Elnour (ed.), *Gum Arabic and Breast Cancer Biology*,
https://doi.org/10.1007/978-981-97-8518-6_7

and treatments remains imperative. Challenges such as affordability, treatment abandonment, and limited access to state-of-the-art medical technology hinder progress in developing countries. Bridging this gap requires addressing economic burdens, enhancing screening programs, and improving access to advanced treatments. By leveraging the IsDB's role in promoting social and economic development, the introduction of cancer-on-a-chip technology in member countries could significantly advance cancer research and personalized treatment approaches, ultimately improving patient outcomes and reducing mortality rates.

Keywords Cancer-on-a-chip · Organ-on-a-chip · Tumor microenvironment · Personalized medicine · Islamic Development Bank (IsDB) · Sustainable development, Renewable energy

Highlights

1. Cancer-on-a-chip technology holds immense potential for advancing personalized medicine and cancer research.
2. No established mechanism currently exists to implement cancer-on-a-chip technology in IsDB member countries.
3. The proposed strategy involves initiating a coordination platform and spinning it off into a company.
4. The growing organ-on-a-chip market, projected to reach **$331** million by **2023**, highlights the technology's potential.
5. Challenges such as affordability and limited access to advanced technology hinder progress in developing countries.
6. Bridging the gap requires addressing economic burdens, enhancing screening programs, and improving treatment access.
7. The asymptomatic stage of cancer presents an opportunity to increase recovery rates by **60–80%**.

7.1 Introduction

Leveraging research on tumor-on-a-chip technology is essential for exploring the potential of introducing cancer-on-a-chip technology to the IsDB member countries and identifying potential implementation barriers. Studies emphasize the technology's capacity for personalized drug screening and modeling tumor microenvironments (Imparato et al., 2022; Young, 2013). By replicating intricate cancer models and integrating vascular components, tumor-on-a-chip systems provide a promising platform for investigating metastasis and drug development (Rodrigues et al., 2020; Wang et al., 2022). These advancements, along with the ability to simulate tumor physiology and interactions within the tumor microenvironment, position cancer-on-a-chip as a valuable tool for advancing cancer research and therapy (Imparato et al., 2022; Komen et al., 2021). This organ-on-a-chip technology can potentially revolutionize cancer research, providing a more accurate and efficient means of studying the disease at a cellular level (Ingber, 2022). Therefore, introducing cancer-on-a-chip technology in IsDB member countries could significantly advance personalized medicine and cancer research.

The Research and Innovation Platform for Micro-nano-Bio Systems (BERCESTE) Project, focusing on Bio, Environmental, and Renewable Chemistry, aligns well with the global trend toward renewable energy options (Gannan et al., 2023). This initiative is crucial for sustainable development and addressing climate change concerns by promoting the use of renewable energy sources. Additionally, financial inclusion efforts in IsDB member countries can benefit from opening bank branches in rural areas to enhance financial service accessibility (Xia et al., 2020). Thus, the BERCESTE Project and improved financial inclusion can significantly contribute to sustainable development in IsDB member countries.

Although the 3D cancer-on-a-chip model could be a highly capable and affordable technique to detect and control cancer, there is currently no established mechanism to implement the technology in IsDB member countries, despite the availability of relevant infrastructure and human resources (Azizipour et al., 2020; Liu et al., 2021). The main strategy to achieve this research's purpose will be to initiate cancer-on-a-chip as a coordination platform among universities and relevant governmental organizations, eventually spinning it off into a registered international company (Cànaves Llabrés, 2022). Under certain time constraints, this company would provide continuous quality services to customers, offering a one-stop solution for diagnosing cancer with individualized treatment concepts in the region (Baka et al., 2022). Hence, establishing a coordinated platform and international company could facilitate the implementation and success of cancer-on-a-chip technology in the region.

The projected growth of the organ-on-a-chip market to approximately $331 billion by 2023, with a compound annual growth rate (CAGR) of 9.7% from 2020 to 2023, reflects the increasing relevance of this technology in biomedical research and drug development. This growth is attributed to advancements in microfluidics and the rising demand for personalized medicine, which causes more efficient and ethical alternatives to traditional drug testing methods (Silva, 2023; Sunildutt et al., 2023). This technology offers precise control over microenvironment conditions, enabling tailored drug discovery and disease modeling (Khalid et al., 2017). Organ-on-a-chip systems provide a platform for understanding tissue complexity, aiding in the development of future therapies (Shah et al., 2024). The miniaturized human organ models created through tissue engineering advancements are instrumental in drug discovery processes (Esch et al., 2015). These advancements in organ-on-a-chip technology hold promise for personalized treatment approaches and more efficient drug screening methods (Kulkeaw & Pengsart, 2021). As a result, the growing organ-on-a-chip market is poised to enhance drug discovery and personalized treatment approaches significantly.

However, addressing the gap between developed and developing countries in accessing cancer technologies and treatments across various healthcare stages is imperative. Challenges such as affordability, treatment abandonment, and limited access to state-of-the-art medical technology hinder progress in developing countries (Sharma et al., 2011). Cancer incidence and mortality discrepancies underscore the need to address barriers to radiotherapy delivery in low- and middle-income countries (LaVigne et al., 2017). The lack of infrastructure, financial resources, and cancer literacy contribute to advanced cancer presentations and poor outcomes in developing nations (Srinivasan et al., 2015). Bridging this gap requires addressing economic burdens, enhancing screening programs, and improving access to advanced treatments. Therefore, addressing these barriers is crucial for improving cancer care and outcomes in developing countries.

For instance, the asymptomatic stage of cancer presents a crucial opportunity to increase the recovery rate by approximately 60–80%. Adequate screening can identify cancers at early stages where treatment regimens are effective, uncomplicated, and cheaper. While this approach has been explored in developed countries, extending these efforts to developing nations is essential. Early detection through screening programs is vital for improving survival rates, as many cancers are diagnosed at advanced stages in developing countries due to limited access to healthcare and screening facilities (Akinyemiju, 2012). Bridging the gap in early detection and treatment access is imperative to enhance outcomes and reduce mortality rates, especially in resource-constrained settings where cancer diagnosis often occurs at later stages (Sobri et al., 2021). Thus, enhancing early detection and treatment access is critical for improving cancer outcomes in resource-constrained settings.

7.1.1 Background of Cancer-on-a-Chip Technology

The IsDB plays a crucial role in promoting social and economic development in member states and Muslim communities by providing interest-free financing to alleviate poverty through project development, aligning with Islamic principles (Sennanda et al., 2023). IsDB's focus on foreign aid is essential to prevent disparities and promote sustainable development (Mustafa & Razak, 2012). By emphasizing education and scientific advancements, IsDB's efforts resonate with the Quran's call for a scientific approach to all affairs. This is exemplified in research on the Isd system in Staphylococcus aureus and the importance of strengthening research capacity in countries like Tanzania (Sennanda et al., 2023). This commitment to addressing contemporary needs and promoting knowledge aligns with the Quran's emphasis on education (Sennanda et al., 2023). Therefore, IsDB's initiatives align with Islamic principles and focus on sustainable development through education and research.

A modern "Cancer-on-a-Chip" (CN chip) approach involves creating living organisms on a chip that closely mimic their natural form, enabling the reproduction, monitoring, and evaluation of specific biological functions (Duzagac et al., 2021). This technology allows for the culture of patient-derived tumor samples and the assessment of their response to therapies (Bērziņa et al., 2021). By utilizing different cell populations within a gelatin hydrogel, such as cancer cells, monocytes, and endothelial cells, these platforms can simulate the 3D growth of angiogenic blood vessels toward cancerous organoids (Haque et al., 2021). Organs-on-chips have been explored for drug efficacy testing, including a micro-engineered 3D assay of epithelial-mesenchymal transition during cancer progression (Esch et al., 2015). These chips aim to replicate the structural tissue arrangements and functional complexity of living organs, providing physiologically relevant models for studying organ physiology and disease mechanisms (Huh et al., 2012). Hence, "Cancer-on-a-Chip" technology offers a realistic and advanced platform for studying cancer and testing therapies.

The "Cancer-on-a-Chip" (CN chip) technology integrates microfluidics, wireless sensor networks, stem cells, and 3D bioprinting to create a biophysical microenvironment mimicking native tumor conditions (Caytiles & Park, 2014). This approach enables applications in drug development, personalized testing, and disease modeling, potentially revolutionizing medical research and reconstructive embryology (Pandya et al., 2017). Microfluidic platforms have been instrumental in cancer research, offering high sensitivity, low-cost, and precise control over the tumor microenvironment (Chaudhuri et al., 2016). By simulating tumor microenvironments on chips, these technologies facilitate drug testing, disease diagnosis, and the study of cancer metastasis (Liu et al., 2022). The combination of microfluidic chips and cell biology has advanced multi-organ systems, enhancing research capabilities in understanding complex biological processes. As a result, the integration of advanced technologies in "Cancer-on-a-Chip" platforms significantly enhances cancer research and therapeutic development.

7.1.2 Importance of Organ-on-a-Chip Technology Transfer in Healthcare Innovation to Islamic Development Countries (IsDB)

The advancement of OoC technology significantly impacts healthcare innovation by transforming drug development, personalized medicine, disease modeling, regenerative medicine, and addressing economic and ethical considerations. OoC technology involves the development of microdevices that replicate organ structures and functions, allowing for the examination of organ responses to various conditions and compounds (Low et al., 2020). By integrating multiple organ compartments, OoC systems can simulate the human body, aiding in the prediction of drug pharmacokinetics and reducing the need for animal testing in drug trials (Doost, 2024; Takebe et al., 2017). This technology provides a foundation for more precise disease modeling, personalized drug responses, and the creation of predictive preclinical models for drug discovery (Park et al., 2019). Additionally, OoC technology has the potential to advance virology research, enhance toxicology assessments, and improve sustainable drug screening practices (Feitor et al., 2023; Tang et al., 2022; Yi et al., 2017). Collaborations between pharmaceutical companies and academic institutions are propelling the adoption of OoC technology, positioning it at the forefront of healthcare innovation (Esch et al., 2015). Therefore, OoC technology is pivotal in advancing healthcare innovation by improving drug development, personalized medicine, and disease modeling.

The implementation of OoC technology in IsDB member countries has the potential to significantly impact healthcare innovation by addressing specific healthcare challenges and improving medical outcomes. By leveraging OoC technology, these countries can enhance cancer research (Lewison et al., 2020), align with the digital economy paradigm to boost productivity (Alshubiri et al., 2022), utilize Islamic finance for economic growth (Yasirwan, 2024), and strengthen healthcare systems to manage non-communicable diseases (Fadhil et al., 2022). Integrating OoC technology with these initiatives can advance medical research, improve healthcare delivery, and foster sustainable development in the region. Hence, OoC technology can drive healthcare innovation and sustainable development in IsDB member countries.

Cancer-on-a-chip pathways are innovative tools that address critical issues in cancer research. Studies have shown that CHIP, a U-box-type E3 ligase, plays a significant role in cancer progression by regulating various proteins involved in oncogenesis (Cheng et al., 2018; Gohno et al., 2022; Hiyoshi et al., 2014; Jang et al., 2011; Li et al., 2013; Sarkar et al., 2017; Won et al., 2017; Xiao et al., 2017; Xu et al., 2018; Zhang et al., 2016). These pathways are closely linked to cancer stem cells, which are crucial for cancer recurrence and resistance to treatment (Chen et al., 2015; Yan et al., 2013). The use of microsystems like Cancer-on-a-chip offers a promising approach to studying cancer hallmarks and addressing challenges in cancer research (Young, 2013). By targeting specific proteins and pathways, such as those regulated by CHIP, novel therapeutic strategies can be developed to combat cancer progression (Parrales et al., 2017; Yao et al., 2010). Integrating ChIP-seq data with gene expression analysis provides a powerful tool to identify functional sub-pathways in cancer, offering insights into oncogenic processes (Atsalaki et al., 2020). Thus, Cancer-on-a-chip pathways offer novel therapeutic strategies for addressing cancer progression and treatment resistance.

For example, by 2035, it is projected that there will be around 14.8 million cancer survivors among economically developed countries, with the number of cancer survivors increasing significantly over the years (Zwemer et al., 2011). This growing population of cancer survivors will require ongoing medical care to address both short- and long-term effects of cancer and its treatments (Yabroff et al., 2011). The term “cancer survivor” refers to individuals who have been diagnosed with cancer, completed curative treatment, and are disease-free (Leeuwen et al., 2018). With advancements in early detection and treatment, the number of cancer survivors is expected to continue rising, necessitating the development of effective survivorship care programs (Harten et al., 2013). The increase in cancer survivors is a global trend, with over 32 million cancer survivors worldwide, a number that is projected to rise further in the coming years (García-Vivar, 2022). Survival rates for cancer patients have indeed significantly improved over the years due to advancements in treatments and supportive care, leading to a larger population of long-term survivors (Bhakta et al., 2017). However, this increase in survival rates has also brought about challenges such as the financial burden on cancer survivors, including high out-of-pocket expenditures and financial hardship (Ekwueme et al., 2019). The costs associated with cancer care can indeed negatively impact treatment outcomes and quality of life for patients (Ryan et al., 2017). Additionally, the long-term costs of cancer treatment, especially with the introduction of new and costly drugs, have become a significant concern (Corral et al., 2016). As a result, the rising number of cancer survivors underscores the need for effective survivorship care programs and addressing the financial burden of cancer treatment.

The share of the world's largest commercial company in Turkey, its turnover for 2018 reached $400 billion and the middle- and southeast-income group, the number of cancer patients is around 2.2 million and growing because of lifestyle changes (Lin et al., 2021). In addition to the cost of the drugs, the costs of supportive therapies, which include liquid food products, painkillers, platelet boosters, blood and red cell enhancers, and patient follow-up, are included (Mariotto et al., 2020). The increasing number of cancer patients in Turkey, influenced by lifestyle changes, highlights the growing need for various supportive therapies beyond drug costs, such as liquid food products, painkillers, and blood enhancers (Çelik et al., 2020). Studies have shown a rise in breast cancer incidence and mortality in developing countries like Turkey due to lifestyle changes and population growth (Özmen et al., 2019). Understanding the costs associated with cancer treatments, like metastatic colon cancer, is crucial for healthcare planning in Turkey (Kockaya et al., 2013). Additionally, the impact of breast cancer on Turkish women underscores the importance of tailored treatment decisions and histopathological analysis (Özmen, 2014; Özmen et al., 2016). Therefore, addressing the growing incidence of cancer in Turkey requires understanding treatment costs and the need for supportive therapies.

The potential benefits of cancer-on-a-chip systems designed and presented in this project, supported by the IsDB-MCS grant in Islamic Development Bank, are briefly described in the current Regulation, Market, and Training documents (Osman & Sulieman, 2023). Investigator Treasurer: The market needs and the projects that will unlock these markets. They can be produced in the long-term because of investments. However, estimates may increase once again to $250 million or more. Therefore, to increase the value of this research and development (R&D) product, it is necessary to prove the concept, validation, and the first functional model/data produced with a minimum amount, still sufficiently significant for the private sector and venture investors (Zhang et al., 2022). Besides the university and other academic partners, two private firms and venture investors with known in-market medical device applications will be included in the project chain that will guide technology transfer and commercialization activities (Lerner, 2022). Hence, the commercialization of cancer-on-a-chip systems involves collaboration with academic and private sector partners to validate and market the technology.

7.1.3 Islamic Development Bank (IsDB-MCS)

The Islamic Development Bank Member Country Partnership Strategy (IsDB-MCS) aims to foster economic growth through good governance and strong institutions, contributing to human capital development in member countries (Mustafa & Razak, 2012). By providing foreign aid and financing aid, the IsDB-MCS plays a crucial role in supporting socio-economic development (Pericoli, 2020). This strategy has a direct impact on poverty reduction and economic growth in member countries, aligning with the goal of reinforcing a virtuous cycle of development (Mustafa & Razak, 2012; Pericoli, 2020). Through its initiatives, the IsDB-MCS contributes to the overall well-being and advancement of its member countries by addressing key socio-economic challenges and promoting sustainable development. Therefore, the IsDB-MCS strategy significantly aids in socio-economic development and poverty reduction.

The IsDB has implemented the MedCo Science, Technology and Innovation Cooperation (TIC) Roadmap, focusing on priority areas such as food security, energy, disease control, materials, water supply, and Islamic science revival. This initiative involves strategic destinations in Islamic countries and utilizes instruments like high-tech finance and capacity-building. However, many least developed countries lack resources for comprehensive science, technology, and innovation systems (de Fonseca & Veloso, 2018). Collaboration between governments and the private sector is crucial as research and innovation are increasingly network-based (de Fonseca & Veloso, 2018). Capacity-building for innovation and sustainability should prioritize individual behavior change to drive lasting changes (Paiva et al., 2022). Science and technology finance can enhance green productivity, drive development in peripheral regions, and improve financing efficiency (Lu et al., 2023). Financial development significantly boosts resource allocation efficiency (Liu et al., 2022). Hence, the MedCo TIC Roadmap by IsDB is vital for fostering science and technology innovation, especially through public–private collaboration and capacity-building.

The IsDB is a multilateral financial institution founded in 1975 with 56 member countries (Kamal & Ray, 2020). IsDB operations are designed to empower people to bolster their economies, developing their productive capacity, and overcoming both natural and man-induced disasters (Twiggs, 2020). In addition, IsDB supports its member countries in their development projects in education, health, science, technology, and innovation, and other sectors aimed at the advancement of their communities, all consistent with broader economic and social development plans (Twiggs, 2020). To attain its strategic objectives, IsDB generates useful knowledge products to help its members overcome their development challenges. Such products are generated by different groups of IsDB, including collaboration with medical experts of Medically Advanced Countries and Medical Cooperation (MACMC), and other professionals of IsDB (Twiggs, 2020). To address this issue, IsDB introduced the MedCo Science, TIC Roadmap (Elzahi Saaid Ali, 2022). As a result,

IsDB plays a crucial role in supporting and empowering its member countries through various development projects and knowledge products.

7.2 Fundamentals of Cancer-on-a-Chip Technology

Cancer-on-a-chip technology integrates microfluidics, tissue engineering, and biomaterials to mimic tumor microenvironments, enabling high-throughput personalized drug screening and therapy testing (Fetah et al., 2019; Tsai et al., 2017). These systems advance cancer research by replicating human cancer responses more accurately than animal models, aiding in drug discovery and reducing reliance on animal testing (Komen et al., 2021; Young, 2013). By incorporating microfabrication technologies and single-cell omics, tumor-on-a-chip models offer a more precise representation of solid tumors, enhancing our understanding of cancer biology (Bērziņa et al., 2021; Santiago et al., 2019). The technology's ability to simulate the tumor microenvironment, including vascularization, contributes to improved preclinical drug screening and pathophysiological relevance in cancer research (Lim et al., 2021; Wang et al., 2022). Therefore, cancer-on-a-chip platforms hold promise for revolutionizing cancer studies and advancing treatment.

Most established cancer-on-a-chip systems are single- or few-organ systems showing heterotypic interactions in two-dimensional (2D) and 3D dynamics (Zhang et al., 2022). Nevertheless, cancer multifacets are still limited because of communication mediated by short-range paracrine signals toward distant organs (Banales et al., 2020). Therefore, an array of multiple organ models on a chip that can create a condition called body-on-a-chip for tumoral disease is still indispensable for the reasons discussed in the text (Rodrigues et al., 2020). The demand for designing an organ-on-a-chip array that is closely related to different parts of the body and each organ has a relationship with one another is essential, either for basic medicine or drug effects and toxicity studies (Cong et al., 2020). Hence, developing comprehensive body-on-a-chip systems is crucial for advancing cancer research and therapy testing.

Cancer-on-a-chip technology is an engineering approach to tumoral disease modeling, where the cells, tissues, biochemical, and mechanical microenvironment of a living organism are mimicked on a glass or polymeric chip in vitro (Caballero et al., 2020). It provides an extraordinary opportunity to study the complex and multi-faceted relationships of the three-dimensional (3D) real tumor with its environment relying on interdisciplinary approaches (Paradiso et al., 2021). Recently, the concept of a tumor-on-a-chip is weaving fields that aim to take advantage of strategies employed in microfluidic engineering, microfabrication technology (Liu et al., 2021), and 3D cell biology to design and construct a reliable, real-time, and functional cellular tumor model applicable to cancer drug research, diagnostics, imaging, and therapeutics, which play important roles in both advancements and shortcomings of the fight against cancer disease (Habanjar et al., 2021). As a result, cancer-on-a-chip technology is an invaluable tool for cancer research, diagnostics, and therapeutic development.

7.2.1 Principles and Working Mechanisms of Organ-on-a-Chip Technology

Organ-on-a-chip technology integrates microfabricated cell culture platforms that mimic human organ functions by modulating multitissue organization, biochemical cues, vascular perfusion, cellular crosstalk, and mechanical forces (Roye et al., 2021). These systems aim to replicate tissue interactions within the body by delivering nutrients and reagents precisely through microfluidic channels (Ribeiro et al., 2021). By combining microsystem technology, cell biology, and microfluidics, organ-on-a-chip models create in vitro representations of organ functionality (Busek et al., 2022). The technology allows for the strategic integration of multiple organ compartments to simulate the human body, aiding in predicting drug pharmacokinetics (Takebe et al., 2017). Through precise control of spatiotemporal factors, organ-on-a-chip systems expedite the generation of patient-specific organoids, enhancing their potential for personalized biomedical applications (Zheng et al., 2021). Therefore, organ-on-a-chip technology significantly enhances the ability to mimic human organ functions and predict drug responses.

Microphysiological systems are a significant part of organ-on-a-chip technology, blending human biology, biophysics, electronics, and other disciplines (Kang, 2022). While the contributing disciplines provide theoretical knowledge, ideas, techniques, etc., each applies a mix of general principles and becomes familiar with the specific civilization comprising the user and the application (Leung et al., 2022). Cancer-on-a-chip technology moving into the IsDB-MCS can achieve three key objectives: finding cancer, identifying its location, and treating it (Wood et al., 2022). The MSC and MCS work well together due to membrane-containing nano-carrier entities (PNC) that can help detect cancer on the chip by simultaneously attaching to many molecularly different cancer cells of the same type (Akgönüllü et al., 2021). Both MSC and MCS possess similar multiple capabilities, both structural and functional, and the ability to adjust their working features, such as mechanical, electrical, electromagnetic, chemical, optical, and magnetic properties on-the-fly (Wu et al., 2020). Hence, the integration of microphysiological systems in cancer-on-a-chip technology offers versatile and adaptive solutions for cancer detection and treatment.

7.2.2 *Advantages and Limitations of Organ-on-a-Chip Technology*

The widespread adoption of organ-on-a-chip technology in biomedical research and clinical applications faces several limitations and challenges. These include the use of primary cells derived from normal tissue to form systems more representative of in vivo organ systems (Wang et al., 2013), the need for convincing data demonstrating the advantages of human organ chips over animal models to gain acceptance in laboratories (Ingber, 2022), the complexity of 3D culture configurations as the technology develops further (Arık et al., 2018), and the challenges regarding throughput, assay multiplexing, and experimental complexity that need to be addressed for organ-on-a-chip models to become a routine methodology (Shoemaker et al., 2020). Additionally, the lack of systematic academic literature on the economic impact of applying organs-on-chips in pharmaceutical settings poses a challenge (van den Berg et al., 2019). When added to small planar bilayer lipid bilayers (BLM), ionic current or fluorimetry techniques have been used to get real-time data from low-density ion channels (Walewska et al., 2022). However, channel lifetime and deterministic behavior remain as major challenges (Salariseddigh et al., 2023). Patch-by-patch BLMs sometimes offer the solution, but if lifetime is a problem for a small membrane, the problem will be amplified when increasing the number of BLMs is required (Sugihara, 2018).

Recently, novel chip-scale pressure-assisted methods demonstrated successful BLM formation in seconds rather than minutes. Such an improved model should allow more affordable drug screening and a platform to reveal the complex action of certain drugs targeting these channels (Kiriiri et al., 2020). These more favorable conditions offer the opportunity to perform high-quality research of a class that will be useful to render such a biological system more useful for the development of AI models (Tajeddin & Mustafaoglu, 2021). Compared with related technologies, C-OC has several advantages. As an in vitro model, C-OC technology can provide real-time and dynamic response data (Sequeira et al., 2020). It has an elaborate microenvironment and can simulate the characteristics of cells and tissues (Bojin et al., 2021). Biological tissues differ from cells cultured in a monolayer on a plate, and 3D cells can be tested simultaneously, which is closer to the physiological state (Habanjar et al., 2021). This model has the advantage of being low-cost and conducive to high-throughput testing (Rienzo et al., 2021). It can simulate physiological and chemical stimulation conditions and is conducive to immune cell research (Miller et al., 2020). However, C-OC also has certain limitations. When adding biological samples or injecting reagents, the C-OC's microstructure is relatively brittle and easily deformed or broken, making it challenging to transport and preserve (Nazir et al., 2021). The external structure and pipeline system are imperfect, and integration is difficult (Li et al., 2021). The current C-OC can only simulate one or two organ models, and the nutrient supply is not ideal (Leung et al., 2022). Constructing a multi-organ microfluidic chip is also difficult and time-consuming, requiring a high degree of cross-disciplinary and professional knowledge from the workers (Zhou et al., 2024).

7.3 Technology Transfer Process of Organ-on-a-Chip Technology

Key factors influencing the successful technology transfer of organ-on-a-chip innovations from academic research to commercial and clinical applications include the ability of these technologies to mimic in vivo physiology and enhance drug discovery research (Huh et al., 2012). The convergence of microengineering with microfluidics and cell biology has led to the development of organs-on-chips that replicate organ functionality, providing a platform for drug screening and preclinical testing (Hess & Siegwart, 2012). Successful transfer is facilitated by cooperative approaches between academia and industry partners, improving technology assessment and potential (Middelkamp et al., 2016). Involving stakeholders early on in health technology assessment is crucial for the development and application of organs-on-chips in drug development (Ongaro et al., 2020). Additionally, the choice of materials, such as polylactic acid, impacts the performance and sustainability of organ-on-a-chip platforms. Key factors influencing the successful technology transfer of organ-on-a-chip innovations from academic research to commercial and clinical applications include the ability of these technologies to mimic in vivo physiology and enhance drug discovery research (Huh et al., 2012). The convergence of microengineering with microfluidics and cell biology has led to the development of organs-on-chips that replicate organ functionality, providing a platform for drug screening and preclinical testing (Hess & Siegwart, 2012). Successful transfer is facilitated by cooperative approaches between academia and industry partners, improving technology assessment and potential (Middelkamp et al., 2016). Involving stakeholders early on in health technology assessment is crucial for the development and application of organs-on-chips in drug development (Ongaro et al., 2020). Additionally, the choice of materials, such as polylactic acid, impacts the performance and sustainability of organ-on-a-chip platforms.

However, it is regrettable that a significant gap exists between global technology and products (Wang & Feng, 2021). The costs of research, development, and commercialization of the Chinese IOCHIPs are not within the reach of most patient and research groups (Azizipour et al., 2020). The Arab world has a large cancer population and can provide many demands markets but has very limited scientific achievements and model platforms in the field due to blockade and sanction constraints (Roemer, 2021). In addition, the chip costs and applications urgently need human resources and institutional construction services (Miller, 2022). How to effectively and quickly inject China's high-level cancer-on-a-chip technology achievements into the Arab market and promote clinical application to meet Arab demand, not to let high-end foreign products enter the Chinese market unchecked, has become the problem to be solved (Wang et al., 2023). In this chapter, the method of science and technology commercialization was presented in detail, including the main idea of cancer-on-a-chip technology and the model innovation system of the Belt and Road Inflammation & Cancer Multifunctional Chip Consortium (BRIC-MCS) (Geyer, 2023). Material methods are divided into hardware-equipment-reagent-channel-gene-co-culture model setup, multicellular-tumor type-on-chip simulation tests, sensitivity and specificity comparison tests, sample collection, analysis services, etc. (Ahmed, 2022). Software methods are related to consortium-establishment-connection-construction-answers. Special answers include consortium open, government-executive, and pharmaceutical support (National Academies of Sciences, Engineering, and Medicine, 2023). PREM (RPRAM Refugees, Poor people, Elderly people, Pregnant woman, Minors)'s funds, medical insurance-aid coupons, channel-group-share. These methods establish the technical advantage of a new business model and give advisory suggestions (Ranta et al., 2021). In the late half of the twentieth century, most of the leading discoveries in the field were made by cancer researchers themselves (Sonnenschein & Soto, 2020). From the beginning of the twenty-first century, cancer research quietly began to be outsourced to physical engineering and other groups of people (da Silva, 2024). The reason for finding outsiders is that regardless of the time or reason, they have always adhered to the type of social action of refusing rebirth, staying away from society, and not conforming to the conventional rules of social games (Isosävi, 2020). These groups of people use basic theories and technologies in the fields of microfluidics, biosystems, semiconductors, and lithography, as vanguards of the reformation of cancer research (Wu et al., 2021). After a century of exploration, people worldwide have deeply grasped the functions and application prospects of cell- and tumor-on-a-chip (Ruiz-Espigares et al., 2021). Oncology has been recognized as an important strategic field for chip technology to provide S&T guarantees for personalized medicine, new targeted drugs, and cellular intelligence systems (Yang et al., 2022).

7.3.1 *Identification of Key Stakeholders of Organ-on-a-Chip Technology*

Key stakeholders of organ-on-a-chip technology include pharmaceutical companies partnering with academic institutions (Esch et al., 2015), regulatory bodies like the FDA (Marx et al., 2020), organizations focusing on rare diseases such as AstraZeneca (Blumenrath et al., 2020), and societies like EUROoCS overseeing technology implementation (Mastrangeli & van den Eijnden-van Raaij, 2021). Academic and industrial stakeholders worldwide are increasingly adopting organ-on-a-chip models (Shoemaker et al., 2020). The technology offers a promising approach for early validation of drug efficacy and toxicity detection (Middelkamp et al., 2016). Stakeholder identification is crucial for the success of organ-on-a-chip technology, involving various entities from pharmaceuticals to regulatory bodies and academic societies (Kumar et al., 2016).

7.3.2 *Intellectual Property Considerations of Organ-on-a-Chip Technology*

Key intellectual property considerations and challenges in the development and commercialization of Organ-on-a-Chip technology include the need to address patenting strategies for these innovative systems (Sung, 2022). Organ-on-a-chips, also known as tissue chips or microphysiological systems (MPS), offer advantages over conventional cell-based in vitro models (Peck et al., 2019). Challenges also arise in integrating multiple organs on a single platform, ensuring accurate biosensing technologies, and addressing the complexity of multi-organ interactions (Sunildutt et al., 2023). To date, Organ-on-a-Chips technology has mostly been developed to improve the predictability of drug discovery and preclinical development, to provide experimental data for the development of improved in silico models, and to support the reduction, refinement, and replacement of animal experimentation (Peck et al., 2019). Furthermore, the commercialization of Organ-on-a-Chip technology requires alignment with 3D printing and faces challenges related to stakeholder involvement and early health technology assessment (Zheng et al., 2021). Overall, the development and commercialization of Organ-on-a-Chip technology involves navigating complex intellectual property landscapes while addressing technological and regulatory challenges to bring these innovative systems to market successfully.

7.3.3 *Regulatory and Compliance Aspects*

Critical regulatory and compliance challenges facing the development, approval, and adoption of OoC technology in biomedical research and clinical applications include the need for convincing data to demonstrate the advantages over traditional models (Ingber, 2022), optimized storage protocols for broad use outside specialized labs (Gröger et al., 2017), guidelines from regulatory agencies for technology validation (Singh et al., 2022), incorporation of OoC into regulatory processes by agencies like the FDA (Thakar & Fenton, 2023), and the development of novel models reflecting human physiology for safety assessment (Heringa, 2019). Additionally, challenges involve the validation of in vitro findings for nanoparticle research (Kang & Park, 2021), the need for physiologically relevant connections between organ systems (Veliz et al., 2023), and the integration of different organ compartments to mimic human physiology (Lee & Sung, 2013).

7.4 Case Studies in Organ-on-a-Chip Technology Transfer

The successful transfer and commercialization of organ-on-a-chip technologies from academic research to industrial applications are influenced by several key factors. These include addressing challenges related to throughput, assay multiplexing, and experimental complexity to ensure wider acceptance by academic and industrial stakeholders (Shoemaker et al., 2020). Developing new on-chip functions and reducing device costs and material requirements also present opportunities for innovation and interdisciplinary collaboration (Chan et al., 2013). Pharmaceutical companies are increasingly partnering with academic institutions to explore and leverage organ-on-a-chip technology for drug discovery (Esch et al., 2015). Furthermore, integrating multiple organs-on-a-chip systems funded by governmental agencies is crucial for advancing drug development and toxicity studies (Wikswo et al., 2013).

This technology transfer oversaw the new mid-2019 established not-for-profit joint US (Harvard Bioengineered Kidney Microscale Collaborative) and Denmark (Herlev-Gentofte Hospital) initial unique effort aimed at cancer-on-a-chip in the human immune system-on-a-chip (Canton, 2021). The infusion of live, non-bone-marrow-damaging macrophages was implemented using a similar approach used by the Israël Weizmann Institute to fully animalize mice's human immune system immune cells, such as monocyte-derived cells, by allogeneically transplanting and abbreviating freshly picked human monocytes or hematopoietic stem cells (HS transcripts) in subsequent weeks (Biram et al., 2022). Humanized mice were used in cancer vaccine immune system engagement studies (Guil-Luna et al., 2021). Allogeneic monocyte-derived cells' selectivity and effective migration into infrared (IR) illuminated solid tumors that were transcutaneously scrutinized to be present were confirmed by in vivo fluorescent yellow/green tracking (Toffoli et al., 2023). Up to 5 weeks after the end of the ultra-microscope EVOS on boarded manipulations (Bhan et al., 2022), single cycles of immune cells were sacrificed for immune organs by above skin macrophages' patch getImage temporarily keeps kept essential desired natural-looking immune cells' clinical shape-image as a safer, less-demanding alternative to stuck human lymphocyte samples' sample buffer supernatant (Bērziņa et al., 2021). Immune cells should be acquired antigen-based in vivo dual staining after instant isolation to identify immune-specific surface markers as easily.

7.4.1 US-Denmark Cancer-on-a-Chip Technology Transfer Case

Collaborative efforts between the USA and Denmark significantly impact the technology transfer and commercialization of cancer-on-a-chip innovations. Governments' initiatives to promote collaborative innovation, universities' roles in technology transfer, and firms' shift toward collaborative research and development

(R&D) partnerships are crucial (Bertello et al., 2023). Engaging in R&D collaborations and adopting IP sharing policies influence technology transfer performance (Hemmatian et al., 2022). Establishing formalized relationships between anemic committees and technology transfer offices enhances the readiness of innovations for commercialization (Rose et al., 2017). Collaborations between firms and universities are essential to effective national innovation systems (Huang et al., 2021). The cooperation between industry and academia facilitates continuous technological advancements and bidirectional knowledge transfer (Lee & Sung, 2013). Effective collaboration among scientific institutions, R&D centers, and technology-oriented enterprises is vital for developing technological entrepreneurship (Badzińska, 2021).

7.4.2 Successful Examples

Successful examples of cancer-on-a-chip technology transfer from research institutions to commercial applications involve interdisciplinary collaboration in designing and optimizing these systems for cancer research and drug discovery (Tsai et al., 2017). Integrating solid tumor tissue formation techniques with microvasculature-on-a-chip technologies has advanced preclinical drug screening with pathophysiological relevance (Lim et al., 2021). Novel microfabrication technologies, precise instrumentation, single-cell omics, and systems biology approaches enhance tumor-on-a-chip systems for cancer research (Santiago et al., 2019). By mimicking human cancer responses, cancer-on-a-chip technology aims to surpass the limitations of animal testing (Komen et al., 2021). Leveraging advances in organ-on-a-chip and cancer-on-a-chip technology can help bridge the gap between in vitro and in vivo studies (Komen et al., 2021). These strategies align intending to translate research advancements into practical applications in cancer treatment and drug development.

A consortium aiming to advance common approaches and technologies in healthcare platforms should integrate microfluidic technology for continuous monitoring (Sequeira-Antunes & Ferreira, 2023). This technology, with its applications in point-of-care diagnostics and biological monitoring, aligns with the need for true compound platform assessment (Sequeira-Antunes & Ferreira, 2023). Additionally, the use of microfluidics in biosensors and energy harvesting technologies showcases its versatility in healthcare (Luka et al., 2015; Mahmud et al., 2022). Leveraging telecommunication platforms for data monitoring and analysis, as seen in the integration of microfluidics with mobile phones, can revolutionize medical research toward a knowledge-driven approach (Sequeira-Antunes & Ferreira, 2023). The advantages and limitations of microsystems research projects, such as those at Eureka Samart Systems, will shed light on the practical implications of these technologies in healthcare (Juang & Chiu, 2022).

The outcomes of the projects in the field of microfabricated human organ systems of MPS Consortium look very promising and could be a good foundation for the creation of a more complicated platform (Taute et al., 2021). The conclusions of all projects confirm that microfabricated human organ systems (or the so-called organ-on-a-chips) mimic the complex functions of human organs (Kurth et al., 2020). Not only the expected cell type-specific functions were reached, but also the induction and modeling of pathology states which are suitable for therapy and drug development (Qu et al., 2023). The results from the WP02 show the excellent possibilities to monitor standard ChimPAbs-based applications to profile tissue-organ function, including inflammation. The different platforms developed within the consortium-cross projects (lung, liver, blood–brain-barrier, healthy aging with heart-kidney-urethra functions) could be a basis for a corrugated portal for personalized tissue models' implementation (Kunkle, 2021). The question is about special financial translational procedures for successful products of syncretic goals.

7.4.3 Challenges Faced and Lessons Learned During Organ-on-a-Chip Technology Transfers

The challenges encountered during the transfer of organ-on-a-chip technologies from research to commercialization include the inability of a single Organ-Chip to fully model all aspects of an organ's function, necessitating modifications to address specific research questions (Ingber, 2020). Additionally, challenges arise in transferring thousands of micro LEDs and bonding them with their drivers due to technical obstacles related to transfer rate, alignment accuracy, bonding strength, and reliability (Chen et al., 2021). Lessons learned from these challenges emphasize the importance of adapting Organ-Chip systems based on research needs and focusing on technological innovations to enhance platform capabilities (Chen et al., 2021; Ingber, 2020).

Private partners can enhance scientific research through public–private partnerships (PPPs) by engaging in joint innovation projects and supporting the innovation process (Holovko & Sakevych, 2020). These partnerships can provide a framework for better project scientific proof and protection of private investments during implementation (Zapivakhin et al., 2018). By securing follow-on investments and fostering international cooperation, private partners can contribute to the sustainability and success of grandiose scientific ideas (Bosanac, 2022). PPPs offer a model that mitigates weaknesses associated with profit-maximizing motives, allowing for long-term cooperation between public and private sectors (Trucco & Petrenj, 2017). The involvement of private businesses in PPPs can be incentivized by a combination of project financing methods and a shared risk-reward structure (Snyder et al., 2014). Through PPPs, private partners can play a crucial role in advancing scientific endeavors while ensuring mutual benefits and long-term collaboration.

The challenges faced in cancer research projects are multi-faceted, encompassing market-related obstacles such as securing sufficient funding for long and costly development processes. The specificity of cancer research projects, with high entry barriers and intense competition, necessitates substantial financial support for success (Vogel et al., 2019). Public investments play a crucial role in enhancing the translational strategy and increasing the likelihood of commercial success in this highly competitive market (Neta et al., 2015). Ensuring a steady flow of funds is essential for project execution, with team quality and capability being vital for effective research project conduct within specified constraints (Tuffaha et al., 2018). Discrepancies in research funding allocation highlight the need for evidence-based strategies to maximize the impact of research investments on cancer control (Coronado et al., 2018).

7.5 Potential Impact on Healthcare

Cancer comes second in causes of mortality after cardiovascular diseases, and together with other non-communicable diseases, it increasingly supersedes infectious diseases, injuries, and accidents as a main target of healthcare efforts (Bray, 2023). This is particularly true in some of the IsDB's countries with comparatively high economic and scientific potential (Ali & Ali, 2023). Therefore, the implementation of the anti-cancer local treatment in the form of implantable drug depots for localized anti-cancer chemotherapy in suitable patients represents not only a necessity but also an attractive business opportunity (Wang et al., 2021). Despite some evident challenges and risks connected with the project, considering its phase and the limited therapeutic alternatives, it would be more attractive than many other health and science-related investment proposals (Ramadan & Zourob, 2020). Furthermore, the potential for capitalizing on the project's success by commercializing the technology and therapeutic products also appeals to global and regional stakeholders and investors (Bettanti et al., 2021).

7.5.1 Improving Drug Development and Personalized Medicine Using Organ-on-a-Chip Technology

Organ-on-chip technology offers significant advancements in drug development and personalized medicine by providing insights into drug action mechanisms, toxicity, and efficacy in preclinical and clinical settings. These platforms enable the exploration of new therapeutic options for rare diseases and facilitate personalized medicine through models that represent patient diversity (Carvalho et al., 2023). By mimicking organ structures and interactions, organ-on-chip systems enhance the predictability of drug discovery, support the development of in silico models, and reduce the reliance on animal experimentation (Komen et al., 2021). Additionally, these technologies allow for the integration of multiple organ compartments to simulate the human body, aiding in predicting drug pharmacokinetics and improving drug success rates (Bulboacă et al., 2020). Overall, organ-on-chip platforms hold promise in enhancing drug efficacy, safety, and cost-effectiveness in personalized medicine and drug development (Duzagac et al., 2021). In drug development and personalized medicine, microfluidic platforms mimic the tumor microenvironment manifestations, providing a more predictive assessment of the efficiency of anti-cancer drugs and personalized medicine (Ayuso et al., 2022). Microfluidic systems allow precise control over combination drug administration and measurement of the concentrations within and around tumor cells in response to drug transport, metabolism, or other cell processes (Ayuso et al., 2022). Additionally, sophisticated tests can be performed on cancer-on-a-chip platforms to probe how cancer cells respond to certain types of drugs and to seek predictive correlations for personalized chemotherapy treatment for the most effective agents or regions within a pharmacokinetic and pharmacodynamic parameter space (Liu et al., 2021). Furthermore, the capability to generate reliable in vitro parameters, e.g., concentration threshold determination, would improve drug clearance definitions (Wang et al., 2021).

7.5.2 Enhancing Cancer Research and Therapeutic Approaches

In this context, reducing the negative societal and economic costs of cancer and reaching sustainable development justifies the mutual efforts to make these advanced technologies accessible and attractive for the Muslim and the rest of the developing societies (Driver & Mishra, 2023). Proposing several programs and activating sustainable partnerships with the IsDB, together with the academic, clinical, industrial, and socio-economic sectors of each MC (Loewenson et al., 2021), we aim that their citizens can soon benefit from an affordable and personalized cancer-on-a-chip platform adapted to their pathological/disease problems and thus have access to advance in vitro cancer research models as well as powerful high-throughput therapeutic screening systems (Imparato et al., 2022). Establishing the basis for the development of tailored cancer-on-a-chip plat and to enable the testing and validation of advanced therapeutic approaches in the countries and environments of growth economic and technological development should not be limited to the pioneers but increased and involved in the benefit of the underprivileged populations (Rodrigues et al., 2020). Cancer has become one of the leading causes of death worldwide, notably in Islamic and developing countries where access to diagnosis, treatment, care, and palliative care remains insufficient (Abdullah et al., 2020).

The cancer-on-a-chip technology, presents a groundbreaking opportunity for enhancing healthcare services and national development in IsDB Member Countries. This innovative platform, spearheading the commercialization of cancer-on-a-chip technology, holds immense promise for revolutionizing the global health industry. By leveraging microfluidics and biofabrication techniques, such as those discussed in the references on circulating tumor cells isolation (Lemaire et al., 2018), this technology enables early cancer diagnosis, monitoring of metastasis, and molecular biomarker analyses. Furthermore, the integration of organs-on-chips in drug discovery, as highlighted in (Tao et al., 2021), underscores the transformative potential of this technology in precision oncology and personalized medicine, ultimately improving the lives and well-being of individuals worldwide.

7.6 Economic and Social Benefits of Organ-on-a-Chip Technology

Organ-on-chip technology has the potential to significantly impact healthcare by transforming drug discovery, development, and personalized medicine. These chips simulate human body systems, improving drug efficacy testing, toxicity assessment, and disease modeling, which could lead to higher drug success rates and reduced costs associated with failed trials (Ewart et al., 2021). Major pharmaceutical companies are investing in this technology to advance drug development and form partnerships with academic institutions, positioning themselves at the forefront of innovation (Blumenrath et al., 2020). The integration of multiple organs-on-a-chip enables personalized medicine applications, enhancing predictions of drug safety and efficacy (Low et al., 2020). Overall, organ-on-chip technology shows promise in advancing healthcare, reducing medical costs, and improving patient outcomes.

The economic effect of the organ-on-a-chip project will be as follows: replacement of experts with a specialized device leading to cost savings (Balafoutis et al., 2020). This device can also be used in the pharmacy market to test the efficacy and safety of new drugs (Ma et al., 2021; Zhu, 2020). From a strategic point of view, organ-on-a-chip technology has the potential to revolutionize drug testing and personalized medicine (Ingber, 2022). This organ-on-a-chip technology replicates the functions of human organs on a microchip, allowing for more accurate and efficient testing of drugs and their effects on the human body (Ingber, 2022). Additionally, organ-on-a-chip technology has the potential to reduce the need for animal testing in the pharmaceutical industry project's products will enable the existing infrastructure to enter a new niche capable of implementing research outcomes (van Berlo et al., 2021). The results of the organ-on-a-chip technology project can be applied in the further development of biosensors and biochips, leading to the cannabis industry modernization (Bezelya et al., 2023). The organ-on-a-chip technology project work demonstrating the available successful results should be used to attract additional financial resources (Li et al., 2023). An agreement with investors may be formalized. They will be offered the project feasibility study and the initiative products for the project commercialization (Battaglia et al., 2021). The cost approach will be applied to the prospect projects. Materials and technologies will be protected by patents. There are also other sources of financing the organ-on-a-chip technology project in agriculture and bioindustry, including the Chinese market (Taute et al., 2021). Finally, the synergy between these industries will allow future profits from the organ-on-a-chip technology project work to outweigh the costs (Taute et al., 2021).

The social significance and effect of organ-on-a-chip technology project work are as follows: one in three peoples suffers from cancer and the scale of this problem is constantly growing. Every 5 s a person dies of malignant neoplasm (Saini et al., 2020). At least 70% of clinical cases are diagnosed with cancer at the final stage, when effective treatment is very difficult (Crosby et al., 2022). Today, cancer diagnostic accuracy, non-invasiveness, high speed, and accuracy are insufficient requirements for widespread use of methods currently in existence (Aggarwal, 2023). The development of a device incorporating many, if not all, of these features, will be of great value in point-of-care testing (Shrivastava et al., 2020).

From an economic point of view, the organ-on-a-chip technology transfer will be of much value. Currently, the total figure of expenses on malignant tumor management can reach as high as $1.3 trillion (Lai, 2022). This amount is determined by a sum of expenses on primary health care, oncologic care, specific screening regimen, and total therapy. Even high-income countries spend 70% of this sum on oncologic care (Aggarwal et al., 2022). Nevertheless, each country expends about 1.5% of its economic costs and in some cases these figures reach 5% (Pramesh et al., 2022). Usually, both high- and low-income countries have significant expenses for all elements of this sum (Edejer et al., 2020). These points validate our intentions to work on the project.

7.6.1 Economic Growth and Job Creation Using Organ-on-Chip Technology

The implementation of organ-on-chip technology can stimulate economic growth and job creation in member countries of the Islamic Development Bank by revolutionizing drug development, enhancing manufacturing processes, and fostering innovation. Organ-on-chip devices offer early validation of drug efficacy and toxicity testing (Middelkamp et al., 2016), while advancements in mass production methods for these chips improve design and manufacturability (Lantada et al., 2018). Additionally, the integration benefits of monolithic 3D labs-on-chips enhance efficiency and throughput (Lantada et al., 2018). Such technological progress can lead to economic growth by promoting investment in high-tech sectors (Ahmad & Mamun, 2020), fostering entrepreneurship through technology incubation centers (Akhuemonkhan et al., 2014), and creating job opportunities in industries utilizing these cutting-edge technologies (Liu et al., 2019). Islamic banking systems can further support this growth (Caporale & Helmi, 2016), emphasizing the potential synergy between technological innovation and financial mechanisms in driving economic development.

In this connection, IsDB-MCS have a significant potential to use cross-hatching facilitation between OIC researchers and their foreign counterparts to secure access to technology design drivers, whether for general or specific purposes reinforcing their economies and addressing societal issues, especially the area of healthcare, among which cancer is challenging developed and developing ones (Elzahi Saaid Ali, 2022). The reason for this gravity is obvious when comparing the R&D and production capabilities of cancer diagnostics and therapy facilities in Islamic countries with the leading power in this field (Alshehri et al., 2020). However, the current situation has a serious drawback, as the gridlock of access to technology vital for reinforcing production in oncology leads to disproportionate expenditure returns related to the ensuing sustained survival of the relatively sufficient individuals belonging to developed anamnestical population in Islamic countries (Abdul-Sater et al., 2020). Key drivers for reaching the goal of self-reliance in this field could be access to an arsenal of innovative technological and industrial bases, as well as the biomedical research community's interest in utilitarian research aimed at the provision of technology-transfer-relevant R&D, such as ethnopharmaceutical cancer treatments, and eradication of lung cancer and breast cancer, which is critical together with prostate cancer for social and political reasons (Bonvillian, 2021). In this connection, a public–private partnership focusing on aspects of industrial relevance to IsDB-MCS healthcare production through technology apps and transfer procedures designed to achieve the machine-dreamed end is surely welcome (Ali, 2022).

Islamic countries need to create new jobs for their fast-growing, young population to sustain future growth. In addition, to deliver these jobs, they need to grow and diversify their economic base to accommodate the rise of an educated and skilled workforce, instead of public-sector jobs or low-skilled service-driven employment (Mohamed et al., 2021). It is essential to diversify Islamic countries' economies so that they gradually reduce their reliance on revenues from natural resources, especially in oil and gas, which, besides being finite, do not generate sufficient employment, particularly for educated and skilled youngsters (Muhammad et al., 2021). Islamic countries are lagging considerably behind advanced or even emerging-market countries in terms of productivity, and rising public-sector employment makes it harder for the economies to raise living standards (İlhan, 2020). In support of job creation, Islamic countries could increase their ability to attract foreign investment and transfer of technology of affected sectors, as well as enhance their efficient use of available resources, including via technology-driven plans (Mahbub et al., 2022).

7.6.2 Access to Advanced Healthcare Technologies in Developing Countries

Organ-on-chip technology holds promise in advancing healthcare in developing countries by providing a platform for studying human physiology, disease mechanisms, and drug responses, ultimately leading to improved healthcare outcomes (Dasgupta et al., 2024). This technology has been utilized to enhance predictions of drug safety and efficacy, which can be particularly beneficial in addressing rare diseases (Blumenrath et al., 2020). By enabling early validation of drug efficacy and toxicity detection, organ-on-chip models serve as valuable tools in drug development (Middelkamp et al., 2016). These models, such as liver-on-a-chip, offer high-fidelity representations of organ-specific physiology and pathology (Kim et al., 2023). Regulatory guidelines are essential to validate and standardize the use of organ-on-chip technology for various applications, including disease modeling (Singh et al., 2022). Overall, the integration of organ-on-chip technology into healthcare systems in developing countries can significantly contribute to advancing healthcare accessibility and outcomes.

Overall, the development of the neuropharmacological research laboratory in Qatar with UMR-magnetic resonance imaging and genetically engineered animal study tools created by experts from the M.M. Shemyakin and Yu. A. Ovchinnikov Institute of Bioorganic Chemistry is of particular interest in the context of a comprehensive approach to the implementation of genome editing techniques for understanding the factors contributing to the development of modern thought processes of mankind and intellectual abilities from such mental diseases as Asperger's Syndrome. Islamic banking funds today remain the main source of scientific research and development, independently or through private equity funds, and not only in developing countries (Wani et al., 2021). In the global market, sovereign funds play a declaratory role of a long-term investor who finds it attractive to co-invest or change the current vector of scientific research and development based on public–private partnership and objective scientific information due to the relatively low market value of intellectual rights in the production of state-of-the-art medical technologies compared to the infrastructure for new materials (Sejko, 2023).

Nowadays, many LLMCs have a great need for high-technology equipment, but due to budget constraints, experts predict a considerable challenge to affording those devices (Hinrichs-Krapels et al., 2022). Solutions to conducting cancer research and developing treatment programs are also crucial for the Zanzibar archipelago (Jørgensen et al., 2022). This research is in parallel with the development of cooperation within the Arab Republic of Egypt Study and Research Laboratory of Information and Control Complex Systems and Qatar Computing Research Institute (part of Hamad Bin Khalifa University) (Abdel-Latif, 2023). Qatar, having an Islamic banking interest in transferring knowledge from experimental studies to practical implementation, got an opportunity to use the so-called convertible medical equipment to equip them (Waqas et al., 2021). This complex system can be used to study new anti-cancer drugs and personalized medicine on 3D cancer models with a high simulation of the tumor environment based on tissue engineering as a cancer-on-a-chip technology created to model human cancer in laboratory conditions that can mimic cancer, such as its microenvironment, made with multinozzles 3D-bioprinting (Zhang et al., 2022).

7.7 Collaborative Opportunities with IsDB-MCS

The Cancer-on-a-Chip project aims to leverage micro-nano-technology platforms for cancer research in IsDB member countries and China, addressing the unmet need for advanced diagnostic technologies in developing nations. Collaborative efforts with IsDB-MCS can lead to the development of innovative devices integrating nano-features and advanced biointerfaces, enhancing cancer diagnostics and treatment efficacy (Li et al., 2017). This initiative aligns with the trend toward miniaturization in fabrication technologies, crucial for applications like microdevices and nano sensors (Xie et al., 2021). By utilizing micro/nano-based devices for cancer monitoring and diagnostics, the project can tap into the potential of circulating biomarkers in cancer detection and treatment (Ko et al., 2016). Such advancements in micro/nanotechnologies hold promise for revolutionizing cancer research and healthcare delivery globally.

The Cancer-on-Chip project, funded by the Islamic Development Bank-Merieux Connaught Technology Transfer Center (IsDB-MCS), aims to revolutionize patient-centric oncology by utilizing micro-technology solutions to enhance the Biopsy-Assay-Assessment-Treatment chain through interconnected micro- and nano-systems, known as "Cancer-on-a-Chip" (Mertz et al., 2018). This initiative aligns with the growing trend of patient-centered approaches in oncology, supported by regulatory bodies and research (Farah et al., 2023). Leveraging recent biotechnological advancements, such as ChIP-seq technology (McFarlane et al., 2019), can provide valuable insights into epigenetic mechanisms, transcriptional regulation, and gene expression in cancer (Audia & Campbell, 2016). Integrating patient advocates and incorporating patient experiences in clinical development can further enhance the patient-centric focus of such projects (Lovrić et al., 2022). Cancer-on-chip technology, utilizing lab-on-a-chip models, provides a promising approach to bridge the gap between tumor molecular signatures and clinical data, facilitating the prediction of tumor biology and treatment responses (Fetah et al., 2019). These models accurately replicate the tumor microenvironment, enabling real-time analysis of cell interactions and high-throughput screening (Michna et al., 2018). By offering precise control over the microenvironment, these devices can impact treatment decisions based on individual resistance mechanisms before therapy initiation (Hachey & Hughes, 2018). The advantages of microfluidics, including miniaturization, automation, and rapid results, improve the feasibility of predicting patient responses and customizing treatments accordingly (Xu & Cheng, 2022). This technology has the potential to revolutionize cancer research by providing a more efficient and effective platform for studying tumor progression and treatment responses (Goenka et al., 2023).

7.7.1 Existing Initiatives and Programs

Many positive stories are emerging from the work of these scientists and researchers with special emphasis on joining forces, both internal and external collaborations, and creating international teams (Ramadan & Zourob, 2020). A special workshop in 2018 highlighted some of these special cases and provided reports and updates on the achieved progress presenting publications, patents, start-ups, innovative methods, and devices (Busek et al., 2022). The main MCS area was awarded in biotechnology, followed by 9% in health, 10% in pharmacy, 5% in biophysics, and 32% in chemistry and conduct (Djisalov et al., 2021). Within the context of its IsDB Re-shaping the Future initiative, the IsDB has been actively mobilizing and further developing the Medical and Chemical Sciences (IsDB-MCS) programs to address challenges arising from the current pandemic (Elzahi Saaid Ali, 2022). There were many questions about the success of this initiative. The same set of questions is asked regarding its future, especially after the completion of the current pandemic and the potential return to the previous levels of funding (Malik et al., 2020). In this direction, testimonials from previous award winners might give inspiration and information about the former successes and potential for future expansion (Ma et al., 2020). During these 25 years, around 750 scientists and researchers from the 57 IsDB member countries have received this award. More than 150 million USD in prize money and other contributions have been allocated (Droesse, 2022).

7.7.2 Potential Funding and Support Mechanisms

The IsDB aims to support tumor-on-a-chip technologies through grants allocated to institutions and organizations with the capacity to integrate these technologies effectively (Bruneel et al., 2012). While the potential of tumor-on-a-chip systems for cancer research is evident, challenges remain in translating this technology to clinical practice (Tsai et al., 2017). Organ-on-a-chip advancements, such as personalized models for toxicity prediction, hold promise for more efficient human trials (Esch et al., 2015). Technology business incubators play a crucial role in supporting innovation and entrepreneurship, providing resources and services to nurture technology-based ventures (Li & North, 2017). The integration of 3D bioprinting with microfluidic chips enhances their potential for modeling cancers (Guo et al., 2021). The IsDB offers support for projects through the MCS, financial rewards, and visibility, which can enhance international recognition for research (Fonner et al., 2014). This visibility is crucial for the scientific community and can benefit the OIC, IsDB, and various institutions interested in oncology solutions. By forging alliances and providing advocacy, IsDB contributes to the development of oncology solutions and strengthens international partnerships (Fonner et al., 2014).

7.8 Conclusion and Future Directions

The advance of knowledge and the international network with institutions with advanced technology transfer systems would lower the technology transfer risks for the IsDB for partnerships with member countries, one of which would be the promotion of international research projects aimed at solving these issues with an aim to raise the already existing number of identified validated de-risked PoCs ready for commercialization (Scuotto et al., 2020). Funding of international workshops devoted to this topic could further stimulate an international network to create connections among potential commercial development investors in academia and start-up companies (Ramadan & Zourob, 2020). The participation and the generation of the data that would be part of the technology transfer potential readiness should be stimulated among the Islamic Development Bank member countries by funding calls for proof-of-concept (PoC) projects from the host university and industry matching grants (Pericoli, 2020). In conclusion, the time window for commercialization in the fields of cancer drug discovery and therapeutics present Ex Horizons 2020 and the European Budget 2018–2019 offers an opportunity for preparatory actions aimed at the technology transfer with the promotion of international research calls to the IsDB (Teixeira & Tavares-Lehmann, 2022). In conclusion, combining microfluidics and three-dimensional tissue culture models has created potential in-vitro systems for anti-cancer drug discovery and therapeutic response (Badr-Eldin et al., 2022). These organ-on-a-chip systems have started to find their way from academic laboratories to small bio-technology companies and are presently confined to the step of development to preclinical tests (Ma et al., 2021). The high-technology readiness level (TRL) of these systems creates an opportunity for the immediate technology transfer into Islamic Development Bank Merck Collaboration Scheme member countries, among which many organs with advanced technology transfer centers exist (Ramadan & Zourob, 2020) The existing potential for cancer drug discovery and therapeutics will create the time window for the preparatory phase of the technology transfer (Rodriguez et al., 2022). These technologies also possess the potential in the fields of veterinary medicine and toxicology, which can increase the market of the technology offered for the transfer and consequently the total return on investment for the TRL increase (Joseph et al., 2022).

7.8.1 Key Takeaways and Recommendations

To address technology knowledge transfer barriers in advanced cancer care, a multidisciplinary approach involving academic healthcare institutions, microfabrication technology facilities, and multisector networking is crucial (Nguyen et al., 2023). Collaboration and innovation are needed to overcome technical, clinical, and commercial obstacles (Nguyen et al., 2023). The integration of advanced technologies like AI and blockchain in healthcare supply chains requires careful evaluation for long-term viability (Oluwadare et al., 2024). Initiatives focusing on compliance, such as hand hygiene programs, benefit from evaluation, feedback, and monitoring to ensure adherence to guidelines (Jeanes et al., 2019). Strategic healthcare innovation planning should consider factors like technical, investment, clinical, regulatory, and socio-economic risks (Apell & Eriksson, 2021). By fostering collaborative innovation networks and leveraging technical knowledge, healthcare organizations can enhance their innovation capabilities (Najafi-Tavani et al., 2018).

The IsDB can leverage the latest advancements in oncology and microfabrication technologies through a multidisciplinary approach (D'Haene et al., 2015). The commercialization of these technologies has become more accessible, benefiting clinicians and researchers (Ahn & Lee, 2024). The use of technologies like antibody–drug conjugates (ADCs) in oncology has gained prominence, indicating a shift in treatment approaches (Flynn et al., 2016). Nanopharmaceuticals have shown promise in overcoming obstacles in oncology, such as regulatory challenges and development costs (Salama et al., 2020). As technology progresses, the financial position of firms can improve significantly through disruptive technologies (Maxwell & Metz, 2021). The successful implementation of these technologies requires considering various technical, clinical, and economic aspects (Chen et al., 2019).

7.8.2 *Emerging Trends in Technology Transfer and Healthcare Innovation*

Cancer-on-a-chip technology enables the simulation of human cancer on a microplatform in animal-free environments to develop personalized medicines and nutraceuticals (Farooqi et al., 2023). This chapter reveals the potential of cancer-on-a-chip technology to be transferred into the Islamic Development Bank copy-tournament laboratory to journey through the proof-of-concept stage of cancer-on-a-chip into their commercial entities (Baka et al., 2022). Establishing Islamic Development Bank's technology transfer ecosystem could provide the supporting supply chain to enable cancer-on-a-chip technology applications to facilitate drug discovery and development tailored to local preferences (Imparato et al., 2022). Such drug development will enable formulations targeted to the common ailments among the Islamic Development Bank member countries, particularly in the areas of skin, orthopedic, degenerative conditions, and others (Ma et al., 2021). Industry 4.0 influences innovation in healthcare and technology transfer from academia into markets (Pozzi et al., 2023). This section presents how the fastest-growing competitive research projects can take advantage of the disruptive nature of the Fourth Industrial Revolution to engage in the Islamic Development Bank and to apply new Knowledge Copy-Tournament Laboratory for the development of healthcare innovation (Al Madani et al., 2020). The research is focused on projects that utilize the microfluidic model technology with relevance in tomorrow's medicine (Santos Rosalem et al.,

2020). Our findings give a new perspective to research and development projects conducted in academia, industry, and societies. As institutions responsible for promoting science, education, and the transfer and application of technology, universities are particularly challenged to also transform into drivers of the fourth Industrial Revolution (Spöttl & Windelband, 2021).

Acknowledgments The authors thank the Chemical and Process Engineering Technology at the University of Malaysia, Pahang-UMPSA, Malaysia, for supporting this research.

Fund This research was funded by Eastern Unity Technology under Grant No. UIC 190806.

References

Abdel-Latif, A. F. (2023). *The central science and technology organization in Egypt*. Worldwide Science and Technology Advice. [HTML].

Abdullah, R., Guo, P., & Harding, R. (2020). Preferences and experiences of Muslim patients and their families in Muslim-majority countries for end-of-life care: A systematic review and thematic analysis. *Journal of Pain and Symptom Management, 60*, 1223. sciencedirect.com

Abdul-Sater, Z., Kobeissi, E., Menassa, M., Telvizian, T., & Mukherji, D. (2020). Research capacity and training needs for cancer in conflict-affected MENA countries. *Annals of Globalization and Health, 86*(1). nih.gov

Aggarwal, A. K. (2023). Thermal imaging for cancer detection. *Imaging and Radiation Research, 6*. enpress-publisher.com

Aggarwal, A., Lievens, Y., Sullivan, R., & Nolte, E. (2022). What really matters for cancer care–health systems strengthening or technological innovation? *Clinical Oncology, 34*(7), 430–435. clinicaloncologyonline.net

Ahmad, S. M., & Mamun, A. A. (2020). Opportunities of Islamic FinTech: The case of Bangladesh and Turkey. *CenRaPS Journal of Social Sciences, 2*(3), 412–426. https://doi.org/10.46291/cenraps.v2i3.39

Ahmed, T. (2022). Organ-on-a-chip microengineering for bio-mimicking disease models and revolutionizing drug discovery. *Biosensors and Bioelectronics: X*. sciencedirect.com

Ahn, S., & Lee, H. S. (2024). Applicability of spatial technology in cancer research. *Cancer Research and Treatment, 56*(2), 343–356. https://doi.org/10.4143/crt.2023.1302

Akgönüllü, S., Bakhshpour, M., Pişkin, A. K., & Denizli, A. (2021). Microfluidic systems for cancer diagnosis and applications. *Micromachines, 12*. mdpi.com

Akhuemonkhan, I., Raimi, L., Patel, A., & Fadipe, A. O. (2014). Harnessing the potentials of technology incubation centers (TICs) as tools for fast-tracking entrepreneurship development and actualization of the vision 20:2020 in Nigeria. *Humanomics, 30*(4), 349–372. https://doi.org/10.1108/h-11-2013-0069

Akinyemiju, T. (2012). Socio-economic and health access determinants of breast and cervical cancer screening in low-income countries: Analysis of the world health survey. *PLoS One, 7*(11), e48834. https://doi.org/10.1371/journal.pone.0048834

Al Madani, H., Alotaibi, K. O., & Alhammadi, S. (2020). The role of Sukuk in achieving sustainable development: Evidence from the Islamic Development Bank. *Banks and Bank Systems, 15*, 36. academia.edu.

Ali, A. E. S. (2022). *Empowering the poor through financial and social inclusion in Africa*. [HTML].

Ali, J., & Ali, T. (2023). *Circular economy and agriculture: Mapping scientific productivity, research pattern and future research direction* (pp. 1–46). Environment.

Alshehri, S., Imam, S. S., Rizwanullah, M., Akhter, S., Mahdi, W., Kazi, M., & Ahmad, J. (2020). Progress of cancer nanotechnology as diagnostics, therapeutics, and theranostics nanomedicine: Preclinical promise and translational challenges. *Pharmaceutics, 13*(1), 24. mdpi.com

Alshubiri, F., ALmaashani, A., & Thuaar, S. (2022). The impact of the digital economy paradigm on the productivity and monetary system of Oman. *Journal of Science and Technology Policy Management, 14*(5), 830–858. https://doi.org/10.1108/jstpm-07-2021-0097

Apell, P., & Eriksson, H. (2021). Artificial intelligence (AI) healthcare technology innovations: The current state and challenges from a life science industry perspective. *Technology Analysis & Strategic Management, 35*(2), 179–193. https://doi.org/10.1080/09537325.2021.1971188

Arık, Y. B., van der Helm, M. W., Odijk, M., Segerink, L., Passier, R., van den Berg, A., et al. (2018). Barriers-on-chips: Measurement of barrier function of tissues in organs-on-chips. *Biomicrofluidics, 12*(4), 042218. https://doi.org/10.1063/1.5023041

Atsalaki, X., Koumakis, L., Potamias, G., & Tsiknakis, M. (2020). Chip-seq and gene expression data for the identification of functional sub-pathways: A proof-of-concept in lung cancer. https://doi.org/10.1101/2020.06.15.151712

Audia, J. E., & Campbell, R. M. (2016). Histone modifications and cancer. *Cold Spring Harbor Perspectives in Biology, 8*(4), a019521. https://doi.org/10.1101/cshperspect.a019521

Ayuso, J. M., Virumbrales-Muñoz, M., Lang, J. M., & Beebe, D. J. (2022). A role for microfluidic systems in precision medicine. *Nature Communications, 13*(1), 3086. nature.com

Azizipour, N., Avazpour, R., Rosenzweig, D. H., Sawan, M., & Ajji, A. (2020). Evolution of biochip technology: A review from lab-on-a-chip to organ-on-a-chip. *Micromachines, 11*(6), 599. mdpi.com

Badr-Eldin, S. M., Aldawsari, H. M., Kotta, S., Deb, P. K., & Venugopala, K. N. (2022). Three-dimensional in-vitro cell culture models for efficient drug discovery: Progress so far and future prospects. *Pharmaceuticals, 15*(8), 926. mdpi.com

Badzińska, E. (2021). Exploring critical factors for academic start-ups toward the development of technological entrepreneurship: Preliminary research findings. *European Research Studies Journal, XXIV*(Special Issue 5), 30–47. https://doi.org/10.35808/ersj/2702

Baka, Z., Stiefel, M., Figarol, A., Godier, C., Mallick, A., Joubert, O., et al. (2022). Cancer-on-chip technology: Current applications in major cancer types, challenges and future prospects. *Progress in Biomedical Engineering, 4*(3), 032001. hal.science

Balafoutis, A. T., Evert, F. K. V., & Fountas, S. (2020). Smart farming technology trends: Economic and environmental effects, labor impact, and adoption readiness. *Agronomy*. mdpi.com

Banales, J. M., Marin, J. J., Lamarca, A., Rodrigues, P. M., Khan, S. A., Roberts, L. R., et al. (2020). Cholangiocarcinoma 2020: The next horizon in mechanisms and management. *Nature Reviews Gastroenterology & Hepatology, 17*(9), 557–588. nature.com

Battaglia, D., Paolucci, E., & Ughetto, E. (2021). The role of proof-of-concept programs in facilitating the commercialization of research-based inventions. *Research Policy, 50*, 104268. [HTML].

Bertello, A., Bernardi, P. D., & Ricciardi, F. (2023). Open innovation: Status quo and quo vadis—An analysis of a research field. *Review of Managerial Science, 18*(2), 633–683. https://doi.org/10.1007/s11846-023-00655-8

Bērziņa, S., Harrison, A., Taly, V., & Xiao, W. (2021). Technological advances in tumor-on-chip technology: From bench to bedside. *Cancers, 13*(16), 4192. https://doi.org/10.3390/cancers13164192

Bettanti, A., Lanati, A., & Missoni, A. (2021). Biopharmaceutical innovation ecosystems: A stakeholder model and the case of Lombardy. *The Journal of Technology Transfer*, 1–26.

Bezelya, A., Küçüktürkmen, B., & Bozkır, A. (2023). Microfluidic devices for precision nanoparticle production. *Micro*. mdpi.com

Bhakta, N., Liu, Q., Ness, K., Baassiri, M., Eissa, H., Yeo, F., et al. (2017). The cumulative burden of surviving childhood cancer: An initial report from the St Jude Lifetime Cohort Study (SJLIFE). *The Lancet, 390*(10112), 2569–2582. https://doi.org/10.1016/s0140-6736(17)31610-0

Bhan, C., Dipankar, P., Dash, S. P., Chakraborty, P., Dalpati, N., & Sarangi, P. P. (2022). Recent advances in imaging and analysis of cellular dynamics in real-time. In *Biotechnological advances for microbiology, molecular biology, and nanotechnology* (pp. 311–350) [HTML].

Biram, A., Liu, J., Hezroni, H., Davidzohn, N., Schmiedel, D., Khatib-Massalha, E., et al. (2022). Bacterial infection disrupts established germinal center reactions through monocyte recruitment and impaired metabolic adaptation. *Immunity, 55*(3), 442–458. cell.com

Blumenrath, S. H., Lee, B. Y., Low, L. A., Prithviraj, R., & Tagle, D. A. (2020). Tackling rare diseases: Clinical trials on chips. *Experimental Biology and Medicine, 245*(13), 1155–1162. https://doi.org/10.1177/1535370220924743

Bojin, F., Robu, A., Bejenariu, M. I., Ordodi, V., Olteanu, E., Cean, A., et al. (2021). 3D bioprinting of model tissues that mimic the tumor microenvironment. *Micromachines, 12*(5), 535. mdpi.com

Bonvillian, W. B. (2021). *Emerging industrial policy approaches in the United States*. itif.org

Bosanac, M. (2022). Analysis of public: Private partnerships in education. *Univerzitetska Misao - Casopis Za Nauku, Kulturu I Umjetnost, Novi Pazar, 21*, 126–139. https://doi.org/10.5937/univmis2221126b

Bray, E. A. (2023). Supporting the transition of children and young people with spinal cord injury from pediatric to adult healthcare services.

Bruneel, J., Ratinho, T., Clarysse, B., & Groen, A. J. (2012). The evolution of business incubators: Comparing demand and supply of business incubation services across different incubator generations. *Technovation, 32*(2), 110–121. https://doi.org/10.1016/j.technovation.2011.11.003

Bulboacă, A. E., Boarescu, P., Melincovici, C. S., & Mihu, C. M. (2020). Microfluidic endothelium-on-a-chip development, from in-vivo to in-vitro experimental models. *Romanian Journal of Morphology and Embryology, 61*(1), 15–23. https://doi.org/10.47162/rjme.61.1.02

Busek, M., Aizenshtadt, A., Martinez, M., Delon, L., & Krauss, S. (2022). *User-view on organ-on-a-chip technology*. https://doi.org/10.20944/preprints202202.0032.v1

Caballero, D., Reis, R. L., & Kundu, S. C. (2020). Engineering patient-on-a-chip models for personalized cancer medicine. In *Biomaterials and microfluidics-based tissue engineered 3D models* (pp. 43–64) [HTML].

Cànaves Llabrés, A. I. (2022). *Tumor-on-a-chip: New strategies of in-vitro cancer cell culture*. udg.edu.

Canton, H. (2021). European Union—EU. In *The Europa directory of international organizations 2021* (pp. 552–605). Routledge. [HTML].

Caporale, G. M., & Helmi, M. H. (2016). Islamic banking, credit and economic growth: Some empirical evidence. *SSRN Electronic Journal*. https://doi.org/10.2139/ssrn.2714816

Carvalho, M. R., Yan, L. P., Li, B., Zhang, C. H., He, Y. L., Reis, R. L., & Oliveira, J. M. (2023). Gastrointestinal organs and organoids-on-a-chip: Advances and translation into the clinics. *Biofabrication, 15*(4), 042004.

Caytiles, R. D., & Park, S. W. (2014). A study of the design of wireless medical sensor network-based u-healthcare system. *International Journal of Bio-Science and Bio-Technology, 6*(3), 91–96. https://doi.org/10.14257/ijbsbt.2014.6.3.11

Çelik, H., Varışlı, B., Yardım, A., Yıldırım, S., & Kisli, M. (2020). A clinical and cost comparison of cancer and non-cancer patients in palliative care. *Cumhuriyet Medical Journal*. https://doi.org/10.7197/cmj.vi.692645

Chan, C., Huang, P. H., Guo, F., Ding, X., Hudson, P. J., John, D., et al. (2013). Accelerating drug discovery via organs-on-chips. *Lab-on-a-Chip, 13*(24), 4697. https://doi.org/10.1039/c3lc90115g

Chaudhuri, P. K., Warkiani, M. E., Jing, T., Kenry, K., & Lim, C. T. (2016). Microfluidics for research and applications in oncology. *The Analyst, 141*(2), 504–524. https://doi.org/10.1039/c5an00382b

Chen, X., Jung, J., Shajahan-Haq, A., Clarke, R., Shih, I., Wang, Y., et al. (2015). Chip-bit: Bayesian inference of target genes using a novel joint probabilistic model of chip-seq profiles. *Nucleic Acids Research, 44*(7), e65. https://doi.org/10.1093/nar/gkv1491

Chen, Q., Wang, J., Ou, R., Sun, J., & Chang, L. (2019). Disruptive technologies and career transition strategies of middle-skilled workers. *Career Development International, 24*(5), 475–490. https://doi.org/10.1108/cdi-06-2018-0172

Chen, X., Zhang, Y. S., Zhang, X., & Liu, C. (2021). Organ-on-a-chip platforms for accelerating the evaluation of nanomedicine. *Bioactive Materials, 6*(4), 1012–1027.

Cheng, L., Zang, J., Dai, H., Li, F., & Guo, F. (2018). Ubiquitin ligase chip functions as an oncogene and activates the AKT signaling pathway in prostate cancer. *International Journal of Oncology*. https://doi.org/10.3892/ijo.2018.4377

Cong, Y., Han, X., Wang, Y., Chen, Z., Lu, Y., Liu, T., et al. (2020). Drug toxicity evaluation based on organ-on-a-chip technology: A review. *Micromachines, 11*(4), 381. mdpi.com

Coronado, A., Finley, C., Badovinac, K., Han, J., Niu, J., & Rahal, R. (2018). Discrepancies between Canadian cancer research funding and site-specific cancer burden: A spotlight on ten disease sites. *Current Oncology, 25*(5), 338–341. https://doi.org/10.3747/co.25.4230

Corral, J., Castells, X., Molins, E., Chiarello, P., Borras, J., & Cots, F. (2016). Long-term costs of colorectal cancer treatment in Spain. *BMC Health Services Research, 16*(1), 56. https://doi.org/10.1186/s12913-016-1297-6

Crosby, D., Bhatia, S., Brindle, K. M., Coussens, L. M., Dive, C., Emberton, M., et al. (2022). Early detection of cancer. *Science, 375*(6586), eaay9040. ucl.ac.uk.

D'Haene, N., Mercier, M. L., Nève, N. D., Blanchard, O., Delaunoy, M., Housni, H. E., et al. (2015). Clinical validation of targeted next-generation sequencing for colon and lung cancers. *PLoS One, 10*(9), e0138245. https://doi.org/10.1371/journal.pone.0138245

da Silva, R. G. L. (2024). The advancement of artificial intelligence in biomedical research and health innovation: Challenges and opportunities in emerging economies. *Globalization and Health, 20*. springer.com

Dasgupta, I., Rangineni, D., Abdelsaid, H., Ma, Y., & Bhushan, A. (2024). Tiny organs, big impact: How microfluidic organ-on-chip technology is revolutionizing mucosal tissues and vasculature. *Bioengineering, 11*(5), 476. https://doi.org/10.3390/bioengineering11050476

de Fonseca, R. S., & Veloso, A. P. (2018). The practice and future of financing science, technology, and innovation. Foresight and STI Governance, 12(2), 6–22. https://doi.org/10.17323/2500-2597.2018.2.6.22

Djisalov, M., Knežić, T., Podunavac, I., Živojević, K., Radonic, V., Knežević, N. Ž., et al. (2021). Cultivating multidisciplinarity: Manufacturing and sensing challenges in cultured meat production. *Biology, 10*(3), 204. mdpi.com

Doost, N. (2024). A comprehensive review of organ-on-a-chip technology and its applications in disease diagnostics. https://doi.org/10.20944/preprints202403.0965.v1

Driver, R., & Mishra, S. (2023). Organ-on-a-chip technology: An in-depth review of recent advancements and future of whole body-on-chip. *BioChip Journal, 17*, 1. [HTML].

Droesse, G. (2022). *International financial institutions: Paradigms of organizational structures, funding structures and innovative funding modalities*. Funding International Development Organizations. brill.com

Duzagac, F., Saorin, G., Memeo, L., Canzonieri, V., & Rizzolio, F. (2021). Microfluidic organoids-on-a-chip: Quantum leap in cancer research. *Cancers, 13*(4), 737. https://doi.org/10.3390/cancers13040737

Edejer, T. T. T., Hanssen, O., Mirelman, A., Verboom, P., Lolong, G., Watson, O. J., et al. (2020). Projected health care resource needs for an effective response to COVID-19 in 73 low-income and middle-income countries: A modeling study. *The Lancet Globalization and Health, 8*(11), e1372–e1379. thelancet.com

Ekwueme, D., Zhao, J., Rim, S., Moor, J., Zheng, Z., Khushalani, J., et al. (2019). Annual out-of-pocket expenditures and financial hardship among cancer survivors aged 18–64 years—United States, 2011–2016. *MMWR Morbidity and Mortality Weekly Report, 68*(22), 494–499. https://doi.org/10.15585/mmwr.mm6822a2

Elzahi Saaid Ali, A. (2022). The challenges of Islamic trade finance in promoting SMEs in IsDB member countries. In *Empowering the poor through financial and social inclusion in Africa: An Islamic perspective* (pp. 103–117). Springer. [HTML].

Esch, E. W., Bahinski, A., & Huh, D. (2015). Organs-on-chips at the frontiers of drug discovery. *Nature Reviews Drug Discovery, 14*(4), 248–260. https://doi.org/10.1038/nrd4539

Ewart, L., Αποστόλου, A., Briggs, S. A., Carman, C. V., Chaff, J. T., Heng, A. R., et al. (2021). *Qualifying a human liver-chip for predictive toxicology: Performance assessment and economic implications*. https://doi.org/10.1101/2021.12.14.472674

Fadhil, I., Ali, R., Al-Raisi, S., Belaila, B., Galadari, S., Javed, A., et al. (2022). Review of national healthcare systems in the gulf cooperation council countries for non-communicable diseases management. *Oman Medical Journal, 37*(3), e370. https://doi.org/10.5001/omj.2021.96

Farah, E., Kenney, M., Kica, A., Haddad, P., Stewart, D. J., & Bradford, J. (2023). Beyond participation: Evaluating the role of patients in designing oncology clinical trials. *Current Oncology, 30*(9), 8310–8327. https://doi.org/10.3390/curroncol30090603

Farooqi, M. A., Kang, C. U., & Choi, K. H. (2023). Organ-on-chip: Advancing nutraceutical testing for improved health outcomes. *ACS Omega, 8*, 31632. acs.org

Feitor, J., Brazaca, L., Lima, A., Ferreira, V., Kassab, G., Bagnato, V., et al. (2023). Organ-on-a-chip for drug screening: A bright future for sustainability? A critical review. *ACS Biomaterials Science & Engineering, 9*(5), 2220–2234. https://doi.org/10.1021/acsbiomaterials.2c01454

Fetah, K., DiPardo, B. J., Kongadzem, E. L., Tomlinson, J. S., Elzagheid, A., Elmusrati, M., et al. (2019). Cancer modeling-on-a-chip with future artificial intelligence integration. *Small, 15*(50), e1901985. https://doi.org/10.1002/smll.201901985

Flynn, M. J., Zammarchi, F., Tyrer, P. C., Akarca, A. U., Janghra, N., Britten, C. E., et al. (2016). ADCT-301, a pyrrolobenzodiazepine (PBD) dimer–containing antibody–drug conjugate (ADC) targeting CD25-expressing hematological malignancies. *Molecular Cancer Therapeutics, 15*(11), 2709–2721. https://doi.org/10.1158/1535-7163.mct-16-0233

Fonner, B. A., Tripet, B., Eilers, B. J., Stanisich, J. J., Sullivan-Springhetti, R. K., Moore, R. S., et al. (2014). Solution structure and molecular determinants of hemoglobin binding of the first NEAT domain of IsdB in Staphylococcus aureus. *Biochemistry, 53*(24), 3922–3933. https://doi.org/10.1021/bi5005188

Gannan, I., Kubaji, H., Siwale, W., Frodeson, S., & Venkatesh, G. (2023). Streamlined social footprint analysis of the nascent bio-pellet sub-sector in Zambia. *Sustainability, 15*(6), 5492.

García-Vivar, C. (2022). The use and misuse of the concept of cancer survivor. *Cancer Nursing, 45*(6), 419–420. https://doi.org/10.1097/ncc.0000000000001167

Geyer, M. (2023). *Development of novel microfluidic-based pancreatic cancer-on-a-chip models*. kobv.de

Goenka, A., Khan, F., Verma, B., Sinha, P., Dmello, C., Jogalekar, M. P., et al. (2023). Tumor microenvironment signaling and therapeutics in cancer progression. *Cancer Communications, 43*(5), 525–561. https://doi.org/10.1002/cac2.12416

Gohno, T., Hanamura, T., Kurosumi, M., Yamaguchi, Y., & Hayashi, S. (2022). One DNA methylation regulates chip gene expression of human breast cancer and predicts recurrence. *Anticancer Research, 42*(2), 759–766. https://doi.org/10.21873/anticanres.15534

Gröger, M., Dinger, J. E., Kiehntopf, M., Peters, F. T., Rauen, U., & Mosig, A. S. (2017). Preservation of cell structure, metabolism, and biotransformation activity of liver-on-chip organ models by hypothermic storage. *Advanced Healthcare Materials, 7*(2). https://doi.org/10.1002/adhm.201700616

Guil-Luna, S., Sedlik, C., & Piaggio, E. (2021). Humanized mouse models to evaluate cancer immunotherapeutics. *Annual Review of Cancer Biology, 5*, 119–136. annualreviews.org

Guo, Q., Zhang, L., Liu, J., Li, Z., Li, J., Zhou, W., et al. (2021). Multifunctional microfluidic chip for cancer diagnosis and treatment. *Nanotheranostics, 5*(1), 73–89. https://doi.org/10.7150/ntno.49614

Habanjar, O., Diab-Assaf, M., Caldefie-Chezet, F., & Delort, L. (2021). 3D cell culture systems: Tumor application, advantages, and disadvantages. *International Journal of Molecular Sciences, 22*(22), 12200. mdpi.com

Hachey, S. J., & Hughes, C. C. (2018). Applications of tumor chip technology. *Lab on a Chip, 18*(19), 2893–2912.

Haque, M. R., Rempert, T., Al-Hilal, T. A., Wang, C., Bhushan, A., & Bishehsari, F. (2021). Organ chip models: Opportunities for precision medicine in pancreatic cancer. *Cancers, 13*(17), 4487. https://doi.org/10.3390/cancers13174487

Harten, W., Paradiso, A., & Beau, M. (2013). The role of comprehensive cancer centers in survivorship care. *Cancer, 119*(S11), 2200–2201. https://doi.org/10.1002/cncr.28054

Hemmatian, I., Ponzio, T. A., & Joshi, A. M. (2022). Exploring the role of R&D collaborations and non-patent IP policies in government technology transfer performance: Evidence from U.S. federal agencies (1999–2016). *PLoS One, 17*(5), e0268828. https://doi.org/10.1371/journal.pone.0268828

Heringa, M. B. (2019). The value of organs-on-chip for regulatory safety assessment. *Altex*. https://doi.org/10.14573/altex.1910111

Hess, S., & Siegwart, R. (2012). R&D venture: Proposition of a technology transfer concept for breakthrough technologies with R&D cooperation: A case study in the energy sector. *The Journal of Technology Transfer, 38*(2), 153–179. https://doi.org/10.1007/s10961-012-9249-8

Hinrichs-Krapels, S., Ditewig, B., Boulding, H., Chalkidou, A., Erskine, J., & Shokraneh, F. (2022). Purchasing high-cost medical devices and equipment in hospitals: A systematic review. *BMJ Open, 12*(9), e057516. bmj.com

Hiyoshi, H., Gotô, N., Tsuchiya, M., Iida, K., Nakajima, Y., Hirata, N., et al. (2014). 2-(4-hydroxy-3-methoxyphenyl)-benzothiazole suppresses tumor progression and metastatic potential of breast cancer cells by inducing ubiquitin ligase chip. *Scientific Reports, 4*(1). https://doi.org/10.1038/srep07095

Holovko, O., & Sakevych, L. (2020). Analysis of the factors that determine the features of the organization of public-private partnership in the innovation activity of Ukraine. *Green, Blue and Digital Economy Journal, 1*(2), 39–44. https://doi.org/10.30525/2661-5169/2020-2-7

Huang, C. Y., Yang, M., Li, J. F., & Chen, H. (2021). A DANP-based NDEA-MOP approach to evaluating the patent commercialization performance of industry–academic collaborations. *Mathematics, 9*(18), 2280. https://doi.org/10.3390/math9182280

Huh, D., Torisawa, Y., Hamilton, G. A., Kim, H. J., & Ingber, D. E. (2012). Micro-engineered physiological biomimicry: Organs-on-chips. *Lab-on-a-Chip, 12*(12), 2156. https://doi.org/10.1039/c2lc40089h

İlhan, B. (2020). Stock market liberalization: Effects on stock market development in the emerging Islamic countries. *International Journal of Islamic Economics and Finance Studies, 6*(1), 30–56. dergipark.org.tr

Imparato, G., Urciuolo, F., & Netti, P. (2022). Organ-on-chip technology to model cancer growth and metastasis. *Bioengineering, 9*(1), 28. https://doi.org/10.3390/bioengineering9010028

Ingber, D. E. (2020). Is it time for reviewer three to request human organ-chip experiments instead of animal validation studies? *Advanced Science, 7*(22). https://doi.org/10.1002/advs.202002030

Ingber, D. E. (2022). Human organs-on-chips for disease modeling, drug development and personalized medicine. *Nature Reviews Genetics*. nature.com

Isosävi, J. (2020). Cultural outsiders' reported adherence to Finnish and French politeness norms. *Journal of Pragmatics, 155*, 177. academia.edu.

Jang, K., Lee, K., Kim, S., Jin, T., Choi, E., Jeon, H., et al. (2011). Ubiquitin ligase CHIP induces TRAF2 proteasomal degradation and NF-κB inactivation to regulate breast cancer cell invasion. *Journal of Cellular Biochemistry, 112*(12), 3612–3620. https://doi.org/10.1002/jcb.23292

Jeanes, A., Coen, P. G., Gould, D., & Drey, N. (2019). Validity of hand hygiene compliance measurement by observation: A systematic review. *American Journal of Infection Control, 47*(3), 313–322. https://doi.org/10.1016/j.ajic.2018.08.004

Jørgensen, J. M. A., Christensen, D. L., Nielsen, K. K., Sadiq, H. S., Khan, M. Y., Jusabani, A. M., & Walker, R. (2022). Incidence and characteristics of stroke in Zanzibar—A hospital-based prospective study in a low-income Island population. *Frontiers in Neurology, 13*, 931915. frontiersin.org

Joseph, X., Akhil, V., Arathi, A., & Mohanan, P. V. (2022). Comprehensive development in organ-on-a-chip technology. *Journal of Pharmaceutical Sciences, 111*(1), 18–31. jpharmsci.org

Juang, Y., & Chiu, Y. (2022). Fabrication of polymer microfluidics: An overview. *Polymers, 14*(10), 2028. https://doi.org/10.3390/polym14102028

Kamal, R., & Ray, R. (2020). *A connected and sustainable future–comparing lessons from Southern-led regional banks and networks, CAF and the Islamic Development Bank compared.* Southern-Led Development Finance. [HTML].

Kang, S. M. (2022). Recent advances in microfluidic-based microphysiological systems. *BioChip Journal, 16*(1), 13–26.

Kang, S., & Park, S. E. (2021). Organ-on-a-chip technology for nanoparticle research. *Nano Convergence, 8*(1). https://doi.org/10.1186/s40580-021-00270-x

Khalid, N., Kobayashi, I., & Nakajima, M. (2017). Recent lab-on-chip developments for novel drug discovery. *WIREs Systems Biology and Medicine, 9*(4). https://doi.org/10.1002/wsbm.1381

Kim, M. K., Paek, K., Woo, S., & Kim, J. A. (2023). Bone-on-a-chip: Biomimetic models based on microfluidic technologies for biomedical applications. *ACS Biomaterials Science & Engineering, 9*(6), 3058–3073. https://doi.org/10.1021/acsbiomaterials.3c00066

Kiriiri, G. K., Njogu, P. M., & Mwangi, A. N. (2020). Exploring different approaches to improve the success of drug discovery and development projects: A review. *Future Journal of Pharmaceutical Sciences, 6*, 1–12. springer.com

Ko, J., Carpenter, E. L., & Issadore, D. (2016). Detection and isolation of circulating exosomes and microvesicles for cancer monitoring and diagnostics using micro-/nano-based devices. *The Analyst, 141*(2), 450–460. https://doi.org/10.1039/c5an01610j

Kockaya, G., Polat, M. G., Wertheimer, A. I., Özet, A., Malhan, S., Vural, I. M., et al. (2013). Treatment cost of metastatic colon cancer in Turkey. *Farmeconomia. Health Economics and Therapeutic Pathways, 14*(1), 19–25. https://doi.org/10.7175/fe.v14i1.472

Komen, J., van Neerven, S. M., van den Berg, A., Vermeulen, L., & van der Meer, A. D. (2021). Mimicking and surpassing the xenograft model with cancer-on-chip technology. *eBioMedicine, 66*, 103303. https://doi.org/10.1016/j.ebiom.2021.103303

Kulkeaw, K., & Pengsart, W. (2021). Progress and challenges in the use of a liver-on-a-chip for hepatotropic infectious diseases. *Micromachines, 12*(7), 842. https://doi.org/10.3390/mi12070842

Kumar, V., Rahman, Z., & Kazmi, A. A. (2016). Stakeholder identification and classification: A sustainability marketing perspective. *Management Research Review, 39*(1), 35–61. https://doi.org/10.1108/mrr-09-2013-0224

Kunkle, W. E. (2021). *Investigating the user experience with a 3D virtual anatomy application.* Doctoral dissertation, Nova Southeastern University.

Kurth, F., Györvary, E., Heub, S., Ledroit, D., Paoletti, S., Renggli, K., et al. (2020). Organs-on-a-chip engineering. In *Organ-on-a-chip* (pp. 47–130). Academic Press.

Lai, B. F. L. (2022). *Development of a well plate-based multiplexed organ-on-a-chip platform with perfusable vasculature*. [HTML].

Lantada, A. D., Pfleging, W., Besser, H., Guttmann, M., Wissmann, M., Plewa, K., et al. (2018). Research on the methods for the mass production of multi-scale organs-on-chips. *Polymers, 10*(11), 1238. https://doi.org/10.3390/polym10111238

LaVigne, A., Triedman, S., Randall, T., Trimble, E., & Viswanathan, A. (2017). Cervical cancer in low-and middle-income countries: Addressing barriers to radiotherapy delivery. *Gynecologic Oncology Reports, 22*, 16–20. https://doi.org/10.1016/j.gore.2017.08.004

Lee, J. B., & Sung, J. H. (2013). Organ-on-a-chip technology and microfluidic whole-body models for pharmacokinetic drug toxicity screening. *Bio-technology Journal, 8*(11), 1258–1266. https://doi.org/10.1002/biot.201300086

Leeuwen, M., Husson, O., Alberti, P., Arrarás, J., Chinot, O., Costantini, A., et al. (2018). Understanding the quality of life (QOL) issues in survivors of cancer: Toward the development of an EORTC QOL cancer survivorship questionnaire. *Health and Quality of Life Outcomes, 16*(1). https://doi.org/10.1186/s12955-018-0920-0

Lemaire, C. A., Liu, S. Z., Wilkerson, C., Ramani, V. C., Barzanian, N., Huang, K., et al. (2018). Fast and label-free isolation of circulating tumor cells from blood: From a research microfluidic platform to an automated fluidic instrument, VTX-1 liquid biopsy system. *SLAS Technology, 23*(1), 16–29. https://doi.org/10.1177/2472630317738698

Lerner, J. (2022). *Venture capitalists and the oversight of private firms*. Venture Capital. qcconference.com

Leung, C. M., De Haan, P., Ronaldson-Bouchard, K., Kim, G. A., Ko, J., Rho, H. S., et al. (2022). A guide to the organ-on-a-chip. *Nature Reviews Methods Primers, 2*(1), 33. nature.com

Lewison, G., Hussain, S., Guo, P., Harding, R., Mukherji, D., Sittah, G., et al. (2020). Cancer research in the 57 Organization of Islamic Cooperation (OIC) countries, 2008–17. *Ecancermedicalscience, 14*. https://doi.org/10.3332/ecancer.2020.1094

Li, X., & North, D. (2017). The role of technological business incubators in supporting business innovation in China: A case of regional adaptability? *Entrepreneurship & Regional Development, 30*(1–2), 29–57. https://doi.org/10.1080/08985626.2017.1364789

Li, Z., Ma, L., Wen, Z., Hu, Z., Wu, F., Li, W., et al. (2013). Cancerous inhibitor of PP2A is targeted by natural compound celastrol for degradation in non-small-cell lung cancer. *Carcinogenesis, 35*(4), 905–914. https://doi.org/10.1093/carcin/bgt395

Li, M., Liu, L., Xi, N., & Wang, Y. (2017). Applications of micro/nano automation technology in detecting cancer cells for personalized medicine. *IEEE Transactions on Nanotechnology, 16*(2), 217–229. https://doi.org/10.1109/tnano.2017.2654320

Li, X., Zhang, Y., Abbassi, R., Yang, M., Zhang, R., & Chen, G. (2021). Dynamic probability assessment of urban natural gas pipeline accidents considering integrated external activities. *Journal of Loss Prevention in the Process Industries, 69*, 104388. sciencedirect.com

Li, Q., Wang, C., Li, X., Zhang, J., Zhang, Z., Yang, K., et al. (2023). Epidermis-on-a-chip system to develop skin barrier and melanin mimicking model. *Journal of Tissue Engineering, 14*, 20417314231168529. sagepub.com

Lim, J. Y., Ching, H., Yoon, J., Jeon, N. L., & Kim, Y. T. (2021). Microvascularized tumor organoids-on-chips: Advancing preclinical drug screening with pathophysiological relevance. *Nano Convergence, 8*(1). https://doi.org/10.1186/s40580-021-00261-y

Lin, L., Li, Z., Yan, L., Liu, Y., Yang, H., & Li, H. (2021). Global, regional, and national cancer incidence and death for 29 cancer groups in 2019 and trends analysis of the global cancer burden, 1990–2019. *Journal of Hematology & Oncology, 14*, 197. springer.com

Liu, F., Park, K., & Whang, U. (2019). Organizational capabilities, export growth and job creation: An investigation of Korean SMEs. *Sustainability, 11*(14), 3986. https://doi.org/10.3390/su11143986

Liu, X., Fang, J., Huang, S., Wu, X., Xie, X., Wang, J., et al. (2021). Tumor-on-a-chip: From bioinspired design to biomedical application. *Microsystems & Nanoengineering, 7*(1), 50. nature.com

Liu, Y., Li, S., & Liu, Y. (2022). Machine-learning-driven multiobjective optimization: An opportunity of microfluidic platforms applied in cancer research. *Cells, 11*(5), 905. https://doi.org/10.3390/cells11050905

Loewenson, R., Godt, S., & Kapata, P. C. (2021). *Framings of and priorities for action on the commercial determinants of health in sub-Saharan Africa*. tarsc.org

Lovrić, B., Sajnic, A., Pleština, S., Jakopović, M., & Kurtović, B. (2022). Croatian national cancer patient experience survey. *International Journal of Environmental Research and Public Health, 19*(14), 8285. https://doi.org/10.3390/ijerph19148285

Low, L. A., Mummery, C. L., Berridge, B. R., Austin, C. P., & Tagle, D. A. (2020). Organs-on-chips: Into the next decade. *Nature Reviews Drug Discovery, 20*(5), 345–361. https://doi.org/10.1038/s41573-020-0079-3

Lu, Y., Ahmad, M., Zhang, H., & Guo, J. (2023). Effects of science and technology finance on green total factor productivity in China: Insights from an empirical spatial Durbin model. *Journal of the Knowledge Economy, 15*, 7280. https://doi.org/10.1007/s13132-023-01306-9

Luka, G., Ahmadi, A., Najjaran, H., Alocilja, E. C., DeRosa, M. C., Wolthers, K. R., et al. (2015). Microfluidics integrated biosensors: A leading technology toward lab-on-a-chip and sensing applications. *Sensors, 15*(12), 30011–30031. https://doi.org/10.3390/s151229783

Ma, Y., Mukherjee, S., & Uzzi, B. (2020). Mentorship and protégé success in STEM fields. *Proceedings of the National Academy of Sciences of the United States of America, 117*(25), 14077–14083. pnas.org

Ma, C., Peng, Y., Li, H., & Chen, W. (2021). Organ-on-a-chip: A new paradigm for drug development. *Trends in Pharmacological Sciences, 42*, 119. nih.gov

Mahbub, T., Ahammad, M. F., Tarba, S. Y., & Mallick, S. Y. (2022). Factors encouraging foreign direct investment (FDI) in the wind and solar energy sector in an emerging country. *Energy Strategy Reviews, 41*, 100865. sciencedirect.com

Mahmud, A., Bazaz, S. R., Dabiri, S., Mehrizi, A. A., Asadnia, M., Warkiani, M. E., et al. (2022). Advances in mems and microfluidics-based energy harvesting technologies. *Advanced Materials Technologies, 7*(7). https://doi.org/10.1002/admt.202101347

Malik, K., Meki, M., Morduch, J., Ogden, T., Quinn, S., & Said, F. (2020). COVID-19 and the future of microfinance: Evidence and insights from Pakistan. *Oxford Review of Economic Policy, 36*(Supplement_1), S138–S168. khazar.org

Mariotto, A. B., Enewold, L., Zhao, J., Zeruto, C. A., & Yabroff, K. R. (2020). Medical care costs associated with cancer survivorship in the United States. *Cancer Epidemiology, Biomarkers & Prevention, 29*(7), 1304–1312. nih.gov

Marx, U., Akabane, T., Andersson, T., Baker, E., Beilmann, M., Beken, S., et al. (2020). Biology-inspired microphysiological systems to advance medicines for patient benefit and animal welfare. *Altex*. https://doi.org/10.14573/altex.2001241

Mastrangeli, M., & van den Eijnden-van Raaij, J. (2021). Organs-on-chip: The way forward. *Stem Cell Reports, 16*(9), 2037–2043. https://doi.org/10.1016/j.stemcr.2021.06.015

Maxwell, J. A., & Metz, K. (2021). How technology life cycles increase the financial position of the firm. *American Journal of Management, 21*(3). https://doi.org/10.33423/ajm.v21i3.4357

McFarlane, S., Orr, A., Roberts, A. P. E., Conn, K. L., Iliev, V., Loney, C., et al. (2019). The histone chaperone HIRA promotes the induction of host innate immune defenses in response to HSV-1 infection. *PLoS Pathogens, 15*(3), e1007667. https://doi.org/10.1371/journal.ppat.1007667

Mertz, D., Ahmed, T., & Takayama, S. (2018). Engineering cell heterogeneity into organs-on-a-chip. *Lab-on-a-Chip, 18*(16), 2378–2395. https://doi.org/10.1039/c8lc00413g

Michna, R., Gadde, M., Ozkan, A., DeWitt, M., & Rylander, M. (2018). Vascularized microfluidic platforms to mimic the tumor microenvironment. *Biotechnology and Bioengineering, 115*(11), 2793–2806.

Middelkamp, H. H., van der Meer, A. D., Hummel, J. M., Stamatialis, D., Mummery, C. L., Passier, R., et al. (2016). Organs-on-chips in drug development: The importance of involving stakeholders in early health technology assessment. *Applied in-Vitro Toxicology, 2*(2), 74–81. https://doi.org/10.1089/aivt.2015.0029

Miller, C. (2022). *Chip war: the fight for the world's most critical technology*. mincotur.gob.es
Miller, C. P., Shin, W., Ahn, E. H., Kim, H. J., & Kim, D. H. (2020). Engineering microphysiological immune system responses on chips. *Trends in Bio-technology, 38*, 857. nih.gov
Mohamed, B. H., Ari, I., Al-Sada, M. S., & Koç, M. (2021). Strategizing human development for a country in transition from a resource-based to a knowledge-based economy. *Sustainability*. mdpi.com
Muhammad, F., Khan, A., Razzaq, N., & Karim, R. (2021). Influence of tourism, governance, and foreign direct investment on energy consumption and CO 2 emissions: A panel analysis of Muslim countries. *Environmental Science and Pollution Research, 28*, 416–431.
Mustafa, D., & Razak, N. (2012). Islamic Development Bank, foreign aid and economic growth in Africa: A simultaneous equations model approach. *International Journal of Economics and Finance, 4*(6). https://doi.org/10.5539/ijef.v4n6p94
Najafi-Tavani, S., Najafi-Tavani, Z., Naudé, P., Oghazi, P., & Zeynaloo, E. (2018). How collaborative innovation networks affect new product performance: Product innovation capability, process innovation capability, and absorptive capacity. *Industrial Marketing Management, 73*, 193–205. https://doi.org/10.1016/j.indmarman.2018.02.009
National Academies of Sciences, Engineering, and Medicine. (2023). Equity in innovation within NIH and FTC: Examination of agency approaches and implementation. In *Toward equitable innovation in health and medicine: A framework*. National Academies Press (US). nih.gov
Nazir, F., Iqbal, M., Khan, A. N., Mazhar, M., & Hussain, Z. (2021). Fabrication of robust poly L-lactic acid/cyclic olefinic co-polymer (PLLA/C-OC) blends a study of physical properties, structure, and cytocompatibility for bone tissue engineering. *Journal of Materials Research and Technology, 13*, 1732–1751. sciencedirect.com
Neta, G., Sánchez, M., Chambers, D. A., Phillips, S. M., Leyva, B., Cynkin, L., et al. (2015). Implementation science in cancer prevention and control: A decade of grant funding by the national cancer institute and future directions. *Implementation Science, 10*(1), 4. https://doi.org/10.1186/s13012-014-0200-2
Nguyen, T. T., Nguyen, C. M., Huynh, M. A., Vu, H. H., Nguyen, T., & Nguyen, N. (2023). Field effect transistor based wearable biosensors for healthcare monitoring. *Journal of Nanobiotechnology, 21*(1), 411. https://doi.org/10.1186/s12951-023-02153-1
Oluwadare, A., Akintayo, B. D., Babatunde, O. M., & Olanrewaju, O. A. (2024). Assessing the impact of healthcare 4.0 technologies on healthcare supply chain management: A multi-criteria evaluation framework. *Logistics, 8*(2), 44. https://doi.org/10.3390/logistics8020044
Ongaro, A. E., Giuseppe, D. D., Kermanizadeh, A., Crespo, A. M., Mencattini, A., Ghibelli, L., et al. (2020). Polylactic is a sustainable, low absorption, low autofluorescence alternative to other plastics for microfluidic and organ-on-chip applications. *Analytical Chemistry, 92*(9), 6693–6701. https://doi.org/10.1021/acs.analchem.0c00651
Osman, O. A., & Sulieman, A. M. E. (2023). *Halal and kosher food: Integration of quality and safety for global market trends*. [HTML].
Qu, Y., Ye, J., Lin, B., Luo, Y., & Zhang, X. (2023). Organ mimicking technologies and their applications in drug discovery. *Intelligent Pharmacy, 1*(2), 73–89.
Özmen, V. (2014). Breast cancer in Turkey: Clinical and histopathological characteristics (analysis of 13.240 patients). *The Journal of Breast Health, 10*(2), 98–105. https://doi.org/10.5152/tjbh.2014.1988
Özmen, V., Atasoy, A., Gökmen, E., Özdoğan, M., Güler, N., Uras, C., et al. (2016). Impact of oncotype DX recurrence score on treatment decisions: Results of a prospective multi-center study in Turkey. *Cureus, 8*, e522. https://doi.org/10.7759/cureus.522
Özmen, V., Çakar, B., Gökmen, E., Özdoğan, M., Güler, N., Uras, C., et al. (2019). Cost-effectiveness of gene expression profiling in patients with early-stage breast cancer in a middle-income country, Turkey: Results of a prospective multi-center study. *European Journal of Breast Health, 15*(3), 183–190. https://doi.org/10.5152/ejbh.2019.4761

Paiva, T., Ribeiro, M. P., & Coutinho, P. (2022). Capacity-building model to promote innovation and sustainability in the Portuguese agro-industrial sector. *Sustainability, 14*(23), 15873. https://doi.org/10.3390/su142315873

Pandya, H., Dhingra, K., Prabhakar, D., Chandrasekar, V., Natarajan, S., Vasan, A., et al. (2017). A microfluidic platform for drug screening in a 3D cancer microenvironment. *Biosensors and Bioelectronics, 94*, 632–642. https://doi.org/10.1016/j.bios.2017.03.054

Paradiso, F., Serpelloni, S., Francis, L. W., & Taraballi, F. (2021). Mechanical studies of the third dimension in cancer: From 2D to 3D model. *International Journal of Molecular Sciences, 22*(18), 10098. mdpi.com

Park, S., Georgescu, A., & Huh, D. (2019). Organoids-on-a-chip. *Science, 364*(6444), 960–965. https://doi.org/10.1126/science.aaw7894

Parrales, A., Thoenen, E., & Iwakuma, T. (2017). The interplay between mutant p53 and the mevalonate pathway. *Cell Death and Differentiation, 25*(3), 460–470. https://doi.org/10.1038/s41418-017-0026-y

Peck, R., Hinojosa, C. D., & Hamilton, G. A. (2019). Organs-on-chips in clinical pharmacology: Putting the patient into the center of treatment selection and drug development. *Clinical Pharmacology & Therapeutics, 107*(1), 181–185. https://doi.org/10.1002/cpt.1688

Pericoli, A. (2020). Islamic finance and charity in the Muslim world. The role of the Islamic Development Bank in financing aid. *JEBA (Journal of Economics and Business Aseanomics), 5*(2), 113–133. yarsi.ac.id. https://doi.org/10.33476/j.e.b.a.v5i2.1656

Pozzi, R., Rossi, T., & Secchi, R. (2023). Industry 4.0 technologies: Critical success factors for implementation and improvements in manufacturing companies. *Production Planning & Control*. [HTML].

Pramesh, C. S., Badwe, R. A., Bhoo-Pathy, N., Booth, C. M., Chinnaswamy, G., Dare, A. J., et al. (2022). Priorities for cancer research in low-and middle-income countries: A global perspective. *Nature Medicine, 28*(4), 649–657. nature.com

Ramadan, Q., & Zourob, M. (2020). Organ-on-a-chip engineering: Toward bridging the gap between lab and industry. *Biomicrofluidics*. nih.gov

Ranta, V., Aarikka-Stenroos, L., & Väisänen, J. M. (2021). Digital technologies catalyzing business model innovation for circular economy—Multiple case study. *Resources, Conservation and Recycling, 164*, 105155. sciencedirect.com

Ribeiro, M., Ali, P., Metcalfe, B., Moschou, D., & Rocha, P. (2021). Microfluidics integration into low-noise multi-electrode arrays. *Micromachines, 12*(6), 727. https://doi.org/10.3390/mi12060727

Rienzo, M., Jackson, S. J., Chao, L. K., Leaf, T., Schmidt, T. J., Navidi, A. H., et al. (2021). High-throughput screening for high-efficiency small-molecule biosynthesis. *Metabolic Engineering, 63*, 102–125. sciencedirect.com

Rodrigues, R. O., Sousa, P. C., Gaspar, J., Bañobre-Lopez, M., Lima, R., & Minas, G. (2020). Organ-on-a-chip: A preclinical microfluidic platform for the progress of nanomedicine. *Small, 16*(51), 2003517. https://doi.org/10.1002/smll.202003517. [HTML].

Rodriguez, F., Caruana, P., De la Fuente, N., Español, P., Gamez, M., Balart, J., et al. (2022). Nano-based approved pharmaceuticals for cancer treatment: Present and future challenges. *Biomolecules, 12*(6), 784. mdpi.com

Roemer, I. (2021). *Breast cancer in the Gaza strip: The impact of the medical permit regime on public health*. georgetown.edu.

Rose, L. M., Wills, F., Bourassa-Shaw, C., Butler, T. L., Ennis, J. G., Emmons, K., et al. (2017). Collaborative academic medical product development: An 8-year review of commercialization outcomes at the institute of translational health sciences. *Journal of Clinical and Translational Science, 1*(4), 229–234. https://doi.org/10.1017/cts.2017.13

Roye, Y., Bhattacharya, R., Mou, X., Zhou, Y., Burt, M., & Musah, S. (2021). A personalized glomerulus chip engineered from stem cell-derived epithelium and vascular endothelium. *Micromachines, 12*(8), 967. https://doi.org/10.3390/mi12080967

Ruiz-Espigares, J., Nieto, D., Moroni, L., Jimenez, G., & Marchal, J. A. (2021). Evolution of metastasis study models toward metastasis-on-a-chip: The ultimate model? *Small, 17*(14), 2006009. wiley.com

Ryan, G., Kirchhoff, A., Fair, D., Rabin, J., Hyland, K., Kuhlthau, K., et al. (2017). Financial burden in survivors of childhood cancer: A report from the childhood cancer survivor study. *Journal of Clinical Oncology, 35*(30), 3474–3481. https://doi.org/10.1200/jco.2016.71.7066

Saini, A., Kumar, M., Bhatt, S., Saini, V., & Malik, A. (2020). Cancer causes and treatments. *International Journal of Pharmaceutical Sciences and Research, 11*, 3121–3134. researchgate.net

Salama, L., Pastor, E. R., Stone, T., & Mousa, S. A. (2020). Emerging nanopharmaceuticals and nanonutraceuticals in cancer management. *Biomedicine, 8*(9), 347. https://doi.org/10.3390/biomedicines8090347

Salariseddigh, M. J., Jamali, V., Pereg, U., Boche, H., Deppe, C., & Schober, R. (2023). Deterministic identification for molecular communications over the Poisson channel. *IEEE Transactions on Molecular, Biological and Multi-Scale Communications*. [PDF].

Santiago, G. T., Flores-Garza, B. G., Negrete, J. A. T., Lara-Mayorga, I. M., González-Gamboa, I., Zhang, Y. S., et al. (2019). The tumor-on-chip: Recent advances in the development of microfluidic systems to recapitulate the physiology of solid tumors. *Materials, 12*(18), 2945. https://doi.org/10.3390/ma12182945

Santos Rosalem, G., Gonzales Torres, L. A., de Las Casas, E. B., Mathias, F. A. S., Ruiz, J. C., & Carvalho, M. G. R. (2020). Microfluidics and organ-on-a-chip technologies: A systematic review of the methods used to mimic bone marrow. *PLoS One, 15*(12), e0243840. plos.org

Sarkar, S., Brautigan, D., & Larner, J. (2017). Aurora kinase A promotes AR degradation via the E3 ligase chip. *Molecular Cancer Research, 15*(8), 1063–1072. https://doi.org/10.1158/1541-7786.mcr-17-0062

Scuotto, V., Beatrice, O., Valentina, C., Nicotra, M., Di Gioia, L., & Briamonte, M. F. (2020). Uncovering the micro-foundations of knowledge sharing in open innovation partnerships: An intention-based perspective of technology transfer. *Technological Forecasting and Social Change, 152*, 119906. fardapaper.ir.

Sejko, D. (2023). Sovereign investors as ICSID claimants: Lessons from the drafting documents and the case law. *Vanderbilt Journal of Transnational Law*. researchgate.net

Sennanda, M., Ahmed, M., Gerald, K., & Ismail, K. (2023). Project financing and poverty trends in the Islamic Development Bank member countries. *International Journal for Multidisciplinary Research, 5*(2). https://doi.org/10.36948/ijfmr.2023.v05i02.1911

Sequeira, R. C., Criswell, T., Atala, A., & Yoo, J. J. (2020). Microfluidic systems for assisted reproductive technologies: Advantages and potential applications. *Tissue Engineering and Regenerative Medicine, 17*(6), 787–800. nih.gov

Sequeira-Antunes, B., & Ferreira, H. A. (2023). Urinary biomarkers and point-of-care urinalysis devices for early diagnosis and management of disease: A review. *Biomedicine, 11*(4), 1051. https://doi.org/10.3390/biomedicines11041051

Shah, M., Prakashan, D., & Gandhi, S. (2024). Microfluidics, organs-on-a-chip, and 3D printing. In *Human organson-a-chip technology* (pp. 91–112). Academic Press.

Sharma, V., Kerr, S., Kawar, Z., & Kerr, D. (2011). Challenges of cancer control in developing countries: Current status and future perspective. *Future Oncology, 7*(10), 1213–1222. https://doi.org/10.2217/fon.11.101

Shoemaker, J. T., Zhang, W., Atlas, S. I., Bryan, R. A., Inman, S. W., & Vukasinovic, J. (2020). A 3D cell culture organ-on-a-chip platform with a breathable hemoglobin analog augments and extends primary human hepatocyte functions in-vitro. *Frontiers in Molecular Biosciences, 7*. https://doi.org/10.3389/fmolb.2020.568777

Shrivastava, S., Trung, T. Q., & Lee, N. E. (2020). Recent progress, challenges, and prospects of fully integrated mobile and wearable point-of-care testing systems for self-testing. *Chemical Society Reviews*. [HTML].

Silva, R. (2023). Organ chip research in Europe: Players, initiatives, and policies. *Frontiers in Bioengineering and Biotechnology, 11*. https://doi.org/10.3389/fbioe.2023.1237561

Singh, D. B., Mathur, A., Arora, S., Roy, S., & Mahindroo, N. (2022). Journey of organ-on-a-chip technology and its role in future healthcare scenario. *Applied Surface Science Advances, 9*, 100246. https://doi.org/10.1016/j.apsadv.2022.100246

Snyder, H. M., Kim, H., Bain, L. J., Egge, R., & Carrillo, M. C. (2014). Alzheimer's disease public-private partnerships: Update 2014. *Alzheimer's & Dementia, 10*(6), 873–880. https://doi.org/10.1016/j.jalz.2014.06.014

Sobri, F. B., Bachtiar, A., Panigoro, S. S., Ayuningtyas, D., Gustada, H., Yuswar, P. W., et al. (2021). Factors affecting delayed presentation and diagnosis of breast cancer in Asian developing countries women: A systematic review. *Asian Pacific Journal of Cancer Prevention, 22*(10), 3081–3092. https://doi.org/10.31557/apjcp.2021.22.10.3081

Sonnenschein, C., & Soto, A. M. (2020). Over a century of cancer research: Inconvenient truths and promising leads. *PLoS Biology, 18*. plos.org

Spöttl, G., & Windelband, L. (2021). The 4th industrial revolution—Its impact on vocational skills. *Journal of Education and Work, 34*, 29. tandfonline.com

Srinivasan, A., Tiwari, K., Scott, J., Ramachandran, P., & Ramakrishnan, M. (2015). Impact of cancer support groups on childhood cancer treatment and abandonment in a private pediatric oncology center. *Indian Journal of Palliative Care, 21*(1), 68. https://doi.org/10.4103/0973-1075.150192

Sugihara, K. (2018). *Black lipid membranes: Investigation of their properties and their application in the characterization of active membrane components*. access.archive-ouverte.unige.ch.

Sung, J. H. (2022). From organ-on-a-chip toward body-on-a-chip. *Biocell, 46*(5), 1177–1180. https://doi.org/10.32604/biocell.2022.019055

Sunildutt, N., Parihar, P., Salih, A. R. C., Lee, S. H., & Choi, K. H. (2023). Revolutionizing drug development: Harnessing the potential of organ-on-chip technology for disease modeling and drug discovery. *Frontiers in Pharmacology, 14*. https://doi.org/10.3389/fphar.2023.1139229

Tajeddin, A., & Mustafaoglu, N. (2021). Design and fabrication of organ-on-chips: Promises and challenges. *Micromachines*. mdpi.com

Takebe, T., Zhang, B., & Radišić, M. (2017). Synergistic engineering: Organoids meet organs-on-a-chip. *Cell Stem Cell, 21*(3), 297–300. https://doi.org/10.1016/j.stem.2017.08.016

Tang, B., Bendas, S., Krajka, V., May, T., Moritz, A., Constantinou, I., & Dietzel, A. (2022). Self-loading microfluidic platform with ultra-thin nanoporous membrane for organ-on-chip by wafer-level processing. *Frontiers in Sensors, 3*, 974895.

Tao, Y., Lin, C., Pan, M., Zhu, F., & Zhu, D. (2021). Tailored biosensors for drug screening, efficacy assessment, and toxicity evaluation. *ACS Sensors, 6*(9), 3146–3162. https://doi.org/10.1021/acssensors.1c01600

Taute, F., Homs-Corbera, A., & Gaudriault, P. (2021). The challenges and considerations for emerging or future entrepreneurial researchers in microphysiological systems. *Open Research Europe, 1*, 38. europa.eu.

Teixeira, J. E., & Tavares-Lehmann, A. T. C. (2022). Industry 4.0 in the European Union: Policies and national strategies. *Technological Forecasting and Social Change, 180*, 121664. [HTML].

Thakar, R., & Fenton, K. N. (2023). Bioethical implications of organ-on-a-chip on modernizing drug development. *Artificial Organs, 47*(10), 1553–1558. https://doi.org/10.1111/aor.14620

Toffoli, E. C., van Vliet, A. A., Verheul, H. W., van der Vliet, H. J., Tuynman, J., Spanholtz, J., & de Gruijl, T. D. (2023). Allogeneic NK cells induce monocyte-to-dendritic cell conversion, control tumor growth, and trigger a pro-inflammatory shift in patient-derived cultures of primary and metastatic colorectal cancer. *Journal for Immunotherapy of Cancer, 11*(12). nih.gov

Trucco, P., & Petrenj, B. (2017). Resilience of critical infrastructures: Benefits and challenges from emerging practices and programs at local level. *NATO Science for Peace and Security Series C: Environmental Security*, 225–286. https://doi.org/10.1007/978-94-024-1123-2_8

Tsai, H., Trubelja, A., Shen, A. Q., & Bao, G. (2017). Tumor-on-a-chip: Microfluidic models of tumor morphology, growth and microenvironment. *Journal of the Royal Society Interface, 14*(131), 20170137. https://doi.org/10.1098/rsif.2017.0137

Tuffaha, H., El-Saifi, N., Chambers, S. K., & Scuffham, P. A. (2018). Directing research funds to the right research projects: A review of criteria used by research organizations in Australia in prioritizing health research projects for funding. *BMJ Open, 8*(12), e026207. https://doi.org/10.1136/bmjopen-2018-026207

Twiggs, S. I. (2020). *Reconstruction: An assessment of the long-term disaster risk reduction capacity of post-conflict communities*. proquest.com

van Berlo, D., van de Steeg, E., Amirabadi, H. E., & Masereeuw, R. (2021). The potential of multi-organ-on-chip models for assessment of drug disposition as alternative to animal testing. *Current Opinion in Toxicology, 27*, 8–17. sciencedirect.com

van den Berg, A., Mummery, C. L., Passier, R., & van der Meer, A. D. (2019). Personalized organs-on-chips: Functional testing for precision medicine. *Lab-on-a-Chip, 19*(2), 198–205. https://doi.org/10.1039/c8lc00827b

Veliz, D. S., Lin, K., & Sahlgren, C. (2023). Organ-on-a-chip technologies for biomedical research and drug development: A focus on vasculature. *Smart Medicine, 2*(1), e20220030. https://doi.org/10.1002/smmd.20220030

Vogel, A. L., Freeman, J., Duncan, K., Alaro, J., Welch, J. J., Rodrigues, B., et al. (2019). Advancing cancer research in Africa through early-career awards: The big cat initiative. *Journal of Global Oncology, 5*, 1–8. https://doi.org/10.1200/jgo.18.00223

Walewska, A., Krajewska, M., Stefanowska, A., Buta, A., Bilewicz, R., Krysiński, P., et al. (2022). Methods of measuring mitochondrial potassium channels: A critical assessment. *International Journal of Molecular Sciences, 23*(3), 1210.

Wang, M., & Feng, C. (2021). Revealing the pattern and evolution of global green development between different income groups: A global meta-frontier by-production technology approach. *Environmental Impact Assessment Review, 89*, 106600.

Wang, Y., Ahmad, A. A., Shah, P. K., Sims, C. E., Magness, S. T., & Allbritton, N. L. (2013). Capture and 3D culture of colonic crypts and colonoids in a microarray platform. *Lab-on-a-Chip, 13*(23), 4625. https://doi.org/10.1039/c3lc50813g

Wang, H., Brown, P. C., Chow, E. C., Ewart, L., Ferguson, S. S., Fitzpatrick, S., et al. (2021). 3D cell culture models: Drug pharmacokinetics, safety assessment, and regulatory consideration. *Clinical and Translational Science, 14*(5), 1659–1680. wiley.com

Wang, R., Zhang, C., Li, D., & Yao, Y. (2022). Tumor-on-a-chip: Perfusable vascular incorporation brings new approach to tumor metastasis research and drug development. *Frontiers in Bioengineering and Bio-technology, 10*. https://doi.org/10.3389/fbioe.2022.1057913

Wang, Y., Gao, Y., Pan, Y., Zhou, D., Liu, Y., Yin, Y., et al. (2023). Emerging trends in organ-on-a-chip systems for drug screening. *Acta Pharmaceutica Sinica B, 13*(6), 2483–2509. sciencedirect.com

Wani, I., Khanam, M., & Almonifi, Y. S. A. (2021). A brief review of growth and development in Islamic banking. Dr. Shakeel ul Rehman, et al. "A brief review of growth and development in Islamic banking". *IOSR Journal of Economics and Finance (IOSR-JEF), 12*(3). researchgate.net

Waqas, H., Aqeeq, M. A., & Ghaffar, A. (2021). Re-thinking Ribā and the 'Islamic' banking experimentation. Waqas Ali Haider, Muhammad Arsalan Aqeeq, & Dr. Abdul Ghaffar. (2021). ENGLISH: Re-thinking Ribā and the 'Islamic' banking experimentation. *Rahatulquloob, 5*(2), 81–103. rahatulquloob.com

Wikswo, J. P., Curtis, E. L., Eagleton, Z. E., Evans, B. C., Kole, A., Hofmeister, L. H., et al. (2013). Scaling and systems biology for integrating multiple organs-on-a-chip. *Lab-on-a-Chip, 13*(18), 3496. https://doi.org/10.1039/c3lc50243k

Won, K., Im, J., Ban, H., Jung, Y., & Won, M. (2017). Stability of the cancer target DDIAS is regulated by the CHIP/HSP70 pathway in lung cancer cells. *Cell Death and Disease, 8*(1), –e2554. https://doi.org/10.1038/cddis.2016.488

Wood, L. D., Canto, M. I., Jaffee, E. M., & Simeone, D. M. (2022). Pancreatic cancer: Pathogenesis, screening, diagnosis, and treatment. *Gastroenterology, 163*, 386. nih.gov

Wu, Q., Liu, J., Wang, X., Feng, L., Wu, J., Zhu, X., et al. (2020). Organ-on-a-chip: Recent breakthroughs and future prospects. *Biomedical Engineering Online, 19*, 1–19. springer.com

Wu, J., Dai, B., Li, Z., Pan, T., Zhang, D., & Lin, F. (2021). Emerging optofluidic technologies for biodiagnostic applications. *View*. wiley.com

Xia, E., Nazir, M. R., & Bilal, A. (2020). Does financial inclusion enhance economic growth? Empirical evidence from the IsDB member countries. *International Journal of Finance & Economics, 26*(4), 5235–5258. https://doi.org/10.1002/ijfe.2063

Xiao, M., Yan, M., Zhang, J., Xu, Q., Qi, S., Wang, X., et al. (2017). Cancer stem-like cell related protein CD166 degrades through E3 ubiquitin ligase chip in head and neck cancer. *Experimental Cell Research, 353*(1), 46–53. https://doi.org/10.1016/j.yexcr.2017.03.005

Xie, J., Yan, J., Zhu, D., & He, G. (2021). Atomic-level insight into the formation of subsurface dislocation layer and its effect on mechanical properties during ultrafast laser micro/nano fabrication. *Advanced Functional Materials, 32*(15). https://doi.org/10.1002/adfm.202108802

Xu, H., & Cheng, C. (2022). Recent research advances of the biomimetic tumor microenvironment and regulatory factors on microfluidic devices: A systematic review. *Electrophoresis, 43*(7–8), 839–847. https://doi.org/10.1002/elps.202100360

Xu, J., Zhou, J., Dai, H., Liu, F., Li, W., Wang, W., et al. (2018). Chip functions as an oncogene by promoting colorectal cancer metastasis via activation of mitogen-activated protein kinase (MAPK) and AKT signaling and suppression of e-cadherin. *Journal of Translational Medicine, 16*(1), 169. https://doi.org/10.1186/s12967-018-1540-5

Yabroff, K., Lund, J., Kepka, D., & Mariotto, A. (2011). Economic burden of cancer in the United States: Estimates, projections, and future research. *Cancer Epidemiology Biomarkers & Prevention, 20*(10), 2006–2014. https://doi.org/10.1158/1055-9965.epi-11-0650

Yan, M., Yang, X., Wang, L., Clark, D., Zuo, H., Ye, D., et al. (2013). Plasma membrane proteomics of tumor spheres identify CD166 as a novel marker for cancer stem-like cells in head and neck squamous cell carcinoma. *Molecular & Cellular Proteomics, 12*(11), 3271–3284. https://doi.org/10.1074/mcp.m112.025460

Yang, J., Yang, J., Gong, X., Zheng, Y., Yi, S., Cheng, Y., et al. (2022). Recent progress in microneedles-mediated diagnosis, therapy, and theranostic systems. *Advanced Healthcare Materials, 11*(10), 2102547.

Yao, Z., Fenoglio, S., Gao, D., Camiolo, M., Stiles, B., Lindsted, T., et al. (2010). TGF-β IL-6 axis mediates selective and adaptive mechanisms of resistance to molecular targeted therapy in lung cancer. *Proceedings of the National Academy of Sciences of the United States of America, 107*(35), 15535–15540. https://doi.org/10.1073/pnas.1009472107

Yasirwan. (2024). The impact of Islamic finance, economic growth, and globalization on CO_2 emissions. *Jurnal Ekonomi, 29*(1), 127–146. https://doi.org/10.24912/je.v29i1.2094

Yi, H., Lee, H., & Cho, D. (2017). 3D printing of organs-on-chips. *Bioengineering, 4*(1), 10. https://doi.org/10.3390/bioengineering4010010

Young, E. (2013). Cells, tissues, and organs-on-chips: Challenges and opportunities for the cancer tumor microenvironment. *Integrative Biology, 5*(9), 1096. https://doi.org/10.1039/c3ib40076j

Zapivakhin, I., Ilin, I., & Levina, A. (2018). Public-private partnership as city project management technology. *MATEC Web of Conferences, 170*, 01037. https://doi.org/10.1051/matecconf/201817001037

Zhang, H., Zeng, L., He, Q., Tao, W., Zha, Z., & Hu, C. (2016). The E3 ubiquitin ligase chip mediates ubiquitination and proteasomal degradation of PRMT5. *Biochimica Et Biophysica Acta (BBA) – Molecular Cell Research, 1863*(2), 335–346. https://doi.org/10.1016/j.bbamcr.2015.12.001

Zhang, X., Karim, M., Hasan, M. M., Hooper, J., Wahab, R., Roy, S., & Al-Hilal, T. A. (2022). Cancer-on-a-chip: Models for studying metastasis. *Cancers, 14*(3), 648. mdpi.com

Zheng, F., Xiao, Y., Liu, H., Fan, Y., & Dao, M. (2021). Patient-specific organoid and organ-on-a-chip: 3D cell-culture meets 3D printing and numerical simulation. *Advanced Biology, 5*(6). https://doi.org/10.1002/adbi.202000024

Zhou, C., Li, Z., Lu, K., Liu, Y., Xuan, L., Mao, H., & Wang, X. (2024). Advances in human organs-on-chips and applications for drug screening and personalized medicine. *Fundamental Research*. sciencedirect.com

Zhu, J. (2020). Application of organ-on-chip in drug discovery. *Journal of Biosciences and Medicines*. scirp.org

Zwemer, E., Mahler, H., Werchniak, A., & Recklitis, C. (2011). Sun exposure in young adult cancer survivors on and off the beach: Results from project reach. *Journal of Cancer Survivorship, 6*(1), 63–71. https://doi.org/10.1007/s11764-011-0201-y

Chapter 8
Biological Applications of Acacia Gums (AGs) SME on Breast Cancer Cell Lines

Ahmed A. M. Elnour, Abdurahmani Hamid Nour, Aiman A. Bin Mokaizh, and Mohamed E. S. Mirghani

Abstract This chapter examines the potential anti-tumor effects of solvent fractions from AGs on the human breast cancer cell lines MCF-7 and MDA-MB-231. This chapter observed significant apoptotic effects, including upregulation of Bax and caspases 3 and 9, along with morphological changes. Furthermore, the fractions showed anti-angiogenic effects by downregulating vascular endothelial growth factor (VEGF) expression. Non-toxic concentrations of the fractions also downregulated cyclin D1 and upregulated p21 and p27, indicating their potential as anti-tumor agents. This is the first report to support the use of AGs in traditional Sudanese medicine for BC treatment. These findings also establish a model for evaluating other plant secondary metabolites in the Sudanese flora. The methanol AGs fraction inhibited BC cell proliferation, migration, and colony formation. They arrested cells in the G0/G1 phase, induced apoptosis via the intrinsic pathway, and demonstrated VEGF-mediated anti-proliferative effects, suggesting their therapeutic value. This chapter also explored optimizing doxorubicin conjugation with AGs to potentially reduce adverse effects while maintaining therapeutic efficacy in BC treatment. AGs known as stabilizing properties are essential in making it relevant for pharmaceutical and cosmetic applications because they contain secondary metabolites with various pharmacological effects.

A. A. M. Elnour (✉) · A. H. Nour
Faculty of Chemical and Process Engineering Technology, Universiti Malaysia Pahang Al-Sultan Abdullah (UMPSA), Pahang Gambang, Malaysia

Centre of Excellence for Advanced Research in Fluid Flow (CARIFF), University Malaysia, Pahang Gambang, Malaysia
e-mail: adamhassan@ump.edu.my; abrahman@ump.edu.my

A. A. Bin Mokaizh
Faculty of Chemical and Process Engineering Technology, Universiti Malaysia Pahang Al-Sultan Abdullah (UMPSA), Pekan, Pahang, Malaysia

M. E. S. Mirghani
National Oilseed Processing Research Institute (NOPRI), University of Gezira, Medani, Sudan

A. A. M. Elnour (ed.), *Gum Arabic and Breast Cancer Biology*,
https://doi.org/10.1007/978-981-97-8518-6_8

Keywords Gum arabic · *Acacia senegal* · *Acacia seyal* · *Acacia polyacantha* · Breast cancer · Apoptosis · Anti-angiogenesis

Highlights

1. The AGs fractions showed anti-tumor effects in BC cells.
2. The fractions induced apoptosis, morphological changes, and anti-angiogenic effects.
3. Non-toxic concentrations downregulated cyclin D1 and upregulated p21 and p27 expression.
4. This is the first report on the anti-tumor potential of AGs fractions.
5. This study modeled the evaluation of other Sudanese plant metabolites.
6. Optimizing doxorubicin conjugation with AGs may reduce side effects while maintaining its efficacy.
7. AGs-secondary metabolites have pharmacological effects relevant to pharmaceuticals and cosmetics.

8.1 Introduction

AGs are heterogeneous, complex polysaccharides traditionally composed of highly branched macromolecules and are mainly found in the semi-desert regions of sub-Saharan Africa and adjacent areas of the Middle East (Li et al., 2021). It is classified, in the international market, as "food-grade," where it is used to stabilize flavors and emulsify essential oils in soft drinks, and as "pharmaceutical grade" or "prebiotic grade," where it is used in pharmaceutical and nutraceutical products as a matrix for lyophilized substances (Rhazi et al., 2020). The secondary metabolites of AGs have also drawn attention for their potential use as alternative preservatives in cosmetic and food products (Desam & Al-Rajab, 2021). In fact, AGs are known to produce AGs, and each species produces unique AGs types with different physicochemical properties (Rhazi et al., 2020). These properties are influenced by the composition and structure of the associated phytochemicals, which vary as the species, seasons, and regions from which the AGs exudates are collected (Hasheminya & Dehghannya, 2023). Distinct differences in the types of AGs produced have been reported (Jebeso et al., 2023). Therefore, the unique properties of AGs highlight their potential for diverse applications in food, pharmaceuticals, and cosmetics.

AGs are produced by some species of Acacia (Jebeso et al., 2023). These AGs have long been used as a natural delivery system for various compounds, among other uses in pharmaceutical and nutraceutical products (Amiri et al., 2021). Additionally, AGs and their associated phytochemicals, which possess various bioactivities (antioxidative, antibacterial, prebiotic, anti-inflammatory, etc.), are good candidates for novel bioactive complexes (Manzoor et al., 2020). However, increased anti-cancer activities of phytochemicals when combined with specific bioactives have been reported, and AGs can optimize the delivery of such combinations (Kyriakoudi et al., 2021). This chapter summarizes information on the phytochemicals in AGs derived from the stated species and their potential anti-cancer activities, hoping to encourage their biological applications (Ali et al., 2021). Thus, the phytochemicals in AGs from specific AGs highlight their potential anti-cancer activities, encouraging further exploration of their biological applications.

8.2 Acacia Gums (AGs)-Secondary Metabolites Extract (SME)

However, AGs are not limited to food-related applications (Elnour et al., 2020). Their inherent AGs biological activities and potential as carriers of bioactive compounds also make them relevant for pharmaceutical and cosmetic purposes (Culas et al., 2023). The secondary metabolites in AGs have been studied and have been found to constitute a major class of phytochemicals with various pharmacological effects (Elnour et al., 2018). In this chapter, we focused on the biological applications of AGs-secondary metabolites, particularly anti-cancer, and their potential to act as carriers for existing therapeutic substances to develop novel AGs-based pharmaceuticals (Alyafii, 2021). Organic solvent systems were used to dissolve the AGs. Gum study disk diffusion was performed against the representative BC cell lines, T47D and MCF7, followed by bioassay-guided fractionation to prepare test samples to evaluate their biological activities (Barkeer et al., 2024). The conjugation of doxorubicin with AGs was also optimized using a physical mixing method to combine the chemotherapeutic drug with the natural polymer at a certain weight ratio, which may give rise to a formulation that reduces the adverse effects of doxorubicin while maintaining its therapeutic efficacy in the treatment of BC (Hassani et al., 2020). Hence, this chapter underscores the potential of AGs as carriers for anti-cancer drugs and highlights their broader pharmaceutical and cosmetic applications.

AGs are dried exudates harvested from trees of the genus Acacia and predominantly consist of complex polysaccharides and their glycoprotein derivatives (Rawi et al., 2021). AGs, also known as *Acacia senegal* gum, are the most famous and harvested from Acacia senegal trees, mainly found in Africa's Sudan and sub-Sahelian regions (Tadesse et al., 2020). Subsequently, it has been extensively used in the food industry, particularly as a stabilizing agent for beverage emulsions, owing to its highly branched arabinogalactan structure with various molecular weights (Wafari & Ta'awu, 2021). As natural dietary fibers, AGs also possess a prebiotic effect and may benefit the human digestive system by increasing probiotic activity while decreasing pathogenic bacterial populations (Al-Baadani et al., 2021). These properties make AGs a valuable material for the formulation of functional food products such as dietary supplements or low-calorie, high-fiber beverages (Gupta et al., 2023). Thus, AGs' stabilizing and prebiotic properties of AGs make them valuable for developing functional food products such as dietary supplements and high-fiber beverages.

AGs are closely related to African plant species. Gummiferous AGs produce a unique blend of complex polysaccharides as exudates from their stems and branches, eventually forming gummy materials (Satti et al., 2020). These plant AGs exudates have been used for centuries in traditional medicine for their various biological activities, including antibacterial, antioxidant, hepatoprotective, hypoglycemic, anti-inflammatory, and anti-cancer effects (Barkeer et al., 2024). The secondary metabolites present in AGs can be considered beneficial natural additives and thus potentially used in developing value-added products, such as pharmaceuticals and cosmetics, from these widely available and cost-effective plant gums (Madni et al., 2021). Accordingly, the diverse biological activities of AGs make them promising natural additives for the development of value-added products in pharmaceuticals and cosmetics.

8.3 Breast Cancer: An Overview

Natural products have long been considered a source of new chemical entities for developing therapeutic agents for treating various diseases, including cancer (Huang et al., 2021). More than 80% of the world's population relies on traditional medicine for primary healthcare, and a major part of traditional medicine involves the use of plant extracts and their bioactive components (Mao et al., 2022). Traditional remedies involve long-term treatment; hence, compounds derived from such remedies are often presumed to be relatively non-toxic, at least in the short-term (Gahtori et al., 2023). In addition, many phytoconstituents have shown promising multimodal activities (selective cytotoxicity, effects on multidrug resistance and specific biochemical targets, and hormone-modulating activity), which are essential for the prevention and treatment of BC (Chavda et al., 2023). Many plant gums contain secondary metabolites and have a variety of bioactivities (Kumar et al., 2021). A previous biochemical analysis revealed that AGs contained appreciable amounts of proteins, phenolic compounds, and flavonoids (Tiwari et al., 2023). Thus, the biochemical richness and bioactivity of AGs make them promising candidates for developing therapeutic agents, particularly in the prevention and treatment of cancer.

In recent years, there has been a significant increase in interest in exploring plant-derived natural compounds for their potential anti-cancer properties (Kubczak et al., 2021). This growing interest stems from the desire to understand better and utilize the healing properties of plant-based compounds in cancer treatment (Samuel et al., 2021). As scientists delve deeper into this field of research, they are uncovering a wealth of promising compounds with the potential to revolutionize cancer treatment (Azhari Rad et al., 2024). Some potential compounds, including AGs, significantly affect BC cell lines (Elnour et al., 2020). These natural products have sparked interest in their biological applications and potential applications in cancer therapy (Huang et al., 2021). BC is a major health problem among women, and it is the most frequently diagnosed malignant tumor in women from economically less-developed populations (Prusty et al., 2020). At present, among various cancers, BC is one of the identified cancers that can largely be controlled through detection at an early-stage and through healthy dietary modifications or the use of chemo-preventives to reduce cancer risk (Li et al., 2020). Several drugs have been used clinically for BC (Barzaman et al., 2020). For example, tamoxifen is widely used for the treatment of women with estrogen receptor-positive breast cancer, whereas trastuzumab and lapatinib are used for HER2-positive patients (Prat et al., 2020). However, the clinical use of these drugs is limited because of their side effects and the development of drug resistance (Ward et al., 2020). Therefore, developing new drugs (or alternative therapeutic approaches) for preventing and treating BC is crucial (Britt et al., 2020). Figure 8.1 shows a diagram of HER2/EGFR positivity in breast cancer.

8.3.1 Epidemiology and Impact

Cancer is a major global public health problem, particularly in developed and developing countries (Soerjomataram & Bray, 2021). It is the cause of 7.6 million deaths, and 12.4 million new cases occur every year (Cao et al., 2022). BC is the most diagnosed cancer and the leading cause of death among women worldwide, accounting for approximately 1.38 million new cases and 458,000 deaths (Lakkis et al., 2024). These statistics sufficiently elucidate the need to invest in research on BC (Britt et al., 2020). Because of the side effects of chemotherapy and the excessively high cost, patients tend to seek alternatives offered by traditional medicine (Mwaka et al., 2021). The impact of cancer on individuals and their families is devastating. It is a medical issue with significant emotional and financial burden (Kayser et al., 2021). Patients often experience a range of physical and psychological symptoms because of their condition and many struggles with the financial implications of treatment (Hawke et al., 2023). Furthermore, the long-term effects of cancer can affect a person's ability to work and maintain a normal quality of life (Nekhlyudov et al., 2022). Therefore, continued research and investment in finding effective treatments and, ultimately, a cure for cancer is crucial (Barrios, 2022). Furthermore, the anti-cancer properties of AGs have been extensively studied. Studies have shown that these gums exhibit potent cytotoxic effects on BC cell lines, making them potential candidates for future cancer treatments (Elnour et al., 2020). Additionally, the bioactive compounds in these gums suggest their potential to inhibit tumor growth and metastasis in BC is extremely crucial. Further research is warranted to explore the therapeutic potential of these gums for BC treatment (Waldman et al., 2020). The risk of tumor growth and development of metastatic opportunities in the brain, for example, is illustrated in Fig. 8.2.

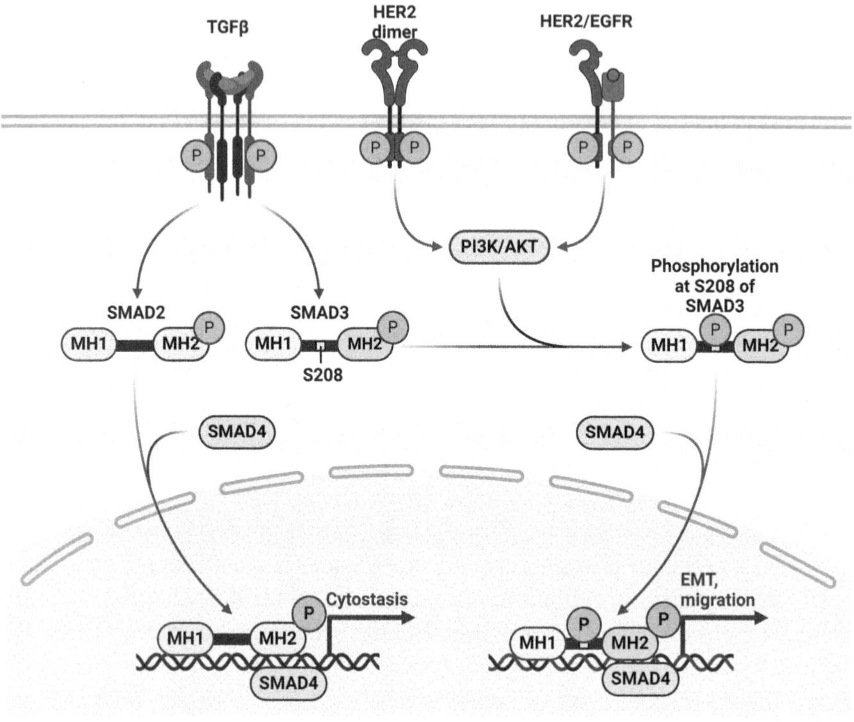

Fig. 8.1 A diagram illustrating HER2/EGFR positivity in breast cancer

8.4 Secondary Metabolites in AGs

The combination of secondary metabolites found in AGs contributes to various biological properties, including tumoricidal effects, which serve as dietary phytochemicals (Batiha et al., 2022). Moreover, this information on secondary metabolites is required to standardize phytochemical-containing compounds based on their respective plant sources (Atiya et al., 2022). Currently, the estimation of quercetin, myricetin, and rutin and their subsequent quantification have been reported in *Acacia catechu* and *Acacia nilotica* using ultra-high-performance liquid chromatography and a photodiode array detector (Batiha et al., 2022). However, to date, there have been no reports on identifying and quantifying quercetin, myricetin, or their glycosides present in AGs (Tiwari et al., 2023). Further research is needed to identify and quantify quercetin, myricetin, and their glycosides in AGs, as their combination of secondary metabolites contributes to various beneficial biological properties.

AGs belong to the Acacia classified under the subfamily Mimosoideae (Batiha et al., 2022). The AGs of these plants are endowed with valuable secondary

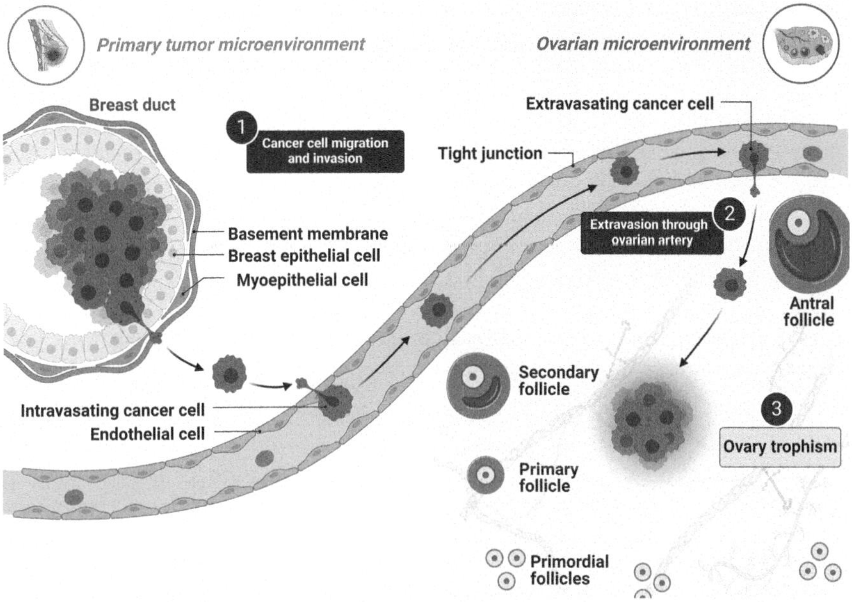

Fig. 8.2 The risk of breast cancer metastasis to the ovary depends on various factors, including the histological type of breast cancer, whether ductal or lobular. The size of the original tumor, especially if it is >5 cm, also plays a significant role in determining the likelihood of ovarian metastasis in breast cancer patients. Additionally, the presence of inflammatory carcinomas is a critical factor that influences the risk of ovarian metastasis. Hormonal changes within a patient can further affect the probability of breast cancer spreading to the ovary. Finally, the time elapsed since the initial diagnosis of breast cancer is an important factor in assessing the risk of ovarian metastasis

metabolites to survive their habitats and ecological needs (Batiha et al., 2022). AGs mainly contain sugar and moderate amounts of protein, but polyphenolic compounds are important dietary fiber constituents (Tiwari et al., 2023). The polyphenolic compounds in AGs are secondary metabolites of AGs and comprise mainly flavonoids, including quercetin, myricetin, and their glycosides, kaempferol and its glycosides, taxifolin, and catechin, as well as non-flavonoid phenolic compounds (Batiha et al., 2022). As dietary phytochemicals, secondary metabolites of AGs contribute to various biological properties, including cytotoxicity (Drabo et al., 2022). Advocated by the biological properties of secondary metabolites found in AGs, this study investigated their cytotoxicity and cellular death pathways in BC cell lines (Tiwari et al., 2023). This study explored the cytotoxicity of secondary metabolites in AGs, emphasizing their potential as valuable dietary phytochemicals.

8.4.1 Chemical Composition and Classification of AGs-Secondary Metabolites

The hydroalcoholic solutions of these three AGs powders were used for a chemical assay of biological organic compounds after freeze-drying, and the resulting AGs matrices were weighed (Atiya et al., 2022). This investigation showed that the composition is dominated by the same organic molecules, whose structural families (carbohydrates, sugars, and manno-proteins) have been previously described in the literature for other AGs (Priya et al., 2023). This study reports on the fractions of the composite fraction's present-day hydroalcoholic solutions, designed to reproduce mixtures dominated by such molecules, and on the AGs powder composition after the freeze-drying process (Sharma, 2021). The organic molecules in AGs powders reinforce their potential utility in various biotechnological applications after freeze-drying.

AGs encompass various species (Amoussa et al., 2020). Secondary metabolites are also extremely diverse because they include flavonoids, sterols, triterpenes, and their derivatives as well as complex tannins exclusively found in AGs (Batiha et al., 2022). These metabolites have a wide range of biotechnological applications that exploit the molecular and sometimes the pharmacological properties of single molecules (D'Amelia et al., 2021). Furthermore, the biological activities of these metabolites as bioregulators in several model organisms have been characterized (Seidi et al., 2021). This study provides deeper insight into the molecular composition of AGs hydroalcoholic solutions that are commonly employed in laboratory settings as organic coatings or scaffolding matrices in the design of nanomaterials for drug delivery and biomedical applications (Moutaharrik et al., 2024). For more information about the chemical composition of AGS-SME (Chap. 3, Sect. 3.5).

8.5 In Vitro Studies of AGs-Secondary Metabolites on Breast Cancer Cells

In vitro investigations have revealed the potential of AGs auriculiformis extract to inhibit the proliferation of BC cells (Sundari & Wijayanti, 2021). Secondary metabolites from plant extracts have demonstrated anti-cancer properties by triggering DNA damage and apoptosis (Khan et al., 2019). Studies on mistletoe extracts have examined their interactions with chemotherapeutic drugs and their cytostatic and cytotoxic effects on various cancer cell lines, including BC cells (Weissenstein et al., 2014). Acacia honey induces apoptosis in human breast adenocarcinoma cell lines (Salleh et al., 2017). These findings underscore the promising anti-cancer potential of natural extracts, highlighting their role in inhibiting cancer cell proliferation and inducing apoptosis.

In the present study, the cytotoxic effects of ethanol extracts of AGs (which are rich in polyphenolics) were evaluated using four BC cell lines (MCF-7, T-47D,

HCC-1937, and MDA-MB-231) and one normal breast cell line (MCF-10A) (Pham et al., 2020). The AGs extract had the highest selectivity toward BC cell lines, with the most sensitive cell line being HCC-1937 (Wawruszak et al., 2021). The AGs extract showed selectivity and good potency toward the estrogen-receptor-positive cell lines (MCF-7 and T-47D), with IC50 values of 64.12 and 54.15 μg/mL, respectively (Elnour et al., 2020). AGs extract may act as an anti-proliferative agent through modulation of the cell cycle (where it inhibits the Wnt/β-catenin pathway) and activates the mitochondria-mediated apoptotic pathway (Wani et al., 2024). The AGs extract also showed moderate activity against MCF-7, T-47D, and MDA-MB-231 cell lines, whereas the AGs extract showed moderate activity against HCC-1937 and MDA-MB-231 cell lines (Maher et al., 2021). These findings suggest that the secondary metabolites of AGs can be developed as chemo-preventive agents for BC (Wani et al., 2024). Thus, these findings suggest that secondary metabolites of AGs can be developed as promising chemo-preventive agents for breast cancer.

BC is the most common malignant disease and a major cause of death among women worldwide. In the search for novel and selective therapeutic and chemo-preventive agents for breast cancer, natural secondary metabolites of AGs can be considered as potential candidates (Alyafii, 2021). In the present study, ethanol extracts of AGs were prepared to investigate their cytotoxic effects on the BC cell lines MCF-7, T-47D, HCC-1937, and MDA-MB-231 and the normal breast cell line MCF-10A (Elnour et al., 2020). The AGs extract showed the highest selectivity toward BC cell lines, particularly the estrogen-receptor-positive MCF-7, T-47D, and HCC-1937 cell lines, with IC_{50} values of 64.12, 54.15, and 58.22 μg/mL, respectively (Campoccia et al., 2021). The AGs extract showed moderate activity against MCF-7, T-47D, and MDA-MB-231 BC cell lines, whereas the AP gum extract showed moderate activity against HCC-1937 and MDA-MB-231 cell lines (Saidu, 2023). An example of in vitro studies of AGs-secondary metabolites and their effects on BC cells is comprehensively discussed in Chap. 9.

8.5.1 Methodologies and Techniques Using Secondary Metabolites in Cancer Cell Lines

In the bioactivity assay, some extracts and fractions showed potential bioactivity against BC cells, with selectivity toward primary BC cells (BB1) (Wang et al., 2023). The bioactive fractions caused the downregulation of BCL-2 and ERBB2 genes associated with breast cancer (Khan et al., 2022). Treatment with the bioactive fraction affects the expression of the E-cadherin gene, which is associated with the loss of invasion and metastasis of breast cancer (Mehralikhani et al., 2021). The trastuzin-like effect of the bioactive fraction suggests that it may be useful for the development of anti-tumor drugs (Sidde et al., 2022). The bioactive fraction induced early and late apoptosis and caused nuclear condensation and DNA fragmentation in MCF-7 cells (Alkahtani et al., 2022). The bioactive fraction is localized in both the cytoplasm and the nucleus of breast cancer cells (Muhammad et al., 2023). These results demonstrate the potential of AGs as secondary metabolites for the treatment of breast cancer (Elnour et al., 2020). Therefore, these results demonstrate the potential of AGs as secondary metabolites for the development of effective treatments against breast cancer.

The methods used in this study included extraction, purification, and analysis of secondary metabolites in the gums (Elnour et al., 2018). Cell culture assays were also conducted to evaluate their effects on breast cancer cell lines. The results showed a significant inhibition of cell growth and proliferation (Maroufi et al., 2020). Apoptosis assays were performed to assess the induction of programmed cell death (Kari et al., 2022). These findings revealed a dose-dependent increase in the population of apoptotic cells (Won et al., 2020). Western blotting was used to examine the expression levels of key proteins involved in cell cycle regulation (Jiang et al., 2020). The results demonstrated the downregulation of cyclin D1 and upregulation of p21, suggesting cell cycle arrest at the G1 phase (Engeland, 2022). Flow cytometry analysis has been used to investigate its impact on cell cycle progression (Semaan et al., 2020). The data indicated a significant accumulation of cells in the G0/G1 phase, further supporting the observed cell cycle arrest (Yin et al., 2020). This study highlights the significant potential of gum-derived secondary metabolites in inhibiting breast cancer cell growth and inducing cell cycle arrest.

Scratch wound assays were used to assess the effects on cell migration and invasion (Luo et al., 2020). The results demonstrated a marked decrease in cell migration and invasion capabilities. Colony formation assays were performed to evaluate the impact on cell clonogenicity (Sulaiman et al., 2021). The results showed a significant reduction in colony formation ability, indicating impaired tumor cell survival and proliferation (Jiang et al., 2020). Additionally, gene expression analysis using quantitative real-time PCR was performed to investigate alterations in the expression of genes related to cancer progression (Zhao et al., 2020). These findings revealed downregulation of oncogenes and upregulation of tumor suppressor genes, suggesting potential anti-cancer activity (Gao et al., 2021). The results also indicated decreased cell proliferation and increased apoptosis in BC cell lines treated with secondary metabolites of AGs (Alyafii, 2021). The extracts and fractions of AGs were tested for bioactivity against BC cell lines (Elnour et al., 2020). The cell lines included primary BC (BB1), which is a ductal carcinoma with an invasive nature, and secondary BC (BT-474 and MCF-7), both of which are lobular carcinomas with a less invasive nature (Safe & Zhang, 2022). The MTT *[3-(4,5-dimethylthiazol-2-yl)-2,5-diphenyl tetrazolium bromide]* cell viability assay was used to determine the percentage of living cells upon treatment with different samples (Kamiloglu et al., 2020). Gene expression was assessed using quantitative real-time polymerase chain reaction (qRT-PCR). Morphological changes and active compound localization were observed using an inverted fluorescence microscope (Papari et al., 2020). Flow cytometry was used to measure apoptosis (Zhu et al., 2021). The most common methods and techniques for employing secondary metabolites to assess cytotoxicity in cancer cells are shown in Fig. 8.3.

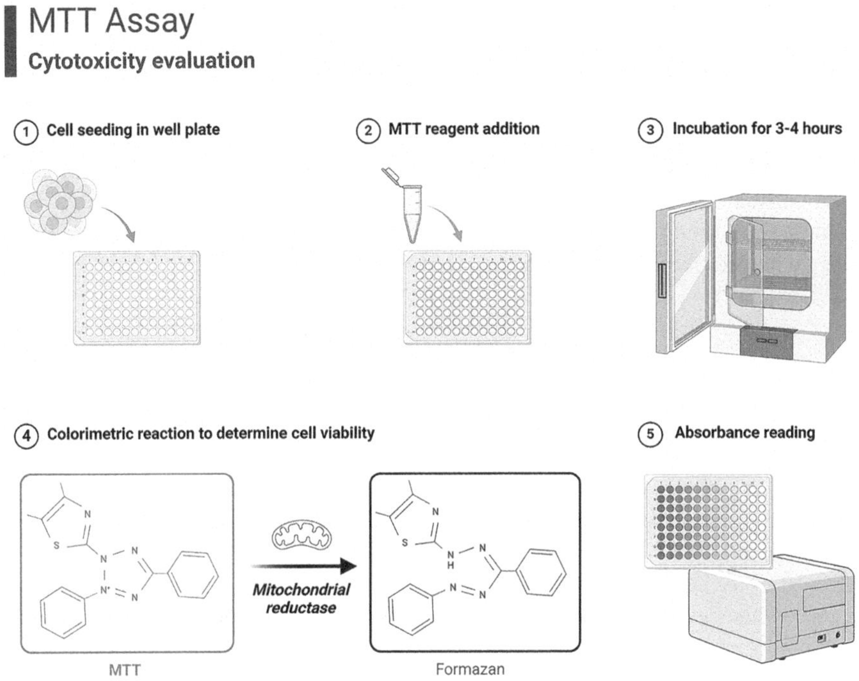

Fig. 8.3 The leading methodologies and techniques for using secondary metabolites in the cytotoxicity evaluation of cancer cell lines

8.6 In Vivo Studies of AGs-Secondary Metabolites on Breast Cancer

In vivo studies on the anti-proliferative effects of the secondary metabolites of AGs on BC cells have been conducted (Elnour et al., 2020). AGs have been found to possess bioactive compounds with pharmacological actions, including secondary metabolites that could potentially impact BC (Batiha et al., 2022). Additionally, the role of *Lactobacillus* plantarum as a probiotic and AGs as a prebiotic in strengthening the gastrointestinal tract could have implications in reducing procarcinogenic enzyme activity in BC (Chundakkattumalayil et al., 2019). Collectively, these studies suggest the potential of AGs and their secondary metabolites to influence BC growth in vivo. These findings highlight the potential of AGs in targeting BC cells. Further research in this area may lead to the development of new therapeutic approaches to combat breast cancer. Overall, the unique properties of AGs and their secondary metabolites make them attractive candidates for further cancer research.

BC is one of the most common types of cancers diagnosed in women. It is the leading cause of cancer death among women, accounting for 23% of all reported cancer cases and 14% of cancer deaths in adults (Afifi et al., 2020). In the search for new chemotherapeutic agents, several plant secondary metabolites have been reported to exert strong anti-carcinogenic activities against BC (Khan et al., 2022). AGs are traditionally used in Sudan to treat tumors (Elnour et al., 2020). The bioactivity of AGs has been proven to be associated with the content of flavanols, phenolic acids, and tannins (Elnour et al., 2022). Additionally, combining the biosynthesized AgNPs with each of the AGs demonstrated a significant enhancement in anti-cancer activity compared to the nanosilver itself (Karade et al., 2021). As a consequence, AGs, with their rich content of flavanols, phenolic acids, and tannins, show significant potential as anti-cancer agents against breast cancer, especially when combined with biosynthesized AgNPs.

8.6.1 Animal Models and Experimental Design

AGs are sources of phytochemicals with a broad-spectrum of biological activities (Atiya et al., 2022). We identified phenolic acids, flavonoids, tannins, and triterpenoids among their secondary metabolites in AGs (Cavazos et al., 2021). Some of these AGs phytochemicals have shown interest in their in vitro cytotoxic activity (Ahmed et al., 2022). Thus, AGs and/or their related secondary metabolites have been hypothesized to inhibit cancer growth (Elnour et al., 2020). To determine this, it was induced BC with adenocarcinoma cells and treated the animals with the compounds (Jiang & Fan, 2020). A volume of 70 μL MCA cells (4646 cells) was injected into the left fourth mammary pad of mice. The tumor cell suspension was loaded into a 0.3-mL insulin syringe, which had been inserted previously in the mammary fat pad and was then slowly and carefully injected (Jacoberger-Foissac et al., 2023). This cell concentration produced visible palpable tumors in 100% of the mice 30 days after tumor cell inoculation (Pan et al., 2022). Treatment of mice with A. macrostachya stem bark for its antibacterial, antibiofilm, and efflux pump inhibitory properties (Barfour et al., 2021). Treatment of mice with *A. polyacantha* 50% MeOH-H_2O crude extract and fractions commenced 1 week after the inoculation of MCA cells.

8.7 Mechanistic Insights into the Anti-cancer Activity

One potential mechanism underlying the anti-cancer activity of AGs-secondary metabolites is their ability to induce apoptosis in BC cell lines (Alyafii, 2021). The anti-cancer activity of a new drug involves various molecular mechanisms (Puah et al., 2021). Drugs interact with specific cancer-related proteins or receptors, triggering downstream effects that lead to cancer cell death. These interactions include inhibiting cancer cell proliferation, induction of apoptosis, and disruption of tumor growth pathways. For instance, the drug may induce apoptosis by phosphorylating specific proteins such as p68 RNA helicase (Dey & Liu, 2012). Additionally, the drug may regulate signaling pathways such as the Janus kinase (JAK)-signal transducer and activator of transcription (STAT) pathway to inhibit cancer cell growth and aggressiveness (Duan et al., 2018). Furthermore, the drug can induce apoptosis via the caspase pathway, a common mechanism associated with the anti-tumor efficacy of many chemotherapeutic agents (Hosseinzadeh et al., 2013). These interactions at the cellular level contribute to the effectiveness of drugs in targeting cancer cells. Additionally, this treatment has shown promise in enhancing innate and adaptive immune responses, further contributing to its effectiveness in combating cancer. Moreover, studies have indicated that the drug exhibits anti-angiogenic properties, which can help prevent cancer growth and spread. The multi-faceted approach of a drug in targeting various aspects of cancer biology makes it a promising candidate for further development and clinical use. An understanding of the anti-cancer mechanisms involving cytokine signaling through the JAK-STAT pathway is presented in Fig. 8.4.

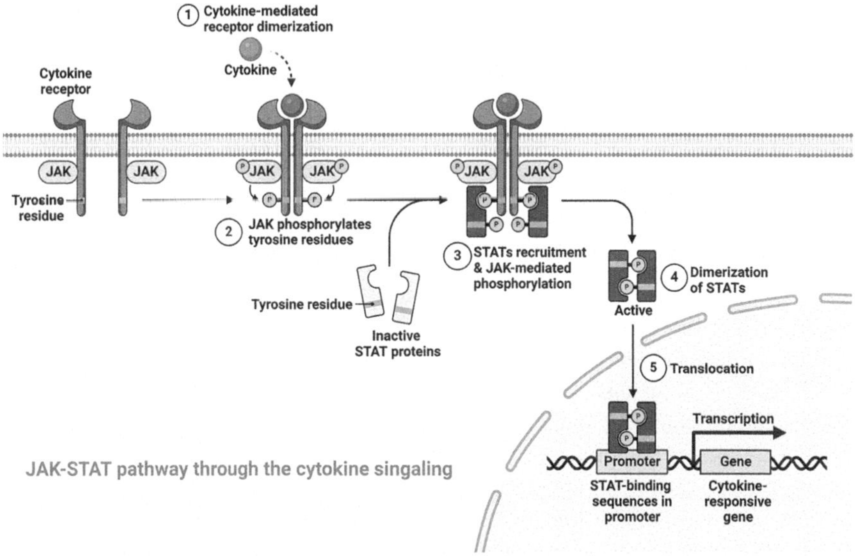

Fig. 8.4 Mechanistic insights into anti-cancer activity through cytokine signaling via the JAK-STAT pathway. In immunology, the JAK-STAT pathway is a crucial signaling mechanism that is activated by cytokines. Cytokines are small proteins released by immune cells that communicate with each other and regulate the immune response. When cytokines bind to their receptors on target cells, receptor-associated Janus kinase (JAK) enzymes are activated, leading to the phosphorylation of STAT proteins. Phosphorylated STATs form dimers and translocate to the nucleus, thereby regulating gene expression and influencing various immune functions

8.7.1 Cell Signaling Pathways and Molecular Targets

Intricate cell signaling pathways involve interactions between key signaling molecules, such as proteins, receptors, and enzymes, leading to specific cellular responses such as proliferation, differentiation, and apoptosis (Zhao et al., 2021). Hormones, growth factors, and cytokines bind to cell surface receptors and initiate intracellular events (Hopkins et al., 2020). For instance, PU.1 regulates cellular communication in the immune system by targeting genes involved in cellular communication (Watt et al., 2021). Inter-organellar calcium signaling also regulates cell metabolism and affects cellular processes (Rimessi et al., 2020). Mito nuclear communication is crucial for homeostasis and stress responses, involving anterograde and retrograde signaling (Walker & Moraes, 2022). The complexity and precision of cell signaling mechanisms emphasize the importance of molecular interactions in cellular regulation (Bludau & Aebersold, 2020). The fidelity and specificity of cell signaling pathways are essential for proper cellular function and the maintenance of tissue homeostasis (de Morree & Rando, 2023). Disturbances in these signaling pathways can lead to various diseases, including cancer, immune disorders, and metabolic diseases (Mehta et al., 2020). Understanding the intricate mechanisms of cell signaling is critical for developing targeted therapeutic interventions and for advancing medical treatments (Qin et al., 2022). Furthermore, advancements in understanding cell signaling are essential for identifying new drug targets and creating innovative treatment strategies (Yip & Papa, 2021). Overall, a comprehensive understanding of cell signaling is paramount for furthering medical research and for improving human health.

8.8 Clinical Relevance and Potential Therapeutic Applications

Moreover, the biological activities observed at diluted concentrations (cytotoxic, modulation of estrogen signaling) of the fractions suggest that their components could serve as adjuvants in designing specific therapies for BC (Alyafii, 2021). Positive interactions only in the P53-mutated MDA-MB-468 cells between A. polyacantha and *A. seyal* components indicate the importance of natural extract combinations for polychemotherapy design (Alyafii, 2021). Finally, this study highlights the growing concern regarding the sustainable management of AGs in African countries. This is the only way to ensure long-term supply of natural products for BC treatment or other therapeutic applications. This chapter presents the synergistic interactive effects of AGs-secondary metabolites on modulation of cancer progression (Elnour et al., 2020). This justifies the traditional use of AGs exudates for cancer treatment. Additionally, the growth-inhibitory effects of the enriched and fractionated *A. polyacantha* + *A. seyal* acetone and Re-Elu-APAc extracts should be considered in more detailed preclinical studies from the perspective of drug

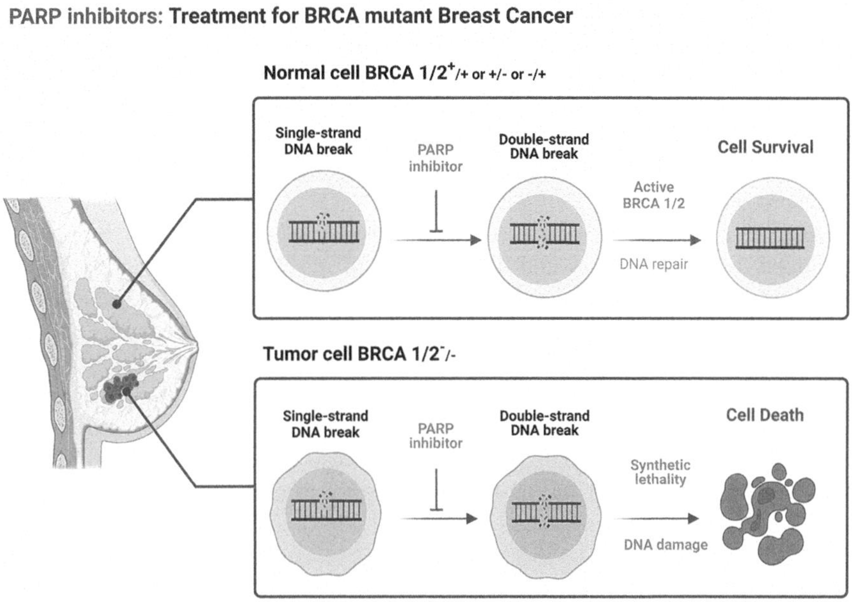

Fig. 8.5 The proposed clinical significance and potential therapeutic uses of AGs in breast cancer through the application of Poly (ADP-ribose) polymerase (PARP) and BReast CAncer gene (BRCA) techniques

development (Elnour et al., 2020). Furthermore, it should be noted that these fractions showed high selectivity for ER-positive phenotypes and P53wt cells, indicating less toxicity to normal non-transformed cells (Alyafii, 2021). The proposed clinical relevance and potential therapeutic applications of AGs in BReast CAncer using poly (ADP-ribose) polymerase (PARP) and BReast CAncer gene (BRCA) techniques are shown in Fig. 8.5.

8.8.1 Challenges and Opportunities

Phytochemical studies of AGs and their derived exudates revealed the presence of a wide variety of biologically active secondary metabolites (Elnour et al., 2022). The challenges and opportunities for utilizing these natural compounds for BC treatment are still being researched and understood (Deepak et al., 2020). The potential for integrating AGs-secondary metabolites into mainstream cancer therapies is an exciting opportunity for future medical treatment (Chavda et al., 2023). These metabolites include terpenes (triterpenoids and sterols), flavonoids, phenolic compounds with their derivatives, and other phytochemicals (Hilal et al., 2024). The identified chemical compounds showed diverse bioactivities, including antioxidant, antimicrobial, anti-inflammatory, and analgesic activities (Tabassum et al., 2023).

Furthermore, the anti-cancer potential of AGs-derived phytochemicals has also been examined (Elnour et al., 2020). Crosstalk and intracellular signaling pathways of A. senegal phytochemicals in breast cancer, both as a single and combined treatment (Shaban et al., 2022). The opportunities and challenges faced by AGs-derived phytochemicals in BC have also been outlined (Alyafii, 2021). The present review will hopefully provide a direction for further research on AGs in Sudan and elsewhere.

Currently, biologically active secondary metabolites of plant origin provide directions for the discovery and development of novel medicines, especially for plant parts that, after their regular biological purposes, become unnecessary, while de facto is waste (Twaij & Hasan, 2022). Some natural compounds have been shown to exhibit anti-cancer activity (Lin et al., 2020). Despite overwhelming success both in understanding progress and in available therapies, cancer continues to be a major cause of mortality and morbidity (Patra et al., 2024). The increasing resistance of cancer cells to currently used drugs further complicates this situation (Sharma et al., 2022). Available therapies and new drug discovery can be improved by screening compounds from natural sources (Najmi et al., 2022). AGs (Fabaceae) are well-known tree species in Sudan that produce gums as exudates from their stems and branches (Jebeso et al., 2023). Gums are complex compounds containing sugars, proteins, and their derivatives that produce antimicrobial peptides after hydrolysis (de Oliveira et al., 2022). Hence, exploring natural compounds from plant waste, like AGs gums, offers significant potential for advancing cancer therapy and overcoming drug resistance.

8.9 Conclusion and Future Directions

This chapter showed that the investigated AGs are soluble fiber secondary metabolites that can be used as potential anti-cancer agents to inhibit the growth of a BC cell line. Low concentrations of the compounds cause cell cycle arrest at the G1 and G2M phases, whereas high concentrations result in massive apoptosis. The pro-apoptotic effect could be regulated by p53 and Bcl-2, since both wild and homozygous disrupted expression cell embryos were reportedly resistant to apoptosis. The ability of the compounds to regulate the mRNA expression of p53 and Bcl-2 was also observed. The potential use of these secondary metabolites as chemo-preventive agents for BC can be explored.

Nearly all the available published reports on AGs point to the taste of *A. senegal* and *A. seyal* gums as the major hindrance to their application in the food and pharmaceutical industries as emulsifying agents. In this study, we report for the first time that after-taste compounds include soluble condensed tannins and cyclitols. The presence of the two groups of compounds explains the "AGs paradox" (biological activity versus no impact on taste). This provides an opportunity to develop techniques for removing taste compounds without affecting the emulsifying activity of the gums. The detection of soluble condensed tannins and cyclitols in *A.*

polyacantha gum suggests that it can also be utilized in food and pharmaceutical industries. In addition, the presence of these compounds is consistent with the traditional use of *A. polyacantha* trees as a source of medicinal compounds.

8.9.1 Summary of Findings

This chapter presents the potential anti-tumor effects of AGs on BC cells and the biological activities of several Sudanese plant secondary metabolites, which are rich in medicinal plants in the kingdom of Sudan's flora. AGs are a complex mixture of polysaccharides and glycoproteins that constitute exudates collected from the bark of the aforementioned AGs trees. In Sudan, AGs have been used in traditional medicine for wound-healing as antioxidants. This chapter evaluated the anti-proliferative effects of n-butanol gum fractions of Sudanese AGs on the human BC cell lines MCF-7 and MDA-MB-231. Significant apoptotic effects were observed, as indicated by the upregulation of Bax and caspases 3 and 9, and morphological changes. The fractions also exhibited anti-angiogenic effects by downregulating VEGF expression. Furthermore, non- toxic concentraions of the AGs fractions. downregulated cyclin D1 and upregulated p21 and p27 expression. These findings indicate the potential anti-tumor effects of AGs solvent fractions on BC cells. This is the first report of such an evaluation, which may further validate the use of AGs in traditional Sudanese medicine to treat BC. Its setup is also a model to evaluate other plant secondary metabolites in Sudanese flora.

The methanol fractions of Sudanese AGs contained bioactive secondary metabolites that inhibited proliferation, migration, and colony formation in the human BC cell lines MCF-7 and MDA-MB-231. The fractions arrested the cells in the G0/G1 cell cycle phase. They induce apoptosis, as demonstrated by morphological changes. An increase in the percentage of cells in the sub-G0/G1 phase, mitochondrial membrane depolarization, and activation of caspase-3 and -9 initiates the intrinsic apoptosis pathway and upregulates Bax. The fractions inhibited angiogenesis, displaying VEGF-mediated anti-proliferative effects non-toxic concentraions of the AGs fractions. downregulated cyclin D1 and upregulated p21 and p27 expression.

Acknowledgments The authors thank the Chemical and Process Engineering Technology at the University of Malaysia, Pahang-UMP, Malaysia, for supporting this research.

Fund This research was funded by Eastern Unity Technology under Grant No. UIC 190806.

References

Afifi, A. M., Saad, A. M., Al-Husseini, M. J., Elmehrath, A. O., Northfelt, D. W., & Sonbol, M. B. (2020). Causes of death after breast cancer diagnosis: A US population-based analysis. *Cancer, 126*(7), 1559–1567.

Ahmed, M., Ahmad, S., Aati, H. Y., Sherif, A. E., Ashkan, M. F., Alrahimi, J., & Korma, S. A. (2022). Phytochemical, antioxidant, enzyme inhibitory, thrombolytic, antibacterial, antiviral, and in silico studies of Acacia jacquemontii leaves. *Arabian Journal of Chemistry, 15*(12), 104345.

Al-Baadani, H. H., Al-Mufarrej, S. I., Al-Garadi, M. A., Alhidary, I. A., Al-Sagan, A. A., & Azzam, M. M. (2021). The use of gum Arabic as a natural prebiotic in animals: A review. *Animal Feed Science and Technology, 274*, 114894.

Ali, I., Ahmed, S. B., Elhaj, B. M., Ali, H. S., Alsubaie, A., & Almalki, A. S. (2021). Enhanced anti-cancer activities of curcumin-loaded green gum acacia-based AgNPs against melanoma and breast cancer cells. *Applied Nanoscience, 11*, 2679–2687.

Alkahtani, S. A., Alshabi, A. M., Shaikh, I. A., Orabi, M. A., Abdel-Wahab, B. A., Walbi, I. A., & Hoskeri, J. H. (2022). In-Vitro cytotoxicity and spectral analysis-based phytochemical profiling of methanol extract of B. hochstetteri and the molecular mechanisms underlying its apoptosis-inducing effect on breast and lung cancer cell lines. *Separations, 9*(10), 298.

AlYafii, S. Y. H. (2021). Molecular Mechanisims Underlying the Anti-Cancer Activity of Gum Arabic from Acacia Sp. In Triple-Negative Breast Cancer Cells. (Master of Science (MS) Thesis). UAEU, https://scholarworks.uaeu.ac.ae/all_theses/941. (941).

Amiri, M. S., Mohammadzadeh, V., Yazdi, M. E. T., Barani, M., Rahdar, A., & Kyzas, G. Z. (2021). Plant-based gum and mucilage applications in pharmacology and nanomedicine: A review. *Molecules, 26*(6), 1770.

Amoussa, A. M. O., Sanni, A., & Lagnika, L. (2020). Chemical diversity and pharmacological properties of the Acacia genus. *Asian Journal of Applied Sciences, 13*(2), 40–59.

Atiya, A., Majrashi, T., Akhtar, S., Khan, A. A., Asiri, A. M. S., Al-Zahrania, H. J., & Saeedhabtar, S. (2022). The value of the genus Acacia in arid and semi-arid environments for the treatment of chronic inflammatory diseases. *Phytomedicine Plus, 2*(3), 100315.

Azhari Rad, R., Naghdi, Y., Majidi Jamalabadi, M., Masoumi, S., Rezakhani, L., & Alizadeh, M. (2024). Tissue engineering scaffolds loaded with a variety of plant extracts: Novel model in breast cancer therapy. *Breast Cancer: Basic and Clinical Research, 18*, 11782234241236358.

Barfour, A. F., Mensah, A. Y., Asante-Kwatia, E., Danquah, C. A., Anokwah, D., Adjei, S., & Mensah, M. L. (2021). Antibacterial, antibiofilm, and efflux pump inhibitory properties of the crude extract and fractions of Acacia macrostachya stem-bark. *Scientific World Journal, 2021*, 5381993.

Barkeer, S., Pothuraju, R., Malakar, P., Pimentel, T. C., Siddiqui, J. A., & Nair, S. A. (2024). Gum acacia dietary fiber: Significance in immunomodulation, inflammatory diseases, and cancer. *Phytotherapy Research, 38*(3), 1509–1521.

Barrios, C. H. (2022). Global challenges in breast cancer detection and treatment. *The Breast, 62*(Suppl 1), S3–S6.

Barzaman, K., Karami, J., Zarei, Z., Hosseinzadeh, A., Kazemi, M. H., Moradi-Kalbolandi, S., & Farahmand, L. (2020). Breast cancer biology, biomarkers, and treatments. *International Immunopharmacology, 84*, 106535.

Batiha, G. E., Akhtar, N., Alsayegh, A. A., Abusudah, W. F., Almohmadi, N. H., Shaheen, H. M., & Waard, M. D. (2022). Bioactive compounds, pharmacological actions, and pharmacokinetics of acacia. *Molecules, 27*(21), 7340.

Bludau, I., & Aebersold, R. (2020). Proteomic and interactomic insights into the molecular basis of cell functional diversity. *Nature Reviews Molecular Cell Biology, 21*(6), 327–340.

Britt, K. L., Cuzick, J., & Phillips, K. A. (2020). Key steps for effective breast cancer prevention. *Nature Reviews Cancer, 20*(8), 417–436.

Campoccia, D., Ravaioli, S., Santi, S., Mariani, V., Santarcangelo, C., De Filippis, A., & Daglia, M. (2021). Exploring the anti-cancer effects of standardized extracts of poplar-type propolis:

In-vitro cytotoxicity toward cancer and normal cell lines. *Biomedicine & Pharmacotherapy, 141*, 111895.

Cao, M., Li, H., Sun, D., He, S., Yan, X., Yang, F., & Chen, W. (2022). Current cancer burden in China: Epidemiology, etiology, and prevention. *Cancer Biology & Medicine, 19*(8), 1121.

Cavazos, P., González, D., Lanorio, J., & Ynalvez, R. (2021). Secondary metabolites, antibacterial, and antioxidant properties of *A. rigidula* leaf extracts, and *Acacia berlandieri* Benth. *SN Applied Sciences, 3*, 522.

Chavda, V. P., Nalla, L. V., Balar, P., Bezbaruah, R., Apostolopoulos, V., Singla, R. K., & Uversky, V. N. (2023). Advanced phytochemical-based nanocarrier systems for breast cancer treatment. *Cancers, 15*(4), 1023.

Chundakkattumalayil, H. C., Kumar, S., Narayanan, R., & Raghavan, K. T. (2019). Role of L. plantarum kx519413 as a probiotic and acacia gum as a prebiotic in gastrointestinal tract strengthening. *Microorganisms, 7*(12), 659.

Culas, M. S., Popovich, D. G., & Rashidinejad, A. (2023). Recent advances in encapsulation techniques for cinnamon bioactive compounds: A review of stability, effectiveness, and potential applications. *Food Bioscience, 57*, 103470.

D'Amelia, V., Docimo, T., Crocoll, C., & Rigano, M. M. (2021). Specialized metabolites and valuable molecules in crops and medicinal plants: The evolution of their use and strategies for their production. *Genes, 12*(6), 936.

de Morree, A., & Rando, T. A. (2023). Regulation of adult stem cell quiescence and its function in tissue integrity maintenance. *Nature Reviews Molecular Cell Biology, 24*, 334–354.

de Oliveira, W. F., Albuquerque, P. B. S., dos Santos Silva, P. M., Coelho, L. C. B. B., & dos Santos Correia, M. T. (2022). Gums. In *Polysaccharides of microbial origin: Biomedical applications* (pp. 43–77). Springer.

Deepak, K. G. K., Vempati, R., Nagaraju, G. P., Dasari, V. R., Nagini, S., Rao, D. N., & Malla, R. R. (2020). Tumor microenvironment: Challenges and opportunities in targeting metastasis in triple-negative breast cancer. *Pharmacological Research, 153*, 104683.

Desam, N. R., & Al-Rajab, A. J. (2021). Importance of natural products in cosmetics. In *Bioactive natural products for pharmaceutical applications* (pp. 643–685).

Dey, H., & Liu, Z. R. (2012). Phosphorylation of p68 RNA helicase by p38 MAP kinase contributes to colon cancer cells apoptosis induced by oxaliplatin. *BMC cell biology, 13*, 1–8.

Drabo, M. S., Shumoy, H., Savadogo, A., & Raes, K. (2022). Inventory of human-edible products from native Acacia sensulato in Africa, America, and Asia: Spotlight on Senegalia seeds, overlooked wild legumes in arid regions. *Food Research International, 159*, 111596.

Duan, M., Du, X., Ren, G., Zhang, Y., Zheng, Y., Sun, S., & Zhang, J. (2018). Obovatol inhibits the growth and aggressiveness of tongue squamous cell carcinoma through regulation of the EGF-mediated JAK-STAT signaling pathway. *Molecular Medicine Reports, 18*(2), 1651–1659.

Elnour, A. A. M., Mirghani, M. E. S., Musa, K. H., Kabbashi, N. A., & Alam, M. Z. (2018). Challenges of extraction techniques of natural antioxidants and their potential application opportunities as anti-cancer agents. *Health Science Journal, 12*(5), 596.

Elnour, A. A., Mirghani, M. E. S., Kabbashi, N. A., Daddiouaissa, D., Musa, K. H., Alam, Z., et al. (2020). Active fractions of crude methanol were obtained from Acacia seyal gum and their antiproliferative effects against human breast cancer cell lines were investigated. *Global Journal of Science Frontier Research, 20*, 51–64.

Elnour, A. A., Mirghani, M. E., Hamid, K. N. A., Musa, F., Shahabipour, N., Ashammakhi, N. H., & Abdurahman, N. (2022). Comparative study of the characterization and extraction techniques of polyphenolic compounds from Acacia seyal gum. *Food Quality and Safety, 6*, fyab034.

Engeland, K. (2022). Cell cycle regulation: p53-p21-RB signaling. *Cell Death and Differentiation, 29*(5), 946–960.

Gahtori, R., Tripathi, A. H., Kumari, A., Negi, N., Paliwal, A., Tripathi, P., & Upadhyay, S. K. (2023). Anti-cancer plant derivatives: Deciphering their oncopreventive and therapeutic potential in molecular terms. *Future Journal of Pharmaceutical Sciences, 9*(1), 14.

Gao, L., Wu, Z. X., Assaraf, Y. G., Chen, Z. S., & Wang, L. (2021). Overcoming anti-cancer drug resistance by restoring tumor suppressor gene function. *Drug Resistance Updates, 57*, 100770.

Gupta, A., Sanwal, N., Bareen, M. A., Barua, S., Sharma, N., Olatunji, O. J., & Sahu, J. K. (2023). Trends in functional beverages: Functional ingredients, processing technologies, stability, health benefits, and consumer perspective. *Food Research International, 170*, 113046.

Hasheminya, S. M., & Dehghannya, J. (2023). Physicochemical, thermal, and rheological characterization of novel biopolymer gum exudate from Astragalus sarcocolla. *Journal of Polymers and the Environment, 31*(3), 965–975.

Hassani, A., Azarian, M. M. S., Ibrahim, W. N., & Hussain, S. A. (2020). Preparation, characterization, and therapeutic properties of gum Arabic-stabilized gallic acid nanoparticles. *Scientific Reports, 10*(1), 17808.

Hawke, L. D., Nguyen, A. T., Sheikhan, N. Y., Strudwick, G., Rossell, S. L., Soklaridis, S., et al. (2023). Swept under the carpet: A qualitative study of patient perspectives on Long COVID, treatments, services, and mental health. *BMC Health Services Research, 23*(1), 1088.

Hilal, B., Khan, M. M., & Fariduddin, Q. (2024). Recent advancements have been made in deciphering the therapeutic properties of plant secondary metabolites, such as phenolics, terpenes, and alkaloids. *Plant Physiology and Biochemistry, 211*, 108674.

Hopkins, B. D., Goncalves, M. D., & Cantley, L. C. (2020). Insulin–PI3K signaling: An evolutionarily insulated metabolic driver of cancer. *Nature Reviews Endocrinology, 16*(5), 276–283.

Hosseinzadeh, R., Khorsandi, K., & Hemmaty, S. (2013). Study of the effect of surfactants on extraction and determination of polyphenolic compounds and antioxidant capacity of fruits extracts. *PloS one, 8*(3), e57353.

Huang, M., Lu, J. J., & Ding, J. (2021). Natural products in cancer therapy: Past, present, and future. *Natural Products and Bioprospecting, 11*(1), 5–13.

Jacoberger-Foissac, C., Allard, B., Allard, D., & Stagg, J. (2023). Assessment of efficacy of immune checkpoint inhibitors in preclinical tumor models. In *The tumor microenvironment: Methods and protocols* (pp. 151–169). Springer.

Jebeso, G., Gebeyehu, D., & Abere, M. (2023). Characteristics and quality of gum Arabic derived from naturally grown Acacia seyal Willdenow trees in the lowland areas of Ethiopia. *Food Chemistry Advances, 2*, 100243.

Jiang, Y., & Fan, L. (2020). Evaluation of anti-cancer activities of Poria cocos ethanol extract in breast cancer: In-vivo and in-vitro identification and mechanism. *Journal of Ethnopharmacology, 257*, 112851.

Jiang, Z., Li, S., Han, M. J., Hu, G. M., & Cheng, P. (2020). High NSUN5 expression promotes cell proliferation via cell cycle regulation in colorectal cancer. *American Journal of Translational Research, 12*(7), 3858.

Kamiloglu, S., Sari, G., Ozdal, T., & Capanoglu, E. (2020). Guidelines for cell viability assays. *Food Frontiers, 1*(3), 332–349.

Karade, V. C., Patil, R. B., Parit, S. B., Kim, J. H., Chougale, A. D., & Dawkar, V. V. (2021). Insights into shape-based silver nanoparticles: A weapon to cope with pathogenic attacks. *ACS Sustainable Chemistry & Engineering, 9*(37), 12476–12507.

Kari, S., Subramanian, K., Altomonte, I. A., Murugesan, A., Yli-Harja, O., & Kandhavelu, M. (2022). Programmed cell death detection methods: A systematic review and categorical comparison. *Apoptosis, 27*(7), 482–508.

Kayser, K., Smith, L., Washington, A., Harris, L. M., & Head, B. (2021). Living with the financial consequences of cancer: A life-course perspective. *Journal of Psychosocial Oncology, 39*(1), 17–34.

Khan, M. I., Bouyahya, A., Hachlafi, N. E., Menyiy, N. E., Akram, M., Sultana, S., & Elebiyo, T. C. (2022). Anti-cancer properties of medicinal plants and their bioactive compounds against breast cancer: A review of recent investigations. *Environmental Science and Pollution Research, 29*(17), 24411–24444.

Khan, T., Ali, M., Khan, A., Nisar, P., Jan, S. A., Afridi, S., & Shinwari, Z. K. (2019). Anticancer plants: A review of the active phytochemicals, applications in animal models, and regulatory aspects. *Biomolecules, 10*(1), 47.

Kubczak, M., Szustka, A., & Rogalińska, M. (2021). Molecular targets of natural compounds with anti-cancer properties. *International Journal of Molecular Sciences, 22*(24), 13659.

Kumar, M., Prakash, S., Radha, K., Pundir, A., Punia, S., & Mekhemar, M. (2021). Beneficial role of secondary antioxidant metabolites from medicinal plants in maintaining oral health. *Antioxidants, 10*(7), 1061.

Kyriakoudi, A., Spanidi, E., Mourtzinos, I., & Gardikis, K. (2021). Innovative delivery systems loaded with plant bioactive ingredients: Formulation approaches and applications. *Plants (Basel), 10*(6), 1238.

Lakkis, N. A., Abdallah, R. M., Musharrafieh, U. M., Issa, H. G., & Osman, M. H. (2024). Epidemiology of breast, corpus uteri, and ovarian cancers in banks with emphasis on breast cancer incidence trends and risk factors compared to regional and global rates. *Cancer Control, 31*, 10732748241236266.

Li, J., Guan, X., Fan, Z., Ching, L. M., Li, Y., Wang, X., & Liu, D. X. (2020). Non-invasive biomarkers for the early detection of breast cancer. *Cancers, 12*(10), 2767.

Li, Z., Zhang, L., Mao, C., Song, Z., Li, X., & Liu, C. (2021). Preparation and characterization of konjac glucomannan and gum Arabic composite gels. *International Journal of Biological Macromolecules, 183*, 2121–2130.

Lin, S. R., Chang, C. H., Hsu, C. F., Tsai, M. J., Cheng, H., Leong, M. K., et al. (2020). Natural compounds as potential adjuvants for cancer therapy: Preclinical evidence. *British Journal of Pharmacology, 177*(6), 1409–1423.

Luo, C., Wang, Y., Wei, C., Chen, Y., & Ji, Z. (2020). Anti-migration and anti-invasion effects of Bruceine D in human triple-negative breast cancer MDA-MB-231 cells. *Experimental and Therapeutic Medicine, 19*(1), 273–279.

Madni, A., Khalid, A., Wahid, F., Ayub, H., Khan, R., & Kousar, R. (2021). Preparation and application of guar gum composites in the biomedical, pharmaceutical, food, and cosmetics industries. *Current Nanoscience, 17*(3), 365–379.

Maher, T., Ahmad Raus, R., Daddiouaissa, D., Ahmad, F., Adzhar, N. S., Latif, E. S., & Mohammed, A. (2021). Medicinal plants with anti-leukemic effects: A review. *Molecules, 26*(9), 2741.

Manzoor, M., Singh, J., Bandral, J. D., Gani, A., & Shams, R. (2020). Food hydrocolloids: Functional, nutraceutical, and novel applications for the delivery of bioactive compounds. *International Journal of Biological Macromolecules, 165*, 554–567.

Mao, J. J., Pillai, G. G., Andrade, C. J., Ligibel, J. A., Basu, P., Cohen, L., et al. (2022). Integrative oncology: Addressing the global challenges of cancer prevention and treatment. *CA: A Cancer Journal for Clinicians, 72*(2), 144–164.

Maroufi, N. F., Vahedian, V., Akbarzadeh, M., Mohammadian, M., Zahedi, M., Isazadeh, A., & Nouri, M. (2020). Apatinib inhibits the breast cancer cell line MDA-MB-231 in-vitro by inducing apoptosis, cell cycle arrest, and regulating the nuclear factor-κB (NF-κB) and MAPK signaling pathways. *Breast Cancer, 27*, 613–620.

Mehralikhani, A., Movahedi, M., Larypoor, M., & Golab, F. (2021). Evaluation of the effect of Foeniculum vulgare on the expression of E-cadherin, dysadherin and Ki-67 in BALB/C mice with 4T1 model of breast cancer. *Nutrition and Cancer, 73*(2), 318–328.

Mehta, M., Dhanjal, D. S., Paudel, K. R., Singh, B., Gupta, G., Rajeshkumar, S., & Dua, K. (2020). Cellular signaling pathways mediating the pathogenesis of chronic inflammatory respiratory diseases: An update. *Inflammopharmacology, 28*, 795–817.

Moutaharrik, S., Meroni, G., Soggiu, A., Foppoli, A., Cerea, M., Palugan, L., & Maroni, A. (2024). Guar gum is a microbially degradable component of an oral colon delivery system based on a combination strategy: Formulation and in-vitro evaluation. *Drug Delivery and Translational Research, 14*(3), 826–838.

Muhammad, S. N. H., Safuwan, N. A. M., Yaacob, N. S., & Fauzi, A. N. (2023). Regulatory mechanism on anti-glycolytic and anti-metastatic activities induced by Strobilanthes crispus in breast cancer in-vitro. *Pharmaceuticals, 16*(2), 153.

Mwaka, A. D., Achan, J., Adoch, W., & Wabinga, H. (2021). From their own perspectives: A qualitative study exploring the perceptions of traditional health practitioners in northern Uganda regarding cancers, their causes, and .... *BMC Family Practice, 22*(1), 155.

Najmi, A., Javed, S. A., Al Bratty, M., & Alhazmi, H. A. (2022). Modern approaches to the discovery and development of plant-based natural products and their analogs are potential therapeutic agents. *Molecules, 27*(2), 349.

Nekhlyudov, L., Campbell, G. B., Schmitz, K. H., Brooks, G. A., Kumar, A. J., Ganz, P. A., Von, A., & D. (2022). Cancer-related impairments and functional limitations among long-term cancer survivors: Gaps and opportunities for clinical practice. *Cancer, 128*(2), 222–229.

Pan, L., Huang, Z., Li, G., Zhan, Q., Zheng, W., Chen, L., & Zhang, X. (2022). A novel and feasible mouse model of a modified inoculation method by subcutaneous EMT6 cell injection for subclinical breast cancer. *Journal of Radiation Research and Applied Sciences, 15*(3), 232–237.

Papari, E., Noruzinia, M., Kashani, L., & Foster, W. G. (2020). Identification of candidate microRNA markers of endometriosis with the use of next-generation sequencing and quantitative real-time polymerase chain reaction. *Fertility and Sterility, 113*(6), 1232–1241.

Patra, S., Pareek, D., Gupta, P. S., Wasnik, K., Singh, G., Yadav, D. D., & Paik, P. (2024). Progress in the treatment and diagnostics of infectious diseases with polymers. *ACS Infectious Diseases, 10*(2), 287–316.

Pham, D. C., Nguyen, H. C., Le Nguyen, T. H., Ho, H. L., Trinh, T. K., Riyaphan, J., & Weng, C. F. (2020). Optimization of ultrasound-assisted extraction of flavonoids from Celastrus hindsii leaves using response surface methodology and evaluation of their antioxidant and antitumor activities. *BioMed Research International, 2020*, 3497107.

Prat, A., Pascual, T., De Angelis, C., Gutierrez, C., Llombart-Cussac, A., Wang, T., & Rimawi, M. F. (2020). HER2-enriched sub-type and ERBB2 expression in HER2-positive breast cancer treated with a dual HER2 blockade. *JNCI: Journal of the National Cancer Institute, 112*(1), 46–54.

Priya, A., Prasath, N. H., Malligarjunan, N., & Pandian, S. K. (2023). In-vitro and in-vivo efficacy of phytoactive-based oral care regimens in the prophylaxis of oral biofilms and augmentation of oral hygiene. In *Pharmacological studies in natural oral care* (pp. 723–773).

Prusty, R. K., Begum, S., Patil, A., Naik, D. D., Pimple, S., & Mishra, G. (2020). Knowledge of symptoms and risk factors of breast cancer among women: A community-based study in a low socio-economic area of Mumbai, India. *BMC Women's Health, 20*, 106.

Puah, B. P., Jalil, J., Attiq, A., & Kamisah, Y. (2021). New insights into the molecular mechanisms underlying the anti-cancer activities of lycopene. *Molecules, 26*(13), 3888.

Qin, S., Xiao, W., Zhou, C., Pu, Q., Deng, X., Lan, L., & Wu, M. (2022). Pseudomonas aeruginosa: Pathogenesis, virulence factors, antibiotic resistance, interaction with host, technological advances, and emerging therapeutics. *Signal Transduction and Targeted Therapy, 7*(1), 199.

Rawi, M. H., Abdullah, A., Ismail, A., & Sarbini, S. R. (2021). Manipulation of the gut microbiota using acacia gum polysaccharides. *ACS Omega, 6*(28), 17782–17797.

Rhazi, L., Lakahal, L., Andrieux, O., Niamba, N., Depeint, F., & Guillemet, D. (2020). Relationship between the molecular characteristics and functional properties of acacia gum. *Food Chemistry, 328*(126), 860.

Rimessi, A., Pedriali, G., Vezzani, B., Tarocco, A., Marchi, S., Wieckowski, M. R., et al. (2020). Interorganellar calcium signaling in the regulation of cell metabolism: A cancer perspective. In *Seminars in cell & developmental biology* (Vol. 98, pp. 167–180). Academic Press.

Safe, S., & Zhang, L. (2022). Role of the aryl hydrocarbon receptor (AhR) and its ligands in breast cancer. *Cancers, 14*(22), 5574.

Saidu, A. (2023). Evaluation of the anticancer potential of stem bark extracts/fraction (s) of prosopis africana and tamarindus indica against mcf-7 cell line.

Salleh, M. A. M., Eshak, Z., & Ismail, W. I. W. (2017). Acacia honey induces apoptosis in human breast adenocarcinoma cell lines (MCF-7). *Jurnal Teknologi, 79*(4).

Samuel, S. M., Kubatka, P., & Büsselberg, D. (2021). Treating cancers using natural medicine: Significance and challenges. *Biomolecules, 11*(11), 1698.

Satti, N. M. E., Ahmed, F. A. M., Bawadekji, A., & Eltahir, S. E. H. (2020). Gum Arabic (Acacia gum): A review. *Journal of the North for Basic and Applied Sciences, 5*(1).

Seidi, F., Yazdi, M. K., Jouyandeh, M., Dominic, M., Naeim, H., Nezhad, M. N., & Mozafari, M. (2021). Chitosan-based blends for biomedical applications. *International Journal of Biological Macromolecules, 183*, 1818–1850.

Semaan, J., El-Hakim, S., Ibrahim, J. N., Safi, R., Elnar, A. A., & El Boustany, C. (2020). Comparative effect of sodium butyrate and sodium propionate on proliferation, cell cycle, and apoptosis in human breast cancer cells MCF-7. *Breast Cancer, 27*, 696–705.

Shaban, N. Z., Sleem, A. A., Abu-Serie, M. M., Maher, A. M., & Habashy, N. H. (2022). Regulation of the NF-κB signaling pathway and IL-13 in asthmatic rats by aerosol inhalation of the combined active constituents of Punica granatum juice and peel. *Biomedicine & Pharmacotherapy, 155*(113), 721.

Sharma, P. (2021). Investigation of the pharmacology of katha powder and its in-vitro cytotoxicity. *Asian Journal of Pharmaceutical and Clinical Research, 14*, 133–140.

Sharma, M., Bakshi, A. K., Mittapelly, N., Gautam, S., Marwaha, D., Rai, N., & Mishra, P. R. (2022). Recent updates on innovative approaches to overcome drug resistance to improve cancer outcomes. *Journal of Controlled-Release, 346*, 43–70.

Sidde, L., Akhil, C., & Manikanta, C. (2022). Assessment of novel peptidomimetics. *Journal of Integral Sciences, 5*(2), 1–8.

Soerjomataram, I., & Bray, F. (2021). Planning for tomorrow: Global cancer incidence and the role of prevention 2020–2070. *Nature Reviews Clinical Oncology, 18*, 663–672.

Sulaiman, S., Arafat, K., Al-Azawi, A. M., AlMarzooqi, N. A., Lootah, S. N. A. H., & Attoub, S. (2021). The combination of butein and frondoside-A exhibited additive anti-cancer effects on tumor cell viability, colony growth, and invasion, and synergism on endothelial cell migration. *International Journal of Molecular Sciences, 23*(1), 431.

Sundari, T. D., & Wijayanti, N. (2021). Antioxidant and anti-proliferative potential of acacia auriculiformis methanol leaves extract against breast cancer cell model. *Jurnal Biologi Tropis, 21*(2), 600–606.

Tabassum, S., Ahmad, S., Ali, B., Usman, F., Jabeen, Q., Sajid-ur-Rehman, M., & Basit, A. (2023). Chemical profiling and evaluation of the toxicological, antioxidant, anti-inflammatory, antinociceptive, and tyrosinase inhibitory potential of Portulacaria afra using in-vitro, in-vivo, and in silico studies. *Arabian Journal of Chemistry, 16*(6), 104784.

Tadesse, W., Dejene, T., Zeleke, G., & Desalegn, G. (2020). Underutilized natural gum and resin resources in Ethiopia for future research and commercial use. *World Journal of Agricultural Research, 8*(2), 32–38.

Tiwari, M., Panghal, A., Mittal, V., & Gupta, R. (2023). Bioactive compounds of acacia, health benefits, and their utilization in the food processing industry: A critical review. *Nutrition and Food Science, 53*(7), 1125–1146.

Twaij, B. M., & Hasan, M. N. (2022). Bioactive secondary metabolites from plant sources: Type, synthesis, and therapeutic uses. *International Journal of Plant Biology, 13*(1), 4–14.

Wafari, U., & Ta'awu, K. (2021). Biotechnological application and the importance of plant gum exudates acacia senegal and acacia seyal to the food industry.

Waldman, A. D., Fritz, J. M., & Lenardo, M. J. (2020). A guide to cancer immunotherapy: From T cell basic science to clinical practice. *Nature Reviews Immunology, 20*(11), 651–668.

Walker, B. R., & Moraes, C. T. (2022). Nuclear–mitochondrial interactions. *Biomolecules, 12*(3), 427.

Wang, H., Chen, L., Yang, B., Du, J., Chen, L., Li, Y., & Guo, F. (2023). Structures, sources, identification/quantification methods, health benefits, bioaccessibility, and products of isorhamnetin glycosides as phytonutrients. *Nutrients, 15*(8), 1947.

Wani, A. K., Prakash, A., Sena, S., Akhtar, N., Singh, R., Chopra, C., & Rahayu, F. (2024). Unraveling molecular signatures in rare bone tumors and navigating the cancer pathway landscapes for targeted therapeutics. *Critical Reviews in Oncology/Hematology, 196*, 104291.

Ward, R. A., Fawell, S., Floch, N., Flemington, V., McKerrecher, D., & Smith, P. D. (2020). Challenges and opportunities for drug resistance in cancer. *Chemical Reviews, 121*(6), 3297–3351.

Watt, S., Vasquez, L., Walter, K., Mann, A. L., Kundu, K., Chen, L., & Soranzo, N. (2021). Genetic perturbation of PU. 1 binding and chromatin looping of neutrophil enhancers are associated with autoimmune diseases. *Nature Communications, 12*(1), 2298.

Wawruszak, A., Borkiewicz, L., Okon, E., Kukula-Koch, W., Afshan, S., & Halasa, M. (2021). Vorinostat (Saha) and breast cancer: An overview. *Cancers, 13*(18), 4700.

Weissenstein, U., Kunz, M., Urech, K., & Baumgartner, S. (2014). Interaction of standardized mistletoe (Viscum album) extracts with chemotherapeutic drugs on cytostatic and cytotoxic effects in-vitro. *BMC Complementary and Alternative Medicine, 14*(1), 6.

Won, Y. S., Kim, J. H., Lizardo, R. C. M., Min, H. J., Cho, H. D., Hong, S. M., & Seo, K. I. (2020). Flavonol isoquercitrin promotes mitochondrial-dependent apoptosis in SK-Mel-2 melanoma cells via the PI3K/AKT/mTOR pathway. *Nutrients, 12*(12), 3683.

Yin, S., Yang, H., Zhao, X., Wei, S., Tao, Y., Liu, M., & Li, J. (2020). The anti-malarial agent artesunate induced G0/G1 cell cycle arrest and apoptosis by increasing intracellular ROS levels in normal liver cells. *Human & Experimental Toxicology, 39*(12), 1681–1689.

Yip, H. Y. K., & Papa, A. (2021). Signaling pathways in cancer: Therapeutic targets, combinatorial treatments, and new developments. *Cells, 10*(3), 659.

Zhao, Z., Song, J., Tang, B., Fang, S., Zhang, D., Zheng, L., & Ji, J. (2020). CircSOD2 induced epigenetic alteration drive hepatocellular carcinoma progression by activating the JAK2/STAT3 signaling pathway. *Journal of Experimental & Clinical Cancer Research, 39*, 1–16.

Zhao, H., Wu, L., Yan, G., Chen, Y., Zhou, M., Wu, Y., & Li, Y. (2021). Inflammation and tumor progression: Signaling pathways and targeted interventions. *Signal Transduction and Targeted Therapy, 6*(1), 263.

Zhu, C., Zhang, C., Cui, X., Wu, J., Cui, Z., & Shen, X. (2021). Trichosanthin inhibits cervical cancer by regulating oxidative stress-induced apoptosis. *Bioengineered, 12*(1), 2779–2790.

Chapter 9
Virtual Screening of Gum Arabic Compounds for Breast Cancer-Related Receptors: Target Selection and Therapeutic Potential

Ahmed A. M. Elnour, Abdurahmani Hamid Nour, Yousif H. Y. Ahmed, and Kourwel Kuai Kourwel

Abstract This chapter examines the virtual screening of gum arabic (GA) compounds targeting breast cancer (BC)-related receptors. It highlights the innovative use of virtual screening to identify potential therapeutic compounds from natural sources, with a focus on GA, for breast cancer treatment. BC is a significant global health concern, and the limitations and side-effects associated with conventional treatments have prompted the exploration of alternative therapeutic avenues. This chapter highlights the potential of GA compounds to reshape BC treatment through pivotal receptor interactions, emphasizing the role of natural compounds and computational methods in drug discovery for BC treatment. The virtual screening process involved strategic target selection, compound library creation, molecular docking, and absorption, distribution, metabolism, and excretion (ADME) assessment. This process identifies GA compounds with potential interactions with BC-related receptors such as HER2, ERα, and ERβ. The calculated docking scores enabled the ranking of GA compounds based on predicted affinities, leading to the validation of binding stability via molecular dynamics and the identification of lead compounds with robust interactions. Furthermore, this chapter underscores the importance of experimental validation in bridging the gap between

A. A. M. Elnour (✉) · A. H. Nour
Faculty of Chemical and Process Engineering Technology, Universiti Malaysia Pahang Al-Sultan Abdullah (UMPSA), Pahang Gambang, Malaysia

Centre of Excellence for Advanced Research in Fluid Flow (CARIFF), University Malaysia, Pahang Gambang, Malaysia
e-mail: adamhassan@umpsa.edu.my; abrahman@umpsa.edu.my

Y. H. Y. Ahmed
National University, Khartoum, Sudan

K. K. Kourwel
Department of Chemistry, School of Applied and Industrial Sciences, University of Juba, Juba, South Sudan

A. A. M. Elnour (ed.), *Gum Arabic and Breast Cancer Biology*,
https://doi.org/10.1007/978-981-97-8518-6_9

virtual screening and clinical translation, solidifying predictions for potential clinical applications. Experimental validation is crucial to ensure the selective binding and therapeutic efficacy of GA-based compounds and assesses their pharmacokinetics, biodistribution, and safety. This chapter also discusses the morphological changes induced by GA compounds in BC cells, highlighting their potential impact on cell survival and proliferation. Additionally, the safety profile of GA, as confirmed by regulatory agencies, supports its potential as a safe and well-tolerated treatment. In conclusion, the comprehensive research process involved in targeting and screening GA compounds for BC highlights their promising potential for BC treatment. This study provides a robust foundation for future research and potential clinical applications in BC therapeutics, emphasizing the need for rigorous experimental validation to ensure the efficacy and safety of GA-based compounds.

Keywords Breast cancer · Gum arabic compounds · Drug discovery · Molecular docking · Virtual screening

Highlights

1. GA compounds target breast cancer receptors via virtual screening.
2. GA holds promise as a breast cancer therapy with natural compounds.
3. Validation is crucial for confirming the efficacy and safety of GA.
4. Explored the influence of GA on breast cancer cell growth.
5. GA is safe and effective against triple-negative breast cancer.

9.1 Introduction

Breast cancer (BC) is a major health issue that has a significant impact on millions of people across the world. Every year, countless individuals and their families are faced with the devastating news of BC diagnosis. The urgency of addressing BC arises from its widespread impact on individuals and communities. The incidence of this type of cancer in women is on the rise, making it the most frequent form of cancer affecting them. Despite advancements in medical science, BC remains a formidable challenge, with conventional treatments often accompanied by limitations and side-effects (Gupta et al., 2023). Currently, conventional BC treatment options consist of surgery, chemotherapy, radiation therapy, and hormone therapy. Although these treatments have been effective in many cases, they have drawbacks (Winer et al., 2013). For example, chemotherapy delivers drugs systemically to combat the uncontrolled proliferation of cancer cells (Liu et al., 2018b). Nonetheless, the indiscriminate targeting of anti-cancer drugs can result in a host of side-effects and suboptimal drug delivery (Alven & Aderibigbe, 2020). In recent years, there has been a shift toward more targeted therapies for BC treatment (Masood, 2019). These therapies are designed to target critical receptors that are instrumental in the growth and advancement of BC (Abbas et al., 2021). One such type of receptor-targeted therapy is hormone therapy, which targets hormone receptors such as the estrogen and progesterone receptors to inhibit the growth of hormone-receptor-positive BC (Chegini et al., 2023). Another targeted therapy is the use of monoclonal antibodies against human epidermal growth factor receptor 2 (HER2) (Tokunaga et al., 2006). These antibodies and chemotherapeutic agents have significantly improved the therapeutic outcomes (Pickard & Williams, 2014). However, despite advancements in receptor-targeted therapies, their efficacy still requires improvement. One of the main challenges is heterogeneity within BC itself (Fukano et al., 2021). This heterogeneity makes it difficult to accurately target all cancer cells with receptor-specific therapies, leading to incomplete responses and the development of resistance (Londoño et al., 2018).

GA, also known as Acacia gum, is not only a dietary fiber, but also a promising avenue in the fight against BC (Elnour et al., 2020). Its unique chemical composition, consisting of complex carbohydrates and glycoproteins, has been studied for its potential therapeutic effects (Babiker et al., 2012). In addition, GA has a long history of medicinal use in various cultures, further highlighting its potential as an asset in modern oncology (Al-Fahad & Al-Mashhdani, 2023).

Through virtual screening, we explored the potential of GA as a novel receptor-targeted therapy for BC (Dauqan & Abdullah, 2013). BC remains a pressing global health concern and is a constantly challenging medical science for the development of more effective treatments. One approach that has shown promise in improving therapeutic outcomes in patients with BC is the use of receptor-targeted therapies (Yin et al., 2020). Therapies, such as anti-estrogen and monoclonal antibodies against HER2 receptors, have significantly improved treatment responses (Sun et al., 2021a). However, despite these advancements in receptor-targeted therapies, their efficacy (Shaw et al., 2014). Some of these limitations include incomplete targeting of all cancer cells due to the heterogeneity within BC itself, leading to incomplete responses and the development of resistance to these therapies (Marusyk et al., 2012). Addressing these limitations and improving the efficacy of receptor-targeted therapies are crucial for the fight against BC. One potential avenue for enhancing the effectiveness of receptor-targeted therapies is GA (Elnour et al., 2020).

GA, often referred to as Acacia gum, is a natural polysaccharide derived from the sap of several species of Acacia trees (Dickinson, 2003). Its unique chemical composition, consisting of complex carbohydrates and glycoproteins, provides a range of therapeutic properties (Atgié et al., 2019). GA has a long history of medicinal use, primarily as a dietary fiber and natural emulsifier (Leonida et al., 2023). Recent studies have also suggested that AGs may have anti-cancer properties (Elnour et al., 2020). Virtual screening, a cutting-edge tool in drug discovery, can be utilized to explore the potential of GA as a receptor-targeted therapy for BC (Hussein AlYafii, 2021). By utilizing virtual screening algorithms, researchers can analyze the molecular structure of GA and predict its binding affinity to specific targets involved in BC progression (Krasoulis et al., 2022). This method enables the prioritization and selection of substances more likely to interact with the desired target.

9.2 Morphological Changes of BC-Cells

9.2.1 Gum Arabic Extract Induces Morphological Changes in BC Cells MCF-7

Elnour et al (2020) studied ASG–AF and PTC-AF samples produced highly significant results ($p \leq 0.5$) when tested on the MCF-7 human cell line. Specifically, the IC_{50} value for PTC-AF was almost twice the mean for MCF-7 cell lines (18.58 ± 0.03 μg/mL), while ASG–AF had a mean IC_{50} value of 12.17 ± 0.08 μg/mL for the same cells as shown in Fig. 9.1. Furthermore, Fig. 9.1 displayed images of surviving MCF-7 cells after a 24-h incubation period, illustrating noticeable

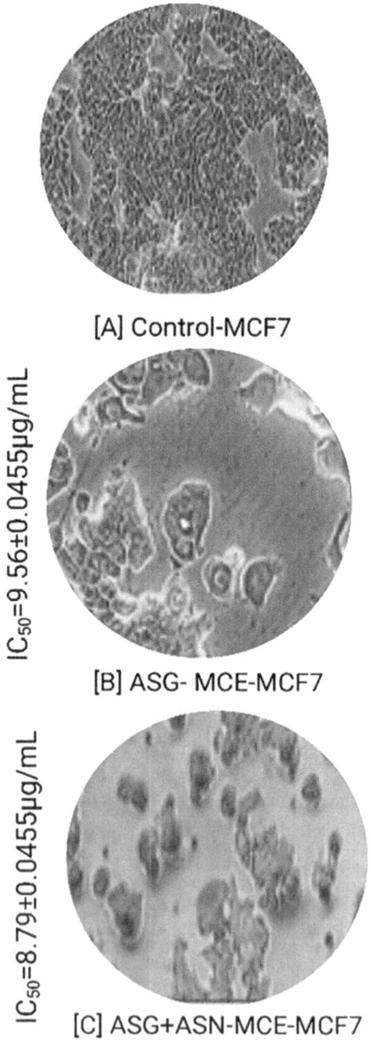

Fig. 9.1 Viable MCF-7 cells after a 24-h incubation period, showing the effects of Acacia seyal gum, Prebio-T crude extract from methanol, and their active fractions. (Source: Adapted from Elnour et al. (2020))

differences between the effects of the two samples on cell survival rates. The study suggested that solvent partitioning could positively impact the screening and characterization of various human cell lines.

9.2.2 *Gum Arabic Extract (GAE) Causes Morphological Changes in MCF-7 compared to MDA-MB-231 Breast Cancer Cells*

The study evaluated the morphological changes in MCF7 cells consider stages of breast cancer cell morphological changes exposed to GAE over the time (Elnour etal. 2020). As depicted in Fig. 9.2, the untreated MCF-7 cells (control) maintained their healthy state and original morphology, characterized by an elongated-spindle shape, adherence, and cell-to-cell contact throughout the treatment periods. This indicated a regular rate of proliferation (Fig. 9.2). In contrast, the MCF-7 cells that were treated underwent changes in their morphology based on both concentration and time. At high concentrations (125 and 500 µg/mL) of GAE treatment, the cells began to lose their original shape after 24 h (Elnour et al., 2020). At concentrations of 62µg/mL and below, treatment resulted in cells losing their elongated-spindle shape and fine membrane extensions after 48 h. In contrast, untreated cells appeared enlarged and flattened and cell-to-cell contact diminished (Elnour et al., 2020). At higher concentrations of 125 and 500 µg/mL, there was a noticeable reduction in cell size and number and membrane blebbing, with floating cells (dead) being more apparent after extending the incubation time to 72 h. Interestingly, (Hussein AlYafii, 2021) studied MDA-MB-231 cells, which found significant reduction in cell size and number at 70 and 100 µg/mL using AGE at 48h. These observed morphological changes using MCF7 cells are typical signs of cell death, possibly through apoptosis, as shown in Fig 9.2.

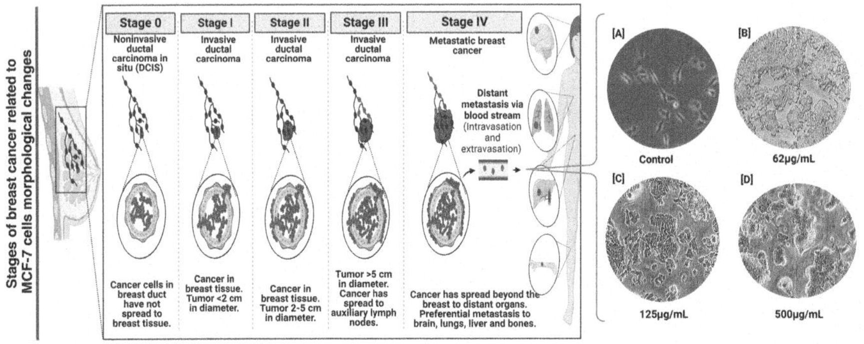

Fig. 9.2 Visual representations of MCF7 cells' morphology after being treated with escalating concentrations of GAE for 48 h are depicted below. These images were captured using the EVOS XL Core Cell Imaging System (Life Technologies), with a magnification of 40×. The white dashed arrows indicate dying cells, the yellow arrows denote elongated-spindle-shaped cells, and the white arrows point to flattened, enlarged cells (Elnour et al., 2020)

9.3 Breast Cancer (BC)-Related Receptors as Therapeutic Targets

BC is a devastating disease affecting millions of lives worldwide (Borgquist et al., 2018). Conventional therapies, while effective to some extent, often lack specificity for tumor tissue, leading to intolerable side-effects and the development of resistance (Hedhili, 2024). In recent years, receptor-targeted therapies have remarkably advanced to deliver highly specific treatments directly to cancer cells (Haque & Desai, 2019). These therapies have significantly improved treatment response, particularly in hormone-receptor-positive and HER2-positive BC (Slamon et al., 2011). Despite these advancements, there are still limitations to current receptor-targeted therapies (Schneider et al., 2008). One limitation is the heterogeneity within BC itself, which poses a challenge in developing therapies that can effectively target all cancer cell types (Keller et al., 2011). Another limitation is the potential for resistance to develop over time, as cancer cells adapt and find ways to evade targeted treatments (An et al., 2021). Receptor-targeted therapies are often associated with significant side-effects. These limitations of conventional treatments highlight the need for alternative approaches to improve efficacy while minimizing side-effects (McIntosh, 2021).

One of the most intriguing aspects of GA is its potential as a receptor-targeted therapy for BC (Elnour et al., 2020 and Hussein AlYafii, 2021). Virtual screening, a cutting-edge technology in drug discovery, has been utilized to analyze the molecular structure of GA and predict its binding affinity to specific targets involved in BC (Ohue et al., 2017). This method uncovered that GA comprises bioactive molecules that are more likely to connect with the intended target. These bioactive compounds in GA have shown promise in inhibiting cell proliferation, inducing apoptosis (cell death), and suppressing the growth of BC cells in preclinical studies (Elnour et al., 2020). Virtual screening has allowed researchers to identify specific compounds within GA that have the potential to selectively target BC cells while minimizing harm to healthy cells (Hussein AlYafii, 2021).

Moreover, GA has shown the potential to enhance the efficacy of conventional treatments by acting as a synergistic agent (Hassani et al., 2020). Virtual screening, a cutting-edge technology in drug discovery, has been utilized to analyze the molecular structure of GA and predict its binding affinity to specific targets in BC. This method uncovered the presence of bioactive compounds in GA that have a higher likelihood of interacting with the desired target (Singthong et al., 2014). Furthermore, using GA as a therapeutic agent is particularly exciting because of its low toxicity and minimal side-effects (Mortensen et al., 2017a). The unique chemical composition of GA, which includes polysaccharides, proteins, and polyphenols, offers a new avenue for developing targeted therapies for BC (Hassani et al., 2020). By harnessing the potential of GA, researchers are paving the way for more effective and safer treatments for BC (Elnour et al., 2020). However, it is important to acknowledge that further research is needed to fully understand the molecular mechanisms of action and optimize the therapeutic potential of GA in BC.

The virtual screening methodology of GA compounds for BC-related receptors encompasses a systematic exploration to assess the anti-cancer potential of GA extract (Hassani et al., 2020). The primary goal of this process is to address the difficulties associated with Triple-Negative BC (TNBC), which is a highly aggressive and treatment-resistant form of cancer that does not respond to conventional hormonal therapy or chemotherapy (Paul et al., 2014). TNBC cells are distinguished by the absence of estrogen receptors (ER), progesterone receptors (PR), and human epidermal growth factor receptor 2 (HER2), which are typically present in other types of BC cells, render them unresponsive to hormonal treatments and resistant to standard chemotherapies (Wei et al., 2021).

This chapter carefully examined the anti-cancer properties of GA extract (GAE) against the MCF7 cell line, a renowned triple-negative human breast cancer model (Barnawi, 2021). According to the findings, the study revealed that GAE suppressed cell growth in a concentration- and time-dependent manner (Begum et al., 2018). The G1/S phase inhibition leads to the arrest of the cell cycle and triggers apoptosis (Begum et al., 2018). The activation of caspase 3/7 and the detection of cleaved caspase 3 and cleaved poly(ADP-ribose) polymerase (PARP) confirmed the existence of significant apoptotic processes, supporting the effects observed (Hannan et al., 2020). The administration of GAE caused the MCF-7 cells to undergo a G1/S cell cycle arrest after cells were exposed to GAE at specific concentrations for 48 h and then examined using cell analyzer techniques. (Hussein AlYafii, 2021). The distribution of cells throughout the cell cycle was analyzed after gum arabic (GAE)-induced G1/S cell cycle arrest, revealing the proportion of cells in each phase. A significant difference between the control and treated groups was observed, with a p-value of less than 0.05.

Further investigations have identified molecular targets and elucidated the mechanism underlying the anti-cancer effects of GAE (Babiker et al., 2012). It was found that gum arabic (GAE) hinders the Wnt/β-catenin signaling pathway in MDA-MB-231 cells by reducing cell cycle regulatory proteins and inhibiting Wnt-targeted genes (Li et al., 2016). This provides a clear understanding of the molecular mechanisms by which GAE its effects, positioning it as a potential chemotherapeutic agent against highly chemoresistant TNBC (Gashua et al., 2015). It emphasized GA positive safety record, acknowledging its designation as a safe food supplement and feed additive by regulatory authorities such as the FDA, Food and Agriculture Organization (FAO)/WHO, and the European Union (Mortensen et al., 2017b). This safety assurance supports the potential of GA as a viable and well-tolerated treatment option for BC (Hassani et al., 2020).

This chapter also, invistigated that GA-derived from AGs exhibits anti-cancer properties against TNBC cells by regulating the cell cycle and promoting apoptosis (Al-Zharani et al., 2023). The objectives were to evaluate the anti-cancer effects of GAE on selected human cancer cell lines, investigate growth-inhibitory mechanisms, and identify biochemical events and biomarkers related to the cells' response to GAE (Elnour et al., 2020). Overall, the virtual screening methodology for GAE compounds has laid a foundational framework for further research and potential clinical applications in BC treatment (Hussein AlYafii, 2021). Figure 9.3 presents a

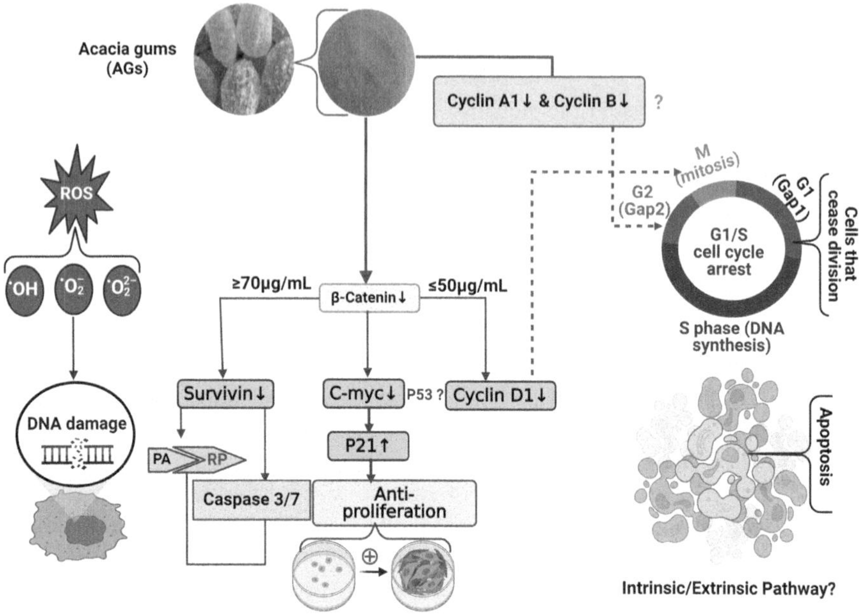

Fig. 9.3 A model illustrating the mechanism of gum arabic extract (GAE)-induced anti-cancer effects in MCF-7- BC cells

model showing the mechanism behind GAE-induced anti-cancer effects in MCF-7 -BC cells.

9.4 Virtual Screening Methodology (VSM)

Virtual screening method (VSM) is a cutting-edge technology that has revolutionized the field of drug discovery. It allows researchers to analyze the molecular structures of various compounds and predict their binding affinity to specific targets involved in disease processes such as BC (Bai et al., 2013). By utilizing VSM, we can identify bioactive compounds within GAE that are more likely to bind to the target of interest in BC (Elnour et al., 2020). This groundbreaking approach saves time and resources compared with traditional laboratory experiments and broadens the scope of potential drug candidates by exploring a vast array of chemical structures (Elnour et al., 2020). Furthermore, VSM bridges the gap between traditional knowledge and contemporary drug discovery by integrating ancient medicinal uses of GAE with modern scientific techniques (Akahane et al., 2020). Through VSM, researchers have discovered that GAE contains bioactive compounds that can potentially target and treat BC. Transitioning GA from traditional use to modern oncology is an exciting prospect that holds promise for improving BC treatment (Nishi & Jayakrishnan, 2004). VSM offers a significant advantage in optimizing the physicochemical properties of GA compounds (Yin et al., 2021). This optimization can improve the effectiveness and reduce potential adverse effects (Rezwendy et al., 2018). VSM can help researchers identify GA compounds with favorable properties for direct tumor targeting while minimizing damage to non-tumor cells (Hussein AlYafii, 2021). The unique chemical composition of GA further enhances its potential as an ally in the fight against BC (Elnour et al., 2020). It comprises a complex mixture of polysaccharides and glycoproteins, contributing to its therapeutic properties (Liang et al., 2020). These components exhibit anti-inflammatory, antioxidant, and immunomodulatory properties, essential for combating cancerous growth and bolstering the body's natural defense mechanisms (Elnour et al., 2018).

One of the significant benefits of VSM technology is that it allows for the concealment of the unfavorable physicochemical properties of the incorporated molecules (Maia et al., 2020). This can be especially beneficial in BC treatment, as it allows for efficient and specific delivery of anti-cancer drugs to the tumor site while minimizing damage to healthy cells (Lionta et al., 2014). This approach allows the discovery of novel receptor-targeted therapies that have the potential to revolutionize BC treatment (Saha & Nanda, 2016). Furthermore, GA holds promise in oncology because of its unique chemical composition and historical medicinal uses (Ashour et al., 2022). Its rich history of use in traditional medicine, coupled with its potential as a targeted therapy in modern oncology, positions GA as a promising candidate for BC treatment (Elnour et al., 2020). Treating brain tumors takes advantage of this characteristic because of the efficient and specific brain delivery of anti-cancer drugs (Wei et al., 2014). Although GA has demonstrated promise in BC treatment, its mechanisms of action and potential side-effects require further investigation through additional research and clinical trials.

The virtual VSM was implemented in the research examining the molecular mechanisms underlying the anti-cancer activity of GA and its potential as an anti-cancer agent for BC-related receptors, specifically focusing on triple-negative breast cancer (TNBC) (Hussein AlYafii, 2021). This chapter focused primarily on TNBC, which is well-known for its highly aggressive behavior and resistance to both traditional chemotherapy and hormonal treatments (Hussein AlYafii, 2021). TNBC is distinguished by the absence of hormone receptors, including ER, PR, and human epidermal growth factor receptor 2 (HER2) (Khosravi-Shahi et al., 2019). This lack of receptors renders them unresponsive to hormonal treatments and highly resistant to standard chemotherapy regimens (Foulkes et al., 2010). Thus, the choice of TNBC underscores the critical demand for new and effective treatment modalities for this particularly difficult sub-type of BC.

To evaluate the anti-cancer activity, VSM were employed to assess the effectiveness of GAE against the human BC model known as MCF-7 or MDA-MB-231, which exhibits triple-negative characteristics and have been widely used as an experimental cell line (Phung et al., 2020). The study uncovered that GAE diminished cell growth in a manner that depended on both concentration and time (Hussein AlYafii, 2021). It led to a halt in the cell cycle at the G1/S phase transition and activated apoptosis. These cellular effects were verified by activating caspase 3/7 and detecting cleaved forms of caspase 3 and poly (ADP-ribose) polymerase (PARP) (Hussein AlYafii, 2021). This study provided a detailed examination of the molecular targets and mechanisms underlying the anti-cancer effects of GAE. GAE inhibited the Wnt/β-catenin signaling pathway in MDA-MB-231 cells (Hussein AlYafii, 2021). This inhibition results in the downregulation of proteins that regulate the cell cycle and suppression of genes targeted by the Wnt pathway. These findings highlight a specific molecular mechanism through which GAE exerts its anti-cancer effects and underscore its potential utility as a chemotherapeutic agent, particularly against TNBC (Hussein AlYafii, 2021).Therefore, they are resistant to conventional therapies.

Additionally, the study addressed the safety and regulatory status of GA, emphasizing its established use as a food supplement and feed additive (Anderson, 1986). Recognized by regulatory authorities such as the FDA, FAO/WHO, and the European Union, GA's safety profile is well documented, suggesting that it could be a viable and well-tolerated treatment option for BC therapies (Anderson & Eastwood, 1989). The objectives outlined in this study were to evaluate the anti-cancer effects of GAE on selected human BC cell lines, investigate the mechanisms behind its growth-inhibitory effects, and identify biochemical markers and downstream responses associated with the cellular impact of the extract (Elnour et al., 2020). In summary, this study's VSM approach methodically assessed the anti-cancer capabilities of GAEs. This study aimed to clarify the molecular targets of the extract, elucidate its mechanism of action, and evaluate its safety and efficacy as a therapeutic option for TNBC (Hussein AlYafii, 2021). These insights lay the groundwork for further research and potential clinical applications in BC treatment, offering promising directions for future research. Table 9.1 summarizes the fundamental

Table 9.1 Key principles and steps of the virtual screening process used in drug discovery for breast cancer (BC) treatment

Step	Description	References
Target identification	Defining the biological target involved in the disease process may be modulated by a drug	Sun et al. (2021b)
Database selection	Choosing a database of chemical compounds that will be screened against the target. This database can contain millions of potential compounds	Andricopulo et al. (2009)
Ligand preparation	Preparing the chemical compounds in the database for screening. This can involve computing the optimal 3D structure and predicting the ligands' charge distribution and tautomeric states	Vita et al. (2019)
Docking	Calculating how well the compounds fit into the binding site of the target. This involves predicting the correct orientation of the ligand within the active site (docking) and estimating the binding strength or affinity (scoring)	Cavasotto et al. (2005)
Scoring	The compounds are ranked based on their predicted affinity or how well they interact with the target. The scoring functions can consider various factors, such as hydrogen bonding, hydrophobic effects, and conformational entropy	Empereur-mot et al. (2015)
Hit identification	Identifying the "hits" or compounds that show a significant level of interaction with the target. These hits are considered potential leads for further drug development	Zhu et al. (2013)
Validation	Experimentally testing the best-scoring compounds through in vitro or in vivo assays to confirm their biological activity	Montes-Grajales et al. (2016)

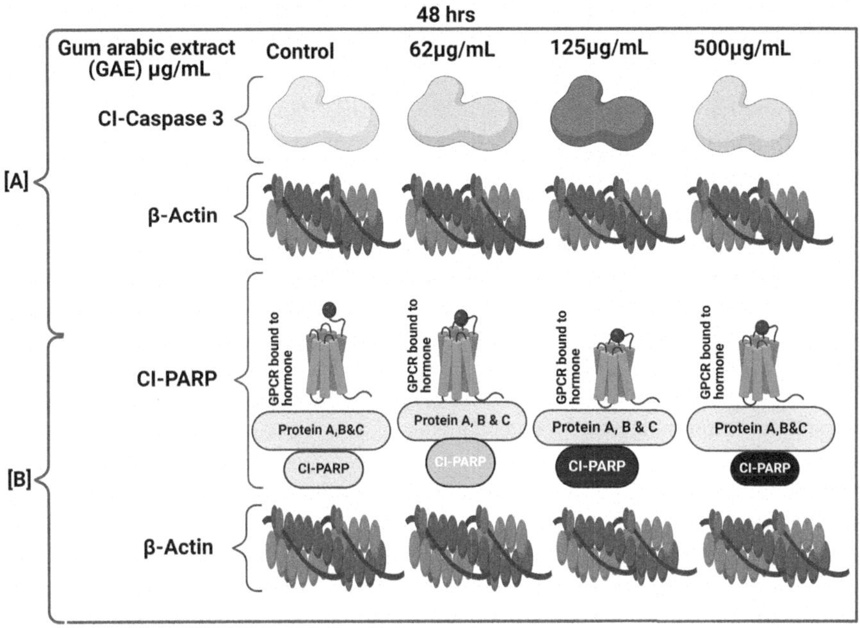

Fig. 9.4 Western blot analysis was performed to assess the levels of cleaved cysteine-aspartic acid protease 3 (caspase3), which is defined as a critical enzyme in the apoptotic protease cascade, marks the terminal step and is firmly associated with cellular demise (Grütter, 2000). It accomplishes this by severing essential proteins that participate in cellular restoration processes (Porter & Jänicke, 1999). Therefore, both caspase 3 and Poly (ADP-ribose) polymerase (PARP) in MCF-7 cells have been cleaved after cells treatment with GAE for 48 h. Thus, proteins were extracted from cells treated with different concentrations of 62, 125, and 500µg/mL GAE or vehicle (DMSO) and subjected to Western blot analysis using antibodies against cleaved caspase 3 and PARP. Finaly, β-actin was used as a loading control. The results are shown in panels A and B, where cleaved caspase 3 and PARP levels are depicted (Adapted by Elnour et al., 2020). Fig 9.4 proposes the mechanism during cleaved caspase 3 and PARP in MCF-7 cells treated with gum arabic extract for 48 hrs

principles and steps of the VSM process used in drug discovery for breast cancer treatment (Fig. 9.4).

9.5 Target Selection and Screening Process

Another essential advantage of GA is its ability to mask unfavorable physicochemical characteristics of the incorporated molecules (Rashima et al., 2016). This makes GA an ideal candidate for drug delivery systems, especially for treating brain tumors where efficient and specific delivery is crucial (Nayak et al., 2012). Furthermore, GA has a long history of use in traditional medicine (Ashour et al., 2022). It has been used for centuries to treat various ailments, including gastrointestinal disorders, respiratory infections, and inflammation (Ashour et al., 2022). In recent years, there has been growing interest in exploring the potential anti-cancer properties of GA (Elnour et al., 2020). Studies have shown that GA exhibits inhibitory effects against certain bacterial species such as *Pseudomonas aeruginosa*, *Staphylococcus aureus*, and *Escherichia coli* (Babiker et al., 2012). Therefore, this highlights the potential of GA not only as a dietary fiber but also as a therapeutic agent in the fight against BC and other diseases.

Virtual screening methodology (VSM) is an innovative tool that bridges the gap between traditional knowledge and contemporary drug discovery (Miles & Ross, 2020). This allows researchers to analyze large databases of chemical compounds and predict their potential interactions with target receptors (Wu et al., 2023). This can significantly accelerate the drug discovery process and identify potential candidates for further investigation, such as GA. Moreover, VSM can provide valuable insights into the molecular interactions of GA with BC cells and receptors, helping researchers better understand its mechanisms of action and efficacy in BC treatment (Hussein AlYafii, 2021). Using VSM, researchers can identify specific receptors targeted by GA and determine the potential side-effects associated with its use in BC treatment (Hussein AlYafii, 2021). GA shows promise as a receptor-targeted therapy for BC treatment. It offers unique advantages, such as its ability to mask the unfavorable characteristics of incorporated molecules and its historical use in traditional medicine(Hussein AlYafii, 2021). Another important advantage of GA in drug delivery systems is its ability to mask unfavorable physicochemical characteristics of the incorporated molecule (Banerjee & Chen, 2007). Additionally, GA has the potential to deliver anti-cancer drugs efficiently and specifically to brain tumors, thereby minimizing damage to healthy brain cells (Lockman et al., 2010). In a study evaluating the role of encapsulant materials on the stability of bioactive compounds in red ginger extract powder, it was found that increasing the amount of GA in the encapsulant material resulted in improved stability and release of bioactive compounds (Böger et al., 2018). This suggests that GA can serve as a therapeutic agent for BC treatment and enhance the stability and efficacy of other bioactive compounds (Elnour et al., 2020). Overall, GA shows promising potential for BC therapy by targeting receptors specifically.

Research on GAE and its potential in combating BC focuses on the challenging sub-type known as triple-negative BC (TNBC)(Dent et al., 2007). Identified due to its aggresive nature of tripil- negative breast cancer (TNBC) (Nedeljković & Damjanović, 2019). This set the stage for rigorous examination of GAE's effects of GAE on cancer cells (Elnour et al., 2020). The anti-cancer properties of GAE were meticulously evaluated using the MDA-MB-231 cell line, which is representative of TNBC (Pélicano et al., 2014). The outcomes demonstrated that GAE effectively suppressed cell growth in both concentration-dependent and time-dependent ways (Hussein AlYafii, 2021). Key findings include the induction of cell cycle arrest at the G1/S phase and the triggering of apoptosis, as evidenced by the activation of caspase 3/7 and the detection of cleaved caspase 3 and PARP (Tyagi et al., 2004). These results underscore GAE's potential efficacy of GAE against this particularly resistant form of BC (Elnour et al., 2020). Further investigation of GA extract's molecular interactions with GAE explained that it impedes the Wnt/β-catenin signaling pathway in TNBC cells (Koval et al., 2018). This disruption leads to downregulating cell cycle regulatory proteins and suppressing Wnt-targeted genes (Ou et al., 2011). The precise delineation of these molecular targets provides valuable insights into GAE's mechanism of action of GAE, bolstering its profile as a potential chemotherapeutic agent against TNBC (Çınar et al., 2022). The GA safety and regulatory status also support its viability as a cancer treatment option. Recognized by major health authorities such as the FDA, FAO/WHO, and the European Union as safe for use in food and feed, GAE's well-established safety profile of GAE makes it an attractive candidate for further development in clinical settings (Kamal et al., 2021). The study laid out clear hypotheses and objectives, proposing that GA from the AGs exerts anti-cancer effects by modulating the cell cycle and promoting apoptosis in TNBC cells (Chen et al., 2021). This study aimed to assess the anti-cancer effects of GAE across various cancer cell lines, delve into its growth-inhibitory mechanisms, and identify key biomarkers associated with cellular responses to treatment (Muangsanit et al., 2020). In conclusion, the comprehensive research process involved in targeting and screening GA compounds for BC highlights their promising potential for TNBC treatment (He et al., 2020a). By establishing TNBC as a target, confirming GAE's anti-cancer activity of GAE, elucidating its molecular basis of action, and affirming its safety, this study provides a robust foundation for future research and potential clinical applications in breast cancer therapeutics (Sarika et al., 2015). Table 9.2 outlines the systematic assessment of potential targets for breast cancer therapy using virtual screening methods.

Table 9.2 Systematic evaluation of potential breast cancer therapy targets using virtual screening techniques

Criteria	Description	References
Biological relevance	The receptor must play a significant role in breast cancer pathogenesis, affecting cell growth, survival, invasion, metastasis, or angiogenesis	Losordo and Isner (2001)
Druggability	The receptor should be "druggable," meaning it is possible to find molecules that can specifically interact with it to induce a therapeutic effect	Hauser et al. (2017)
Clinical evidence	Clinical evidence should exist that modulating the receptor can change the course of the disease. This can include data from observational studies, clinical trials, or existing therapies	Luessen and Conn (2022)
Structural information	Detailed structural information about the receptor. Often, information is obtained through methods such as X-ray crystallography or nuclear magnetic resonance, which aids in virtual screening by providing more precise modeling of ligand–receptor interactions	Ananthan et al. (2009)
Expression patterns	The receptor should ideally be differentially expressed in cancerous versus normal tissues, making it a more specific target with reduced off-target effects	Gan et al. (2012)
Target validation	The receptor should have been validated as a target through previous studies, including biological assays, genetic proof (e.g., knockdown experiments) or evidence from successful targeting in other cancers	Paik et al. (2004)
Availability of ligands	A set of known ligands, including agonists and antagonists, can be very helpful as they provide a starting point for virtual screening and help validate screening methods	Thomsen and Gloyn (2017)

9.6 Analysis of Virtual Screening Results

Virtual screening method (VSM) is a potent computational approach with significant potential for discovering new treatments for BC (Qin et al., 2017). This method allows the efficient evaluation of large compound databases to identify potential candidates that target specific receptors involved in BC (Sultana et al., 2019), offering time and resource savings compared to traditional screening methods (Miles & Ross, 2020). Furthermore, VSM expands the exploration of chemical space, increasing the chances of finding novel and effective treatments while potentially overcoming the limitations and side-effects associated with conventional therapies by specifically targeting cancer cells (Yu et al., 2023). Notably, it can also improve drug delivery and efficacy by overcoming anatomical barriers, such as the blood–brain barrier (Yang et al., 2015). For instance, a polycefin variant conjugated to specific antibodies synthesized using VSM has shown promise in enhancing tumor targeting while minimizing damage to healthy cells (Hudiyanti et al., 2022). Gum-arabic-based carriers identified through VSM has demonstrated several advantages, making them desirable carrier matrices for biopharmaceuticals because of their non-toxic nature, biodegradability, circulation stability, and easy cell uptake (Abu-Serie et al., 2021). Hence, VSM utilizing GA-based carriers holds great promise for developing receptor-targeted therapies against BC.

To fully exploit the potential of GA in virtual BC treatment screening, it is crucial to comprehend the process of selecting suitable compounds (Hussein AlYafii, 2021). This involves using computational methods to identify compounds that can bind to specific receptors in BC. We can assess the binding affinity and potential cytotoxicity of various GA compounds through computational modeling and VSM (Hussein AlYafii, 2021). This enabled us to select compounds that effectively target BC cells while minimizing their side-effects on healthy tissues (Elnour et al., 2020). Additionally, VSM can help optimize drug-nanoparticle systems for targeted delivery by carefully choosing the location of modification to maintain biological activity and ensure proper release of the active drug (Liu et al., 2018a). The unique properties of GA, such as lack of toxicity, the qualities of non-immunogenicity, biodegradability, stability within the bloodstream, and ease of cellular uptake, offer several advantages for developing receptor-targeted therapies for BC (Hassani et al., 2020). This approach can enhance efficacy and safety by enabling personalized medicine tailored to each patient's tumor characteristics (Hassani et al., 2020).

Furthermore, using GA as a carrier matrix provides additional benefits for optimizing drug delivery (Hassani et al., 2020). GA has superior solubility in water and organic solvents. The unique chemical composition of GA makes it an excellent carrier matrix for drug delivery systems because of its blend of complex polysaccharides, such as arabinogalactan and glucuronic acid (Mohamed et al., 2017). This composition provides superior solubility in water and various organic solvents (Adetoro et al., 2020). High water solubility allows for easy incorporation of hydrophobic drugs into the matrix, including those used in BC treatment (Elnour et al., 2020). This ensures effective drug dispersion within the carrier and promotes efficient drug delivery to the cancer cells (Sobulska & Zbiciński, 2020). Furthermore, the structural flexibility of GA enables easy modification and functionalization for attaching targeting ligands or therapeutic molecules, enhancing the specificity of drug delivery to BC cells without significant side-effects (Atgié et al., 2019; Tibbitt et al., 2016). Thus, GA has a long history of medicinal use, supporting its potential use in modern oncology.

VSM employing carriers based on GA can potentially connect traditional expertise with modern drug discovery (Liang et al., 2020). Through VSM of compound libraries against BC-related targets, scientists can identify viable drug candidates that offer increased efficacy and safety in addressing the disease (Salmasfattah et al., 2023; Zhang et al., 2020). This method enables rapid screening of numerous compounds, thereby diminishing the time and expense of conventional drug discovery approaches (Gorgulla et al., 2020). Recent research has illustrated encouraging outcomes in harnessing GA-based nanoemulsions to deliver therapeutic agents specifically to BC cells (Elnour et al., 2020; Hassani et al., 2020).

Applying a VSM workflow to GA compounds involves pre-screening preparation, docking simulations, and post-screening analysis (Klebe, 2006). This approach has facilitated the discovery of potential GA-derived compounds that exhibit binding affinity for BC targets (Elnour et al., 2020). Through VSM, it is possible to identify GA-based compounds that selectively target BC cells while minimizing harm to healthy cells (Hassani et al., 2020). The use of computational models and algorithms to predict interactions between GA compounds and cancer-related receptors involves several methods, such as molecular docking, molecular dynamics simulations, and machine-learning algorithms (Taghizadeh et al., 2022). These computational methods enable the prediction of binding affinity, drug–target interactions, and potential off-target effects, providing valuable insights for selecting primary compounds and developing more effective treatments for BC (Forli et al., 2016). Thus, using VSM, researchers are empowered to accurately predict the interactions between compounds and cancer-related receptors.

Despite significant progress in BC research and treatment, there is a continued need for more effective therapies (Tan et al., 2014). BC poses a significant challenge worldwide, impacting numerous individuals and compelling medical science to seek improved treatments (Sung et al., 2021). An encouraging area of focus in BC research involves the development of therapies that target the receptors (Cetinkaya & Avci, 2022). These treatments aim to specifically target the receptors involved in BC cells, delivering medications directly to the affected area while reducing systemic side-effects (Joseph et al., 2021). GA has traditionally been recognized for its dietary fiber properties. However, it has now emerged as a potential ray of hope in combating BC because of its unique chemical composition, which contains polysaccharides and glycoproteins with anti-cancer properties that have shown promise in inhibiting tumor growth during preclinical studies (Elnour et al., 2020).

The cutting-edge tool known as VSM has played an integral role in unlocking the potential of GAE in treating BC by offering ways to conceal unfavorable physicochemical characteristics associated with incorporating this molecule into treatments

(Hussein AlYafii, 2021). A comparative study of GAE for current BC treatments revealed its potential as a therapeutic option, particularly for aggressive subtypes such as triple-negative BC (Elnour et al., 2020). Through VSM and subsequent experimental validation, it has been demonstrated that GAE hinders the growth of MDA-MB-231 cells, which are a TNBC cell line, with an effect dependent on concentration and time (Khan et al., 2021). This hindrance is linked to significant cell cycle interruption at the G1/S phase, which triggers apoptosis (Han et al., 2021). The latter is evident through the activation of caspase 3/7 and identification of cleaved caspase 3 and PARP (He et al., 2020b). These discoveries emphasize the strong anti-cancer properties of GAE and suggest a unique mechanism distinct from conventional therapies (Elnour et al., 2020).

The study revealed that GAE focuses on the Wnt/β-catenin signaling pathway, which significantly regulates cell growth (Yu et al., 2022). By suppressing this pathway, it is suggested that GAE's ability of GAE to combat cancer is influenced by unique molecular processes that are different from those of typical treatments for BC (Hassani et al., 2020). This targeted strategy may present an advantageous method for treating TNBC, a type of BC characterized by the absence of hormonal receptors and limited reaction to conventional therapies (Sarika et al., 2015).

The safety profile of GA, as confirmed by regulatory agencies such as the FDA and FAO, supports its potential as a safe and well-tolerated treatment option (Kamal et al., 2021). This combination of safety and observed effectiveness positions GAE as a promising candidate for further development in BC therapy (Elnour et al., 2020). This study outlines comprehensive objectives, including evaluating GAE's efficacy of GAE across different types of cancer cells, investigating its mechanisms of action, and identifying key biomarkers that respond to treatment (Hassani et al., 2020). This study hypothesizes that GAE from AGs specifically targets TNBC cells by influencing cell cycle dynamics and promoting apoptosis (Hussein AlYafii, 2021). This chapter aimed to understand further the molecular mechanisms through which GA exerts its anti-cancer effects. These findings confirm the initial VSM results and pave the way for future research, including in vivo studies, to fully comprehend the clinical implications of GA in BC treatment (Hassani et al., 2020). The analysis suggests that compounds found in GA show promise as new anti-cancer agents, especially against TNBC (Atgié et al., 2019). They may offer unique advantages over current treatments by targeting specific molecular pathways involved in cancer cell growth (Hussein AlYafii, 2021). Further exploration is critical to validating these outcomes and uncovering the full therapeutic potential of GA in oncology.

9.7 Therapeutic Potential and Experimental Validation

While VSM has opened new possibilities for identifying potential GA-based compounds, it is crucial to validate the efficacy and safety of these compounds through rigorous in vitro and in vivo experiments (Hussein AlYafii, 2021). These validation steps are necessary to ensure that GA-based compounds exhibit selective binding to

BC cells and demonstrate therapeutic efficacy and minimal toxicity in breast cancer treatment (Elnour et al., 2020). To validate the potential of GA-based compounds, researchers must conduct comprehensive preclinical studies to assess their pharmacokinetics, biodistribution, and therapeutic activity (Hassani et al., 2020). Furthermore, evaluating the stability and scalability of v-based drug delivery systems is crucial to ensure their clinical translation feasibility (Ghasemizad et al., 2022).

GA plays a crucial role in drug delivery systems because of its unique chemical composition and historical medicinal uses (Gerola et al., 2015). Its ability to form stable nanoemulsions and act as a natural surfactant makes it an ideal candidate for encapsulating and delivering therapeutic compounds (Ya-Ali et al., 2020). The natural surfactant properties of GA help stabilize nanoemulsions, allowing for the effective encapsulation and delivery of therapeutic compounds (Böger et al., 2018). Additionally, GA has been shown to have low toxicity in vitro and in vivo, making it a safe option for drug delivery (Kk & Jayakrishnan, 2006). This natural compound also possesses biodegradability and stability in the bloodstream, further enhancing its potential as a drug carrier (Bai et al., 2017). Furthermore, GA can mask the unfavorable physicochemical characteristics of the incorporated molecules, making it an attractive option for targeted drug delivery (Binsi et al., 2017). Thus, GA is a promising material for targeted drug delivery and other applications in the pharmaceutical and food industries.

VSM is an innovative approach that has revolutionized the drug discovery process (Yang et al., 2019). VSM involves computer algorithms and databases to screen large libraries of compound structures for potential interactions with a specific target, such as BC cells (Vyas, 2008). VSM allows researchers to quickly and efficiently identify potential GA-based compounds that may have therapeutic effects on BC cells (Hussein AlYafii, 2021). The use of VSM in conjunction with GA-based compounds offers a promising avenue for the development of receptor-targeted therapies for BC (Hassani et al., 2020). In recent years, VSM has emerged as a powerful tool for drug discovery (Miles & Ross, 2020). Its ability to screen and analyze large libraries of compound structures has significantly accelerated the process of identifying potential drug candidates (Lionta et al., 2014). Thus, promising VSM can lead to further development of compounds, ultimately contributing to the discovery of novel therapeutic agents.

Molecular mechanisms underlying the anti-cancer activity of GA along with its associated content, investigate the therapeutic potential and experimental validation of using GAE compounds in VSM for receptors related to BC (Hussein AlYafii, 2021). A pivotal aspect of this research focused on assessing the anti-cancer capabilities of GAE against the MCF-& and MDA-MB-231 celles line, a triple-negative human BC type (Mirghani et al., 2018). These findings highlighted that GAE effectively inhibited cell proliferation in a concentration- and time-dependent manner (Hussein AlYafii, 2021). This significant anti-proliferative effect was attributed to the arrest of the cell cycle at the G1/S phase and induction of apoptosis (Hussein AlYafii, 2021). The activation of caspase 3/7 further substantiated this and the identification of cleaved caspase 3 and PARP, validating the therapeutic potential of GAE compounds against breast cancer (He et al., 2020b).

GAE has been shown to suppress the growth of BC cells by reducing the expression of c-myc and increasing the expression of p21WAF1 in the MDA-MB-231 cell line, which overexpresses c-myc and promotes cell progression (Xu et al., 2010). On the other hand, a reduction or inactivation of this protein slows down cell division or results in cell growth being halted (Janghorban et al., 2014). The c-myc protein is said to regulate the G1-S transition by activating downstream targets and suppressing the cyclin-dependent kinase (Cdk) inhibitor p21WAF1 (Prall et al., 1998). The p21WAF1 CDK inhibitor functions as a tumor suppressor by preventing the progression of the cell cycle during G1 and S phase transitions (Kim et al., 2010). The objective of assessing the expression of p21WAF1 was to ascertain whether growth inhibition was facilitated by the induction of p21WAF1 (Lee et al., 2006). The graph shown in Fig. 9.5 indicates a considerable rise in p21WAF1 protein levels at concentrations of 65 and 120 μg/mL, while the protein levels were completely eliminated in cells exposed to a concentration of 500 μg/mL GAE. GAE suppresses BC cells by inhibiting the expression of c-myc protein and inducing the expression of p21WAF1 protein, independent of p53 activity as a tumor suppressor protein in these cells (Han et al., 2015). The impact of GAE on protein modifications and cell growth regulation is shown in Fig. 9.5.

This book examined the effectiveness of GAE in treating BC, primarily by investigating its impact on the Wnt/β-catenin signaling pathway in MCF-7 cells, a

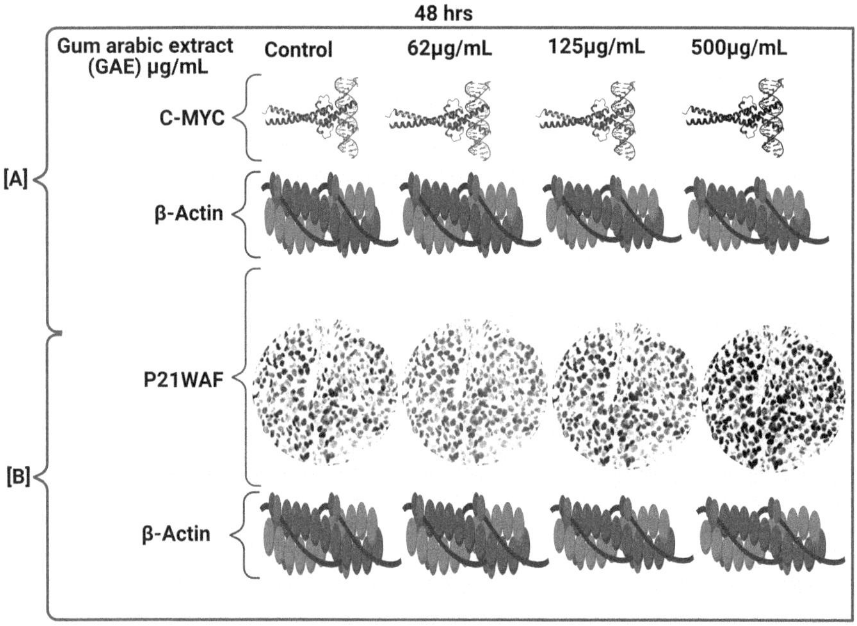

Fig. 9.5 The inhibition of β-catenin signaling by GAE. After treating MCF-7 cells with GAE for 48 h, a concentration-dependent decrease in β-catenin protein was observed. Whole-cell proteins were extracted from cells treated with GAE and vehicle, followed by Western blot analysis for β-catenin with β-actin as a loading control (Adapted from Elnour et al.)

particular type of TNBC cell line (Ziagova et al., 2021). This pathway is crucial for cell proliferation, and its inhibition by compounds derived from GAE suggests a specific molecular mechanism that may underlie their anti-cancer effects (Elnour et al., 2020). This finding provides a targeted approach to cancer therapy, emphasizing the potential of GAE to disrupt cancer growth at the molecular level (Hussein AlYafii, 2021).

Regarding safety and regulatory status, GAE is recognized as safe by major authorities such as the FDA and the FAO (Kaddam & Kaddam, 2020). This recognition, along with its approval as a food supplement and feed additive, supports its potential use as a safe and well-tolerated treatment option for BC (Hassani et al., 2020). This safety profile is significant because it helps integrate GAE into existing cancer treatment paradigms. GAE effectively inhibits the β-catenin signaling pathway. In MCF-7 cells treated with GAE, a reduction in β-catenin protein levels was observed, and this decrease was correlated with the concentration of GAE. After 48 h of treatment with either DMSO or different concentrations of GAE, whole-cell proteins from both cell types were extracted and analyzed by Western blotting for β-catenin, with β-actin as a loading control. These findings demonstrate the potential therapeutic effects of GAE and confirm its ability to inhibit the β-catenin signaling pathway through experimental validation using visual screening as an appropriate technique (Fig. 9.6).

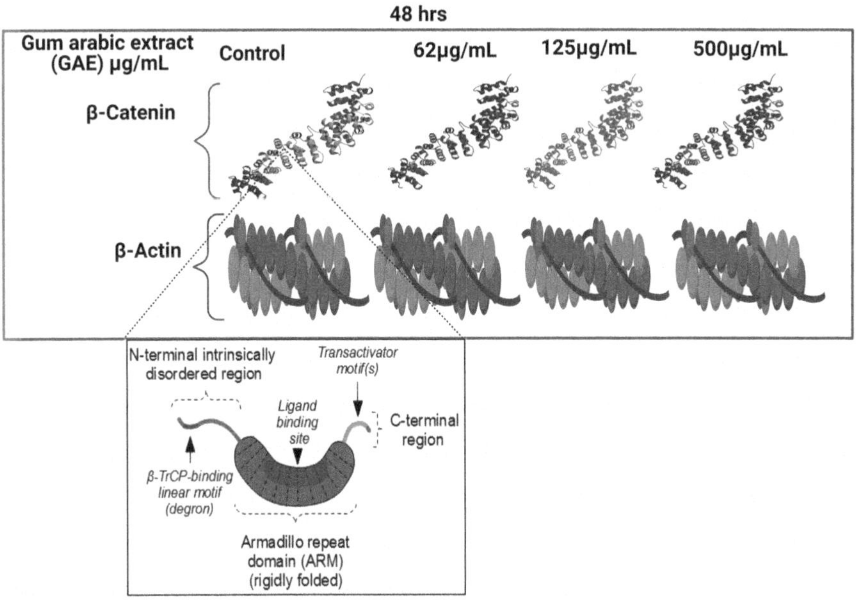

Fig. 9.6 The suppression of the β-catenin signaling pathway by GAE. Treatment with GAE resulted in a decrease in β-catenin protein levels in MCF-7 cells, with the reduction being dependent on the concentration of GAE. To assess β-catenin levels, western blot analysis was conducted on whole-cell proteins extracted after 48 h of treatment with either DMSO (vehicle) or various concentrations of GAE. β-actin served as a loading control in this experiment. (Adapted from the study by Elnour et al., (2020)

9.8 Future Perspectives and Clinical Implications

The future outlook and practical applications of virtual screening to analyze GAE compounds for receptors related to breast cancer include the possibility of discovering new therapeutic targets and creating precise treatments for this type of cancer (Gorgulla et al., 2020). Virtual screening, a computer-based approach commonly used in drug development, can be used to evaluate how GAE compounds interact with specific receptors associated with breast cancer (Hassani et al., 2020). This study has several implications for future research and clinical practice.

1. **Identification of Novel Therapeutic Targets**: Virtual screening of compounds derived from GAE may aid in identifying particular receptors or proteins implicated in the development of breast cancer, which could be potential targets for therapeutic measures (Hussein AlYafii, 2021). By examining how these compounds interact with breast cancer-associated receptors, novel molecular targets for drug discovery may be unveiled, potentially resulting in more precise and potent treatments for breast cancer (Sarika et al., 2015).
2. **Personalized Medicine and Targeted Therapies**: The virtual screening of GAE compounds for receptors related to breast cancer could potentially aid in developing personalized medical strategies (Mavaddat et al., 2019). By understanding the precise interactions between these compounds and specific receptor subtypes or mutations, it may be possible to create custom-tailored therapies that target the molecular features of various breast cancer subtypes (Waks & Winer, 2019). This approach has the potential to improve treatment precision and efficacy in patients.
3. **Reduced Side-Effects and Improved Efficacy**: Virtual screening has the potential to discover compounds derived from GA that can interact with receptors linked to breast cancer (Gorgulla et al., 2020). This could lead to therapies with fewer side-effects and enhanced effectiveness (Liu et al., 2022). By focusing on particular receptors or signaling pathways associated with breast cancer, these compounds may provide a more precise and less harmful treatment method, which could ultimately enhance patient outcomes and quality of life (Wang et al., 2018).
4. **Combination Therapies and Synergistic Effects**: Virtual screening of compounds found in GAE could reveal potential synergistic effects when used alongside current breast cancer treatments (Hussein AlYafii, 2021). By gaining insights into the interactions between GAE compounds and receptors related to breast cancer (Elnour et al., 2020), therefore, we can discover possibilities for combining therapies that improve the effectiveness of standard treatments or combat drug resistance, ultimately leading to more thorough and efficient treatment approaches.
5. **Clinical Translation and Drug Development**: The results of the virtual screening of GA compounds for receptors associated with breast cancer could lead to future preclinical and clinical investigations (Elnour et al., 2020). Compounds showing potential from the virtual screening process may undergo additional

assessments in laboratory tests, animal studies, and eventually in clinical trials to progress these findings into new potential drug candidates for breast cancer treatment.

In summary, the future prospects of virtual screening to identify GAE compounds for breast cancer-related receptors are promising. This approach could contribute to the development of targeted, efficient therapies, advancing personalized medicine and improving treatment outcomes in the fight against breast cancer, potentially advancing personalized medicine and improving treatment results in ongoing efforts against this disease.

9.9 Conclusion

This chapter delves into the therapeutic potential of GAE compounds in BC treatment by leveraging virtual screening to identify promising candidates. Through strategic target selection and molecular docking, GAE interactions with key BC receptors such as HER2, ERα, and ERβ have been unveiled, offering new avenues for treatment. Emphasizing the importance of experimental validation, this study bridges virtual screening with clinical translation and solidifies predictions for potential applications. The effect of GAE on breast cancer cells, including size reduction and apoptosis, underscores its potential as a treatment option. Furthermore, regulatory confirmation of the safety of GAE supports its viability as a therapeutic agent. Particularly in TNBC, known for its aggressiveness, GAE shows promise in selectively targeting cancer cells while sparing healthy cells. In summary, this study lays a strong foundation for the role of GAE in breast cancer therapy, particularly in TNBC treatment. This study paves the way for future clinical applications and promising directions in breast cancer treatment by elucidating the molecular mechanisms and affirming its safety.

Acknowledgments The authors thank the Chemical and Process Engineering Technology at the University of Malaysia, Pahang-UMPSA, Malaysia, for supporting this research.

Fund This research was funded by Eastern Unity Technology under Grant No. UIC 190806.

References

Abbas, A. J., Ibrahim, M. F., & Saifo, M. (2021). Antibody-drug conjugates used in breast cancers. *Journal of Oncology, 30*(7), 6447–6461. https://doi.org/10.1155/2021/9927433

Abu-Serie, M. M., Hamouda, A. F., & Habashy, N. H. (2021). Acacia senegal gum attenuates systemic toxicity in CCl4-intoxicated rats via regulation of the ROS/NF-κB signaling pathway. *Scientific Reports, 11*(1), 20316. https://doi.org/10.1038/s41598-021-99953-y

Adetoro, A. O., Opara, U. L., & Fawole, O. A. (2020). Effect of carrier agents on the physicochemical and technofunctional properties and antioxidant capacity of freeze-dried pomegranate juice (Punica granatum) powder. *Foods, 9*(10), 1388. https://doi.org/10.3390/foods9101388

Akahane, T., Kanomata, N., Harada, O., Yamashita, T., Kurebayashi, J., Tanimoto, A., & Moriya, T. (2020). Targeted next-generation sequencing assays using triplet samples of normal breast tissue. *Primary Breast Cancer, and Recurrent/Metastatic Lesions, 20*(1), 944. https://doi.org/10.21203/rs.2.16536/v5

Al-Fahad, M. E. A., & Al-Mashhdani, H. E. (2023). Evaluation of the addition of different percentages of gum Arabic powder (Acacia senegal) to diets on the productive performance of ross 308 broilers. *IOP Conference Series Earth and Environmental Science., 1262*(7), 072033. https://doi.org/10.1088/1755-1315/1262/7/072033

Al-Zharani, M., Alyami, N. M., Qurtam, A. A., Aljarba, N. H., Alkahtani, S., Mubarak, M., & Daoush, W. M. (2023). Use of multi-walled carbon nanotubes (MWCNTs) stabilized in Arabic gum colloidal solution to induce genotoxicity and apoptosis of human breast and lung cancer cell lines. *Frontiers in Materials, 10*, 1229637. https://doi.org/10.3389/fmats.2023.1229637

Alven, S., & Aderibigbe, B. A. (2020). The therapeutic efficacy of dendrimer and micelle formulations for breast cancer treatment. *Pharmaceutics, 12*(12), 1212. https://doi.org/10.3390/pharmaceutics12121212

An, J., Peng, C., Tang, H., Liu, X., & Peng, F. (2021). New advances in the research of resistance to neoadjuvant chemotherapy in breast cancer. *International Journal of Molecular Sciences, 22*(17), 9644. https://doi.org/10.3390/ijms22179644

Ananthan, S., Zhang, W., & Hobrath, J. V. (2009). Recent advances in structure-based virtual screening of G protein coupled receptors. *The AAPS Journal, 11*(1), 178–185. https://doi.org/10.1208/s12248-009-9094-3

Anderson, D. M. W. (1986). Evidence for the safety of gum Arabic (*Acacia senegal* (L.) Willd.) as a food additive—A brief review. *Food Additives & Contaminants, 3*(3), 225–230. https://doi.org/10.1080/02652038609373584

Anderson, D. M. W., & Eastwood, M. A. (1989). The safety of gum Arabic as a food additive and its energy value as an ingredient: A brief review. *Journal of Human Nutrition and Dietetics, 2*(3), 137–144. https://doi.org/10.1111/j.1365-277x.1989.tb00017.x

Andricopulo, A. D., Salum, L. B., & Abraham, D. J. (2009). Structure-based drug design strategies in medicinal chemistry. *Current Topics in Medicinal Chemistry, 9*(9), 771–790. https://doi.org/10.2174/156802609789207127

Ashour, M. A., Fatima, W., Imran, M., Ghoneim, M. M., Alshehri, S., & Shakeel, F. (2022). A review on the main phytoconstituents, traditional uses, inventions, and patent literature of gum Arabic emphasizing Acacia seyal. *Molecules, 27*(4), 1171. https://doi.org/10.3390/molecules27041171

Atgié, M., Garrigues, J. C., Chennevière, A., Masbernat, O., & Roger, K. (2019). Gum Arabic in solution: Composition and multi-scale structures. *Food Hydrocolloids, 91*, 319–330. https://doi.org/10.1016/j.foodhyd.2019.01.033

Babiker, R., Merghani, T. H., Elmusharaf, K., Badi, R. M., Lang, F., & Saeed, A. (2012). Effects of gum Arabic ingestion on body mass index and body fat percentage in healthy adult females: Two-arm randomized, placebo-controlled, double-blind trial. *Nutrition Journal, 11*, 111. https://doi.org/10.1186/1475-2891-11-111

Bai, F., Xu, Y., Chen, J., Liu, Q., Gu, J., Wang, X., et al. (2013). Free energy landscape for the binding process of huperzine a to acetylcholinesterase. *Proceedings of the National Academy of Sciences of the United States of America, 110*(11), 4273–4278. https://doi.org/10.1073/pnas.1301814110

Bai, L., Huan, S., Li, Z., & McClements, D. J. (2017). Comparison of emulsifying properties of food-grade polysaccharides in oil-in-water emulsions: Gum Arabic, beet pectin, and corn fiber gum. *Food Hydrocolloids, 66*, 144–153. https://doi.org/10.1016/j.foodhyd.2016.12.019

Banerjee, S., & Chen, D.-H. (2007). Magnetic nanoparticles grafted with cyclodextrin for hydrophobic drug delivery. *Chemistry of Materials, 19*(25), 6345–6349. https://doi.org/10.1021/cm702278u

Barnawi, I. O. (2021). Traganum nudatum possesses anti-cancer efficacy against breast cancer MDA-MB-231 cells. *Journal of Pharmaceutical Research International, 33*(47A), 831–838. https://doi.org/10.9734/jpri/2021/v33i47a33081

Begum, S., Chitra, K., Joseph, B., Sundararajan, R., & Hemalatha, S. (2018). Gelidiella acerosa inhibits lung cancer proliferation. *BMC Complementary and Alternative Medicine, 18*(1), 104. https://doi.org/10.1186/s12906-018-2165-1

Binsi, P. K., Nayak, N., Sarkar, P., Joshy, C. G., Ninan, G., & Ravishankar, C. N. (2017). Gelation and thermal characteristics of microwave extracted fish gelatin–natural gum composite gels. *Journal of Food Science and Technology, 54*, 518–530. https://doi.org/10.1007/s13197-017-2496-9

Böger, B. R., Georgetti, S. R., & Kurozawa, L. E. (2018). Microencapsulation of grape seed oil by spray-drying. *Food Science and Technology., 38*(2), 263–270. https://doi.org/10.1590/fst.04417

Borgquist, S., Hall, P., Lipkus, I. M., & Garber, J. E. (2018). Toward prevention of breast cancer: What are the clinical challenges? *Cancer Prevention Research, 11*(5), 255–264. https://doi.org/10.1158/1940-6207.capr-16-0254

Cavasotto, C. N., Kovacs, J. A., & Abagyan, R. (2005). Representing receptor flexibility in ligand docking through relevant normal modes. *Journal of the American Chemical Society, 127*(26), 9632–9640. https://doi.org/10.1021/ja042260c

Cetinkaya, B. D., & Avci, Ç. B. (2022). Molecular perspective on targeted therapy in breast cancer: A review of current status. *Medical Oncology, 39*(10), 149. https://doi.org/10.1007/s12032-022-01749-1

Chegini, A., Gholamzad, A., Khatibi, M., Safvat, M., Nourikhani, M., & Chegini, H. (2023). Beyond diagnosis: Investigating the interplay between estrogen and progesterone receptors in breast cancer patients—A cross-sectional study. https://doi.org/10.21203/rs.3.rs-2921109/v1

Chen, F., Wang, Q., Yu, X., Yang, N., Wang, Y., Zeng, Y., et al. (2021). McPIP1-mediated NFIC alternative splicing inhibits proliferation of triple-negative breast cancer via cyclin D1-Rb-E2f1 axis. *Cell Death and Disease, 12*(4), 370. https://doi.org/10.1038/s41419-021-03661-4

Çınar, V., Hamurcu, Z., Güler, A., Nurdinov, N., & Ozpolat, B. (2022). Serotonin 5-Ht7 receptor is a biomarker poor prognostic factor and induces proliferation of triple-negative breast cancer cells through FOXM1. *Breast Cancer, 29*(6), 1106–1120. https://doi.org/10.21203/rs.3.rs-1200230/v1

Dauqan, E., & Abdullah, A. (2013). Utilization of gum Arabic for industries and human health. *American Journal of Applied Sciences, 10*(10), 1270–1279. https://doi.org/10.3844/ajassp.2013.1270.1279

Dent, R., Trudeau, M., Pritchard, K. I., Hanna, W., Kahn, H. K., Sawka, C., et al. (2007). Triple-negative breast cancer: Clinical features and patterns of recurrence. *Clinical Cancer Research, 13*(15 Pt 1), 4429–4434. https://doi.org/10.1158/1078-0432.ccr-06-3045

Dickinson, E. (2003). Hydrocolloids at interfaces and the influence on the properties of dispersed systems. *Food Hydrocolloids, 17*(1), 25–39.

Elnour, A., Mirghani, M., Kabbashi, N., Md Alam, Z., & Musa, K. (2018). Study of antioxidant and anti-inflammatory crude methanol extract and fractions of Acacia seyal gum. *American Journal of Pharmacology and Pharmacotherapeutics, 5*(1), 3.

Elnour, A. A., Mirghani, M. E., Kabbashi, N., Daddiouaissa, D., Musa, K. H., Alam, M. Z., & Abdurahman, N. H. (2020). Active fractions of methanol crude obtained from Acacia seyal gum and their anti-proliferative effects against human breast cancer cell lines. *Global Journal of Science Frontier Research, 20*(3-C), 51–64.

Empereur-mot, C., Guillemain, H., Latouche, A., Zagury, J. F., Viallon, V., & Montes, M. (2015). Predictiveness curves in virtual screening. *Journal of Cheminformatics, 7*, 52. https://doi.org/10.1186/s13321-015-0100-8

Forli, S., Huey, R., Pique, M. E., Sanner, M. F., Goodsell, D. S., & Olson, A. J. (2016). Computational protein–ligand docking and virtual drug screening with the AutoDock suite. *Nature Protocols, 11*(5), 905–919. https://doi.org/10.1038/nprot.2016.051

Foulkes, W. D., Smith, I., & Reis-Filho, J. S. (2010). Triple-negative breast cancer. *New England Journal of Medicine, 363*, 1938–1948. https://doi.org/10.1056/nejmra1001389

Fukano, M., Park, M., & Deblois, G. (2021). Metabolic flexibility is a determinant of breast cancer heterogeneity and progression. *Cancers, 13*(18), 4699. https://doi.org/10.3390/cancers13184699

Gan, L., He, J., Zhang, X., Zhang, Y., Yu, G., Chen, Y., et al. (2012). Expression profile and prognostic role of sex hormone receptors in gastric cancer. *BMC Cancer, 12*, 566. https://doi.org/10.1186/1471-2407-12-566

Gashua, I. B., Williams, P. A., Yadav, M. P., & Baldwin, T. C. (2015). Characterization and molecular association of Nigerian and Sudanese Acacia gum exudates. *Food Hydrocolloids, 51*, 405–413. https://doi.org/10.1016/j.foodhyd.2015.05.037

Gerola, A. P., Silva, D. C., Jesus, S., Carvalho, R. A., Rubira, A. F., Muniz, E. C., et al. (2015). Synthesis and controlled curcumin supramolecular complex release from pH-sensitive modified gum-arabic-based hydrogels. *RSC Advances, 5*, 94519–94533. https://doi.org/10.1039/c5ra14331d

Ghasemizad, S., Pirsa, S., Amiri, S., & Abdosatari, P. (2022). Optimization and characterization of bioactive biocomposite film based on orange peel incorporated with gum Arabic reinforced by Cr_2O_3 nanoparticles. *Journal of Polymers and the Environment, 30*(1), 1–14. https://doi.org/10.1007/s10924-021-02357-2

Gorgulla, C., Boeszoermenyi, A., Wang, Z., Fischer, P. D., Coote, P., Das, K. M., et al. (2020). An open-source drug discovery platform enables ultra-large virtual screens. *Nature, 580*(7805), 663–668. https://doi.org/10.1038/s41586-020-2117-z

Grütter, M. G. (2000). Caspases: key players in programmed cell death [Review of Caspases: key players in programmed cell death]. *Current Opinion in Structural Biology, 10*(6), 649. Elsevier BV. https://doi.org/10.1016/s0959-440x(00)00146-9

Gupta, N., Kumar, H., Gupta, S., Basavarajaiah, S. M., & Saini, K. S. (2023). A concise review on natural products and their derivatives for breast cancer treatment. *Chemistry & Biodiversity, 20*(8), e202300688. https://doi.org/10.1002/cbdv.202300688

Han, Y. S., Lee, J. H., & Lee, S. H. (2015). Anti-tumor effects of fucoidan on human colon cancer cells via activation of Akt signaling. *Biomolecules & Therapeutics, 23*(3), 225–232. https://doi.org/10.4062/biomolther.2014.136

Han, Y., Wang, J., Meng, J., Lin, J., Yan, C., & Wang, Y. (2021). Shentong Zhuyu decoction inhibits inflammatory response, migration, and invasion and promotes apoptosis of rheumatoid arthritis fibroblast-like synoviocytes via the MAPK P38/PPARγ/CTGF pathway. *Biomed Research International, 2021*, 6187695. https://doi.org/10.1155/2021/6187695

Hannan, M. A., Haque, N., Mohibbullah, M., Dash, R., Hong, Y. K., & Moon, I. S. (2020). Gelidium amansii attenuates hypoxia/reoxygenation-induced oxidative injury in primary hippocampal neurons through suppressing GluN2B expression. *Antioxidants, 9*(3), 223. https://doi.org/10.3390/antiox9030223

Haque, M. M., & Desai, K. V. (2019). Pathways to endocrine therapy resistance in breast cancer. *Frontiers in Endocrinology, 10*, 573. https://doi.org/10.3389/fendo.2019.00573

Hassani, A., Azarian, M. M. S., Ibrahim, W. N., & Hussain, S. A. (2020). Preparation, characterization and therapeutic properties of gum Arabic-stabilized gallic acid nanoparticles. *Scientific Reports, 10*(1), 17808. https://doi.org/10.1038/s41598-020-71175-8

Hauser, A. S., Attwood, M. M., Rask-Andersen, M., Schiöth, H. B., & Gloriam, D. E. (2017). Trends in GPCR drug discovery: New agents, targets and indications. *Nature Reviews Cancer, 16*(12), 829–842. https://doi.org/10.1038/nrd.2017.178

He, J., McLaughlin, R. P., van der Beek, L., Canisius, S., Wessels, L., Smid, M., et al. (2020a). Integrative analysis of genomic amplification-dependent expression and loss-of-function screen identifies ASAP1 as a driver gene in triple-negative breast cancer progression. *Oncogene, 39*(20), 4118–4131. https://doi.org/10.1038/s41388-020-1279-3

He, W., Tao, W., Zhang, F., Qian, J., He, Y., Zhu, W., et al. (2020b). Lobetyolin induces apoptosis of colon cancer cells by inhibiting glutamine metabolism. *Journal of Cellular and Molecular Medicine, 24*(6), 3359–3369. https://doi.org/10.1111/jcmm.15009

Hedhili, L. (2024). Ippia alba essential oil: A promising complementary therapy for breast cancer. *Chemistry & Biodiversity, 21*(4), e202301510. https://doi.org/10.1002/cbdv.202301510

Hudiyanti, D., Khafiz, M. F. A., Anam, K., Siahaan, P., & Christa, S. M. (2022). In-vitro evaluation of curcumin encapsulation in gum Arabic dispersions under different environments. *Molecules, 27*(12), 3855. https://doi.org/10.3390/molecules27123855

Hussein AlYafii, S. Y. (2021). Molecular mechanisims underlying the anti-cancer activity of gum Arabic from Acacia sp. in triple-negative breast cancer cells. 86. Retrieved from https://scholarworks.uaeu.ac.ae/all_theses/1089/

Janghorban, M., Farrell, A., Allen-Petersen, B. L., Pelz, C., Daniel, C. J., Oddo, J., et al. (2014). Targeting C-Myc by antagonizing PP2A inhibitors in breast cancer. *Proceedings of the National Academy of Sciences of the United States of America, 111*(25), 9157–9162. https://doi.org/10.1073/pnas.1317630111

Joseph, C., Daniels, A., Singh, S., & Singh, M. (2021). Histidine-tagged folate-targeted gold nanoparticles for enhanced transgene expression in breast cancer cells in-vitro. *Pharmaceutics, 14*(1), 53. https://doi.org/10.3390/pharmaceutics14010053

Kaddam, L., & Kaddam, A. S. (2020). Effect of gum Arabic (Acacia senegal) on C-reactive protein level among sickle cell anemia patients. *BMC Research Notes., 13*(1), 162. https://doi.org/10.1186/s13104-020-05016-2

Kamal, E., Kaddam, L., Alagib, A., & Saeed, A. (2021). Dietary fibers (gum Arabic) supplementation modulates hepatic and renal profile among rheumatoid arthritis patients, phase II trial. *Frontiers in Nutrition, 8*, 552049. https://doi.org/10.3389/fnut.2021.552049

Keller, P. J., Arendt, L. M., Skibinski, A., Logvinenko, T., Klebba, I., Dong, S., et al. (2011). Defining the cellular precursors to human breast cancer. *Proceedings of the National Academy of Sciences of the United States of America, 109*(8), 2772–2777. https://doi.org/10.1073/pnas.1017626108

Khan, M. A., Siddiqui, S., Ahmad, I., Singh, R. P., Srivastava, A., & Ahmad, R. (2021). Phytochemicals from Ajwa dates pulp extract induce apoptosis in human triple-negative breast cancer by inhibiting AKT/mTOR pathway and modulating bcl-2 family proteins. *Scientific Reports, 11*(1), 10322. https://doi.org/10.1038/s41598-021-89420-z

Khosravi-Shahi, P., Cabezón-Gutiérrez, L., & Salcedo, M. I. A. (2019). State of art of advanced triple-negative breast cancer. *The Breast Journal, 25*(5), 967–970. https://doi.org/10.1111/tbj.13369

Kim, W.-J., Lee, S. J., Choi, Y. D., & Moon, S. J. (2010). Decursin inhibits growth of human bladder and colon cancer cells via apoptosis, G1-phase cell cycle arrest and extracellular signal-regulated kinase activation. *International Journal of Molecular Medicine, 25*(4), 635–641. https://doi.org/10.3892/ijmm_00000386

Kk, N., & Jayakrishnan, A. (2006). Self-gelling primaquine–gum Arabic conjugate: An injectable controlled delivery system for primaquine. *Biomacromolecules, 8*(1), 84–90. https://doi.org/10.1021/bm060612x

Klebe, G. (2006). Virtual ligand screening: Strategies, perspectives and limitations. *Drug Discovery Today, 11*(13–14), 580–594. https://doi.org/10.1016/j.drudis.2006.05.012

Koval, A., Pieme, C. A., Queiroz, E. F., Ragusa, S., Ahmed, K., Blagodatski, A., et al. (2018). Tannins from Syzygium guineense suppress Wnt signaling and proliferation of Wnt-dependent tumors through a direct effect on secreted Wnts. *Cancer Letters, 435*, 110–120. https://doi.org/10.1016/j.canlet.2018.08.003

Krasoulis, A., Antonopoulos, N., Pitsikalis, V., & Theodorakis, S. (2022). DENVIS: Scalable and high-throughput virtual screening using graph neural networks with atomic and surface protein pocket features. *Journal of Chemical Information and Modeling, 62*(19), 4642–4659. https://doi.org/10.1021/acs.jcim.2c01057

Lee, B.-Y., Kim, C.-H., & Moon, S. K. (2006). Honokiol causes the P21WAF1-mediated G(1)-phase arrest of the cell cycle through inducing P38 mitogen-activated protein kinase in vascular smooth muscle cells. *FEBS Letters, 580*(22), 5177–5184. https://doi.org/10.1016/j.febslet.2006.08.064

Leonida, M. D., Kumar, I., Benzecry, A., Song, J., Jean, C., & Belbekhouche, S. (2023). Green synthesis of zein-based nanoparticles encapsulating lupulone: Antibacterial and antiphotoaging agents. *ACS Biomaterials Science & Engineering, 9*(11), 6165–6174. https://doi.org/10.1021/acsbiomaterials.3c01225

Li, N., Yousefi, M., Nakauka-Ddamba, A., Tobias, J. W., Jensen, S. T., Morrisey, E. E., & Lengner, C. J. (2016). Heterogeneity in readouts of canonical Wnt pathway activity within intestinal crypts. *Developmental Dynamics, 245*(8), 822–833. https://doi.org/10.1002/dvdy.24423

Liang, Q., Ma, P., Zhang, Q., Yin, Y., Wang, P., Wang, S., et al. (2020). A gum Arabic assisted sustainable drug delivery system for adult *Drosophila*. *Biology Open, 9*(6), bio052241. https://doi.org/10.1242/bio.052241

Lionta, E., Spyrou, G. M., Vassilatis, D. K., & Cournia, Z. (2014). Structure-based virtual screening for drug discovery: Principles, applications and recent advances. *Current Topics in Medicinal Chemistry, 14*(16), 1923–1938. https://doi.org/10.2174/1568026614666140929124445

Liu, D., Zhang, H., Fontana, F., Hirvonen, J., & Santos, H. A. (2018a). Current developments and applications of microfluidic technology toward clinical translation of nanomedicines. *Advanced Drug Delivery Reviews, 128*, 54–83. https://doi.org/10.1016/j.addr.2017.08.003

Liu, Z., Huang, P., Law, S., Tian, H. Y., Leung, A. W., & Xu, C. (2018b). Preventive effect of curcumin against chemotherapy-induced side-effects. *Frontiers in Pharmacology, 9*, 1374. https://doi.org/10.3389/fphar.2018.01374

Liu, M., Wang, Z., Li, S., Deng, Y., & He, N. (2022). Identification of PHB2 as a potential biomarker of luminal a breast cancer cells using a cell-specific aptamer. *ACS Applied Materials & Interfaces., 14*(46), 51593–51601. https://doi.org/10.1021/acsami.2c12291

Lockman, P. R., Mittapalli, R. K., Taskar, K. S., Rudraraju, V., Gril, B., Bohn, K. A., et al. (2010). Heterogeneous blood–tumor barrier permeability determines drug efficacy in experimental brain metastases of breast cancer. *Clinical Cancer Research, 16*(23), 5664–5678. https://doi.org/10.1158/1078-0432.ccr-10-1564

Londoño, G. A. J., Vicente, A. M. G., Marcos, J. S., Pardo, F. J. P., Amo-Salas, M., Caballero, M. M., et al. (2018). Low-dose radioiodine ablation in patients with low-risk differentiated thyroid cancer. *European Thyroid Journal, 7*(4), 218–224. https://doi.org/10.1159/000489850

Losordo, D. W., & Isner, J. M. (2001). Estrogen and angiogenesis. *Arteriosclerosis Thrombosis and Vascular Biology, 21*, 6–12. https://doi.org/10.1161/01.atv.21.1.6

Luessen, D. J., & Conn, P. J. (2022). Allosteric modulators of metabotropic glutamate receptors as novel therapeutics for neuropsychiatric disease. *Pharmacological Reviews, 74*(3), 630–661. https://doi.org/10.1124/pharmrev.121.000540

Maia, E. H. B., Medaglia, L. R., de Silva, A. M., & Taranto, A. G. (2020). Molecular architect: A user-friendly workflow for virtual screening. *ACS Omega, 5*(12), 6628–6640. https://doi.org/10.1021/acsomega.9b04403

Marusyk, A., Almendro, V., & Polyak, K. (2012). Intra-tumor heterogeneity: A looking glass for cancer? *Nature Reviews Cancer, 12*(5), 323–334.

Masood, S. (2019). The changing role of pathologists from morphologists to molecular pathologists in the era of precision medicine. *The Breast Journal, 26*(1), 27–34. https://doi.org/10.1111/tbj.13728

Mavaddat, N., Michailidou, K., Dennis, J., Lush, M., Fachal, L., Lee, A., et al. (2019). Polygenic risk scores for prediction of breast cancer and breast cancer subtypes. *The American Journal of Human Genetics, 104*(1), 21–34. https://doi.org/10.1016/j.ajhg.2018.11.002

McIntosh, S. (2021). Surgery for good prognosis breast cancers. *Current Breast Cancer Reports., 13*, 125–131. https://doi.org/10.1007/s12609-021-00414-2

Miles, J. A., & Ross, B. P. (2020). Recent advances in virtual screening for cholinesterase inhibitors. *ACS Chemical Neuroscience, 12*(1), 30–41. https://doi.org/10.1021/acschemneuro.0c00627

Mirghani, M. E. S., Elnour, A. A. M., Kabbashi, N. A., Alam, Z., Musa, K. H., & Abdullah, A. (2018). Determination of antioxidant activity of gum Arabic: An exudation from two different locations. *Scienceasia, 44*(3), 179. https://doi.org/10.2306/scienceasia1513-1874.2018.44.179

Mohamed, H. N., Mustafa, S., Fitrianto, A., & Manap, Y. A. (2017). Development of alginate–gum Arabic beads for targeted delivery of protein. *Journal of Biomolecular Research & Therapeutics, 6*, 2. https://doi.org/10.4172/2167-7956.1000155

Montes-Grajales, D., Bernardes, G. J. L., & Olívero-Verbel, J. (2016). Urban endocrine disruptors targeting breast cancer proteins. *Chemical Research in Toxicology, 29*(2), 150–161. https://doi.org/10.1021/acs.chemrestox.5b00342

Mortensen, A., Aguilar, F., Crebelli, R., Domenico, A. D., Frutos, M. J., Galtier, P., et al. (2017a). Re-evaluation of acacia Gum (E 414) as a food additive. *EFSA Journal, 15*(4), e04741. https://doi.org/10.2903/j.efsa.2017.4741

Mortensen, A., Aguilar, F., Crebelli, R., Domenico, A. D., Frutos, M. J., Galtier, P., et al. (2017b). Re-evaluation of tara gum (E 417) as a food additive. *EFSA Journal, 15*(6), 4863. https://doi.org/10.2903/j.efsa.2017.4863

Muangsanit, P., Day, A., Dimiou, S., Ataç, A. F., Kayal, C., Park, H., et al. (2020). Rapidly formed stable and aligned dense collagen gels seeded with Schwann cells support peripheral nerve regeneration. *Journal of Neural Engineering, 17*(4), 046036. https://doi.org/10.1088/1741-2552/abaa9c

Nayak, A. K., Das, B. K., & Maji, R. (2012). Calcium alginate/gum Arabic beads containing glibenclamide: Development and in-vitro characterization. *International Journal of Biological Macromolecules, 51*(5), 1070–1078. https://doi.org/10.1016/j.ijbiomac.2012.08.021

Nedeljković, M., & Damjanović, A. (2019). Mechanisms of chemotherapy resistance in triple-negative breast cancer—How we can rise to the challenge. *Cells, 8*(9), 957. https://doi.org/10.3390/cells8090957

Nishi, K. K., & Jayakrishnan, A. (2004). Preparation and in-vitro evaluation of primaquine-conjugated gum Arabic microspheres. *Biomacromolecules, 5*(4), 1489–1495. https://doi.org/10.1021/bm0499435

Ohue, M., Yamazaki, T., Ban, T., & Akiyama, Y. (2017). *Link mining for kernel-based compound-protein interaction predictions using a chemogenomics approach*. https://doi.org/10.1007/978-3-319-63312-1_48

Ou, C.-Y., LaBonte, M. J., Manegold, P., So, A. Y.-L., Ianculescu, I., Gerke, D. S., et al. (2011). A coactivator role of CARM1 in the dysregulation of B-catenin activity in colorectal cancer cell growth and gene expression. *Molecular Cancer Research, 9*(5), 660–670. https://doi.org/10.1158/1541-7786.mcr-10-0223

Paik, S., Shak, S., Tang, G., Kim, C., Baker, J. B., Cronin, M., et al. (2004). A multigene assay to predict recurrence of tamoxifen-treated, node-negative breast cancer. *New England Journal of Medicine, 351*(27), 2817–2826. https://doi.org/10.1056/nejmoa041588

Paul, A., Gunewardena, S., Stecklein, S. R., Saha, B., Parelkar, N. K., Danley, M., et al. (2014). PKCλ/ι signaling promotes triple-negative breast cancer growth and metastasis. *Cell Death and Differentiation, 21*(9), 1469–1481. https://doi.org/10.1038/cdd.2014.62

Pélicano, H., Zhang, W., Liu, J., Hammoudi, N., Dai, J., Xu, R. H., et al. (2014). Mitochondrial dysfunction in some triple-negative breast cancer cell lines: Role of mTOR pathway and therapeutic potential. *Breast Cancer Research, 16*(5), 434. https://doi.org/10.1186/s13058-014-0434-6

Phung, H. M., Lee, H., Lee, S., Jang, D., Kim, C.-E., Kang, K. S., et al. (2020). Analysis and anti-cancer effects of active compounds from Spatholobi caulis in human breast cancer cells. *Processes, 8*(9), 1193. https://doi.org/10.3390/pr8091193

Pickard, M. A., & Williams, G. T. (2014). Regulation of apoptosis by long non-coding RNA GAS5 in breast cancer cells: Implications for chemotherapy. *Breast Cancer Research and Treatment, 145*(2), 359–370. https://doi.org/10.1007/s10549-014-2974-y

Porter, A. G., & Jänicke, R. U. (1999). Emerging roles of caspase-3 in apoptosis [Review of Emerging roles of caspase-3 in apoptosis]. *Cell Death and Differentiation, 6*(2), 99. Springer Nature. https://doi.org/10.1038/sj.cdd.4400476

Prall, O. W. J., Rogan, E. M., Musgrove, E. A., Watts, C., & Sutherland, R. L. (1998). C-Myc or cyclin D1 mimics estrogen effects on cyclin E-Cdk2 activation and cell cycle reentry. *Molecular and Cellular Biology, 18*(8), 4499–4508. https://doi.org/10.1128/mcb.18.8.4499

Qin, C., Zhao, Y., Gong, C., & Yang, Z. (2017). MicroRNA-154/ADAM9 axis inhibits the proliferation, migration and invasion of breast cancer cells. *Oncology Letters, 14*(6), 6969–6975. https://doi.org/10.3892/ol.2017.7021

Rashima, R. S., Murad, M., Kang, W., Fazilah, A., & Tan, L.-T. (2016). Influence of sodium chloride treatment and polysaccharides as debittering agent on the physicochemical properties, antioxidant capacity and sensory characteristics of bitter gourd (Momordica charantia) juice. *Journal of Food Science and Technology, 54*(1), 228–235. https://doi.org/10.1007/s13197-016-2454-y

Rezwendy, R., Syahdi, R. R., & Yanuar, A. (2018). Indonesian herbal SGLT2 inhibitor discovery through pharmacophore-based virtual screening. *Pharmacognosy Journal, 10*(4), 803–807. https://doi.org/10.5530/pj.2018.4.136

Saha, P., & Nanda, R. (2016). Concepts and targets in triple-negative breast cancer: Recent results and clinical implications. *Therapeutic Advances in Medical Oncology, 8*(5), 351–359. https://doi.org/10.1177/1758834016657071

Salmasfattah, N., Nurulita, N. A., & Dhiani, B. A. (2023). Virtual screening on molecules targeting the interaction between estrogen receptor beta and murine double minute two. *Indonesian Journal of Cancer Chemoprevention, 13*(3), 184. https://doi.org/10.14499/indonesianjcanchemoprev13iss3pp184-194

Sarika, P. R., James, N. R., Kumar, P. R. A., Raj, D. K., & Kumary, T. V. (2015). Gum Arabic-curcumin conjugate micelles with enhanced loading for curcumin delivery to hepatocarcinoma cells. *Carbohydrate Polymers, 134*, 167–174. https://doi.org/10.1016/j.carbpol.2015.07.068

Schneider, B. P., Winer, E. P., Foulkes, W. D., Garber, J. E., Perou, C. M., Richardson, A. L., et al. (2008). Triple-negative breast cancer: risk factors to potential targets. *Clinical Cancer Research, 14*(24), 8010–8018. https://doi.org/10.1158/1078-0432.ccr-08-1208

Shaw, A. T., Ou, S. H. I., Bang, Y. J., Camidge, D. R., Solomon, B., Salgia, R., et al. (2014). Crizotinib in *ROS1*-rearranged non–small-cell lung cancer. *New England Journal of Medicine, 371*, 1963–1971. https://doi.org/10.1056/nejmoa1406766

Singthong, J., Oonsivilai, R., Oonmetta-aree, J., & Ningsanond, S. (2014). Bioactive compounds and encapsulation of yanang (Tiliacora triandra) leaves. *African Journal of Traditional Complementary and Alternative Medicines, 11*(3), 76–84. https://doi.org/10.4314/ajtcam.v11i3.11

Slamon, D. J., Eiermann, W., Robert, N. J., Pieńkowski, T., Martin, M., Press, M. F., et al. (2011). Adjuvant trastuzumab in HER2-positive breast cancer. *New England Journal of Medicine, 365*(14), 1273–1283. https://doi.org/10.1056/nejmoa0910383

Sobulska, M., & Zbiciński, I. (2020). Advances in spray-drying of sugar-rich products. *Drying Technology, 39*, 1774–1799. https://doi.org/10.1080/07373937.2020.1832513

Sultana, R., Islam, M., Haque, M. A., Evamoni, F. Z., Imran, Z. M., Khanom, J., & Munim, A. (2019). Molecular docking-based virtual screening of the breast cancer target NUDT5. *Bioinformation, 15*(11), 784–789. https://doi.org/10.6026/97320630015784

Sun, H., Chen, H., Crespo, J., Tang, G., Robinson, M., & Şahin, A. (2021a). Clinicopathological features of breast cancer with polysomy 17 and its response to neoadjuvant chemotherapy. *European Journal of Breast Health, 17*(2), 128–136. https://doi.org/10.4274/ejbh.galenos.2021.2021-2-9

Sun, J., Prabhu, N., Tang, J., Yang, F., Jia, L., Guo, J., et al. (2021b). Recent advances in proteome-wide label-free target deconvolution for bioactive small molecules. *Medicinal Research Reviews, 41*(6), 2893–2926. https://doi.org/10.1002/med.21788

Sung, H., Ferlay, J., Siegel, R. L., Laversanne, M., Soerjomataram, I., Jemal, A., & Bray, F. (2021). Global Cancer Statistics 2020: GLOBOCAN estimates of incidence and mortality worldwide for 36 cancers in 185 countries. *CA: A Cancer Journal for Clinicians., 71*(3), 209–249. https://doi.org/10.3322/caac.21660

Taghizadeh, M. S., Niazi, A., Moghadam, A., & Afsharifar, A. (2022). Experimental, molecular docking and molecular dynamic studies of natural products targeting overexpressed receptors in breast cancer. *PLoS One, 17*(5), e0267961. https://doi.org/10.1371/journal.pone.0267961

Tan, G., Kasuya, H., Şahin, T. T., Yamamura, K., Wu, Z., Koide, Y., et al. (2014). Combination therapy of oncolytic herpes simplex virus HF10 and bevacizumab against experimental model of human breast carcinoma xenograft. *International Journal of Cancer, 136*(7), 1718–1730. https://doi.org/10.1002/ijc.29163

Thomsen, S. K., & Gloyn, A. L. (2017). Human genetics as a model for target validation: Finding new therapies for diabetes. *Diabetologia, 60*(6), 960–970. https://doi.org/10.1007/s00125-017-4270-y

Tibbitt, M. W., Dahlman, J. E., & Langer, R. (2016). Emerging frontiers in drug delivery. *Journal of the American Chemical Society, 138*(3), 704–717. https://doi.org/10.1021/jacs.5b09974

Tokunaga, E., Oki, E., Nishida, K., Koga, T., Egashira, A., Morita, M., et al. (2006). Trastuzumab and breast cancer: Developments and current status. *International Journal of Clinical Oncology, 11*(3), 199–208. https://doi.org/10.1007/s10147-006-0575-4

Tyagi, A., Agarwal, C., Harrison, G., Glode, L. M., & Agarwal, R. (2004). Silibinin causes cell cycle arrest and apoptosis in human bladder transitional cell carcinoma cells by regulating CDKI–CDK–cyclin cascade, and caspase 3 and PARP cleavages. *Carcinogenesis, 25*(9), 1711–1720.

Vita, S. D., Lauro, G., Ruggiero, D., Terracciano, S., Riccio, R., & Bifulco, G. (2019). Protein preparation automatic protocol for high-throughput inverse virtual screening: Accelerating the target identification by computational methods. *Journal of Chemical Information and Modeling, 59*(11), 4678–4690. https://doi.org/10.1021/acs.jcim.9b00428

Vyas, V. K. (2008). Virtual screening: A fast tool for drug design. *Scientia Pharmaceutica, 76*(3), 333–360. https://doi.org/10.3797/scipharm.0803-03

Waks, A. G., & Winer, E. P. (2019). Breast cancer treatment. *Journal of the American Medical Association, 321*(3), 288–300. https://doi.org/10.1001/jama.2018.19323

Wang, Y., Zhang, W., & Ma, R. (2018). Bioinformatic identification of chemoresistance-associated microRNAs in breast cancer based on microarray data. *Oncology Reports, 39*(3), 1003–1010. https://doi.org/10.3892/or.2018.6205

Wei, X., Chen, X., Ying, M., & Lu, W. (2014). Brain tumor-targeted drug delivery strategies. *Acta Pharmaceutica Sinica B, 4*(3), 193–201. https://doi.org/10.1016/j.apsb.2014.03.001

Wei, D., Wang, H., Zeng, Q., Wang, W., Hao, B., Feng, X., et al. (2021). Discovery of potent and selective CDK9 degraders for targeting transcription regulation in triple-negative breast cancer. *Journal of Medicinal Chemistry, 64*(19), 14822–14847. https://doi.org/10.1021/acs.jmedchem.1c01350

Winer, E. P., Thürlimann, B., Senn, H.-J., Albain, K. S., André, F., Bergh, J., et al. (2013). Personalizing the treatment of women with early breast cancer: Highlights of the St Gallen International Expert Consensus on the Primary Therapy of Early Breast Cancer 2013. *Annals of Oncology, 24*(9), 2206–2223. https://doi.org/10.1093/annonc/mdt303

Wu, J., Wang, W., & Leung, C. H. (2023). Computational strategies for PROTAC drug discovery. *Acta Materia Medica, 2*(1). https://doi.org/10.15212/amm-2022-0041

Xu, J., Chen, Y., & Olopade, O. I. (2010). MYC and breast cancer. *Genes & Cancer, 1*(6), 629–640. https://doi.org/10.1177/1947601910378691

Ya-Ali, P., Yarahmadi, F., & Mehrnia, M. A. (2020). Efficacies of two nano-formulations of tasmanian blue gum essential oil to control Callosobruchus maculatus. *Journal of Economic Entomology, 113*(3), 1555–1562. https://doi.org/10.1093/jee/toaa069

Yang, T., Martin, P., Fogarty, B., Brown, A., Schurman, K., Phipps, R. J., et al. (2015). Exosome delivered anti-cancer drugs across the blood-brain barrier for brain cancer therapy in Danio rerio. *Pharmaceutical Research, 32*(6), 2003–2014. https://doi.org/10.1007/s11095-014-1593-y

Yang, X., Wang, Y., Byrne, R., Schneider, G., & Yang, S. (2019). Concepts of artificial intelligence for computer-assisted drug discovery. *Chemical Reviews, 119*(18), 10520–10594. https://doi.org/10.1021/acs.chemrev.8b00728

Yin, L., Jiang, D., Bian, X. W., & Yu, S. C. (2020). Triple-negative breast cancer molecular subtyping and treatment progress. *Breast Cancer Research, 22*, 61. https://doi.org/10.1186/s13058-020-01296-5

Yin, L., Zhao, B., & Sun, Y. (2021). Virtual screening of drug proteins based on imbalance data mining. *Mathematical Problems in Engineering*. https://doi.org/10.1155/2021/5585990

Yu, S., Han, R., & Gan, R. (2022). The Wnt/B-catenin signaling pathway in hematological neoplasms. *Biomarker Research, 10*, 74. https://doi.org/10.1186/s40364-022-00418-9

Yu, L., He, X., Fang, X., Liu, L., & Liu, J. (2023). Deep learning with geometry-enhanced molecular representation for augmentation of large-scale docking-based virtual screening. *Journal of Chemical Information and Modeling, 63*(21), 6501–6514. https://doi.org/10.1021/acs.jcim.3c01371

Zhang, Z., Zhou, L., Xie, N., Nice, E. C., Zhang, T., Cui, Y., & Huang, C. (2020). Overcoming cancer therapeutic bottleneck by drug repurposing. *Signal Transduction and Targeted Therapy, 5*(1), 113. https://doi.org/10.1038/s41392-020-00213-8

Zhu, T. R., Cao, S., Su, P. C., Patel, R. N., Shah, D., Chokshi, H. B., et al. (2013). Hit identification and optimization in virtual screening: Practical recommendations based on a critical literature analysis. *Journal of Medicinal Chemistry, 56*(17), 6560–6572. https://doi.org/10.1021/jm301916b

Ziagova, M. G., Mavromatidou, C., Samiotis, G., & Amanatidou, E. (2021). Enhancing phenolic content of medicinal aromatic plants extracts-biofunctional foods preparation. *Plants, 11*(1), 76. https://doi.org/10.3390/plants11010076

Chapter 10
Breast-on-a-Chip Technology Application Using Acacia Gums (AGs) Secondary Metabolites Extract (SME)

Ahmed A. M. Elnour, Abdurahmani Hamid Nour, and Nureddin Ashammakhi

Abstract This chapter discusses the application of BoCs technology in studying BC and outlines the challenges, limitations, and future directions in this domain. This emphasizes the limitations of traditional research methodologies in accurately mimicking the complex dynamics of BC and highlights the emergence of OoC technology as a revolutionary tool for replicating the physiological and mechanical functions of human organs, including the breast. The integration of natural compounds with BoC technology, particularly Acacia gums (AGs), has been explored as a novel approach in BC research, offering potential therapeutic avenues for disrupting the metastatic cascade and enriching our understanding of BC pathophysiology. The challenges and limitations in applying BoC technology to study BC are discussed, including the technical complexity of accurately mimicking the tumor microenvironment (TME), maintaining cell viability, and integrating immune system components. However, the future of BoC technology appears promising, with potential advancements in biomaterials, microfluidic design, three-dimensional (3D) printing, nanofabrication, and the integration of immune system components and stem cell technology. These advancements could lead to improved replication of tumor complexity, personalized medicine, high-throughput drug screening, and transformative impacts on BC research, paving the way for groundbreaking discoveries and advancements in treatment methodologies. In conclusion, this chapter underscores the immense potential of BoCs technology in advancing our

A. A. M. Elnour (✉) · A. H. Nour
Faculty of Chemical and Process Engineering Technology, Universiti Malaysia Pahang Al-Sultan Abdullah (UMPSA), Pahang Gambang, Malaysia

Centre of Excellence for Advanced Research in Fluid Flow (CARIFF), University Malaysia, Pahang Gambang, Malaysia
e-mail: adamhassan@umpsa.edu.my; abrahman@umpsa.edu.my

N. Ashammakhi
Department of Biomedical Engineering, Institute for Quantitative Health Science and Engineering, College of Engineering, and College of Human Medicine, Michigan State University, East Lansing, MI, USA
e-mail: ashammak@msu.edu

A. A. M. Elnour (ed.), *Gum Arabic and Breast Cancer Biology*,
https://doi.org/10.1007/978-981-97-8518-6_10

understanding of BC and improving treatment strategies by providing a more accurate and sophisticated platform for studying the complex architecture and cellular interactions of the breast TME.

Keywords Breast cancer · Breast-on-a-chip technology · Organ-on-a-chip technology · *Acacia senegal* gum · *Acacia seyal* gum · *Acacia polyacantha* gum

Highlights

1. The AGs-SME-based breast-on-a-chip technology offers a more precise, efficient, and sustainable solution for studying and treating breast cancer.
2. Organ-on-a-chip technology mimics human organ function and accelerates BC research for better treatment.
3. AGs-SME and breast-on-a-chip technology are used to study breast tissue growth, renewal, and disease progression in a controlled microenvironment.
4. AGs offer an environment-friendly alternative to synthetic surfactants in microfluidic systems, highlighting a sustainable approach to cancer research.
5. Bioactive compounds/secondary metabolites from AGs target specific pathways and molecules involved in BC progression and metastasis.
6. AGs-secondary metabolites (SM) in breast-on-a-chip technology hold promise for improving BC understanding and therapies.

10.1 Introduction

BC is a formidable adversary of global health (Muratov et al., 2020). Traditional research methodologies, such as animal models and in vitro cell cultures, have proven inadequate due to their limitations in accurately mimicking the complex dynamics of the disease (Furlan et al., 2014). OoC technology has emerged as a revolutionary tool, offering microscale devices that replicate the physiological and mechanical functions of human organs, including the breast (Martin et al., 2021). This innovation allows for more accurate study of tissue growth, renewal, and disease processes, marking a significant step forward in understanding BC. Specifically, it facilitates the examination of the metastatic process, providing insights into the mechanisms underlying tumor cell preference for certain metastatic sites and the associated cellular damage (Castaño et al., 2018). The exploration of *A. senegal*, *Acacia seyal*, and *Acacia polyacantha* gums (AGs) is particularly noteworthy in this context. These AGs are recognized for their bioactive compounds/SM, which have shown promise in modulating cellular behaviors and the TME (Stanojković, 2020). By integrating these natural compounds with BoC technology, researchers can investigate the effects on breast tissue growth, renewal, and disease progression within a controlled microenvironment, potentially unveiling new therapeutic avenues for disrupting the metastatic cascade (Ganpisetti et al., 2022). The synergy between BoC technology and SM represents a novel approach to BC research (Gracia-Lor et al., 2017). This combination not only enhances the simulation of physiological conditions for studying metastasis but also opens possibilities for discovering interventions to inhibit or alter the progression of BC (Teiten et al., 2013). Integrating these natural metabolites could significantly enrich our understanding of BC pathophysiology and lead to innovative therapeutic strategies (Yakovenko et al., 2018). AGs offer an environment-friendly alternative to synthetic surfactants in microfluidic systems, highlighting a sustainable approach to cancer research (Maračić et al., 2023). Their biocompatibility and safety profile further underscore the potential of this integrated method to improve the efficacy and functionality of BoC devices, facilitating a deeper investigation into cancer cell behavior and interactions within the TME (Fumagalli et al., 2022). The introduction outlines a future in which the convergence of OoC technology and natural metabolites from AGs could revolutionize BC research. This approach addresses the limitations of traditional research methods. It leverages the therapeutic potential of natural compounds, providing a more accurate, efficient, and sustainable pathway for understanding and combating BC (Islam et al., 2022). As this field evolves, it promises to uncover new insights into BC pathophysiology, foster the development of targeted therapies, and ultimately improve patient outcomes. The intersection of innovative technology with the natural therapeutic potential of AGs sets a new frontier in the fight against BC, paving the way for more effective and personalized treatment modalities.

10.2 AGs-Secondary Metabolites

AGS-SM has various biological activities, including anti-inflammatory, antioxidant, antimicrobial, and anti-cancer properties (Elnour et al., 2020). These SM can potentially target specific pathways and molecules involved in BC progression and metastasis (Hirai et al., 2005). Furthermore, using AGs-based natural surfactants in "OoC" technology could potentially enhance the accuracy and reliability of in vitro models by providing a more physiologically relevant microenvironment for studying BC (Farooqi et al., 2023). Incorporating AGs-SM in BoCs technology is promising for advancing our understanding of BC and developing more effective therapies for this devastating disease. Using AGS-based natural surfactants and incorporating AGs-SME in BoC technology offers exciting opportunities for BC research. AGs are composed of complex polysaccharides and glycoproteins. The chemical composition and bioactive SM present in these gums can vary according to species, geographical location, and environmental conditions (Linder et al., 2012). The key components typically found in AGs are long carbohydrate molecules with various structures and functions (Al-Assaf et al., 2009). AGs often consist of arabinose, galactose, rhamnose, and glucuronic acid. These AGs chemical composition and bioactive SM content of AGs can vary according to species, geographical location, and environmental conditions. The key components typically found in AGs are summarized in Fig. 10.1.

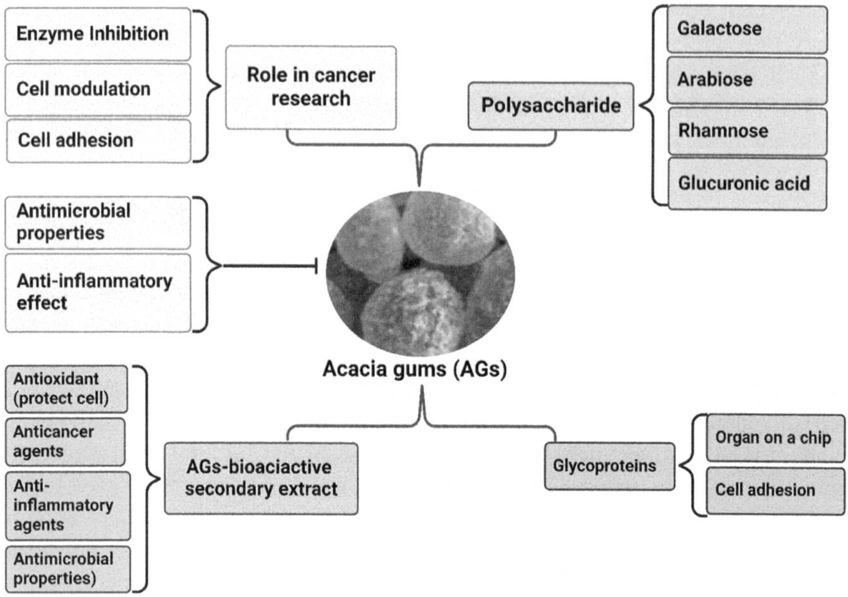

Fig. 10.1 Some AGs possess various biological activities

10.3 Principles and Design of BoCs Technology

BoC technology aims to recreate the structural and functional characteristics of breast tissue within a laboratory setting (Jungreuthmayer et al., 2011). This innovative method involves the development of a microfluidic platform designed to replicate the complex environment of the breast, including its tissue structure, various cell types, and cellular interactions (Michna et al., 2018). The principles underlying the design of BoC technology are founded on a blend of engineering and biological sophistication aimed at replicating human tissue functionality on a microscale (Hayward et al., 2020). Miniaturization is central to this approach, enabling the precise manipulation of tiny fluid volumes that are essential for accurately mimicking the conditions found within biological systems (Ruzycka-Ayoush et al., 2019). This precision is not only crucial for the realistic representation of biological microenvironments but also for ensuring the economical use of valuable biological samples and reagents (Burel et al., 2017). Parallel to miniaturization, Laminar Flow plays a pivotal role, whereby fluids are managed in such a way that they flow in parallel layers without intermixing (Liu et al., 2018). This characteristic is instrumental in achieving precise control over fluid movement and establishing stable gradients, thereby simulating the complex biological interfaces found in living systems (Zhang et al., 2015).

It is equally important to create a controlled microenvironment within these devices that meticulously maintains the necessary chemical and physical conditions, including oxygen, nutrients, and other cellular gradients, closely mirroring the natural environment inhabited by cells (Connell et al., 2014). Advancing further into the realm of tissue engineering, this technology excels in Tissue and Organ Mimicry (Tomov et al., 2019). By incorporating living cells and tissues into chips, these devices are adept at emulating the intricate structure and multifunctionality of actual organs (Liu et al., 2021a). This includes the use of various cell types to replicate specific organ functions and interactions, thereby providing a more accurate and functional model of the organ being studied (Schmeichel & Bissell, 2003). Addressing the dynamic aspect of biological tissues, this technology incorporates dynamic mechanical forces, simulating the physical forces to which organs are subjected, such as shear stress from fluid movement and mechanical strain from tissue expansion and contraction (Larobina et al., 2021). Finally, the aspect of integration and automation brings these devices into the future of biomedical research, with built-in sensors and actuators enabling continuous monitoring and automated adjustments (Szydzik et al., 2017). This facilitates not only high-throughput testing, but also extended duration experiments, opening new avenues in the study of disease and drug development. Together, these principles form the foundation of BoC technology, pushing the boundaries of what is possible in the recreation of human organ systems for research and therapeutic purposes (Bhatnagar et al., 2018). In BC research, BoC technology has proven to be a potent tool (Mankoff, 2008). It is significantly contribute to biomedical research and drug development by providing more accurate disease models for studying pathogenesis and treatment responses, facilitating high-throughput drug screening, enabling personalized medicine, reducing animal testing, and allowing a detailed understanding of biological processes (Thakar & Fenton, 2023). Designing BoC models that accurately simulate the TME involves several steps, as presented in Table 10.1.

BoC models offer several advantages over traditional research methods, including more accurate replication of the breast tissue's complexity, enabling the study of cell–cell and cell–matrix interactions, simulating dynamic conditions such as fluid flow and mechanical forces, and providing controlled experimentation environments (Frankman et al., 2022). These models reduce ethical concerns by potentially replacing animal testing, allowing real-time monitoring and analysis, facilitating drug screening and personalized medicine, reducing costs and resource use, improving reproducibility, and delivering faster results than animal studies (Thakar & Fenton, 2023). Therefore, BoCs models represent a significant advancement in cancer research, offering a more precise, ethical, and effective tool for exploring tumor biology, testing drug efficacy, understanding resistance mechanisms, and developing potential therapeutic interventions. The development and advantages of BoCs technology are summarized in Table 10.2.

Table 10.1 Design and innovation steps in BoCs model development

Steps	Description	References
Selection of cell lines	Appropriate cancer cell lines, including primary tumor cells and metastatic lines that reflect the cancer type and characteristics under study are chosen	Hung et al. (2014)
Incorporating microenvironment cells	Various cell types found in the breast microenvironment, such as fibroblasts and immune cells, must be included to mimic the tumor ecosystem	Yiying Liu et al. (2023)
Microfluidic platform development	A platform that can create interconnected compartments for different cell types is being developed. This platform allows controlled interactions and mimics the spatial organization of tumors	Haeberle and Zengerle (2007)
Use of extracellular matrix	Hydrogels or other analogs provide a scaffold for cell growth and interactions, essential for replicating tissue architecture	Todd et al. (2016)
Controlled conditions	The system is designed to precisely control the biochemical environment, including nutrient supply and drug gradients	Song et al. (2021)
Mechanical cues	Mechanisms to apply natural mechanical forces experienced by breast tissue cells are integrated	Riehl et al. (2021)
Integration of sensors and actuators	For real-time monitoring and automated condition regulation, sensors and actuators are incorporated	Clarke et al. (2021)
Validation	The model's accuracy is validated by comparing cancer cell behavior within the chip to known BC processes	Focaroli et al. (2014), Riethdorf et al. (2007)
Reproducibility and scalability	The design ensures that the model can be reliably reproduced and scaled up for extensive drug testing and personalized medicine	Rubiano et al. (2021), Vidi et al. (2014)
Ethical and safety considerations	Non-hazardous materials are used, and proper biological waste disposal methods are ensured	Kim and Fajardo (2019)

10.4 Breast Tumor Microenvironment (TME)

The TME plays a crucial role in BC initiation, progression, and response to treatment (Fernández-García et al., 2016). The TME consists of various cell types, including cancer cells, immune cells, fibroblasts, endothelial cells, and extracellular components such as extracellular matrix and soluble factors (Sjöberg et al., 2016). This complex network of interactions creates a unique microenvironment that influences tumor growth, invasion, metastasis, and therapeutic resistance (Jiang et al., 2019). BC is a highly heterogeneous disease, with different subtypes characterized by distinct molecular profiles and clinical outcomes (Mitchell et al., 2018). Understanding the TME is essential for developing effective treatments and improving patient outcomes (Liu et al., 2021b). BoCs technology offers a promising approach to studying the BC-TME in a more accurate and controlled manner (Mintz

Table 10.2 Development and advantages of BoCs technology

Category	Subcategory	Description	References
Fundamental principles	Miniaturization	Operate on a microscale to handle small fluid quantities with high precision, crucial for replicating biological volumes and for economical use of samples and reagents	MacQueen et al. (2012)
	Laminar flow	Fluids flow in parallel layers without mixing, enabling precise control and stable gradient creation for biological simulation	Moehlenbrock and Martin (2007)
	Controlled micro-environment	Accurate chemical and physical environment control, including oxygen and nutrient gradients that cells encounter in-vivo	Yiying Liu et al. (2023)
	Tissue and organ mimicry	Incorporation of living cells and tissues to mimic real organs' structure and function, using multiple cell types for organ-specific functionality and interactions	Vidi et al. (2014)
	Dynamic mechanical forces	Simulation of in vivo physical forces, such as shear stress from fluid flow or mechanical strain from tissue movements	Carp et al. (2008)
	Integration and automation	Inclusion of sensors and actuators for monitoring and automation, essential for high-throughput testing and extended experiments	Ausserer et al. (2001)
Design and fabrication	Selection of cell line	Choosing appropriate cancer cell lines that represent the studied BC type and characteristics	Smith et al. (2017)
	Incorporating micro-environment associated cells	Including various cell types present in the breast microenvironment, such as fibroblasts and immune cells, to mimic tumor ecosystem interactions	Kim and Fajardo (2019)
	Microfluidic platform	Developing a platform that creates interconnected compartments for different cell types, allowing for controlled interaction and spatial organization mimicry	Fragoso et al. (2011)
	Extracellular matrix	Using hydrogel or other analogs as a scaffold for cell adhesion, growth, and interactions, essential for tissue architecture mimicry	Yang et al. (2020)
	Controlled conditions	Designing the system for precise biochemical environment control, including nutrient supply and drug gradients	Forry and Locascio (2011)
	Mechanical cues	Applying natural mechanical forces that breast tissue cells experience	Mansoor et al. (2016)
	Sensors and actuators	Integrating sensors for real-time monitoring and actuators for automated environmental regulation	Nguyen et al. (2022)
	Validation	Comparing cancer cell behavior within the chip to known BC processes for model accuracy	Frech et al. (2005)
	Reproducibility and scalability	Ensuring design allows for high-throughput drug testing and personalized medicine applications	Nahak et al. (2022)
	Ethical and safety considerations	Using non-hazardous materials and ensuring proper disposal of biological waste	Thakar and Fenton (2023)

(continued)

Table 10.2 (continued)

Category	Subcategory	Description	References
Advantages	Enhanced accuracy	Close replication of breast tissue's architectural and functional complexity, including the TME, for physiologically relevant data	Ramirez et al. (2012)
	Tissue-specific micro-environment	Co-culture of multiple cell types in defined spatial arrangements for studying crucial cell-cell and cell-matrix interactions	Moccia and Haase (2021)
	Dynamic conditions	Simulation of dynamic mechanical forces and fluid flow experienced by tissues in the body	Santiago et al. (2019)
	Controlled exper-imentation	Precise control over the cellular microenvironment, more challenging in traditional culture systems	Hipolito et al. (2018)
	Reduced ethical concerns	Potential reduction or replacement of animal use in research, aligning with ethical considerations	Thakar and Fenton (2023)
	Real-time monitoring and analysis	Integration with sensors and imaging technologies for continuous cellular behavior and treatment	Bulun et al. (2009)

et al., 2018). By incorporating AGs-SME into BoCs models, the potential for exploring the impact of these natural compounds on BC cells and their microenvironment can be investigated. BoCs technology and AGs present a novel and innovative approach to studying BC.

This chapter explores the application of BoCs technology utilizing AGs-SME in understanding and studying the BC-TME. By leveraging the unique capabilities of OoC technology, specifically BoCs models, researchers can gain valuable insights into the complex interactions among BC cells, immune cells, fibroblasts, and endothelial cells in the TME (Shi et al., 2023). This comprehensive exploration of BoC technology utilizing AGS-SME can revolutionize BC research, leading to an improved understanding of tumor biology, personalized medical approaches, and the development of more effective therapeutic strategies for BC patients (Shen et al., 2023). The use of BoCs technology, combined with the incorporation of AGs-SME, offers a unique opportunity to investigate the intricate interactions within the BC-TME. This exploration can provide valuable insights into the effects of these natural compounds on BC cells and their surrounding healthy cells, ultimately leading to a deeper understanding of the progression and treatment of BC. In recent years, there has been growing interest in developing innovative approaches to study the complex TME of BC more accurately and effectively (Ling et al., 2022). One such approach is the utilization of BoCs technology, which enables researchers to recreate the physiological conditions of the breast tissue in a highly controlled manner (Firatligil-Yildirir et al., 2023). Finally, the TME plays a crucial role in BC initiation, progression, and response to treatment (Zhu et al., 2022). The TME consists of various cell types, including:

1. **Cancer Cells**: The primary cells that constitute the tumor itself (Anderson & Simon, 2020).
2. **Immune Cells**: These include lymphocytes and macrophages, which can either fight against the tumor or sometimes be co-opted by the tumor to facilitate its growth (Anderson & Simon, 2020).
3. **Fibroblasts**: Often referred to as cancer-associated fibroblasts in the TME, they can promote tumor progression and drug resistance (Ortiz-Otero et al., 2020).
4. **Endothelial Cells**: These cells line blood vessels and support angiogenesis, enabling tumors to create their own blood supply, thereby facilitating growth by providing necessary nutrients (Hanahan & Weinberg, 2016).

Additionally, the TME contains non-cellular components such as:

1. **Extracellular Matrix**: A complex network of proteins and other molecules that provide structural support to cells and can influence cell behavior and signaling (Κάρλου et al., 2010).
2. **Soluble Factors**: Including cytokines, growth factors, and hormones that mediate communication between cells and can modulate tumor growth and immune responses (Marincola, 2006).

This complex network of interactions creates a unique microenvironment that influences tumor growth, invasion, metastasis, and therapeutic resistance (Kitadai, 2010). BC is a highly heterogeneous disease, with different subtypes characterized by distinct molecular profiles and clinical outcomes. Understanding the TME is essential for developing effective treatments and improving patient outcomes (Althobiti et al., 2020). BoCs technology offers a promising approach to studying the BC-TME more accurately and controlled (Wu et al., 2020). These models can help explore complex cell interactions, the effects of different drugs, and the mechanisms of resistance and metastasis, potentially leading to more effective therapies (Liverani et al., 2021). An example of a breast tumor microenvironment is shown in Fig. 10.2.

10.5 Recapitulation of BC-TME Using OoC

The recapitulation of the BC-TME using OoC technology involves creating a miniaturized, three-dimensional model that imitates a breast tumor's complex in vivo environment (Choi et al., 2015). This approach is designed to incorporate essential elements of the TME, such as various cell types, extracellular matrix components, and dynamic mechanical and biochemical conditions, to facilitate studies more representative of the actual disease state (Kim et al., 2015). The key points in recapitulating the BC-TME using OoC include the following.

1. **Multiple Cell Types**: To replicate the cellular complexity of the TME, the variety of cells present in it, including cancer cells, immune cells, fibroblasts, endothelial cells, and others, will be incorporated (Castillo-Sánchez et al., 2022).

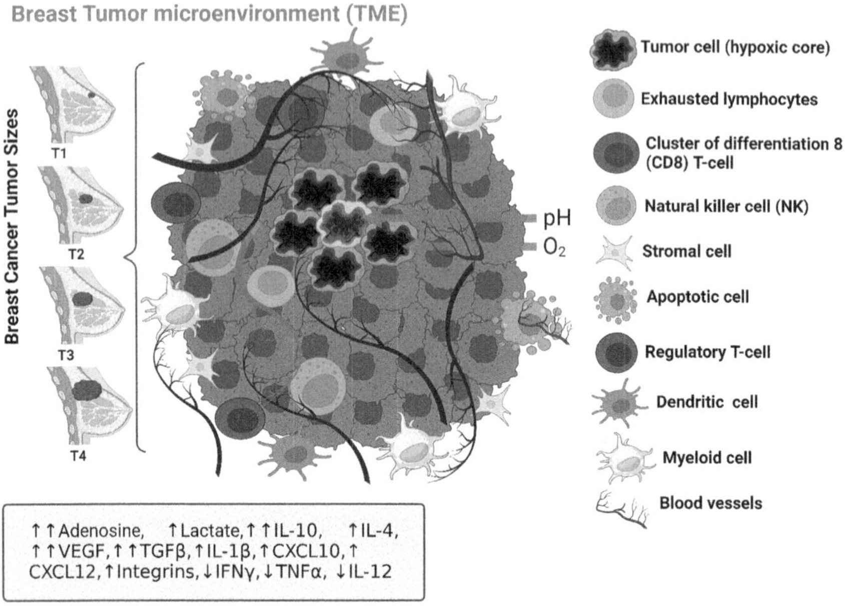

Fig. 10.2 The components of the breast tumor microenvironment (TME), which consists of heterogeneous tumor cells, stromal cells, and diverse immune cells within a network of abnormal vasculature and collagen. This complex ecosystem is characterized by poor perfusion and densely packed glycolytic tumor cells, resulting in reduced oxygen levels, acidic pH, and inadequate nutrient availability. These conditions promote the accumulation of **metabolic** by-products, such as lactate, alongside a milieu enriched with anti-inflammatory cytokines and chemokines, such as **Interleukin** (**IL-10** and **IL-4**), and vascular endothelial growth factor (**VEGF**), while levels of interferon gamma (**IFNγ**), TNFα, and tumor necrosis factor (**IL-12**) are decreased. The altered immune response within the TME plays a crucial role in tumor progression and response to therapy

2. **Extracellular Matrix (ECM)**: Hydrogels or other scaffold materials mimic the ECM found in the breast tissue, which supports cell adhesion and influences cell behavior and signaling (Vong & Kalluri, 2011).
3. **Biochemical Gradients**: Engineering the chip to maintain stable gradients of oxygen, nutrients, and signaling molecules is vital for replicating the TME and studying cancer cell migration and invasion (Ren et al., 2020).
4. **Mechanical Forces**: Mechanical forces such as fluid shear stress and cyclic strain cells experience in the body are applied to replicate the physical cues present in the TME (Narkhede et al., 2017).
5. **Microfluidics**: Utilizing microfluidic technology for precisely manipulating and controlling small volumes of fluids, which is crucial for simulating blood and lymphatic flow patterns (Shang et al., 2019).
6. **High-Throughput and Real-Time Analysis**: Designing OoC platforms to allow for high-throughput drug testing and real-time analysis of cellular responses through integrated sensing technologies (Fetah et al., 2019).

7. **Patient-Derived Models**: Patient-derived cells are used to create personalized OoC models to study individual disease mechanisms and evaluate therapeutic efficacy (Haque et al., 2022).

Using OoC technology to replicate the TME offers a powerful tool for enhancing our understanding of BC biology, evaluating drug responses, and potentially improving the prediction of clinical outcomes in cancer research (Generali et al., 2015). By incorporating advanced techniques, such as 3D bioprinting and microfluidics, researchers are revolutionizing the study of BC and paving the way for more targeted and effective treatments (Ganpisetti et al., 2022). Recapitulation of breast tumor microenvironment using OoC as presented in Table 10.3.

Table 10.3 Demonstrates breast cancer-on-a-chip study using different cell types for different proposes

Study	Cell types used	Conditions	Main findings	References
Drug response				
BCOC-2019-001	Cancer cells, fibroblasts, immune cells	Nutrient gradients, shear stress	Drug A is effective in tumor reduction	Frieboes et al. (2009)
BCOC-2020-002	Cancer cells, endothelial cells	Drug screening, microfluidics	Drug B, less toxic, targets specific tumors	Dimitriou et al. (2021)
Tumor microenvironment (TME)				
TME-Chip-2018	Cancer cells, immune cells	ECM analogs, biochemical gradients	Complex TME recapitulated	Kumar and Varghese (2018)
MicroTME-2019	Cancer cells, fibroblasts	Mechanical forces, 3D culture	Impact of mechanical stress on tumor	Grisendi et al. (2019)
Metastasis exploration				
MetaS-Chip-2021	Cancer cells, immune cells	Microfluidics, cell migration and invasion	New metastasis pathways identified	Cho et al. (2021)
LymphoChip-2020	Cancer cells, endothelial cells	Lymphatic flow, ECM simulation	Tumor cells' intravasation into the lymph system	Barua et al. (2020)
Personalized medicine approaches				
Personal-OoC-2021	Patient-derived cancer cells	Genomic data, personalized conditions	Personalized treatment strategies enabled	Kiyotani et al. (2021)
GenoPheno-Chip-2022	Patient-derived cancer cells	3D culture, drug responses	Genotype-phenotype correlation	Sandhya et al. (2022)

10.6 Application of AGs-SME in BoCs Models

The application of AGs-SME in BoCs models holds great potential for advancing our understanding of BC and developing targeted therapeutic strategies. The SM derived from AGs has been shown to possess significant anti-cancer properties (Elnour et al., 2020). These metabolites have demonstrated anti-proliferative, pro-apoptotic, and anti-migratory effects on BC cells. Furthermore, AGs-SME may have shown potential in modulating the TME by affecting angiogenesis, inflammation, and immune response. Incorporating these AGs-SME into BoCs models can provide a more accurate representation of the complex interactions between cancer cells and the TME, allowing researchers to study the efficacy of targeted therapies and evaluate the mechanisms behind their actions. The preparation and integration of AGs-SME into a BoCs system would typically involve several steps:

1. The investigation into the anti-cancer properties of SME, sourced from species such as *Acacia senegal* or *Acacia seyal* gum, has not yet incorporated a multi-faceted approach, from extraction to evaluation within a BoC system. The initial phase involves the meticulous purification and characterization of these metabolites. Techniques such as chromatography are utilized for purification, followed by various analytical methods to ascertain their structure, purity, and concentration (Chikamai et al., 1995). This ensures that the substances introduced to the BoCs are well-defined and suitable for biological experiments.
2. The next stage centers on integrating these metabolites into the BoC device, an engineered platform to mimic the vascular structure and cellular environment of the breast tissue (Clarke et al., 2021). The chip is seeded with human BC cells and, potentially, other cellular components of the TME, such as immune cells and fibroblasts (Monico et al., 2022). Forming an extracellular matrix (ECM) using materials like hydrogels, collagen, or Matrigel™ is crucial, providing a scaffold that supports cell adhesion and growth, mirroring the natural TME (Hoyle et al., 2022).
3. The application of AGs-SME is a critical step, achieved by either incorporating them into the ECM or infusing them into the microfluidic system. This allows for direct interaction with the BC cells, potentially affecting crucial processes such as proliferation, apoptosis, and migration. Monitoring and analysis of cellular responses are conducted in real-time, utilizing integrated sensors or periodic sampling, focusing on changes in cell viability, apoptosis rates, and migration patterns among other relevant endpoints.
4. Evaluating the therapeutic potential of AGs-SME involves assessing their efficacy as anti-cancer agents, observing drug responses, impacts on the TME, and effects on angiogenesis or metastasis (Elnour et al., 2020). Comparing collected data against controls facilitates a comprehensive understanding of how AGs-SME influence BC cell behavior and modify the TME.
5. Furthermore, the methodological framework includes advanced drug delivery strategies, such as the incorporation method for even distributing metabolites within the ECM, infusion into the microfluidic system for systemic delivery, and

microinjection for localized treatment effects. Dose optimization is conducted by establishing concentration gradients and time-dependent exposure to identify effective doses that minimize cytotoxic effects.

6. Key assays and analyses, including MTT/MTS assays for cell viability, TUNEL assay for apoptosis, and scratch assays for migration, provide quantitative and qualitative data on the metabolites' impact. Additional studies on angiogenesis, inflammation, and immune response further elucidate the potential of AGs-SME in cancer treatment, emphasizing the innovative approach of using BoCs technology to advance BC research.

10.7 Experimental Results and Analysis

A series of experiments was launched to explore the kinetics and mechanisms behind the sorption of butylbenzene and tert-butylbenzene onto sand particles and their biodegradation. This research included assessing the sorption of these hydrocarbons to sand particles and examining the cell surface properties of microbial strains capable of degrading them (Chrzanowski et al., 2008). Additionally, the study investigated the effect of plant extracts, notably those containing saponins, on these processes. Another significant aspect of this research focused on evaluating the impact of AGs-SME on BC cell lines using advanced BoCs technology. This involved a meticulous setup of experiments designed to uncover the therapeutic potential of these metabolites and their interaction with cancer cells and the TME.

The experimental setup for testing AGs-SME on BC cell lines involved several key steps: chip preparation and cell seeding, choosing the right extracellular matrix components, and introducing AGs-SME into the microfluidic environment. These steps were essential in creating a controlled and physiologically relevant platform for studying the effects of natural compounds on cancer cells. The experiments also included maintaining the chip under optimal conditions, establishing dosing parameters, and implementing monitoring protocols to assess the cells' responses to treatment. Endpoint assays and analysis, including various cellular and molecular techniques, were conducted to evaluate the efficacy of AGs-SME in inhibiting cancer cell proliferation, migration, and other critical aspects of tumor biology.

The comprehensive approach adopted in this research extends to examining cell–cell and cell–ECM interactions, molecular signaling pathways, the effects of mechanical forces, and oxygen gradient effects within the TME on a chip (Polacheck et al., 2014). These analyses provide insights into the complex dynamics of cancer progression and the impact of therapeutic agents like AGs-SME. By studying drug responses, resistance, tumor immune microenvironment, metastatic potential, angiogenesis, and feedback loops, a BoC model is a sophisticated tool for unraveling the multi-faceted nature of BC (Castaño et al., 2018). Ultimately, this research aims to pave the way for novel therapeutic strategies and a deeper understanding of BC pathophysiology, leveraging the unique capabilities of OoC technology.

10.8 Challenges, Limitations, and Future Directions

We summarize the challenges, limitations, and future directions associated with applying the breast-on-a-chip (BoC) technology in breast cancer (BC) research as follows:

10.8.1 Challenges and Limitations

The application of BoCs technology in the study of BC presents several challenges and limitations.

1. **Technical Complexity**: Accurately mimicking the complex tumor microenvironment (TME), which consists of diverse cell types, signaling molecules, and extracellular matrix, poses a significant technical challenge (Ganpisetti et al., 2022).
2. **Physiological Replication**: Recreating physiological conditions, such as blood flow and mechanical forces on a microscale, adds complexity to the replication of the TME (Teiten et al., 2013).
3. **Cell Viability and Integration**: Maintaining cell viability over long periods, integrating immune system components, and achieving high-throughput automation are technical challenges in BoCs technology (Teiten et al., 2013).

10.8.2 Future Directions

Despite these challenges, the future of BoCs technology in BC research appears promising, with the potential for significant advancements:

1. **Innovations in Biomaterials and Microfluidic Design**: Advancements in biomaterials and microfluidic design are expected to enhance the physiological relevance of BoC models.
2. **3D Printing and Nanofabrication**: Adoption of 3D printing and nanofabrication techniques, along with the development of automated, high-throughput systems, could revolutionize the use of BoCs in cancer research.
3. **Integration of Immune System Components and Stem Cell Technology**: Integrating immune system components, stem cell technology, and networking multiple BoC models opens new avenues for creating more complex and realistic tumor models.
4. **Personalized Medicine and High-Throughput Drug Screening**: BoC models could offer personalized insights into treatment responses, high-throughput drug screening, and studying drug mechanisms in a context that more closely mimics the human body (Yakovenko et al., 2018).

In conclusion, although there are technical challenges and limitations in applying BoC technology to study BC, future directions involve innovations in technology, biomaterials, and design, as well as the potential for personalized medicine and high-throughput drug screening, which could lead to groundbreaking discoveries and advancements in treatment methodologies.

10.9 Conclusion

In this chapter, we comprehensively explore the application of breast-on-a-chip (BoC) technology in studying BC and the potential of natural compounds, specifically the Acacia gums (AGs), to enhance the functionality and efficacy of BoC models. This chapter underscores the significance of accurately mimicking the complex architecture and cellular interactions of the breast tumor microenvironment (TME) to gain deeper insights into the complexity of BC and its interactions with various components of the microenvironment. The challenges and limitations in applying BoC technology to study BC are thoroughly examined, emphasizing the technical complexity of replicating the TME, maintaining cell viability, and integrating immune system components. However, the future of BoC technology appears promising, with potential advancements in biomaterials, microfluidic design, 3D printing, nanofabrication, and the integration of immune system components and stem cell technology. These advancements could lead to improved replication of tumor complexity, long-term culture stability, and facilitation of regulatory and ethical developments in the field. In conclusion, this chapter highlights the immense potential of BoCs technology to advance our understanding of BC and to improve treatment strategies. By providing a more accurate and physiologically relevant model for studying the disease, BoC technology offers valuable insights into BC biology, including metastasis, co-culture systems, immuno-oncology, biomarker identification, and mechanistic studies. Integrating natural compounds, such as AGs-SME, presents a novel and innovative approach to studying BC, paving the way for groundbreaking discoveries and advancements in treatment methodologies.

Acknowledgments The authors thank the Chemical and Process Engineering Technology at the University of Malaysia, Pahang-UMPSA, Malaysia, for supporting this research.

Fund This research was funded by Eastern Unity Technology under Grant No. UIC 190806.

References

Al-Assaf, S., Sakata, M., McKenna, C., Aoki, H., & Phillips, G. O. (2009). Molecular associations in acacia gums. *Structural Chemistry, 20*, 325. https://doi.org/10.1007/s11224-009-9430-3

Althobiti, M., Muftah, A. A., Aleskandarany, M. A., Joseph, C., Toss, M. S., Green, A., & Rakha, E. A. (2020). The prognostic significance of BMI1 expression in invasive breast cancer is dependent on its molecular subtypes. *Breast Cancer Research and Treatment, 182*, 581. https://doi.org/10.1007/s10549-020-05719-x

Anderson, N., & Simon, M. C. (2020). The tumor microenvironment. *Current Biology, 30*, R921. https://doi.org/10.1016/j.cub.2020.06.081

Ausserer, W. A., Bousse, L., Gallagher, S. J., Kennedy, C. B., & Phan, H. L. (2001). Automated lab-on-a-chip analysis of DNA fragments. *JALA Journal of the Association for Laboratory Automation, 6*, 69. https://doi.org/10.1016/s1535-5535-04-00139-x

Barua, A., Choudhury, P., Mandal, S., Panda, C. K., & Saha, P. (2020). Anti-metastatic potential of a novel xanthone sourced by Swertia chirata against in vivo and in vitro breast adenocarcinoma frameworks. *Asian Pacific Journal of Cancer Prevention, 21*, 2865. https://doi.org/10.31557/apjcp.2020.21.10.2865

Bhatnagar, A., Scott, K., Govender, V., & George, A. (2018). Pushing the boundaries of research on human resources for health: Fresh approaches to understanding health worker motivation. *WHO South-East Asia Journal of Public Health, 7*, 13. https://doi.org/10.4103/2224-3151.228422

Bulun, S. E., Lin, Z., Zhao, H., Lü, M., Amin, S., Reierstad, S., & Chen, D. (2009). Regulation of aromatase expression in breast cancer tissue. *Annals of the New York Academy of Sciences, 1155*, 121. https://doi.org/10.1111/j.1749-6632.2009.03705.x

Burel, J. G., Yu, Q., Arlehamn, C. S. L., Weiskopf, D., Zapardiel-Gonzalo, J., Taplitz, R., et al. (2017). An integrated workflow to assess technical and biological variability of cell population frequencies in human peripheral blood by flow cytometry. *The Journal of Immunology, 198*, 1748. https://doi.org/10.4049/jimmunol.1601750

Carp, S. A., Selb, J., Fang, Q., Moore, R. H., Kopans, D. B., Rafferty, E. A., & Boas, D. A. (2008). Dynamic functional and mechanical response of breast tissue to compression. *Optics Express*. https://doi.org/10.1364/biomed.2008.bsub7

Castaño, Z., Juan, B. P. S., Spiegel, A., Pant, A., DeCristo, M. J., Laszewski, T., et al. (2018). IL-1β inflammatory response driven by primary breast cancer prevents metastasis-initiating cell colonization. *Nature Cell Biology, 20*, 1084. https://doi.org/10.1038/s41556-018-0173-5

Castillo-Sánchez, R., Churruca-Schuind, A., Martinez-Ival, M., & Salazar, E. P. (2022). Cancer-associated fibroblasts communicate with breast tumor cells through extracellular vesicles in tumor development. *Technology in Cancer Research & Treatment, 21*. https://doi.org/10.1177/15330338221131647

Chikamai, B., Osman, M. E., Menzies, A. R., & Banks, W. B. (1995). An evaluation of methods for characterising and monitoring gum Arabic of commerce and related acacia gums. *Wood Science and Technology, 30*, 49. https://doi.org/10.1007/bf00195268

Cho, H.-Y., Choi, J. H., Kim, K.-J., Shin, M., & Choi, J. W. (2021). Microfluidic system to analyze the effects of interleukin 6 on lymphatic breast cancer metastasis. *Frontiers in Bioengineering and Biotechnology, 8*. https://doi.org/10.3389/fbioe.2020.611802

Choi, Y.-C., Hyun, E., Seo, J., Blundell, C., Kim, H. C., Lee, E.-H., et al. (2015). A microengineered pathophysiological model of early-stage breast cancer. *Lab on a Chip, 15*, 3350. https://doi.org/10.1039/c5lc00514k

Chrzanowski, Ł., Bielicka-Daszkiewicz, K., Owsianiak, M., Aurich, A., Kaczorek, E., & Olszanowski, A. (2008). Phenol and N-alkanes (C12 and C16) utilization: Influence on yeast cell surface hydrophobicity. *World Journal of Microbiology and Biotechnology, 24*, 1943. https://doi.org/10.1007/s11274-008-9704-8

Clarke, G. A., Hartse, B. X., Asli, A. E. N., Taghavimehr, M., Hashemi, N., Shirsavar, M. A., et al. (2021). Advancement of sensor integrated organ-on-chip devices. *Sensors, 21*. https://doi.org/10.3390/s21041367

Connell, J. L., Kim, J., Shear, J. B., Bard, A. J., & Whiteley, M. (2014). Analyzing secondary metabolite production by 3D printed bacterial populations using scanning electrochemical microscopy. *Microscopy and Microanalysis, 20*, 1182. https://doi.org/10.1017/s1431927614007648

Dimitriou, P., Li, J., Tornillo, G., McCloy, T., & Barrow, D. A. (2021). Droplet microfluidics for tumor drug-related studies and programmable artificial cells. *Global Challenges, 5*. https://doi.org/10.1002/gch2.202000123

Elnour, A. A., Mirghani, M. E., Kabbashi, N., Daddiouaissa, D., Musa, K. H., Alam, M. Z., & Abdurahman, N. H. (2020). Active fractions of methanol crude obtained from Acacia seyal gum and their antiproliferative effects against human breast cancer cell lines. *Global Journal of Science Frontier Research, 20*(3-C), 51–64.

Farooqi, M. A., Kang, C.-U., & Choi, K. H. (2023). Organ-on-chip: Advancing nutraceutical testing for improved health outcomes. *ACS Omega, 8*, 31632. https://doi.org/10.1021/acsomega.3c03155

Fernández-García, B., Eiró, N., Miranda, M.-A., Cid, S., González, L. O., Domínguez, F., & Vizoso, F. (2016). Prognostic significance of inflammatory factors expression by stroma from breast carcinomas. *Carcinogenesis, 37*, 768. https://doi.org/10.1093/carcin/bgw062

Fetah, K., DiPardo, B. J., Kongadzem, E.-M. L., Tomlinson, J. S., Elzagheid, A., Elmusrati, M., et al. (2019). Cancer modeling-on-a-chip with future artificial intelligence integration. *Small, 15*, e1901985. https://doi.org/10.1002/smll.201901985

Firatligil-Yildirir, B., Yalcin-Ozuysal, Ö., & Nonappa, N. (2023). Recent advances in lab-on-a-chip systems for breast cancer metastasis research. *Nanoscale Advances, 5*, 2375. https://doi.org/10.1039/d2na00823h

Focaroli, S., Mazzitelli, S., Falconi, M., Luca, G., & Nastruzzi, C. (2014). Preparation and validation of low cost microfluidic chips using a shrinking approach. *Lab on a Chip, 14*, 4007. https://doi.org/10.1039/c4lc00679h

Forry, S. P., & Locascio, L. E. (2011). On-chip CO_2 control for microfluidic cell culture. *Lab on a Chip, 11*, 4041. https://doi.org/10.1039/c1lc20505f

Fragoso, A., Latta, D., Laboria, N., von Germar, F., Hansen-Hagge, T. E., Kemmner, W., et al. (2011). Integrated microfluidic platform for the electrochemical detection of breast cancer markers in patient serum samples. *Lab on a Chip, 11*, 625. https://doi.org/10.1039/c0lc00398k

Frankman, Z. D., Jiang, L., Schroeder, J. A., & Zohar, Y. (2022). Application of microfluidic systems for breast cancer research. *Micromachines, 13*, 152. https://doi.org/10.3390/mi13020152

Frech, M. S., Jones, L. P., & Furth, P. A. (2005). Validation of transgenic models of breast cancer: Ductal carcinoma in situ (DCIS) and Brca1-mutation-related breast cancer. *Breast Cancer Online, 8*, e42. https://doi.org/10.1017/s1470903105003135

Frieboes, H. B., Edgerton, M. E., Fruehauf, J. P., Rose, F. R. A. J., Worrall, L. K., Gatenby, R. A., et al. (2009). Prediction of drug response in breast cancer using integrative experimental/computational modeling. *Cancer Research, 69*, 4484. https://doi.org/10.1158/0008-5472.can-08-3740

Fumagalli, F., Calbi, V., Sora, M. G. N., Sessa, M., Baldoli, C., Rancoita, P. M. V., et al. (2022). Lentiviral haematopoietic stem-cell gene therapy for early-onset metachromatic leukodystrophy: Long-term results from a non-randomised, open-label, phase 1/2 trial and expanded access. *The Lancet, 399*, 372. https://doi.org/10.1016/s0140-6736(21)02017-1

Furlan, A., Vercamer, C., Bouali, F., Damour, I., Chotteau-Lelièvre, A., Wernert, N., et al. (2014). Ets-1 controls breast cancer cell balance between invasion and growth. *International Journal of Cancer, 135*, 2317. https://doi.org/10.1002/ijc.28881

Ganpisetti, R., Mounika, M. S., Bag, I., Rana, R., Sanjay, G., & Giribabu, K. (2022). 3D bioprinting models for novel breast cancer strategies. *Research Journal of Pharmacy and Technology, 15*, 5576. https://doi.org/10.52711/0974-360x.2022.00942

Generali, D., Ardine, M., Strina, C., Milani, M., Cappelletti, M. R., Zanotti, L., et al. (2015). Neoadjuvant treatment approach: The Rosetta stone for breast cancer? *JNCI Monographs, 2015*, 32. https://doi.org/10.1093/jncimonographs/lgv019

Gracia-Lor, E., Rousis, N. I., Zuccato, E., Bade, R., Baz-Lomba, J. A., Castrignanò, E., et al. (2017). Estimation of caffeine intake from analysis of caffeine metabolites in wastewater. *The Science of the Total Environment, 609*, 1582. https://doi.org/10.1016/j.scitotenv.2017.07.258

Grisendi, G., Dall'Ora, M., d'Esposito, A., Casaria, G., Spano, C., Rossignoli, F., et al. (2019). Abstract A17: Challenging pancreatic ductal adenocarcinoma and its stroma by a combination of chemo and gene therapy: A preclinical study. *Cancer Research, 79*, A17. https://doi.org/10.1158/1538-7445.panca19-a17

Haeberle, S., & Zengerle, R. (2007). Microfluidic platforms for lab-on-a-chip applications. *Lab on a Chip, 7*, 1094. https://doi.org/10.1039/b706364b

Hanahan, D., & Weinberg, R. A. (2016). The hallmarks of cancer. *Cell, 100*, 57. https://doi.org/10.1093/med/9780199656103.003.0001

Haque, M. R., Wessel, C. R., Leary, D. D., Wang, C., Bhushan, A., & Bishehsari, F. (2022). Patient-derived pancreatic cancer-on-a-chip recapitulates the tumor microenvironment. *Microsystems & Nanoengineering, 8*, 36. https://doi.org/10.1038/s41378-022-00370-6

Hayward, K. L., Kouthouridis, S., & Zhang, B. (2020). Organ-on-a-chip systems for modeling pathological tissue morphogenesis associated with fibrosis and cancer. *ACS Biomaterials Science & Engineering, 7*, 2900. https://doi.org/10.1021/acsbiomaterials.0c01089

Hipolito, J., Peretz-Soroka, H., Moya-Torres, A., Booy, E. P., Yang, K., Gupta, M., et al. (2018). Microfluidic devices for studying the effect of netrin-1 on neutrophil and breast cancer cell migration. *Advanced Biosystems, 2*, 1700178. https://doi.org/10.1002/adbi.201700178

Hirai, M. Y., Klein, M., Fujikawa, Y., Yano, M., Goodenowe, D. B., Yamazaki, Y., et al. (2005). Elucidation of gene-to-gene and metabolite-to-gene networks in Arabidopsis by integration of metabolomics and transcriptomics. *Journal of Biological Chemistry, 280*, 25590. https://doi.org/10.1074/jbc.m502332200

Hoyle, H. W., Stenger, C. M. L., & Przyborski, S. A. (2022). Design considerations of benchtop fluid flow bioreactors for bio-engineered tissue equivalents in vitro. *Biomaterials and Biosystems, 8*, 100063. https://doi.org/10.1016/j.bbiosy.2022.100063

Hung, L. Y., Wang, C. H., Hsu, K. F., Chou, C. Y., & Lee, G. B. (2014). An on-chip cell-SELEX process for automatic selection of high-affinity aptamers specific to different histologically classified ovarian cancer cells. *Lab on a Chip, 14*, 4017. https://doi.org/10.1039/c4lc00587b

Islam, M. R., Islam, F., Nafady, M. H., Akter, M., Mitra, S., Das, R., et al. (2022). Natural small molecules in breast cancer treatment: Understandings from a therapeutic viewpoint. *Molecules, 27*, 2165. https://doi.org/10.3390/molecules27072165

Jiang, E., Yan, T., Xu, Z., & Shang, Z. (2019). Tumor microenvironment and cell fusion. *Biomed Research International, 2019*, 1. https://doi.org/10.1155/2019/5013592

Jungreuthmayer, C., Birnbaumer, G., Zanghellini, J., & Ertl, P. (2011). 3D numerical simulation of a lab-on-a-chip—Increasing measurement sensitivity of interdigitated capacitors by passivation optimization. *Lab on a Chip, 11*, 1318. https://doi.org/10.1039/c0lc00543f

Kim, E. Y., & Fajardo, L. L. (2019). Quality and safety considerations in breast cancer screening. *Proceedings in Obstetrics and Gynecology, 9*, 1. https://doi.org/10.17077/2154-4751.1449

Kim, H. M., Kim, H., & Park, Y. N. (2015). Sarcomatoid cholangiocarcinoma with osteoclast-like giant cells associated with hepatolithiasis: A case report. *Clinical and Molecular Hepatology, 21*, 309. https://doi.org/10.3350/cmh.2015.21.3.309

Kitadai, Y. (2010). Angiogenesis and lymphangiogenesis of gastric cancer. *Journal of Oncology, 2010*, 1. https://doi.org/10.1155/2010/468725

Kiyotani, K., Toyoshima, Y., & Nakamura, Y. (2021). Personalized immunotherapy in cancer precision medicine. *Cancer Biology and Medicine, 18*, 955. https://doi.org/10.20892/j.issn.2095-3941.2021.0032

Kumar, V., & Varghese, S. (2018). Ex vivo tumor-on-a-chip platforms to study intercellular interactions within the tumor microenvironment. *Advanced Healthcare Materials, 8*, e1801198. https://doi.org/10.1002/adhm.201801198

Larobina, D., Pommella, A., Philippe, A.-M., Nagazi, M. Y., & Cipelletti, L. (2021). Enhanced microscopic dynamics in mucus gels under a mechanical load in the linear viscoelastic regime. *National Academy of Sciences of the United States of America, 118*, e2103995118. https://doi.org/10.1073/pnas.2103995118

Linder, H. P., Bykova, O., Dyke, J. G., Etienne, R. S., Hickler, T., Kühn, I., et al. (2012). Biotic modifiers, environmental modulation and species distribution models. *Journal of Biogeography, 39*, 2179. https://doi.org/10.1111/j.1365-2699.2012.02705.x

Ling, L., Aldoghachi, A. F., Chong, Z. X., Ho, W. Y., Yeap, S. K., Chin, R. J., et al. (2022). Addressing the clinical feasibility of adopting circulating miRNA for breast cancer detection, monitoring and management with artificial intelligence and machine learning platforms. *International Journal of Molecular Sciences, 23*, 15382. https://doi.org/10.3390/ijms232315382

Liu, D., Zhang, H., Fontana, F., Hirvonen, J., & Santos, H. A. (2018). Current developments and applications of microfluidic technology toward clinical translation of nanomedicines. *Advanced Drug Delivery Reviews, 128*, 54. https://doi.org/10.1016/j.addr.2017.08.003

Liu, Y., Sun, L., Zhang, H., Shang, L., & Zhao, Y. (2021a). Microfluidics for drug development: From synthesis to evaluation. *Chemical Reviews, 121*, 7468. https://doi.org/10.1021/acs.chemrev.0c01289

Liu, J., Wei, Y., Wu, Y., Li, J., Sun, J., Ren, G., & Li, H. (2021b). ATP2C2 has potential to define tumor microenvironment in breast cancer. *Frontiers in Immunology, 12*, 657950. https://doi.org/10.3389/fimmu.2021.657950

Liu, Y., Liu, R., Liu, H., Lyu, T., Chen, K., Jin, K., & Tian, Y. (2023). Breast tumor-on-chip: From the tumor microenvironment to medical applications. *The Analyst, 148*, 5822. https://doi.org/10.1039/d3an01295f

Liverani, C., Vita, A. D., Spadazzi, C., Miserocchi, G., Cocchi, C., Bongiovanni, A., et al. (2021). Lineage-specific mechanisms and drivers of breast cancer chemoresistance revealed by 3D biomimetic culture. *Molecular Oncology, 16*, 921. https://doi.org/10.1002/1878-0261.13037

MacQueen, L. A., Chebotarev, O., Simmons, C. A., & Sun, Y. (2012). Miniaturized platform with on-chip strain sensors for compression testing of arrayed materials. *Lab on a Chip, 12*, 4178. https://doi.org/10.1039/c2lc40670e

Mankoff, D. A. (2008). Molecular imaging as a tool for translating breast cancer science. *Breast Cancer Research, 10*, S3. https://doi.org/10.1186/bcr2126

Mansoor, M., Haneef, I., Akhtar, S., Rafiq, M. A., Luca, A. D., Ali, S. Z., & Udrea, F. (2016). An SOI CMOS-based multi-sensor MEMS chip for fluidic applications. *Sensors, 16*, 1608. https://doi.org/10.3390/s16111608

Maračić, S., Jakopec, S., Piškor, M., Leventić, M., Lapić, J., Djaković, S., et al. (2023). Mechanochemical synthesis and antiproliferative activity of novel ferrocene quinoline/quinolone hybrids. *Applied Organometallic Chemistry, 37*. https://doi.org/10.1002/aoc.7124

Marincola, F. M. (2006). *Interleukin-10*. CRC Press. https://doi.org/10.1201/9781498713405

Martin, J. D., Miyazaki, T., & Cabral, H. (2021). Remodeling tumor microenvironment with nanomedicines. *Wiley Interdisciplinary Reviews Nanomedicine and Nanobiotechnology, 13*, e1730. https://doi.org/10.1002/wnan.1730

Michna, R. J., Gadde, M., Özkan, A., DeWitt, M. R., & Rylander, M. N. (2018). Vascularized microfluidic platforms to mimic the tumor microenvironment. *Biotechnology and Bioengineering, 115*, 2793. https://doi.org/10.1002/bit.26778

Mintz, R., Gao, M. A., Lo, K., Lao, Y. H., Li, M., & Leong, K. W. (2018). CRISPR technology for breast cancer: Diagnostics, modeling, and therapy. *Advanced Biosystems, 2*, 1800132. https://doi.org/10.1002/adbi.201800132

Mitchell, M. A., Burkett, A., Pen, L., Zhang, T., & Amjadi, K. (2018). Effect of chemotherapy on removal of indwelling pleural catheters in breast cancer patients with malignant pleural effusions. *Respiration, 96*, 552. https://doi.org/10.1159/000491097

Moccia, C., & Haase, K. (2021). Engineering breast cancer on-chip—Moving toward subtype specific models. *Frontiers in Bioengineering and Biotechnology, 9*, 694218. https://doi.org/10.3389/fbioe.2021.694218

Moehlenbrock, M. J., & Martin, R. S. (2007). Development of an on-chip injector for microchip-based flow analyses using laminar flow. *Lab on a Chip, 7*, 1589. https://doi.org/10.1039/b707410g

Monico, D. A., Calori, I. R., Souza, C., Espreafico, E. M., Bi, H., & Tedesco, A. C. (2022). Melanoma spheroid-containing artificial dermis as an alternative approach to in vivo models. *Experimental Cell Research, 417*(1), 113207. https://doi.org/10.1016/j.yexcr.2022.113207

Muratov, S., Canelo-Aybar, C., Tarride, J. É., Alonso-Coello, P., Dimitrova, N., Borisch, B., et al. (2020). Monitoring and evaluation of breast cancer screening programmes: Selecting candidate performance indicators. *BMC Cancer, 20*, 795. https://doi.org/10.1186/s12885-020-07289-z

Nahak, B. K., Mishra, A., Preetam, S., & Tiwari, A. (2022). Advances in organ-on-a-chip materials and devices. *ACS Applied Bio Materials, 5*, 3576. https://doi.org/10.1021/acsabm.2c00041

Narkhede, A. A., Shevde, L. A., & Rao, S. (2017). Biomimetic strategies to recapitulate organ specific microenvironments for studying breast cancer metastasis. *International Journal of Cancer, 141*, 1091. https://doi.org/10.1002/ijc.30748

Nguyen, V. V. T., Ye, S., Gkouzioti, V., van Wolferen, M. E., Yengej, F. Y., Melkert, D., et al. (2022). A human kidney and liver organoid-based multi-organ-on-a-chip model to study the therapeutic effects and biodistribution of mesenchymal stromal cell-derived extracellular vesicles. *Journal of Extracellular Vesicles, 11*(11), e12280. https://doi.org/10.1002/jev2.12280

Ortiz-Otero, N., Marshall, J. R., Lash, B., & King, M. R. (2020). Chemotherapy-induced release of circulating-tumor cells into the bloodstream in collective migration units with cancer-associated fibroblasts in metastatic cancer patients. *BMC Cancer, 20*, 873. https://doi.org/10.1186/s12885-020-07376-1

Polacheck, W. J., German, A. E., Mammoto, A., Ingber, D. E., & Kamm, R. D. (2014). Mechanotransduction of fluid stresses governs 3D cell migration. *Proceedings of the National Academy of Sciences of the United States of America, 111*, 2447. https://doi.org/10.1073/pnas.1316848111

Ramirez, S. I., Scholle, M., Buckmaster, J., Paley, R., & Kowdley, G. C. (2012). Breast cancer tumor size assessment with mammography, ultrasonography, and magnetic resonance imaging at a community based multidisciplinary breast center. *The American Surgeon, 78*, 440. https://doi.org/10.1177/000313481207800435

Ren, X., Hipolito, J., Lin, F., & Kung, S. K. P. (2020). Applications of microfluidic devices in advancing NK-cell migration studies. *Methods in Enzymology, 631*, 357. https://doi.org/10.1016/bs.mie.2019.05.052

Riehl, B. D., Kim, E., Bouzid, T., & Lim, J. Y. (2021). The role of microenvironmental cues and mechanical loading milieus in breast cancer cell progression and metastasis. *Frontiers in Bioengineering and Biotechnology, 8*, 608526. https://doi.org/10.3389/fbioe.2020.608526

Riethdorf, S., Fritsche, H. A., Müller, V., Rau, T., Scheffold, C., Rack, B., et al. (2007). Detection of circulating tumor cells in peripheral blood of patients with metastatic breast cancer: A validation study of the CellSearch system. *Clinical Cancer Research, 13*, 920. https://doi.org/10.1158/1078-0432.ccr-06-1695

Rubiano, A. M., Indapurkar, A., Yokosawa, R., Miedzik, A., Rosenzweig, B. A., Arefin, A., et al. (2021). Characterizing the reproducibility in using a liver microphysiological system for assaying drug toxicity, metabolism, and accumulation. *Clinical and Translational Science, 14*, 1049. https://doi.org/10.1111/cts.12969

Ruzycka-Ayoush, M., Cimpan, M. R., Ríos-Mondragón, I., & Grudziński, I. P. (2019). Microfluidics for studying metastatic patterns of lung cancer. *Journal of Nanobiotechnology, 17*, 71. https://doi.org/10.1186/s12951-019-0492-0

Sandhya, S., Hogenson, T. L., & Fernández-Zapico, M. E. (2022). Patient-derived organoids, creating a new window of opportunities for pancreatic cancer patients. *EMBO Molecular Medicine, 14*, e15707. https://doi.org/10.15252/emmm.202215707

Santiago, G. T., Flores-Garza, B. G., Negrete, J. A. T., Lara-Mayorga, I. M., González-Gamboa, I., Zhang, Y. S., et al. (2019). The tumor-on-chip: Recent advances in the development of microfluidic systems to recapitulate the physiology of solid tumors. *Materials, 12*, 2945. https://doi.org/10.3390/ma12182945

Schmeichel, K. L., & Bissell, M. J. (2003). Modeling tissue-specific signaling and organ function in three dimensions. *Journal of Cell Science, 116*, 2377. https://doi.org/10.1242/jcs.00503

Shang, M., Soon, R. H., Lim, C. T., Khoo, B. L., & Han, J. (2019). Microfluidic modelling of the tumor microenvironment for anti-cancer drug development. *Lab on a Chip, 19*, 369. https://doi.org/10.1039/c8lc00970h

Shen, S., Tu, C., Shen, H., Li, J., Frangou, C., Zhang, J., & Qu, J. (2023). Comparative proteomics analysis of exosomes identifies key pathways and protein markers related to breast cancer metastasis. *International Journal of Molecular Sciences, 24*, 4033. https://doi.org/10.3390/ijms24044033

Shi, W., Mirza, S., Kuss, M., Liu, B., Hartin, A., Wan, S., et al. (2023). Embedded bioprinting of breast tumor cells and organoids using low-concentration collagen-based bioinks. *Advanced Healthcare Materials, 12*, e2300905. https://doi.org/10.1002/adhm.202300905

Sjöberg, E., Augsten, M., Bergh, J., & Jirström, K. (2016). Expression of the chemokine CXCL14 in the tumour stroma is an independent marker of survival in breast cancer. *British Journal of Cancer, 114*, 1117. https://doi.org/10.1038/bjc.2016.104

Smith, S. E., Mellor, P., Ward, A. K., Kendall, S., McDonald, M., Vizeacoumar, F. S., et al. (2017). Molecular characterization of breast cancer cell lines through multiple omic approaches. *Breast Cancer Research, 19*, 65. https://doi.org/10.1186/s13058-017-0855-0

Song, K., Zu, X., Du, Z., Hu, Z., Wang, J., & Li, J. (2021). Diversity models and applications of 3D breast tumor-on-a-chip. *Micromachines, 12*, 814. https://doi.org/10.3390/mi12070814

Stanojković, T. (2020). A marine natural products as modulators of multidrug resistance. *Journal of Cancer Research Updates, 9*, 96. https://doi.org/10.30683/1929-2279.2020.09.11

Szydzik, C., Gavela, A. F., Herranz, S., Roccisano, J., Knoerzer, M., Thurgood, P., et al. (2017). An automated optofluidic biosensor platform combining interferometric sensors and injection moulded microfluidics. *Lab on a Chip, 17*, 2793. https://doi.org/10.1039/c7lc00524e

Teiten, M.-H., Dicato, M., & Diederich, M. (2013). Curcumin as a regulator of epigenetic events. *Molecular Nutrition & Food Research, 57*, 1619. https://doi.org/10.1002/mnfr.201300201

Thakar, R., & Fenton, K. N. (2023). Bioethical implications of organ-on-a-chip on modernizing drug development. *Artificial Organs, 47*, 1553. https://doi.org/10.1111/aor.14620

Todd, J. R., Ryall, K. A., Vyse, S., Wong, J. P., Natrajan, R., Yuan, Y., et al. (2016). Systematic analysis of tumor cell-extracellular matrix adhesion identifies independent prognostic factors in breast cancer. *Oncotarget, 7*, 62939. https://doi.org/10.18632/oncotarget.11307

Tomov, M. L., Gil, C. J., Cetnar, A., Theus, A. S., Lima, B., Nish, J. E., et al. (2019). Engineering functional cardiac tissues for regenerative medicine applications. *Current Cardiology Reports, 21*, 105. https://doi.org/10.1007/s11886-019-1178-9

Vidi, P. A., Maleki, T., Ochoa, M., Wang, L., Clark, S., Leary, J. F., & Lelièvre, S. A. (2014). Disease-on-a-chip: Mimicry of tumor growth in mammary ducts. *Lab on a Chip, 14*, 172. https://doi.org/10.1039/c3lc50819f

Vong, S., & Kalluri, R. (2011). The role of stromal myofibroblast and extracellular matrix in tumor angiogenesis. *Genes & Cancer, 2*, 1139. https://doi.org/10.1177/1947601911423940

Wu, Z., Gong, Z., Ao, Z., Xu, J., Cai, H., Muhsen, M., et al. (2020). Rapid microfluidic formation of uniform patient-derived breast tumor spheroids. *ACS Applied Bio Materials, 3*, 6273. https://doi.org/10.1021/acsabm.0c00768

Yakovenko, A. N., Cameron, M. E., & Treviño, J. G. (2018). Molecular therapeutic strategies targeting pancreatic cancer induced cachexia. *World Journal of Gastrointestinal Surgery, 10*, 95. https://doi.org/10.4240/wjgs.v10.i9.95

Yang, J., Bahcecioglu, G., & Zorlutuna, P. (2020). The extracellular matrix and vesicles modulate the breast tumor microenvironment. *Bioengineering, 7*, 124. https://doi.org/10.3390/bioengineering7040124

Zhang, Y., Xiao, R.-R., Yin, T., Zou, W., Tang, Y., Ding, J., & Yang, J. (2015). Generation of gradients on a microfluidic device: Toward a high-throughput investigation of spermatozoa chemotaxis. *PLoS One, 10*, e0142555. https://doi.org/10.1371/journal.pone.0142555

Zhu, J., Shen, Y., Wang, L., Qiao, J., Zhao, Y., & Wang, Q. (2022). A novel 12-gene prognostic signature in breast cancer based on the tumor microenvironment. *Annals of Translational Medicine, 10*, 143. https://doi.org/10.21037/atm-21-6748

Κάρλου, M., Tzelepi, V., & Efstathiou, E. (2010). Therapeutic targeting of the prostate cancer microenvironment. *Nature Reviews Urology, 7*, 494. https://doi.org/10.1038/nrurol.2010.134

The manufacturer's authorised representative in the EU is Springer Nature Customer Service Centre GmbH, Europaplatz 3, 69115 Heidelberg, Germany. If you have any concerns regarding our products, please contact ProductSafety@springernature.com

Printed and bound by CPI Group (UK) Ltd, Croydon, CR0 4YY
01/10/2026
02225963-0005